DEUCALION

AND OTHER STUDIES

IN

ROCKS AND STONES

BY

JOHN RUSKIN

CONTENTS OF VOLUME XXVI

[1] These are not arranged chronologically, but in accordance with the order given by Ruskin on p. 387.

APPENDIX

LETTERS, ADDRESSES, AND NOTES

MINOR RUSKINIANA : *Continued :—*

LIST OF ILLUSTRATIONS

(From Drawings by the Author[1])

---

[1] Except where otherwise stated.

ILLUSTRATIONS PRINTED IN THE TEXT

FACSIMILES

Note.—Of the drawings reproduced in this volume, the *frontispiece* was No. 163 in the Ruskin Exhibition at Coniston, 1900; No. 391 in the Ruskin Exhibition at the Royal Society of Painters in Water-Colours, 1901; and No. 416 in the Ruskin Exhibition at Manchester, 1904.

The drawing reproduced on Plate A was No. 337 at Manchester, and that on Plate B was No. 110 at Manchester.

No. II. was No. 157 at the Royal Society of Painters in Water-Colours; and No. XIIA. was No. 60 at Coniston, No. 312 at the Water-Colour Society, and No. 220 at Manchester.

One of the drawings of agates, engraved in the *Geological Magazine*, was No. 162 at Coniston, and was sold for the benefit of the Institute.

INTRODUCTION TO VOL. XXVI

THIS volume collects Ruskin's writings on Geology and Mineralogy. *Deucalion*—the principal work here included—was itself intended by Ruskin to collect "the notices of phenomena relating to geology which were scattered through my former works";[1] but the scheme of that book was altered as it advanced, and it came to consist almost entirely of additional studies. Many of "the notices" to which he refers are contained in other volumes; more especially the fourth volume of *Modern Painters*, in the case of geology, and *The Ethics of the Dust*, in that of mineralogy. These are, of course, not here repeated, though references to them are often supplied in editorial notes. With these exceptions, the present volume brings together all the author's papers, letters, lectures, books, and catalogues on the subjects in question.

The arrangement is, as usually in this edition, chronological, and the contents are: I. A paper of 1863, to which the author attached considerable importance, *On the Forms of the Stratified Alps of Savoy*. II. Two papers of 1865, *On the Shape and Structure of some parts of the Alps, with reference to Denudation*. III. Seven papers (1867–1870), *On Banded and Brecciated Concretions*. IV. *Deucalion*, published at intervals between 1875 and 1883. V. A paper of 1884, *On the Distinctions of Form in Silica*. These are in large print.

They are followed, in smaller print, by VI., a series of Catalogues of Minerals (1883–1886), and VII., a *Grammar of Silica* (hitherto unpublished). The *Catalogues* and the *Grammar* are parts of a general scheme, as explained below (p. lx.).

In an Appendix various minor writings are collected. These also are arranged chronologically, namely, I. A *Notice respecting some Artificial Sections illustrating the Geology of Chamouni* (1858). II. A series of letters (1864) on the Conformation of the Alps, with especial reference to glacial action. III. An appreciation (1874) of James David Forbes and his work on glaciers. IV. A report of a lecture on Stones (1876). V. Some letters on the Alpine Club (1878). VI. An

---

[1] See *Deucalion*, vol. ii. ch. ii. § 1 (below, p. 333).

Introduction to Mr. W. G. Collingwood's *Limestone Alps of Savoy* (1884), a work printed by Ruskin as a " Supplement to *Deucalion*."

Lastly, the Appendix includes some fragments (hitherto unpublished) which are here printed from Ruskin's MSS., namely, VII., the beginning of a chapter on "The Garnet," intended for a continuation of *In Montibus Sanctis*, and VIII., "A Geological Ramble in Switzerland"; this is mentioned below, p. xxviii.

Ruskin complained with some emphasis that his contributions to geology and mineralogy attracted little attention, as compared with his writings on art, "though precisely the same faculties of eye and mind are concerned in the analysis of natural and of pictorial forms."[1] For this neglect several explanations may be found. The world is in the habit of applying the formula, "One man, one subject." To Ruskin it looked for criticisms of art and life, and descriptive writing, and did not care to consider him seriously as a geologist. Again, Ruskin in his writings on geology was in the habit, as he says, of "teaching by question, rather than by assertion";[2] his chapters had "sometimes become little more than notes of interrogation";[3] he did not from the first assume the tone of authority which he permitted himself in other subjects. His method of entering upon the territory of the men of science was, it must be admitted, not ingratiating. He did not profess to have studied their subjects very far; but he required them to confess that they were mostly wrong, and that they must begin afresh with new systems of nomenclature and classification of his devising. But the neglect of his geological writings may also be attributed to another cause. They were not considered, because they were little known. They were buried in back numbers, scattered in miscellaneous periodicals, or distributed among scarce pamphlets. It is hoped that the present volume, which for the first time collects Ruskin's studies, will serve to call more attention to a branch of his work, in which he was profoundly interested, which is rich in suggestion, and upon which he spent much labour, both in research and in literary embellishment.

The contents of this volume do not for the most part carry us beyond the date in Ruskin's life which we have already reached;[4] though, in order to bring all his studies of stones together in the same volume, it will be necessary to touch incidentally on later times. In this Introduction we shall go back over past years, tracing Ruskin's

---

[1] See *On Distinctions of Form in Silica*, §§ 3, 29 (below, pp. 373, 386).
[2] *Ibid.*, § 29, p. 386.
[3] *Deucalion*, ii. ch. ii. § 18, p. 342.
[4] His illness of 1878 : see Vol. XXV. p. xxvi.

geological life, so to say, and noting the circumstances in which the various pieces included in the volume were written. He remarks in one of the letters of 1864, printed in this volume, that he had spent "eleven summers and two winters in researches among the Alps,[1] directed solely to the questions of their external form and its mechanical causes" (p. 548). By "solely" he means that his geological researches were thus directed. It was not in Ruskin's nature to devote himself at any time solely to any one subject; but the following pages will show how often, and how long, he had studied the questions discussed in this volume.

Geology and mineralogy, and not painting or literature, were Ruskin's earliest love. No acquisition of later years — not his most radiant Turner or choicest manuscript—gave him pleasure so keen as he felt in the possession of his first box of minerals,[2] and no subsequent possession, he says, had so much influence on his life.[3] The ambition of his boyhood was to connect with his name, not a system of art criticism, but a system of mineralogy.[4] The dream of his early manhood was that he should become, not a master of English, but President of the Geological Society.[5] As a boy he spent many a day at the British Museum, comparing the minerals there with the descriptions of them in Jameson's *Mineralogy*;[6] and for a present on his fifteenth birthday he chose Saussure's *Voyages dans les Alpes*—a book which to the end of his working life was almost always kept at hand and frequently quoted and referred to. Saussure was his master in geology,[7] for this reason among others, that he "had gone to the Alps, as I desired to go myself, only to *look* at them, and describe them as they were, loving them heartily—loving them, the positive Alps, more than himself, or than science, or than any theories of science."[8] The first of his prose[9] writings, which appeared in print, were notices on the Colour of the Rhine and the Twisted Strata of Mont Blanc.[10] His interest was excited, as he says in *Deucalion*, "very early in life by

---

[1] Presumably he omits his earlier tours of 1833 and 1835, and counts 1844, 1846, 1849, 1851, 1854, 1856, 1859, 1860, 1861 (winter also), and 1862 (winter also).

[2] *Fors Clavigera*, Letter 4, § 3.

[3] See below, p. 294 *n.*

[4] See below, pp. 97, 553.

[5] *Ibid.*, p. 97.

[6] *Manual of Mineralogy*, by Robert Jameson, F.R.S.E., Edinburgh, 1821. See *Præterita*, i. § 139.

[7] *Modern Painters*, vol. iv. (Vol. VI. p. 214 *n.*), and vol. v. (Vol. VII. p. 164).

[8] *Ibid.*, vol. iv. (Vol. VI. p. 476). Compare below, p. 560.

[9] A slight piece of verse was actually first: see Vol. II. p. xviii. *n.*

[10] Vol. I. pp. 191–196.

the forms and fractures in the mountain groups of Savoy."[1]   The
paper of 1834, "On the Strata of Mont Blanc," was the result of
observations made during his first continental tour (1833).   His second
tour was in 1835 (his geological studies on that occasion have already
been referred to[2]), and in his versified account of the tour "the dreams
of the geologist" find place.[3]

At Oxford Ruskin received further impetus towards the study of
geology from the attention paid to him by Dr. Buckland, the Reader
in Geology.   He mentions how great a favour he held it to be allowed
to prepare diagrams for Buckland's lectures;[4] and he records how at
a later time Buckland enjoined upon him to read Forbes's *Travels in
the Alps* for the decisive word on the theory of glaciers (p. 134).[5]   He
spent the winter and the spring of 1840–1841 on the Continent, and
sent to Dr. Buckland from Naples a notice of "A Landslip near
Giagnano," which was duly communicated to the Ashmolean Society
(Vol. I. p. 211).   He had become a Fellow of the Geological Society in
1840; he also joined the Mineralogical Society, and contributed to the
Meteorological Society's *Transactions* (Vol. I. p. 206).   He was a frequent
attendant at the meetings of these Societies, and had once heard Darwin
read a paper at the Geological.   Shortly afterwards, at a dinner-party
at Dr. Buckland's, he had met Darwin.   "He and I got together," he
wrote to his father, "and talked all the evening."[6]

In 1842 Ruskin had again spent several months among the Alps—
engrossed, as he said in letters to friends, with "snow and granite."[7]
To what purpose his studies in this sort were directed, appeared, when
the first volume of *Modern Painters* was published (1843), in the sections
"Of Truth in Skies" and "Of Truth of Earth."   His observations of
the phenomena of snow and ice had not, however, as yet been entirely
exact; and he notes in *Deucalion* "a grave error," with regard to the
accumulation of snow, "which, strangely enough, remained undetected,
or at least unaccused, in spite of all the animosity provoked by my
earlier writings" (p. 129).   Similarly, in the fifth volume of *Modern
Painters* (1860) he corrects an account of cloud-phenomena among the
high mountains which, in the first volume, he had accepted from
Saussure without independent consideration.[8]

[1] *Deucalion*, i. ch. xiv. § 5 (below, p. 275 and n.).
[2] Vol. I. p. xxx.
[3] Vol. II. p. 407 (29).
[4] *Præterita*, i. § 225.
[5] Compare *Fors Clavigera*, Letter 34, § 13.
[6] From a letter of April 22, 1837, given in full in a later volume of this edition.
[7] Vol. II. pp. 222–223.
[8] See on this subject *The Storm-Cloud of the Nineteenth Century*, § 52.

Ruskin's study of all such phenomena became closer and more detailed in subsequent years. The first volume of *Modern Painters* being disposed of, he returned in 1844 to the Alps. His tour of that year was memorable among other things for his meeting with Principal Forbes. The occasion remained firmly fixed in Ruskin's memory, and is several times referred to in autobiographical passages—in *Deucalion* (p. 219), in the Preface to *The Limestone Alps of Savoy* (p. 569), and in *Præterita* (ii. § 97). The following is his note of it made at the time:—

> "*July* 16.—I have just had a most uncomfortable chat with Professor Forbes—uncomfortable because my father forced me to show my sketches when they didn't want to see them, and because when I had found him out, which I did not for a long time, I knew nothing of what he had asserted."

The diary then proceeds to describe a mountain excursion:—

> "When I got up this morning I was quite exhausted, limbs aching, and a little feverish feeling altogether. It will be a good warning to me never to walk too quickly up hill again. No walk on the journey knocked me up so completely. After breakfast Mr. Stone and Mr. Anderson came in—a pleasant surprise—but which kept us from going out till ten—all the better, I believe, for me. At ten we started, got leisurely up the hill against strong wind, began to recover a little after an hour's walk, but still felt weak. Worked on —crossing the valley of the Col—up hills on other side, reached the summit on the left of the Col about two; view noble, but Bernese Alps just topped by cloud. Professor Forbes tells me the summit so conspicuous from Martigny is the Bietschhorn—the one with the jagged ridge above Brieg the Aletschhorn—the little glacier coming from it the Ober Aletsch Glacier. The peak seen from the Valais, Simplon way, is the Matterhorn. The broad expanse above the Kaltwasser Glacier, which rose to-day as I climbed to a noble elevation, is the Breithorn, the side of it to the north, remarkable for a perfect line of snow at an angle of 45°, is the Monte Leone. Descended from this point of view by a little lake, and climbed a higher peak, which at length fully rewarded me; it commanded the Valais far down, the Bernese Alps in their whole extent unbroken, and two mountains beyond the Valley of Saas, which I took for the Monte Rosa and Cervin, but Professor Forbes tells me they are no such thing. I think not myself now. The wind was violent, and the slaty ridge, shattered and broken into deep crevices, afforded small footing, so that I stayed not long, but it was a glorious panorama. Descended among steep loose sliding stones to the edge of the

loveliest snow lake I ever saw—crystal water, lightly rippled, edge of snow, which appeared to have melted below and then broken off, giving a sharp edge. The pieces which fell into the water extended much farther under it than above, and being perfectly pure, exhibited its fine green. A vast field extended above it. Altogether a day to be most grateful for, and to remember long."

Other extracts from Ruskin's diary of 1844 have been given in Vol. III. (pp. xxv.–xxvii.), and some were included by him in *Præterita* (ii. §§ 97 *seq.*). His mind was during this tour mainly fixed on the clouds and rocks and snows. After the meeting with Forbes he went up to the Bell Alp.[1] The panorama of the Alps as seen from that place, which he drew at this time and afterwards slightly coloured, is in the Ruskin Museum at Sheffield; he refers to it in this volume as giving trustworthy record of the then state of the snows (p. 222).

The next impulse which Ruskin received was, however, in the direction of pictures, his Italian tour of 1845 being followed by the second volume of *Modern Painters*. The next year found him again for a short time among the Alps, but he went on to Italy and his preoccupations were now with painting and architecture.[2] On the way home, however, in 1845, he halted at the Pass of the St. Gothard in connexion with Turner's drawings. Stones and rocks here again occupied his attention.[3] But his thoughts were principally tending at this time in the direction which was to lead him to write *The Seven Lamps of Architecture* and *The Stones of Venice*. His interest in geology, though for a while overlaid by other studies, was still keen. Thus in 1847 we find him acting as one of the secretaries of the Geological Section at the Oxford meeting of the British Association;[4] and in his diary for 1846–1847 there are some pages in which he classifies, with reference to various drawings of his own, the different forces which seem to have governed mountain forms. He lays particular stress, here as always, on "the muscular or inherent structure" and "the undulatory power."

When *The Seven Lamps* was out of his way, Ruskin set out in the spring of 1849 for Chamouni once more, and it was on his work there and afterwards at Zermatt that, as he says in this volume (*Limestone* Preface), the mountain studies in the fourth volume of *Modern*

---

[1] So Ruskin always wrote this place-name, regarding "Bel Alp" as "a modern vulgarism."
[2] See Vol. VIII. p. xx.
[3] See Vol. V. p. xvi.
[4] See Vol. VIII. p. xxv. and *n.*

THE MER DE GLACE FROM THE MONTANVERT
(1849)

Painters were principally based. The extracts from his diaries of 1849, already given (Vol. V. pp. xvii. *seq.*), show how careful those studies were. The diary of this tour contains many lists of minerals, geological sections, and observations. A passage describing the ruin of the Cascade des Pélerins is in the fourth volume of *Modern Painters.*[1] Another passage of the kind may here be given, in connexion with what he says (p. 552) about his watching of the Alpine watercourses:—

"CHAMOUNI, *Sunday, June* 17*th.*—Half-past five. Pouring still; and fresh snow just down to the level of the pines, all along, from the top of Montanvert bending in above the châlets of Blaitière Dessous; out again and well on to the top of Tapia; then taken up by the lower ridge of La Côte, and the correspondent ridge of Taconay, which I always thought till this moment was much lower. But that notch in both their flanks at the same level is significative; I must examine it. I got out, however, before dinner to-day; during a fair blink which lasted just long enough to let me, by almost running and leaping all the streams, reach the end of the pinewood next the Source of the Arveron, in order to see the waterfall. I had then to turn to the left to the wooden bridge over the Arveron, when, behold, a sight new to me—an avalanche had evidently taken place from the glacier into the very bed of the great cataract, and the consequence was that the stream was as nearly choked as could be with *balls and ellipsoids* of ice, from the size of its common stones to that of a portmanteau, which were rolling down with it wildly, generally swinging out and in of the water as it waved, but when they came to the shallow parts tumbled and tossed over one another, and then plunging back into the deep water like so many stranded porpoises, spinning as they went down, and showing their dark backs with wilder swings after their plunge—white as they emerged—black, owing to their clearness as seen in the water, the stream itself of a pale clay colour, opaque, larger by one-half than ever I saw it, and running, as I suppose, not less than ten miles an hour, the whole mass—water and ice—looking like some thick paste full of plums or ill-made pineapple ice with quantities of fruit in it; and the whole, looking like a solid body (for the nodules of ice hardly changed their relative position during the quarter of a minute they were severally in sight), going down in a mass, thundering and rumbling against the bottom and the shore, and the piles of the bridge, it made one giddy to look at it; and this the more, because on raising the eye there was the great cataract itself—every time it was seen

---

[1] See Vol. VI. pp. 342–344 *n.*

startling one as if it had just begun, or were increasing every instant, like a large avalanche bounding and hurling itself hither and
thither, as if it was striving to dash itself to pieces, not falling
because it could not help it—and behind, there was a fearful storm
coming up by the Breven, its grisly clouds warping up as it seemed
against the river and cataract, and pillars of hail behind."

Two more of his drawings of the time are here introduced—one
(Plate A) of the Mer de Glace at Chamouni, the other of the chain
of Mont Blanc, as seen from the Col de la Seigne, looking towards
Courmayeur (Plate B).   For the time, however, Ruskin laid by his
observations and drawings of the mountains, for he was now to devote
the greater part of three years to *The Stones of Venice*.[1]   In the early
summer of 1854 that long task was over, and he set out again for
Switzerland and Chamouni.   On this tour the observations of 1849
were supplemented, and Ruskin felt himself equipped for the continuation of *Modern Painters*.

The fourth volume of *Modern Painters*, with its close analysis and
description of mountain structure, formed in the author's opinion, as
stated in this volume (p. 568), "the most valuable and least faultful
part of the book."   He began to republish these chapters in 1884
in a series of reprints entitled *In Montibus Sanctis*.[2] Geologists also
account them the most important of his contributions, or aids, to
their science.   "We must not forget his services to our science," said
the President of the Geological Society in an obituary notice of
Ruskin, "in directing the attention of artists and others to the effect
of geological structure, and of the characters of rocks, on scenery";
and the speaker added that the chapters in *Modern Painters* "might
be read with advantage by many geologists."[3] Ruskin himself made
a more modest claim for them; they "should be read," he said, "to
young people by their tutors as an introduction to geological study."[4]

Certainly those chapters are typical of Ruskin's point of view in
approaching geology.   He was little interested in unknown ages and
immeasurable forces.   And so, in *Deucalion*, Ruskin defines "our own
work" as beginning where all theory ceases, and as being the study
of forms which have "actually stood since man was man" (p. 113).

---

[1] See the summary of these years in Vol. V. p. xxxi.

[2] The notes which he appended in that year to the chapters will be found in
Vol. VI. at pp. 116, 121, 122, 124, 126, 128, 131, 132, 133, 135, 136, 138; Post-
scripts at pp. 127, 145; the Preface to the reprints, in Vol. III. p. 678.

[3] Annual address by the President, William Whitaker, F.R.S., May 1900: see
the *Quarterly Journal of the Geological Society*, vol. 56, pp. lx., lxi.

[4] *Modern Painters*, vol. iv. (note of 1885), Vol. VI. p. 128.

FROM THE COL DE LA SEIGNE, LOOKING TO COURMAYEUR
(1849)

Similarly in a letter to his friend, Mr. A. Tylor, who had sent him a paper from the *Geological Magazine*,[1] he wrote (1875):—

> "I am grateful to you for sending me the binomial curve and the Glaciers on Mount Sinai—but it's all much too grand and far away back for me. I don't care three farthings what happened when Mount Sinai was under ice.
>
> "I want to know how long the Staubbach has been falling where it is in the Valley of Lauterbrunnen, and why it hasn't cut itself further back?
>
> "There's a mere nutshell of a question for you geologists. You ought to crack it for me as easily as a squirrel does a beech-nut, and give me my question out of the shell. But I can't get anybody ever to answer about what I want to know."

The bent of Ruskin's mind, in all such studies, was severely practical (p. 166). He turned away from theory, conjecture, speculation, to what could certainly be known, seen, drawn, and measured. In *Deucalion*, and elsewhere, he often speaks with seeming contempt of "science" and "men of science"; but, as one of the acutest of his critics has pointed out, this was a piece of literary *finesse*. "It was on the point of science that issue was joined; and if he did not reproach his adversary in that this adversary was too little, and not too much, a man of science, he reproached him to no purpose."[2] And, in fact, what Ruskin claimed for his own writings on mountain form is that they are of observation, experiment, and verification all compact. What he wrote was founded not on what he had read in books, but on his own "watchings of the Alps"; he "closed all geological books," and set himself "to see the Alps in a simple, thoughtless, and untheorising manner, but to see them, if it might be, thoroughly."[3] Hence, he asserted that the work of *Proserpina*, which was tentative, was "quite different from that of *Deucalion*, which is authoritative as far as it reaches, and will stand out like a quartz dyke, as the sandy speculations of modern gossiping geologists get washed away."[4]

Ruskin, then, in *Modern Painters* limited himself to what could be securely seen in mountain form. But incidentally he touched upon

---

[1] For this paper see p. 290 *n.*, and compare p. 368. For other references to Mr. Tylor, see p. 316 *n.*, Vol. IV. p. 107, Vol. XV. p. 369, and Vol. XXIII. p. liii.

[2] *John Ruskin*, by Mrs. Meynell, p. 245. Mrs. Meynell calls attention, as an example of Ruskin's "exquisite and characteristic wit," to his criticism of Tyndall's phrase "contact with facts," as expressive merely of "occasional collision with them." See below, p. 285.

[3] *Modern Painters*, vol. iv. (Vol. VI. pp. 214 *n.*, 475).

[4] *Proserpina*, ii. ch. i. § 42 (Vol. XXV. p. 413). Compare, below, p. 197.

problems of the past, and put out statements not at the time generally received. He made notes of some of these in his Appendix ("Rock Cleavage"), laying special stress on aqueous, as opposed to igneous, causes of structure in the detail of mountain forms; and though, on the other hand, he "still fancied that ice could drive imbedded rocks and wear down rock surfaces,"[1] he challenged the accepted opinion about the erosive action of rivers, and he questioned generally whether the explanations by geologists of the effects of pressure were adequate to the phenomena to be explained. Perhaps it was of these heresies that Ruskin's father was thinking in a characteristic letter to Mrs. Simon, written at a time when the fourth volume of *Modern Painters* was being much discussed. The father's admiration for his son's work was always tempered by regard for received opinions:—

"7 BILLITER STREET, *19th February*, 1858.

"Mrs. R. named to me your having heard that my Son's meddling with Political Economy might weaken his Influence in matters of Art. I feared this myself, but by his own confession his studies of Political Economy have not encroached much on his time—and on this weary subject a few new ideas will do no harm. The *Times* couples him with Dr. Guthrie, and says they are both in a state of helpless Ignorance of the first principles of Political Philosophy.[2] I might, perhaps, prefer the Simplicity of Dr. Guthrie to the Philosophy of the *Times*; but, if my Son has so greatly committed himself in his last little book, I trust to the talk of Mr. Simon, Mr. Helps, and Mr. Carlyle bringing us an amended Second Edition. I hope the public, as far as it kindly interests itself in my Son's writings, will not suppose his Geology also a deviation from the right path. From Boyhood he has been an artist, but he has been a geologist from Infancy, and his geology is perhaps now the best part of his Art, for it enables him to place before us Rocks and Mountains as they are in Nature, in place of the very bad likenesses of these objects presented to us in most of the old paintings or modern Drawings."

Notice of Artificial Sections at Chamouni.—Ruskin was always learning; and having finished his fourth volume, he set out for Chamouni once more. On this tour (1856), again, he was busily engaged, as we have seen in an earlier volume,[3] in geological studies. At this time he

---

[1] See Vol. VI. p. 116 (note of 1885).

[2] In a review on January 2, 1858, of *The City, its Sins, and its Sorrows*, by Thomas Guthrie, D.D., containing an incidental notice of Ruskin's *Political Economy of Art* (published in 1857): "These are men of great power, of great eloquence, and of great popularity. It is surely instructive to see such men, distinguished in their respective spheres, displaying the most helpless ignorance of the first principles of political economy."

[3] Vol. VII. p. xxi.

followed closely the papers in geological magazines and discussions at the learned societies. In the year 1856 there had been some criticism of the statements made by Forbes with regard to the geological constitution of the chain of Mont Blanc. Forbes had shown in his *Travels*,[1] from observations both on the north and on the south side of the chain, that "there were appearances of limestone dipping under the granite of Mont Blanc." Ruskin had illustrated the same phenomena in the fourth volume of *Modern Painters*.[2] Now, however, the observations had been called in question, and Forbes replied to his critics in a paper read before the Royal Society of Edinburgh.[3] Ruskin determined to utilise his opportunities at Chamouni to investigate the subject yet further; and in order to put the matter beyond all doubt, he employed workmen to excavate. The excavations came to an end with an accident, as he describes in *Deucalion* (p. 288). His diary of 1856 contains several mentions of the "pickaxe and spade" work which he directed, and also a detailed account of the results obtained, with coloured sketches. He afterwards communicated these results to the Royal Society of Edinburgh. This "Notice" comes first in the Appendix (I.) to the present volume (pp. 545–547).

Artistic work—the arrangement and cataloguing of Turner's drawings, *The Elements of Drawing*, *Academy Notes*, and other things—now occupied Ruskin for some years, and there was still the fifth volume of *Modern Painters* to be finished.[4] It was not till this had been done, and *Unto this Last* had been written, that Ruskin found leisure to resume his geological studies. The years 1861–1863 were an important time in what I have called his geological life; and it is curiously characteristic of the workings of his mind, that the same years should be those in which he was much occupied with Political Economy.[5] Wide as was his intellectual range, and scattered as was his work, there was always some inner bond correlating his studies. Thus, as he tells us in *Deucalion*, his studies in the origin and sculpture of mountain form were connected in his own mind "with the practical hope of arousing the attention of the Swiss and Italian mountain peasantry to an intelligent administration of the natural treasures of their woods and streams" (p. 339).

In 1861 Ruskin "went into Savoy and spent two winters on the south slope of the Mont Salève, in order to study the secondary ranges

[1] See ch. xi. pp. 202 *seq.* (edition of 1900).
[2] See Vol. VI. pp. 255, 256.
[3] See the references given below, p. 545 *n.*
[4] See Vol. VII. pp. xxv. *seq.*
[5] See the Introduction to Vol. XVII.

of the Alps and their relation to the Jura" (p. 569). We have seen
some notes, in an earlier volume, of his geological rambles at Mornex
(Vol. XVII. p. lxi.). It was at this time [1] that he read Studer's
Geologie der Schweiz, "learning enough German to translate for himself the parts of his volumes which relate to the Northern Alps,"
and "writing them out carefully, with brilliantly illuminated enlargements of his tiny woodcuts, proposing the immediate presentation of
the otherwise somewhat dull book to the British public in this decorated form" (below, p. 569). The translations from Studer with these
coloured drawings, are preserved at Brantwood in a volume lettered
"Rock Book, 1861–2." In another note-book at Brantwood there
are further extracts from Studer, and several pages describing what
may be called a geological ramble in Switzerland (with a scheme for
further routes of the same kind). Some passages of this are given
in the Appendix to this volume (pp. 577–579). One may surmise
that the proposed presentation of Studer in decorative form would
have contained more of Ruskin than of Studer. This may have been
the beginning of "the book about Switzerland which people would
buy" that he mentioned to his father.[2] "The Scenery of Switzerland, by the Author of *Modern Painters*" would have run, it is safe
to predict, into many editions.

The Stratified Alps of Savoy (1863).—His mountain studies at
Mornex and Annecy were, however, to take a different form. In 1862
Professor Ramsay had published a paper [3] on the action of glaciers
which caused a great stir in the scientific world. It failed to carry
conviction to Ruskin's mind or to accord with his own observations;
and having undertaken to lecture at the Royal Institution, he selected
for his subject "The Forms of the Stratified Alps of Savoy." At that
time, says Mr. Collingwood, "many distinguished foreign geologists
were working at the Alps; but little of conclusive importance had
been published, except in papers embedded in Transactions of various
societies. Professor Alphonse Favre's great work [4] did not appear until

---

[1] And not in 1866, as he states below, p. 569; though the second note-book
of extracts from Studer no doubt belongs to the later year.
[2] Vol. XVII. p. xxxii.
[3] Further particulars of it will be found below, p. lxv.
[4] *Recherches Géologiques dans les parties de la Savoie, du Piémont et de la Suisse,
voisines du Mont-Blanc*, 3 vols. At vol. i. p. 251 of this book, Professor Favre
thus refers to Ruskin's paper (in regard to the cliffs of the Salève): "Dernière-
ment M. Ruskin a nié la présence de ces couches redressés. Il pense que les
divisions que l'on voit dans les roches sont produites par le clivage, et que les
couches presque horizontales de la montagne se prolongent dans celles que l'on
croit verticales (*The Geologist*, 1863, vi. 257)."

1867, and the *Mechanismus der Gebirgsbildung* of Professor Heim not till 1878; so that for an English public the subject was a fresh one."[1] The reports and discussions of Ruskin's paper in the reviews and magazines testify to the interest which his paper aroused. It was founded on close study of the phenomena, and was very carefully written, containing many passages of such eloquence as might be expected from the author of *Modern Painters*. Unfortunately the manuscript is not among Ruskin's papers, nor do the editors know whether it is anywhere extant. It is clear, however, that the writers of the notices in the *London Review* and the *Geologist* had access either to the author's manuscript or to full shorthand notes. In this volume, therefore, Ruskin's abstract of his lecture (pp. 3–11) is supplemented by the account in the *London Review* (pp. 12–17), and by additional passages from the *Geologist* (see the notes at pp. 4, 5, 7, 9). From the latter source two of the diagrams shown by Ruskin at the lecture are also reproduced (Plates I. and III.).

A main object of the lecture was to protest against the extreme application of the glacier theories which Ramsay's paper was bringing into vogue. Ruskin also combated the explanation given by the Swiss geologists of the north-west face of the Salève. This had been " considered to be formed by vertical beds, raised into that position during the tertiary periods"; Ruskin's investigations led him, on the contrary, to conjecture that "the appearance of vertical beds was owing to a peculiarly sharp and distinct cleavage, at right angles with the beds" (p. 6). "I was the first to point out," says Ruskin of this lecture, "the real relation of the vertical cleavages to the stratification, in the limestone ranges belonging to the chalk formation in Savoy" (p. 98). A drawing which is at Brantwood and is here reproduced, with Ruskin's note, illustrates the phenomena in question (Plate II.). He took the Brezon—*brisant*, breaking wave—as a typical illustration of the mighty wave-like action of the force that moulds gigantic rock-masses almost into breakers ready to nod and fall. The structure of the "Limestone Alps of Savoy" was described in further detail twenty years later by Ruskin's pupil, Mr. Collingwood, in a work so entitled, which was published as a "Supplement to *Deucalion*."

Letters on the Conformation of the Alps (1864).—Ruskin continued the controversy in the letters to the *Reader*, which are collected in the Appendix (II. pp. 548–558). These letters show his usual skill in controversy, and contain many fine passages. They were too desultory to make any decisive mark, but they did not lack attentive

---

[1] *Life and Work of John Ruskin*, 1900, p. 205.

readers. In one of the letters Ruskin supported Forbes's theory of glaciers, and referred to the "envious temper" of some of its critics (p. 550). Principal Forbes noticed the reference, and wrote to Ruskin the letter of thanks which is printed below (p. 561). Ruskin's incursion into the geological field was the subject also of conversation and correspondence with Carlyle, who sent the following letter in reply to some gifts from Ruskin :—

"CHELSEA, 22 *February*, 1865.

" DEAR RUSKIN,—You have sent me a munificent Box of Cigars, for which what can I say in answer? It makes me both sad and glad. *Ay de mi!*

> " We are such stuff
> Gone with a puff,
> Then think, and smoke Tobacco!"

The Wife also has had her Flowers; and a letter which has charmed the female mind. You forgot only the first chapter of *Aglaia :*—don't forget; and be a good boy for the future.

" The Geology Book wasn't *Jukes ;* I found it again in the Magazine,— reviewed there : ' Phillips,' is there such a name? It has again escaped me. I have a notion to come out actually some day soon ; and take a serious Lecture from you on what you really know, and can give me some intelligible outline of, about the Rocks ;—*bones* of our poor old Mother ; which have always been venerable and strange to me. Next to nothing of rational could I ever learn of the subject. That of a central fire, and molten sea, on which all mountains, continents, and strata are spread floating like so many hides of leather, knocks in vain for admittance into me these forty years : who of mortals can really believe such a thing! And that, in descending into mines, these geological gentlemen find themselves approaching *sensibly* their central fire by the sensible and undeniable *increase of temperature* as they step down, round after round—has always appeared to me to argue a *length of ear* on the part of those gentlemen which is the real miracle of the phenomenon. Alas, alas : we are dreadful ignoramuses all of us! Answer nothing ; but don't be surprised, if I turn up some day.

> " Yours ever,
> " T. CARLYLE."

This letter may have suggested to Ruskin to write *The Ethics of the Dust*—a " lecture on Crystallography " which Carlyle highly praised.[1]

Papers on " Denudation " (1865).—Meanwhile the letters on the Conformation of the Alps were followed up in the next year by two papers in the *Geological Magazine* on " The Shape and Structure of

---

[1] See Vol. XVIII. p. lxxiv.

some Parts of the Alps, with reference to Denudation" (pp. 21–34). These papers, which were very carefully written, deal destructively with theories of erosion—in a quizzical and questioning way, however; and with regard to theories of elevation, ask the geologists to explain why some rocks were raised bending and others rigid—questions to which he returned in *Deucalion* (p. 108).

"*Breccias of Mornex*" (A Fragment).—Ruskin's next contributions to his favourite subject were the result of another sojourn in Switzerland. He had gone home from Savoy in May 1863 to give the lecture at the Royal Institution, but in September he returned to Mornex, while in the following month he stayed for some time at the Swiss Baden, "where the beach of the Limmat is almost wholly composed of brecciated limestones." He "began to examine them thoughtfully; and perceived, in the end, that they were, one and all, knots of as rich mystery as any poor little human brain was ever lost in."[1] Already at Mornex he had studied the breccias, and one of his note-books (which afterwards goes on to notes on Banded and Concretionary Formations) opens with a careful description of "Breccias of Mornex," thus:—

> "They lie near the base of the Salève on its slope towards the Alps; just above the village of Mornex there are extensive masses of them, passing beneath the uppermost houses of the village. They occur in fragments all the way towards La Mura. They lie immediately on the Jura limestone and come between it and the molasse, apparently resulting from some violent diluvial action before the molasse was deposited.
>
> "A. They are composed of pebbles or knots of the inferior limestone, and of chert, cemented by a sandy paste.
>
> "The limestone pebbles are partially rounded; the chert masses not rounded, but have much the appearance of being partly secreted from the mass of the rock under some such condition as that of chalk flints. They have none of the branched and irregular processes of chalk flints, but as those are in most cases produced by association with organic forces, the simpler shapes of the chert pebbles of this conglomerate do not much check me in my notion of their being partly produced by segregation. If they were originally detached masses, they have certainly undergone some strange modification since their envelopment, but I can't make them out.
>
> "B. Both these kinds of pebbles have undergone warping and rending since their envelopment. The cherts are separated by

---

[1] *Ethics of the Dust*, § 97 (Vol. XVIII. p. 327).

clefts, and the pieces separated have shifted afterwards to one side or the other; and the clefts are filled with the sand-paste, while similar clefts in the sand-paste are filled with the chert.

"The limestone pebbles seem less rent, but show just the same evidence of contraction of substance as the solid rocks beneath.

"C. Both the limestone and chert are traversed by a cleavage in the same direction, which in the freshly broken mass is traceable in the cherts only."

"*Banded and Brecciated Concretions*" (1867–1870).—Pursuing his study of the breccias at Baden, he saw, as Mr. Collingwood says, that the difference between the limestone formations, in their structural aspects, and the hand specimens, which he picked up in the bed of the Limmat or had already in his collection of brecciated minerals at home, "was chiefly a matter of size, and that the resemblances in form were very close." He had for some time past in his mind a hint of Saussure's that the contorted beds of the limestones might possibly be due to some sort of internal action, resembling on a larger scale that separation into concentric or curved bands which is seen in calcareous deposits. The contortions of gneiss were similarly analogous, it was suggested, to those of the various forms of silica. Ruskin had also been much impressed by a paper by Mr. George Maw (p. 44), in which it had been suggested that apparent strata might be due to segregation. One conclusion which Ruskin drew from these hints was that "if the structure of minerals could be fully understood, a clue might be found to the very puzzling question of the origin of mountain structure."[1]

Such were the general ideas, partly in his mind already and presently to be confirmed by consideration, with which Ruskin set himself to the popular and elementary lessons in crystallography (1865), which he entitled *The Ethics of the Dust* (Vol. XVIII.), and to the papers, "On Banded and Brecciated Concretions," which he contributed to the *Geological Magazine* at intervals during the years 1867–1870, and which are now reprinted in this volume (pp. 37–84). The larger questions, suggested by his analogy between minerals and mountains, the nature of which has already been indicated, are stated by him in § 10 (pp. 43, 44), and more fully in the paper of 1884, "On the Distinctions of Form in Silica," § 29 (p. 386). Ruskin's analysis is, apart from the larger theories which he deduced from it, "a most valuable source of facts and suggestions to the student of agates."[2]

---

[1] W. G. Collingwood's *Life and Work of John Ruskin*, 1900, p. 247.
[2] See the passage quoted from Professor Rupert Jones, below, p. 208 *n*.

Ruskin carried these papers forward at intervals up to 1870; his call to be the Professor of Fine Arts at Oxford then left him no time to continue them, and perhaps, too, he felt that any further analysis of the subject called for work with the microscope, to which he was always disinclined. Fourteen years later, however, he returned to the subject, as we shall see (p. xlix.).

The papers are here reprinted from the *Geological Magazine*, with one or two corrections marked by Ruskin in a set of them which he gave to Mr. Allen.

The *manuscript* of the fifth paper is at Brantwood, now bound up with *Deucalion* material. A sheet of it is here given in facsimile (p. 66).

Forbes, Tyndall, and Glacier Motion.—Ruskin's next appearance in the geological field was in connexion with the fierce controversies which were waged around James Forbes's theory of glacier motion. References to this figure occasionally in *Fors Clavigera*, and occupy a considerable portion of the present volume. It may be well, therefore, to give here some general summary of the dispute, with references to the various passages in which Ruskin deals with the subject.[1]

---

[1] To a friend who remonstrated with Ruskin on the controversial tone of the earlier chapters in *Deucalion*, he replied as follows (October 14, 1875):—

"DEAR MR. WILLETT,[1]—Instead of trying for the things that make for *peace*, try only for those, at present, that 'make for truth.'[2]

"There is no connection whatever between the action of tidal waves and glaciers. A receding tide *sucks back*; a 'receding,' so called, glacier no more sucks back than a charging regiment, consumed by shot in its advance. And a glacier, as I told you, has more vigorous *advancing* action when appearing to recede than when appearing to advance. Get this once clearly into your head and you will see that there is no more possibility of using equivocal terms.

"If you really find that *Deucalion* gives you neither new ideas nor information, I can only say I am glad it is so orthodox, and that as either other geologists or I must use very loose language, I am at least so far trustful in my care of expression as to think they had better adopt mine than I theirs. It is at least not *mere* looseness of language in which they describe effects taking place on a sea bottom they never reached.

"There is not the slightest fear of men who are doing good work criticising each other wrongly. Donatello and Ghiberti criticise each other, and that fiercely and mercilessly. Your modern Academician follows in *your* sense the 'things that make for peace,' and would no more find fault with his neighbour's work to his face, than praise it, if he could help it, behind his back.

"Ever faithfully yours,
"J. RUSKIN."

---

[1] For Mr. Willett's own contributions to *Deucalion*, see pp. 206, 212 n.; and for other references to him, see Vol. XIV. p. 351, Vol. XVI. pp. lxvi., 255, Vol. XVIII. p. 203.
[2] Romans xiv. 19.

In *Fors Clavigera*, Letter 34, § 13 (Vol. XXVII.), Ruskin gives a résumé of the state of glacier knowledge as it existed up to 1840. Everybody knew that glaciers moved. Byron had put the fact in lines which Ruskin quotes as illustrative of the poet's "measured and living truth": [1]—

> "The glacier's cold and restless mass
> Moves onward day by day."

Two theories were in the field to account for this motion of glaciers. One was the "Sliding Theory" of Saussure; the other, the "Dilatation Theory" of Charpentier. Ruskin discusses and dismisses these in *Deucalion*, pp. 228, 229. Nobody had as yet conceived of the ice as being "anything but an entirely solid substance."

In 1840 M. le Chanoine Rendu, afterwards Bishop of Annecy, published his *Théorie des Glaciers de la Savoie*.[2] In this work he likened the motion of glaciers to that of rivers: "There is between the Glacier des Bois and a river a resemblance so complete that it is impossible to find in the glacier a circumstance which does not exist in the river. In currents of water the velocity is not uniform throughout their width nor throughout their depth; the friction of the bottom, that of the sides, the action of obstacles, causes a variation in the velocity which is undiminished only towards the middle of the surface. Now, the mere inspection of the glacier is sufficient to prove that the velocity of the centre is greater than that of the sides" (ch. viii. p. 85 of the English edition). But M. Rendu did not profess to have made observations establishing the theory which he here foreshadows. "The fact of motion exists," he writes, "but the mode of motion is entirely unknown. Perhaps with long observations, with experiments upon ice and snow carefully made, we shall succeed in grasping it; but we are still in want of these first elements" (ch. viii. p. 78). And again: "There are a host of facts that would seem to induce the belief that the substance of glaciers enjoys a kind of ductility which allows it to mould itself upon the locality which it occupies—to thin out, to swell, to contract, and to spread as a soft paste would do. Nevertheless, when we deal with a piece of ice, when we strike it, we find in it a rigidity which is in direct opposition to the appearances of which we have just spoken. Perhaps

---

[1] See *Præterita*, i. §§ 172, 173; and compare Vol. I. p. 202 (a passage written by Ruskin in 1836, six years before Forbes published his theory).

[2] Published under this title at Chambéry. Translated into English in 1874 in a volume entitled *Theory of the Glaciers of Savoy, by M. le Chanoine Rendu, translated by Alfred Wills*, etc., etc. (see p. 559 n.).

experiments made upon larger masses would give other results" (ch. vii. p. 71).

Rendu, then, had by a process of intuition come very near to the heart of the mystery, but he had not made experiments or observations to establish any theory. He was, we may say, on a hot scent, but he had not found.

Meanwhile another scientific man had been experimenting on the glaciers. This was Agassiz of Neuchâtel, who had built himself a hut on the Unteraar Glacier, whence he had made many ingenious observations (see *Fors*, § 14). The hut was built of stones in 1840, against a great boulder on the central moraine of the glacier, and was known as the Hôtel des Neuchâtelois (see p. 228).[1] In 1841 Agassiz invited James Forbes, who was then travelling in Switzerland, to visit him there. Forbes went, and on this occasion noticed particularly the veined structure of the glacier which played an important part in the suggestion, and in the proof, of his own theory. Forbes revisited the glaciers in 1842, and he wrote from Switzerland four letters on Glaciers to Dr. Jameson.[2] These contained the original draft of the Plastic or Viscous Theory, which was expounded in 1843 in a more methodical and detailed manner in his *Travels through the Alps of Savoy and other Parts of the Pennine Chain*[3]—the theory being, as Ruskin puts it (*Fors*, 34, § 14), "that glaciers were not solid bodies at all, but semi-liquid ones, and ran down in their beds like so much treacle."

Forbes's theory is stated by himself in the following terms:—

"My theory of glacier motion then is this: a glacier is an imperfect fluid, or a viscous body, which is urged down slopes of a certain inclination by the mutual pressure of its parts" (*Travels*, p. 366, ed. 1900).

His claim to have established this theory rests upon the following several observations: "(1) The motion of the ice is continuous. (2) The centre moves faster than the sides. (3) This change of velocity is continuous. (4) The variation of velocity in the breadth of a glacier is proportional to the absolute velocity, at the time, of the ice under experiment. (5) The velocity of a glacier increases when the steepness of its bed increases. (6) A continuous mass of ice on the Brenva

---

[1] The Hôtel itself was destined in later times to furnish evidence of glacier motion. In August 1884 portions of the great boulder were found lower down on the glacier, proving that since 1840 the rate of travel had been 55 metres a year (see *Alpine Journal*, vol. xii. pp. 177, 229).

[2] The letters are dated July 4, August 10, August 22, and October 5, and first appeared in the *Edinburgh New Philosophical Journal*, October 1842 and January 1843.

[3] Second edition, 1845. Reprinted in 1900, with other papers, under the title *Travels through the Alps*, edited by W. A. B. Coolidge.

became distorted without the formation of any crevasses. (7) The greater the heat and rainfall, the more liquid is the ice, as shown by its greater velocity. (8) The melting of the surface in summer is compensated by a thickening (resulting from diminished fluidity in the viscous mass) in winter. (9) In cases where the tenacity of the ice is insufficient to bear the strain crevasses are formed. (10) These crevasses, and also the broken continuity of the whole mass, are closed up by time and cohesion. (11) The differential motions of a viscous mass account for the veined structure. (12) In every respect, experiments on viscous substances show motions similar to those of glaciers."[1]

The publication of this theory caused a great stir in scientific circles. It encountered the two stages which, as Whewell observed in a letter to Forbes, is the common lot of such things—"the stage when people say they are not true, and the stage when people say they are not new." Rendu did not enter into the controversy, but he was presently, as we shall see, used as a stick to beat Forbes with. Agassiz, who had missed what had been under his eyes for so many years, and whose visitor had come and walked off with what might have been his nugget (*Fors*, Letter 34, § 15), was furious. He wrote indignantly to Forbes accusing him of perfidy;[2] while his companion, M. Desor, declared that Forbes had only "rediscovered after Rendu,"[3] and "found consolation in describing the cowardice of the Ecossais on the top of the Jungfrau."[4]

It was the second stage of the controversy which caused Ruskin's intervention. English men of science had contended that the theory was not new, and that it required emendation by themselves. Some of the leading incidents in the battle must be mentioned.

In 1857 Huxley and Tyndall published conjointly, in the *Philosophical Transactions* of the Royal Society, a paper on Glacier Motion.[5] They

---

[1] This is Professor George Forbes's summary, given in his Introductory Remarks to *Rendu* (as cited on p. xxxiv. *n.*), pp. 13, 14.

[2] See *Excursions et Séjours dans les Glaciers et les Hautes Régions des Alpes de M. Agassiz et de ses compagnons de voyage*, by E. Desor, Neuchâtel, 1844, pp. 438-443.

[3] *Ibid.*, p. 441.

[4] See *Fors*, 34, § 15. M. Desor comments on Forbes's silence, "réserve excessive," and "coldness in presence of the most magnificent scenes of nature" (*Excursions et Séjours*, pp. 396, 436), when on the top of the Jungfrau; but it is not clear that he intended, as is assumed by Ruskin, to impute to Forbes any lack of physical courage.

[5] *Philosophical Transactions of the Royal Society of London for the year 1857*, vol. 147. No. XV. "On the Structure and Motion of Glaciers." By John Tyndall, F.R.S., Professor of Natural Philosophy, Royal Institution; and T. H. Huxley, F.R.S., Fullerian Professor of Physiology, Royal Institution, pp. 327-346.

started, in opposition to Forbes, what is known as the "Regelation Theory," founded on an experiment made by Faraday in 1850 (see p. 127 *n.*), and supported by experiments made by Tyndall himself. According to this theory, as Ruskin sums it, "a glacier advances by breaking itself spontaneously into small pieces, and then spontaneously sticking the pieces together again" (p. 230). On Tyndall's desire to show that his theory explained something which Forbes's did not, Ruskin pours contempt in *Fors*, Letter 34; and to like effect, Professor George Forbes writes (against § 424 of *Forms of Water*), "It appears, then, that the addition made by Dr. Tyndall to Forbes's theory is that ice cracks under tension sufficient to crack it."[1] "The capacity of ice for sticking together," says Ruskin, "no more accounts for the making of a glacier, than the capacity of glass for sticking together accounts for the making of a bottle" (p. 127).

To this paper Forbes replied in much the same sense, finding nothing in the experiments of regelation to modify his theory.[2]

The controversy now entered upon a personal stage, and charges of bad faith were made or insinuated against Forbes. In 1859 it was proposed in the Council of the Royal Society that Principal Forbes should receive the Copley Medal in acknowledgment of his researches upon glaciers. The proposal was made by Whewell; it was opposed by Tyndall's friends (especially Huxley), on the ground that Forbes's theory had been anticipated by Rendu and that Forbes had omitted to give sufficient acknowledgment of Rendu's researches. The medal was ultimately not awarded to Forbes (see the English edition of *Rendu*, pp. 9–12).

In 1859 Forbes collected his contributions to the subject in a volume entitled *Occasional Papers on the Theory of Glaciers, now first Collected and Chronologically arranged, with a Prefatory Note on the Recent Progress and Present Aspect of the Theory.*

In 1860 Tyndall published his *Glaciers of the Alps*. In this book he restated his Regelation Theory; made various criticisms upon Forbes's viscous theory; and suggested that Forbes had insufficiently recognised Rendu's contributions to the subject.[3] Tyndall's quotations from Forbes omitted Forbes's own acknowledgment of Rendu's researches, and the omission was commented upon by Lyon Playfair (see his letter of July 25, 1860, in the English version of *Rendu*, p. 3).

---

[1] *Rendu* (as cited above), p. 178.

[2] See Chapter XX. in Forbes's collection of *Occasional Papers* (pp. 228 *seq.*), the chapter being reprinted from the *Proceedings of the Royal Society of Edinburgh,* April 19, 1858.

[3] *Glaciers of the Alps*, 1860, pp. 299 *seq.*

Later in the year Forbes issued his *Reply to Professor Tyndall's Remarks in his work " On the Glaciers of the Alps" relating to Rendu's " Théorie des Glaciers."*

The case against Forbes was in July 1863 restated by the *Quarterly Review* (vol. 114, pp. 77–125), in an article which gave him much pain (see p. 561 *n.*). The writer disparages Forbes and greatly praises Tyndall; insinuating also that Rendu's work had been ignored by the former. The attack on Forbes was the more marked, as in an earlier number of the *Review* (vol. 74, pp. 39–70) Forbes's work had been appreciatively noticed.

Ruskin had from the first been convinced by Forbes's arguments. He had met Forbes, as we have seen, in 1844; and though he never saw him again, he greatly respected and admired his character, and he accepted Forbes's book as the standard work on its subject.[1] Ruskin took occasion accordingly, in his lecture at the Royal Institution in 1863, to refer to Forbes's theory as firmly established (p. 10); and, in his letters to the press in 1864, to defend Forbes from his assailants (see below, p. 550; and compare above, p. xxix.). Forbes's letter of thanks to Ruskin (p. 561) shows how bitterly Forbes felt the attacks made upon him. He died in 1868.

In 1872 the attacks were renewed. In that year Tyndall published *The Forms of Water in Clouds and Rivers, Ice and Glaciers.* In this he returned to the charge against Forbes, though more by implication than by express statement. He laid great stress on Rendu's work (§§ 397 *seq.*), whilst of Forbes he said that from his "able and earnest advocacy, the public knowledge of this doctrine of glacial plasticity is almost wholly derived" (§ 404), thus leaving it to be inferred that Forbes was only a populariser of another man's theory. Next, Tyndall questioned, without expressly denying, the viscous theory, and put forward once more as an emendation his own theory of regelation (§§ 412 *seq.*).

In 1873 Principal Shairp, Professor Tait, and Mr. Adams-Reilly published their *Life and Letters of James David Forbes.* In this work, "in the interests of truth and justice to the dead," they noticed the renewal of Tyndall's attacks, and vigorously replied to them; reprinting also, in an appendix, Forbes's own reply of 1860. In the number of the *Contemporary Review* for August 1873, Tyndall wrote a rejoinder, which he presently reprinted as a pamphlet—*Principal Forbes and his Biographers.*

---

[1] See the summary of references to it in *Modern Painters*, vol. iv. (Vol. VI. p. 214 *n.*).

The publication of *The Forms of Water* brought Ruskin into the fray. The number of *Fors Clavigera* for October 1873 (Letter 34) was largely devoted to a vigorous onslaught upon Tyndall. He satirises the style and scope of the book (*Fors*, § 12); partly defends Forbes for his method of proceeding towards Agassiz (§ 15); condemns Tyndall's veiled attack on Forbes's theory (§§ 16, 17); and "chastises" the "criminal indulgence" of "the jealousies of the schools" (§ 17).

Forbes's son, and some of his friends, were meanwhile taking independent action in consequence of the publication of Tyndall's book and his pamphlet. Their reply took the shape of the volume (described below, p. 559 *n.*), containing Rendu's memoir together with a translation of it, so that English readers might be able to judge for themselves whether Forbes had in fact been anticipated by Rendu. There were added in the book, "Introductory Remarks" by Professor George Forbes; and a reply to Tyndall, analysing the whole controversy, by Professor P. G. Tait. Ruskin's article was also reprinted from *Fors Clavigera*, and to it he appended some additional remarks, here given in the Appendix (pp. 559–562). In these, he does not deal with the question of priority as between Rendu and Forbes, but delivers an eulogium on the character of the latter.

In the Oxford lectures on glaciers (1874), partly incorporated in *Deucalion*, Ruskin returned to the subject of glacier motion. He asserts that they are "fluid bodies," and reiterates the statement that Forbes first "discovered and proved the fact" (p. 125). He mentions Rendu's lucky guesses, but notes some of his errors (p. 132). He also criticises Tyndall's *Glaciers of the Alps*, a work, he says, in which the author was "eager to efface the memory of Forbes's conclusive experiments" (p. 139).[1] He gives a summary of rival glacier theories, and shows

---

[1] In a conversation with Mr. M. H. Spielmann, published in the *Pall Mall Gazette* of April 21, 1884, Ruskin is thus reported :—

"Another favourite topic of Mr. Ruskin's is the shortcomings of our men of science. On this he descanted with great vehemence. 'The majority of them,' said he, 'have no soul for anything beyond dynamics, the laws of chemistry, and the like. They cannot appreciate the beauties of nature, and they regard the imaginative man—one who can feel the poetry of life—as a donkey regards his rider : as an objectionable person whom he must throw off if he possibly can. Such a man is Tyndall. The *real* scientific man is one who can embrace not only the laws that be, but who can feel to the full the beauty and truth of all that nature has to show, as the Creator made them. Such a man was Von Humboldt, such a man was Linnæus, such a man was Sir Isaac Newton. As regards my opinion of Tyndall, I admire his splendid courage (I am a dreadful physical coward myself: I enjoy my life too much ever to risk the losing of it), and his schoolboy love of adventure ; he has a real and intense interest in the subjects which he takes up : naïve to a degree—incurably

that Forbes alone had satisfactorily explained the facts.[1] While accepting and endorsing Forbes's account of the matter as conclusive, Ruskin adds as a suggestion of his own that the influence of the "subsiding languor" of a glacier's "fainting mass" is itself a constant source of motion (p. 133).

Oxford Lectures on Glaciers (1874).—Ruskin's intervention in defence of Forbes had begun, as we have seen, in *Fors Clavigera* for October 1873, and he returned incidentally to the subject of glacier theories in the following month. He intended to pursue it, but presently came to the conclusion that he had better deal with the subject in Oxford lectures than in *Fors*.[2] These lectures were first announced in March 1874, but Ruskin found it necessary to postpone them[3] till the October term, and went abroad for several months. Further study of the glaciers themselves was one of the objects which he proposed to himself for this tour,[4] and Mr. Allen—his companion on many geological rambles in earlier years—was to meet him at Courmayeur. "I shall go to Courmayeur," he wrote from Assisi (June 8), "and study the Brenva Glacier; it is the riband structure I want to make out.

so; but he has never felt himself to be a sinner against science in the least because of his all-overwhelming vanity. His conduct to James Forbes respecting the Glacier Theory was the outcome of the schoolboy feeling when he sees the Alps for the first time: "Good gracious! no one ever saw this before; and I can tell the world all about it as no one ever did before!" And here is this nuisance of a man who has told the world what is not true, and so, hoping with his whole soul that Forbes is wrong, and hoping and expecting that he is right, he does all he can to get Forbes out of the way and to get people to believe in his theory. Why, he has set back the Glacier Theory twenty years and more! But before long people will find that this theory was all decided before this conceited, careless schoolboy was born. And that is why I always attack him, and shall continue to do so until I die. The whole attitude of the scientific world at present is: "We shall discover everything entirely afresh, no matter who discovered it before—especially James Forbes: we shall believe nothing that Forbes has said if we possibly can help it, and we will believe anything that he has not said if we possibly can;" and as he has said the exact truth in that matter the result has been extremely unfortunate for science in general!'"

[1] Professor Albert Heim, in his *Handbuch der Gletscher Kunde* (Stuttgart, 1885), finds fault, however, with Forbes: "He speaks of a theory of plasticity, but it is rather a case of an abstraction from the facts, than an explanation based on the physical properties of ice. He says in fact little more than that the glacier moves like the stream because the ice must be viscous. His error consisted in always comparing viscous (zähflüssige) substances with the glacier, instead of those whose inner cohesion is less than their internal friction" (pp. 190–191). But Professor Heim's own account of the matter seems to come to much the same result (see pp. 336–337).

[2] See Letter 43, § 16.

[3] See Vol. XXIII. p. xxx.

[4] See, again, *Fors Clavigera*, Letter 43, § 16.

I think Forbes insufficient on this point only "; for as he explained in a later letter (Assisi, June 20):—

> "Forbes has shown perfectly that it forms at right angles to the pressure. But he doesn't show how the pressure forms it. He supposes a series of rents, filled with solid ice. But this would not cause a regularly successive structure at all. There would be small cracks running together into large ones; small solid veins ramifying into large ones. But no successive conditions."

Ultimately Ruskin became absorbed in his *Mornings in Florence*, and Mr. Allen was left to examine the glaciers by himself, as also to prepare for Ruskin a digest of Tyndall's book on the *Glaciers of the Alps*.[1] Ruskin himself on his way home from Italy spent a fortnight at St. Martin and Chamouni, and there began to rewrite his lectures for Oxford on the glaciers.[2] His description of the Valley of Cluse in *Deucalion* (below, pp. 148 *seq.*) refers to this visit. The lectures, delivered in October 1874, were markedly successful, as we have seen;[3] and this fact was no doubt one of the causes which led him to begin collecting his studies in geology and mineralogy, though at one time he intended to publish his lectures on glaciers separately.[4]

"DEUCALION" (1875–1883)

The publication of *Deucalion*, as of *Proserpina*, was spasmodic—the first "Part" appearing in October 1875, the last in May 1883. Though there was a consistent purpose in the book, and though certain leading subjects of inquiry were pursued, yet it is in a manner formless. He had so many irons in the fire, so many balls in the air, that he did not concentrate himself continuously on any one subject. And presently, as we saw in the last Introduction,[5] illness and weakness intervened; but even before that time he could only (as he said) pursue his various

---

[1] After joining Ruskin at Florence, Mr. Allen went to Chamouni, where he made many observations and collected information for Ruskin. On a point mentioned by Ruskin in this volume (p. 132), Mr. Allen's diary contains the following entry:—

"My guide, Joseph Tairraz, says that rain does fall at times right up on the summit of Mont Blanc, and that he has been in a pelting rain on the Grand Plateau. Also that when the sun is hottest at midday snow will melt on rock even at the summit. He also says that water issues from the Glacier des Bois all winter, more or less, no matter how severe the weather is" (August 17, 1874).

[2] See Vol. XXIII. p. lii.

[3] *Ibid.*, p. liii.

[4] See *Fors Clavigera*, Letter 53, § 1 *n.*

[5] Vol. XXV. pp. xxvi.–xxviii.

studies "as a kind of play, irregularly, and as the humour comes upon me."[1] The various books which he had in the press were like drawers into which he flung whatever material happened to come to hand, sometimes without any scrupulous nicety of classification. Thus when he gave a public lecture, it was utilised for a book; and if there was no book into which it precisely fitted at the time, it had to find a place where best it could.

Such was the case, as he frankly explains, with the lecture on Snakes, delivered in 1880, which was made to serve as a chapter in *Deucalion*, though it contained nothing about geology or mineralogy—a serpentine method of ordering his materials which was not very fully covered by calling the lecture "Living Waves." To his description of the movements of snakes he attached some importance. "Actually at this moment (Easter Tuesday, 1880)," he wrote, "I don't believe you can find in any scientific book in Europe a true account of the way a bird flies—or how a snake serpentines. My Swallow lecture was the first bit of clear statement on the one point; and when I get my Snake lecture published, you will have the first extant bit of clear statement on the other; and that is simply because the anatomists can't, for their life, look at a thing till they have skinned it."[2] Ruskin's account of the motion of snakes, which he compares to skating, will be found on pp. 316–318. He had for many years made a study of the colours and movements of Snakes; the occasion of the lecture was, however, accidental. He had happened to hear an address by Huxley on the same subject, which had at once profoundly interested and challenged him. His own lecture on Snakes "became, apparently," says Ruskin, "rather a piece of badinage suggested by Professor Huxley's than a serious complementary statement." As a piece of playful and graceful banter, the chapter is one of Ruskin's happiest and most harmonious pieces; but "nothing," he says, "could have been more seriously intended." It gave his "spiritual version of the development of species," as supplementary to Huxley's discourse on the physical evolution of the snake.

Of the impression made by the lecture at the time of its delivery, a note has been written by Mr. Wedmore:—

"The charm of the later writing—so different both from the exuberance of the earlier, and the measured strength and gravity of some of the middle period, in the *Seven Lamps* for instance—was, so far at least as I can tell and remember, the charm of the later man; the man approaching

---

[1] *Fors Clavigera*, Letter 59 (November 1875), § 1.
[2] "Letters on a Museum or Picture Gallery" (in a later volume of this edition).

old age; claiming some of its privileges; exercising its rights of unfettered affection towards the persons and objects it chose; chiding; encouraging; asking to be accepted implicitly. . . . Only twice did I see him. Once was at a lecture—his lecture on 'Snakes' at the London Institution. Once was at a house in the Prince of Wales's Terrace, or very near it, at Mrs. Bishop's, where for her delight and that of her friends he lectured privately,[1] and charming it was, but the effort was less admirable and less complete than at the London Institution. At both places, what one felt about him was that he was benign and bewitching; but at the London Institution— perhaps owing in part to the greater urgency and reasonableness of his theme (it was a protest, indignant, affectionate, against the evils of cramming [2])—at the London Institution he had the most of force and of depth.

"I remember well his advent—the door opening at the bottom of the theatre—and, with William Morris, I think, and certainly Frederic Leighton and other friends, and patting Leighton on the back (or was it William Morris?) a little nervously, yet bearing himself bravely, the observed of observers, this man of world-wide fame, and, what is so much more impressive and important to those who feel it at all, of extraordinary and magnetic genius—this genius was suddenly amongst us. And, gravely and slowly, with a voice at once of good quality and a rough Cumbrian burr,[3] Ruskin began his discourse. All listened intently; and as the theme developed, and his interest in it gained, and as he felt—for he must have felt—that he held us in the hollow of his hand—the fascination increased, and the power and beauty that justified it. I have heard, with great delight, another impressive genius—Tennyson—read some of his poems. The enjoyment was singular; the experience remarkable. But, in the drawing-room in Manchester Square, the author of 'The Revenge; a Ballad of the Fleet' reached no effect that was quite so much of an enchantment as did John Ruskin, with his voice more and more wonderful and tender, that March afternoon in Finsbury."[4]

Ruskin exhibited at the lecture several drawings and diagrams, of which he distributed an explanatory account. This is here printed as a note to the lecture (p. 330). He subsequently issued six sheets of photographic reproductions for binding up with *Deucalion*. The subjects are all here included (see p. 295 *n.*).

Another lecture, incorporated in *Deucalion*, is of earlier date. This

---

[1] An account of the lecture, which was concerned with Miss Alexander's drawings, appeared in the *Spectator* of June 10, 1883, and is included in a later volume of this edition. An account of it is also given in *St. George*, vol. ii. pp. 55-57.

[2] See below, p. 328.

[3] But there is no such thing, and Ruskin's R was not the Northumbrian burr.

[4] "A Note on Ruskin," by F. Wedmore, in the *Anglo-Saxon Review*, March 1900, p. 136; reprinted in *Whistler and Others*, 1906.

is the lecture on the heraldry of Stones, called "The Iris of the Earth," which he gave at the London Institution in February 1876. Mr. Collingwood has given an amusing account of Ruskin's preparations for this lecture. He bade "one of his younger friends run to various Professors and make inquiries about etymological and mineralogical details:—'What else are the Professors there for?' he would say; and he would be greatly impressed if we could answer his questions without appeal to the higher powers." The day after the first lecture he wrote to Mr. Collingwood :—

> "Those French derivations are like them. No authorities on Heraldry are of the slightest value after the fifteenth century—even Guillim is only good for something in the first edition, the rest nowhere. My pearl is all right—I got it from the Book of St. Albans, 1480 [1]—but my shield is not absolutely in the old terms. I invent 'Colombin,' for the old 'plumby,' and use 'écarlate' for 'tenné'—mine is to be the norma for St. George's heraldry, not a merely historical summary. I hope to be back on Saturday evening. . . . The lecture went well and pleased my audience—and pleased myself better than usual, in that I really got everything said that I intended, of importance." [2]

Another lecture to which Ruskin attached importance was delivered at Kendal (and afterwards at Eton) on "Yewdale and its Streamlets." A full report of this lecture was separately published in pamphlet form, before it was included in *Deucalion;* and some additional passages from the pamphlet are here given in notes (see pp. 260, 261).

These lectures, as incorporated in *Deucalion*, had immediate connexion with the proper theme of the book. The lecture on Stones introduces his chapters on mineralogy, and the lecture on "Yewdale and its Streamlets" carries the studies of erosion from the Alps to the hills of the Lake District. But the order of the chapters is sufficiently indicative of the interruptions which occurred in the book. It begins with general discussions of mountain form, opening up some of the questions which Ruskin desired to submit for revision or reversal, [3] and then passes to theories of glacier motion. This portion of the book (i. chaps. i.–vi.) corresponds with the ground covered in the Oxford Lectures of 1874. Then, in the middle of the discussion of glaciers, Ruskin goes off to mineralogy. The diversion was due to the fact that from 1876 onwards he was deeply interested and occupied with

---

[1] See below, pp. 182, 184, 187.
[2] *Life and Work of John Ruskin*, 1900, p. 329.
[3] See i. ch. xiv. §§ 1, 2 (pp. 273, 274), and p. lxiv.

plans for his St. George's Museum at Sheffield. Chapter VII. (the London Institution lecture of 1876) is a general introduction explanatory of the connexion in his mind between mineralogy and heraldry and art. Chapter VIII. describes the plan which he proposed for the arrangement of the collection of minerals in his Museum; whilst in Chapter IX. he opens up some of the questions in mineralogy which he wished his students to consider. But in the autumn of 1876 he went abroad, and stayed for a few days among the Alps on his way to Venice.[1] This brought him back in Chapter X. (dated "Village of Simplon, 2nd September, 1876") to the glaciers, and we have a further discussion of theories of glacier motion. The occasion passes; and in Chapter XI. he reverts to the subject of crystallisation. He returned home to Brantwood in the summer of 1877; and, recognising that his days of active exploration among the Alps were over, "began, as better suited to my years, the unadventurous rambles by the streams of Yewdale, whose first results were given in my Kendal lecture"[2] (October 1877). This became Chapter XII. The scene is changed, but the argument remains constant, and Ruskin raises by the streams of Yewdale the same questions—of erosion, of cleavage, elevation—as by the cliffs of Uri or aiguilles of Chamouni. A long interruption, caused by his illness of 1878, now intervenes. When the work is resumed, a chapter (XIII.) on the stellar crystallisation of silica comes first, and the first volume closes (Chapter XIV., "Schisma Montium") with a general account of the questions which he desired to raise with regard to geological theories respecting the causes of mountain form.

The further development of such themes was intended to occupy a second volume of *Deucalion*. He had made many experiments of which the results were promised for that volume (see p. 291). Just as at Broadlands he had made "experiments on the glaciers"[3] with butter and honey, so at Brantwood in 1877–1879 he occupied himself with experiments on dough in order to observe the behaviour of ductile substances under pressure. He still intended also to carry forward his investigations into glaciers. In this connexion Mr. Collingwood prints[4] the following letter (July 25, 1879) which he received from Ruskin:—

"Yes, Chamouni is as a desolated home to me—I shall never, I believe, be there more: I could escape the riff-raff in winter, and early spring; but that the glaciers should have betrayed me, and

[1] Vol. XXIV. p. xxxiv.
[2] Preface to *The Limestone Alps of Savoy*, § 6 (below, p. 570).
[3] Vol. XXIV. p. xxi.
[4] *Life and Work of John Ruskin*, 1900, p. 338. Mr. Collingwood made the drawings, but as the plans for *Deucalion* were changed, Ruskin did not use them

their old ways know them no more, is too much. . . . I was gladly surprised to hear of your going to the Aiguille du Tour, if the whole field around it is still pure; but all's so wrecked; perhaps it's all mud and stones by this time.

"However, the thing I want of you is to get as far up the old bed of the Glacier des Bois as you can, and make a good graphic sketch for me of any bit of rock that you can find of the true bottom among the débris. *Graphic*, I say,—as opposed to coloury or shadowy; show me the edges and ins and outs, well—with any notes of direction and effect of former ice on it you can make for yourself. You know I don't believe the ice ever moves at the *bottom* of a glacier at all,—in a general way; but on so steep a slope as that of the Bois, it may sometimes have been dragged a little at the bottom, as it is ordinarily at the sides. Anyhow, sketch me a bit of the rocks, and tell me how the boulders are lodged, whether merely dropped promiscuously, or driven into lines or corners.

"Please give my love to the big old stone under the Breven, a quarter of a mile above the village, unless they've blasted it up for hotels."

When, however, *Deucalion* was resumed other material happened to be ready to hand, and the second volume began with the lecture on Snakes already discussed. Chapter II. ("Revision") contained a general account of Ruskin's faith "in the creating Spirit, as the source of Beauty—in the governing Spirit, as the founder and maintainer of Moral Law." The choice of subject for the next chapter (the last to be published) was conditioned by the severe winter of 1878–1879. This afforded unusual opportunities for studying the crystallisation of ice. He intended to pursue this subject, and the MS. material at Brantwood contains part of an additional chapter, which is here given (p. 363). A few other notes for *Deucalion* are also added, but the book was never finished. Such time as he could spare to these studies was henceforth given, as we shall see, to the arrangement of minerals.

With regard to the title of the book—*Deucalion: Collected Studies of the Lapse of Waves and Life of Stones*—Ruskin dedicated, as his studies of flowers to *Proserpina*, so those of rocks and waves to *Deucalion* —the Noah of Greek tradition, who, when all the world beside perished in the great flood, saved himself and his wife on the top of a high mountain. And when the waters subsided, *Deucalion* consulted the oracle, and was bidden to repair the loss of mankind by throwing behind them the bones of their mother; that is, the stones of the

Earth—"lifeless seed of life," as Ruskin says.[1] He seems, however, to have felt that his title was somewhat cryptic; for in his copy of the book, marked for revision, there is the following note:—

> "Stones. First text: 'God is able of these stones to raise up children upon Abraham.' Second: 'If these should hold their peace, the very stones would immediately cry out.'[2] Then quote the end of the Introduction and the piece about myths [*i.e.*, § 5; see pp. 98–99], explaining the value of the myth of Deucalion as connected also with the story of Lycaon and of Philemon and Baucis."

Unfortunately Ruskin did not carry out the intention of revising *Deucalion*, and one can only roughly guess at the ideas which he intended to develop. The flood of Deucalion occurred, according to some mythographers,[3] as a punishment for the impieties of Lycaon and his sons. The piety of Philemon and Baucis, on the other hand, was rewarded by Zeus taking them to a safe eminence when all the world beside was visited by the flood.[4] Thus the story of Deucalion was, as Ruskin says, the story of the Betrayal and the Redemption. To Ruskin the study of natural phenomena was part of the evidence of natural theology. He found "sermons in stones, and good in everything."

Of the *manuscript* of *Deucalion*, Mr. F. W. Hilliard possesses (with some omissions) that of vol. i. ch. ii. § 5 to ch. iii. § 13. This shows the usual careful revision. Mr. Hilliard also possesses the MS. of a chapter entitled "Pruina Arachne" (see p. 347 *n.*). Mr. A. Macdonald has the MS. of vol. i. ch. xiv. §§ 1–14. No other MS. is known to the editors.

There is at Brantwood a copy of the first volume containing several notes and revisions by the author. Some of these notes were written, as an entry at ch. vi. § 13 shows, at Sallenches in 1882. Occasional use has been made of this copy (see, *e.g.*, pp. 110, 112, 151).

To the first volume of *Deucalion* Ruskin compiled an index to which he attached importance as "classifying the contents so as to enable the reader to collect all notices of importance relating to any

---

[1] Below, p. 555. It appears from a letter to Dr. John Brown (July 24, 1874) that Ruskin originally intended to call the book "Monte Rosa." In the first proof of the title-page (in Mr. Allen's possession) the sub-title reads, "Collected Studies of the Labour of Waters and Life of Stones." On the words *lapse* and *labour*, see *Munera Pulveris*, § 59 (Vol. XVII. p. 183).

[2] Matthew iii. 19; Luke xix. 40.

[3] Apollodorus, iii. 8, 1.

[4] Ovid, *Metamorphoses*, viii. 621 *seq.*

one subject and to collate them with those in my former writings."
The index was, he adds, "made as short as possible" (p. 273). It has
seemed desirable to reprint this index substantially as Ruskin wrote
it; but additions have been introduced. Some of these were noted by
him in his copy for revision; the others supply references to the papers
collected in this volume.

CATALOGUES OF MINERALS (1883–1886)

The division of this volume which follows *Deucalion* brings us to
work which occupied a good deal of Ruskin's time in 1883 and the next
few years. This was the overhauling of his collections of minerals and
precious stones; the formation out of them of collections or specimens
which he presented to schools or public institutions; and the writing
of explanatory catalogues.

He had always assiduously collected minerals during his tours; many
pages in his diaries are occupied with lists of the specimens acquired
on a particular walk, or a particular tour. He also bought largely from
dealers,[1] and his collection at Brantwood, which was especially rich in
siliceous minerals, numbered 3000 specimens, many of them of great
rarity and special interest. It was a favourite diversion to arrange
and rearrange, classify and reclassify, these specimens; and there remain
at Brantwood catalogues in various stages of completion. The work
into which *Fors Clavigera* had drawn him caused him to begin dis-
persing his collection. There was the St. George's Museum at Sheffield
to be equipped; and while selecting specimens for that, he went on to
make up typical collections for various museums and schools in which
he was interested. He had long entertained a design "of making
mineralogy, no less than botany, a subject of elementary education,
even in ordinary parish schools, and much more in our public ones.
With this view," he says, "long before the Guild existed, I arranged
out of my own collection a series of minerals which were found useful
at Harrow; and another for a girls' school at Winnington, Northwich,
where the lectures on mineralogy were given which I afterwards expanded into *The Ethics of the Dust*."[2] The gift of a collection of
minerals was made to Harrow School in 1866, in connexion with two
lectures given there by Ruskin in that year and in the year next

---

[1] One of his purchases involved him in a lawsuit: see *Fors Clavigera*, Letter 76,
§ 18.

[2] *The Guild of St. George: Master's Report*, 1884, § 6.

following.[1] The collection consisted of 237 specimens, and was accompanied by a list, which, however, was not written by Ruskin. The Headmaster (Dr. H. M. Butler) presented a suitable showcase, with drawers underneath, for the housing of the collection, which stood in the Vaughan Library until the opening of the Butler Museum in 1886. The collection, as now arranged in that Museum, follows the order of the original list, and each specimen bears a descriptive label, headed "Ruskin Collection." Of Ruskin's interest in the school at Winnington full account has already been given.[2] The school was afterwards broken up, and the Ruskin collection of minerals was probably dispersed, as were some other gifts of his. "What time I could spare to the subject in later years" was spent, Ruskin says, in "systematizing my knowledge of the forms of Silica."[3] Partly in connexion with the St. George's Museum at Sheffield, and partly from his interest in the rearrangement of the minerals in the British Museum, after their transference from Bloomsbury to South Kensington, Ruskin set himself in the years 1882–1884 to bring his knowledge to bear in the forming of illustrative collections of siliceous and other minerals. A collection of selected examples at the British Museum was to be the central one; subsidiary to it were to be various other collections in different parts of the country; whilst the whole series was to be further systematised and explained, as we shall presently see, in elementary handbooks. He had a double purpose in this work; he wanted to show how in his opinion a museum of minerals should be arranged, and also to illustrate some of his theories of classification of "banded and concretionary formations." This work was "important," he wrote,[4] "as the first practical arrangement ever yet attempted for popular teaching." The paper, with a postscript, "On Distinctions of Form in Silica" (pp. 373–391), which he wrote for the Mineralogical Society in 1884, explains some of his purposes.

Ruskin's broken health prevented the accomplishment of his full

---

[1] The first lecture was given to the School Scientific Society on October 9, 1866, at the house of Mr. Bosworth Smith. "Mr. Ruskin," says the Minute Book, "delivered an exceedingly interesting lecture on the Progress of Natural Science in England. He concluded by pointing out where there was room for much discovery, illustrating his remarks with some beautiful geological specimens which he kindly presented to the Society."

The second lecture was given to the same Society on October 12, 1867, in the Vaughan Library. "Mr. Ruskin addressed the Society," says the Minute, "on the subject of Crystallization, which he illustrated by a few of the specimens from his very handsome present of minerals and crystals."

[2] Vol. XVIII. pp. lxiii. seq.
[3] The Guild of St. George: Master's Report, 1884, § 6.
[4] Letter to Miss Beever (Hortus Inclusus, ed. 1, p. 59).

design; but much of it was carried out, and it is amazing to those who know the extent of his activity in other directions that he should have found time and energy for the catalogues and other work upon mineralogy now collected in this volume. The secret of it all is to be found in his tireless industry, in the enthusiasm and curiosity with which he approached the study of the beautiful in nature, in his passionate desire to communicate to others what he himself saw and believed.

The British Museum Catalogue.—The story of the series of specimens of Silica, which he obtained permission from the Trustees of the British Museum to arrange and catalogue (pp. 397–414), is very characteristic of his keen interest and enthusiasm. He had long known and studied the collections of the Museum in Bloomsbury, and Professor Story-Maskelyne, formerly keeper of the minerals, was among his friends. He had at various times presented specimens to the department—some pink crystals of Fluor in 1850 (Case 7 g.), a specimen of Iceland Spar in 1865 (on a table at the eastern end of the Gallery), and a long branch of Native Copper in 1868 (Wall-case H.). In 1880 the Natural History collections began to be removed to the new Museum in the Cromwell Road, South Kensington, and Ruskin was much interested in the arrangement of the minerals, now under the charge of a new keeper, his friend Mr. L. Fletcher, F.R.S. Ruskin was constantly in and out of the Museum, and Mr. Fletcher would sometimes ask his advice upon the more artistic arrangement of particular cases or specimens. He was full of suggestions. He was greatly delighted with Mr. Fletcher's arrangement of a series of specimens and models to serve as an introduction to the study (see below, p. 458), but he begged that the models in the General Collection should be kept separate from the real specimens. "The public really ought by this time to understand," he wrote to Mr. Fletcher at a later date (Sandgate, 1887), "that a museum is neither a preparatory school, nor a peep-show; but it may be made more delightful than either." "The big topazes," he wrote in a letter of 1882, "I would keep in a dark place and let nobody ever see them but with fumigation and pious offerings,—a guinea a head at least!" The following letters show how pleased Ruskin was at being allowed to take some of the chalcedonies in hand, for an arrangement of his "very own," as the children say:—

"[LONDON, 1882.]

"DEAR MR. FLETCHER,—Forgive me if I snap too like a puppy at the lovely morsel you offered me just as I was going away yesterday—the rearrangement of the chalcedonies. They are such pretty things—such strange ones, and such *findable* ones—that of all minerals, *they*, it seems to me, ought to be most recommended to the public notice. A schoolboy can't pick up diamonds or topazes or rock-crystals on Brighton beach, or even, for the asking, on St. Michael's Mount; but every other flint he breaks may have a bit of chalcedony in it, and I've had more *hunting* pleasure out of it (sponge-saturating or cell-lining) than in any other mineral whatever.

"Now, as you have got the cases at present, I don't remember (as having caught my eye) either the Auvergne or Indian forms, and all the lovely Trevascus ones are put under the table in *another* case!

"Now, could you indeed get out—for arrangement in your room on Saturday?—these Trevascus ones; and from wherever they are, the Indian, Auvergne, and Iceland volcanic forms. I have myself some Cornish dark ones, associated with pyrites as fine as lace, which I don't remember seeing specimens of. I don't want to disturb the present case with its fine branched ones, till we see what we have to do; but if the reserve drawers could be rummaged, I know there are some precious forms of flint chalcedony in them, and we could then get the Flint, Lava, Trap, and Granite chalcedonies each into their own group, and so illustrate their progressive beauty. I'll come early to-morrow in case anything of the kind should be possible, and if not just now, will hope that Mr. Davies will gradually get at what I want for me, and to-morrow I can go on with my prisms without making disturbance, and am,

"Ever gratefully yours,

"J. RUSKIN."

"LAON, 14*th August*, 1882.

"DEAR MR. FLETCHER,—This place is so full of interest that I could not get to work on the chalcedonies till to-day; but I have nearly got the whole now into order, for an illustrated series of 100 specimens, only among them I have included the lovely large Cornish ones out of the table-ends, which I am bent on getting massed with the other beautiful ones, which are now spoiled by nasty flints and hornstones. Meantime, I begin with the flint nodule enclosing another, in that case, and so rise to my lovely quartz with the triangular cleavages—you shall have the MS. catalogue back from Troyes or Dijon before the end of this week. I desired Mr. Nockold to wait upon you with the fibrous heliotrope he has cut for me, and a lovely section of another, that you might see the sort of man he was; but I hope he has waited for a letter of introduction, which, I *believe*, I told him to do.

"This place is very odd in its geology—hard limestone on the top of sands—an outlier—and with lovely springs at the top of it! The surface is not more than three square miles at most, so the springs can't be supplied by any superficial rain. It's the oddest thing to me that ever was, and I'm going to look for some Laonnois Moses in the Cathedral legends. . . .

"Your affectionate
"J. Ruskin.

"Safe address: Avallon, Yonne."

"Rheims, 14th August, 1882.

"Dear Mr. Fletcher,—I send in a separate packet the catalogue of chalcedony series, as I wrote it, since its references are thus straightforward to the specimens as you kindly placed them in the drawers; but supposing they were placed in a separate case for exposition, and the catalogue, revised with your authorization, printed for use, I have written in red, under my black number, the numbers which should be put in the tray of each specimen for its reference in the catalogue. These red numbers run from 1 to 100, and arrange in consecutive order the flints, jaspers, chalcedonies, agates, and mural agates passing into quartz, where this catalogue ceases. Under the jaspers are given the heliotropes, and Numbers 29 and 30 are left for two sards, which I want to connect them with the chalcedonies; also Numbers 38, 39 are left for two Auvergne chalcedonies which are wanted to complete the system. The remaining numbers—if the MS. be cut up, and pasted, each separately, in order of the red numbers—are, I hope, consecutively right; but there may be probably many little improvements to make on the printer's proof, and my long descriptions of the examples of pseudomorphs may be omitted—but I don't mess the MS. with erasures.

"I have next to write the *headings* descriptive of each class, and saying what Flint, Heliotrope, etc., *are*. These I will quickly block out and send for your corrections and additions—or subtractions! Of course I leave all this entirely to your own time and way; only, I don't want to let the plan hang fire for want of any diligence on my part after you have conceded such full opportunity of completing it.

"I mean to make my *own* numbers, unseen by the public, simply consecutive, to avoid confusion through all the species. $+$ 1, $+$ 2, etc., for the first hundred; $+\!\!\!+$ 1, $+\!\!\!+$ 2, etc., for the second hundred; $+\!\!\!+\!\!\!+$ for the third; and other signs for fourth on—but I can't write more to-day, and have bothered you enough already in all conscience.

"Ever affectionately and gratefully yours,
"J. R."

Broken health, and other work, interfered with this chalcedony catalogue; and when the task was resumed in 1884, it took a different
form. Ruskin, it was agreed, should arrange a special case, containing
a series of specimens of the more common forms of Native Silica, and
should write a catalogue of his own, descriptive of them.   He entered
upon the enjoyment of this pleasant morsel with great gusto—working
daily during the summer of 1884 in Mr. Fletcher's room, and supple-
menting specimens already in the Museum with many from his own
collection (see p. 398).   It says much for the keeper's diplomacy that
Ruskin submitted with complete amiability to the suggestions and corrections of superior authority, indulging occasionally in outbursts always
friendly and often playful.   "What's Hemi-morphite?" he wrote (July
18, 1884), "and how dare you use such words to me?"   "*I* shall write
it Half-formite," he declared in a subsequent note.   Mr. Fletcher had
told him that one of the specimens of Silica had been presented to
the Museum by Count Apollos de Moussin Poushkin (No. 3: see p. 399).
Ruskin feared that the whole case was now "sure to be called the
Rouschkin-Pouschkin Case."   Mr. Fletcher helped him greatly with the
Catalogue, though the responsibility for any opinions expressed in it is
the author's alone, and Ruskin took all suggestions in excellent part:—

> "I'm not in the least minded," he wrote (July 24, 1884),
> "to defend myself in this matter of expression, for that sentence
> puzzled me terribly, and I'm entirely content to leave it as you
> have retouched it, without entering into any whys or whethers.
> Don't think there's any pettishness in declining debate.   I would fain
> enter into many questions at length—more to inform myself than
> defend; but I am quite content to let all go as you put it, and
> am extremely desirous of saving as much your time as mine.   I
> will only venture to protest against classing hyalite with opal for
> 'chemical reasons.'   You might as well class a man with a wolf
> because they were both meat.   And I am particularly proud of my
> separation of opal from hyalite by the peculiar structure of opal in
> being straight-banded, transverse to the vein."[1]

On retiring to Brantwood, Ruskin added a Postscript to the Catalogue (p. 412), and sent for revise a Preface in which he paid compliments to Mr. Fletcher.   These were struck out by Mr. Fletcher,
and Ruskin wrote (August 3):—

> "The little expatiation of Epilogue was I think necessary to show
> the reason of the whole thing, and I hope you will be able to pass
> my unscientific analysis of chlorite, of which assuredly vulgar people

[1] On this point, see below, p. 48.

had better learn something, anyhow, than nothing, scientifically. No one will for a moment think the Museum answerable for it.

"It is, I feel, better on the whole that no compliments should be paid you in *this* catalogue; but I'll have my revenge in my 'Grammar of Silica.'"

The collection remains, as Ruskin arranged it, in the Pavilion at the far end of the Mineral Gallery. As Ruskin felt that descriptive labels would diminish the beauty of the exhibit, it was suggested to him that copies of the Catalogue should be placed upon the case for the use of visitors—an arrangement which greatly pleased him:—

"It seems to me," he wrote to Mr. Fletcher (Oxford, October 28, 1884), "that the arrangement authorized by the Trustees is the most complimentary possible to me. Had they merely permitted the sale without laying the books on the case, hundreds would never trouble themselves about the matter. As you have with fine binding,[1] etc., set all forth, the books on the case are an attraction and recommendation at once. I am a great deal more pleased about all this than I'm able to say."

Visitors to the Museum will share the pleasure, if (as Mr. Collingwood suggests[2]) "they will spend a few hours at Cromwell Road with the pamphlet in hand, just as tourists at Venice are seen comparing his notes with the pictures in the Academy. As the shilling catalogue is by no means abstruse, and the specimens are more beautiful than most picture-shows, the unscientific reader would not find his time lost in learning something new about Nature, and something new to most readers, I suspect, about Ruskin."

The hours and days spent by Ruskin himself at the Museum were among the happiest of his later life. The beauty and the mystery of the minerals filled him ever with greater and greater delight. He writes of his enjoyment of "lovely little fights with the gloves on" with Mr. Fletcher, who presently visited him at Brantwood, and of the pleasure of meeting Dr. Günther.[3] "Ever so much love to that dear Dr. Günther," he wrote (Sandgate, December 26, 1887); "I've a long

---

[1] The ordinary copies, first placed upon the case, had vanished; whereupon Mr. Fletcher had the pages mounted on paper too large to be stowed away in visitors' pockets, and bound at his own cost. The "fine binding" is now worn out and replaced by a buckram one, paid for by the State.

[2] *Life and Work of John Ruskin*, 1900, p. 249.

[3] Keeper of the Zoological Department of the British Museum, 1875–1895. Ruskin refers to one of his books below (pp. 296, 310 *n.*), and paid a tribute to his work at the Museum in a letter to the *Pall Mall Gazette* of November 26, 1884 (reprinted in a later volume of this edition).

letter to write to him; meantime please tell him that if he wants alligators or any other sort of Anti-Georgian monster *drawn*, *I* can draw them with all their lovely expressions, and a mile long, if he orders them by the fathom." His friendship with Mr. Fletcher resulted at this time (1887) in the presentation of two valuable specimens to the Museum—one a diamond, the other a ruby. He had some years before placed the diamond in the Museum on loan; and, after some playful manœuvring, he now formally presented both stones. Mr. Fletcher suggested that the diamond should be named "The Ruskin Diamond." "The Diamond," replied Ruskin (Sandgate, December 14, 1887), "is not to be called the Ruskin, nor the Catskin, nor the Yellowskin, diamond. (It is not worth a name at all, for it may be beaten any minute by a lucky Cape digger.) But I will *give* it to the Museum on the condition of their attaching this inscription to it:—

> The Colenso Diamond,
> Presented in 1887 by John Ruskin
> "In Honour of his Friend, the loyal
> And patiently adamantine
> First Bishop of Natal."

The uncut diamond, thus labelled, may be seen in Case 1 g.; it is a large and symmetrical crystal, weighing 130 carats; Ruskin paid £1000 for it, buying it from Mr. Nockold.

A ruby in Case 9 f. bears the following inscription:—

> The Edwardes Ruby,
> Presented in 1887 by John Ruskin
> "In Honour of the
> Invincible Soldiership
> And loving Equity
> Of Sir Herbert Edwardes' Rule
> By the Shores of Indus."

For this stone Ruskin paid £100.

At the same time he wrote the following inscription for the pink crystals of fluor (Case 7 g.) which he had presented many years before (p. xlix.):—

> The Couttet Rose-Fluors,
> Presented in 1850 by John Ruskin
> "In Honour of his Friend, Joseph Couttet,
> The last Captain of Mont Blanc,
> By whom they were found." [1]

---

[1] In finally approving the inscriptions Ruskin wrote to Mr. Fletcher (Sandgate, January 8, 1888): "I think the entire Geological Society should meet at Chamouni

The Catalogue of Silica is here reprinted without alteration from Ruskin's pamphlet (which is still current). The proof-sheets of a small portion of the Catalogue are in the possession of Mr. Wedderburn, and from them a few notes have been added (see pp. 410, 411, 412 *nn.*).

The Sheffield Collection.—Next in importance to the British Museum specimens is the collection which Ruskin sent to the St. George's Museum at Sheffield. The collection was gradually formed, and the Catalogue was written in pieces[1] and never completed. The principles of its arrangement as designed by Ruskin, and the system of numbering the specimens for catalogue purposes, are explained in *Deucalion* (see pp. 165, 197–203). The Catalogue, as now printed (pp. 418–456), is put together from various copies and proofs which exist at the Museum itself and in the possession of Mr. Allen and Mr. Wedderburn (see Bibliographical Note, p. 417).

Unfortunately the Catalogue is only of partial use to a visitor now examining the collection. Even in the classes to which the Catalogue extends, some of the specimens described in it are not in the Museum, and at least half of the specimens in the Museum are not described in the Catalogue. The Sheffield collection was supplemented by a second which Ruskin had originally designed for a separate building at Bewdley, and it has been further enriched by purchases made by or gifts presented to the Trustees of St. George's Guild. Few of the specimens now bear the numbers given in Ruskin's Catalogue, but many may be identified by written labels (generally excerpted from the Catalogue), and the specimens described in the Catalogue may often be found in the corresponding sections of the existing arrangement, of which therefore a brief synopsis should here be given.[2]

In the Mineral Room there are eleven showcases lettered A. to K. Below each case there are drawers, containing further examples. Case A. contains examples of Flint; Case B., of Conglomerates and

this year, and resolve never to return,—till they had found the Home of Rose-Fluor"! For notices of his guide, Couttet, see Vol. IV. p. xxv. *n.*, Vol. XX. p. 371, and *Fors Clavigera*, Letters 4, § 2, and 75, § 10 (where also he is called "Captain of Mont Blanc"). If the Geological Society should ever make the expedition, they should try the Aiguille d'Argentière; for it was there (as Couttet darkly hinted to Mr. Allen) that he found his treasure. Ruskin gave him 600 francs for the specimens.

[1] References to it will be found in *Fors Clavigera*, Letters 65 (§ 22), 68 (§ 12), and 71 (§ 15).

[2] The quotations in the following description are from *A Popular Illustrated Handbook to the Collection of Minerals and Objects of Art*, 1900, published by the City of Sheffield under the direction of the Ruskin Museum Committee.

Chalcedony; Case C., Agates; Case D., Jasper; Case E., Opal. "Besides examples of the noble or precious opal from Queensland, Hungary, Mexico, Brazil, and Honduras, there are the following varieties:— Girasol or Fire Opal, Cachalong, Hydrophane, Wood-opal, Opal-jasper, Semi-opal, or Pitch-opal, Geyserite, Menilite, Common White Opal, and Hyalite or Glass-opal, all of which are to be seen in the natural state." Case F. contains Labradorite (Felspar Class); Case G., Noble or Precious Metals, including fine examples of Gold, Silver, and Platinum; a specimen of Native Silver, purchased by Ruskin for £70, is, he says, entirely unique.[1] Case H., Useful Metals (specimens of the native forms and ores of Copper, Iron, Tin, Lead, Zinc, and Antinomy); Case I., Quartz:—

"The very pure form called *Rock-crystal* is quite colourless, but when stained of a purple or violet colour is known as *Amethyst,* of which a massive block of crystals may be seen; also specimens of it crystallized on chalcedony. Other forms, from dark brown to black tinge, are called *Smoky Quartz* and *Cairngorm,* several examples being present. The most noticeable of these is a large mass of crystals of smoky-quartz associated with mica and albitic granite, having two magnificent crystals of pale blue topaz projecting therefrom; which show very well the angular truncation characteristic of the crystals. . . . The specimen measures thirty-one inches in circumference ($11'' \times 9\frac{1}{2}'' \times 6''$). In the right-hand side compartment of this cabinet is another mass of quartz completely covered with crystals of yellow topaz. Other named varieties are *Rose-quartz, Milky-quartz,* and *Babel-quartz.* There is also a particularly fine series of crystals of quartz containing 'inclosures' or fibres and spangles of minerals, such as Rutile, Tourmaline, Hæmatite, Chlorite, etc., forming very beautiful objects."

Case J. contains Precious Stones. As these are not included in Ruskin's Catalogue, a description is again given:—

"The different varieties of Precious Stones used as Gems have been lavishly supplied with many rich examples, several being the finest known. With few exceptions, they remain embedded in the rock in which they have developed.

"Diamond is represented by several examples, three of these being the perfectly formed octahedral crystals typical of the diamond in its natural crystalline form, the planes retaining their natural lustre; they are from the '*diamantiferous blue ground*' or '*yellow ground*' of the South African fields. There is also an interesting crystal from Brazil which is in the form of a cube with pyramidal termination.

---

[1] *St. George's Guild: Master's Report,* 1881, § 3.

"Of Sapphire and Ruby (*Corundum*), the principal varieties are included, viz. :—the blue, red, and yellow, being *Sapphire*, *Ruby*, and *Oriental Topaz*.

"Topaz: these crystals are altogether exceptional, and in addition to the unique examples mentioned above in association with Quartz, there are other rare specimens, such as *Pink Topaz;* and a large crystal of a dark sherry brown colour, in all probability the largest to be seen anywhere. In shape it is a right prism and measures nearly eight inches in length.

"Beryls and Emeralds: there are choice selections of these, the Emeralds particularly being represented by several crystals of great size, in the matrix. Beryls may be seen embedded in the quartz block covered with yellow topaz crystals before mentioned, and also other fine examples in this cabinet.

"Garnets in the matrix may be seen represented by several good ex-amples of the almandine variety.

"Lapis-lazuli, or azure stone: some selected specimens of this mineral are placed in this cabinet."

Finally, Case K. contains examples of Fluor Spar; the contents of this case, however, are periodically changed, a series of Ambers, or Calcites, or other classes being substituted for exhibition.

Particulars of some of these precious stones at Sheffield will be found in a later volume of this edition.[1] Other choice specimens were either given by Ruskin, at his own cost, or lent by him, on account of the St. George's Guild, to the Ladies' College (Somerville) at Oxford, and to Felsted House (a training college for schoolmistresses), also at Oxford. Ruskin further presented a collection of mixed silicas to Balliol College, Oxford.

Kirkcudbright Collection.—The next Catalogue (pp. 458–486) is of Two Hundred Specimens of Familiar Minerals which Ruskin presented and arranged for the Museum of Kirkcudbright. The gift was due to the fact that Mrs. Severn often stayed at Kirkcudbright with her sister, Mrs. Milroy, and had other friends there who were active in the matter.

The collection remains as arranged by Ruskin, and is kept apart from other collections of minerals in the Museum. Bibliographical details about the Catalogue are given on p. 457.

St. David's School Collection.—The next collection (pp. 491–513) is one which Ruskin presented to "St. David's School, Reigate" in 1883.

---

[1] *The Guild of St. George: Master's Report, 1884,* § 9.

His friend, Miss Constance Hilliard, who has often been mentioned in these Introductions, married the Rev. W. H. Churchill, and it was to her husband's school that this collection, with a carefully written and most instructive catalogue, was presented. Mr. Churchill's school was afterwards removed to Stone House, Broadstairs, and the Ruskin collection, as he desired, was transferred to the new quarters.

Two editions of the Catalogue, showing numerous variations, are known to the editors (see Bibliographical Note, p. 488). Mr. Wedderburn possesses a copy of the Second Edition, with further revisions made upon it by Ruskin; these are here incorporated.

Coniston Collection.—The fifth collection, described by Ruskin, consists of 124 examples presented by him to the Coniston Institute. The descriptive list is here reprinted (pp. 516–518) from the *Catalogue of the Ruskin Museum, Coniston Institute.*

To Mr. Collingwood, Ruskin presented sixty-five specimens from his collection to illustrate the papers in the *Geological Magazine* on "Banded and Brecciated Concretions." This collection has been given by Mr. Collingwood to the Coniston Institute. Many of the specimens are the originals illustrated in the *Magazine*, and they are arranged in sequence, so that reference with the papers in hand is easy. Two more of the originals are at Sheffield (see p. 434, M. 9 and 10).

Edinburgh Selection.—The sixth collection enumerated by Ruskin (see p. 387) was one which he sent to Edinburgh in 1884 to illustrate his paper, "On Distinctions of Form in Silica." He did not, however, carry out his intention of presenting the collection, nor does the Mineralogical Society possess any copy of the Catalogue. No complete copy of it is known to the editors. The text as here printed (pp. 520–526) is made up from (1) a printed copy of ten pages of the Catalogue (containing Nos. 1–60) which is in the possession of Mr. W. G. Collingwood; (2) a sheet of additional MS., found among Ruskin's papers at Brantwood. (For further details, see the Bibliographical Note, p. 519.)

Minor Collections.—Among other institutions to which Ruskin presented smaller collections of minerals was Whitelands Training College at Chelsea. His notes on some specimens in the College are here printed (p. 528) from a pamphlet of 1883, entitled *The Ruskin Cabinet: Whitelands College*, supplemented by MS. notes attached to other specimens also presented by Ruskin. Whitelands was the College selected by

Ruskin for the institution of his " May Queen" Festival, of which account will be found in a later volume. Shortly afterwards he instituted a similar " Rose Queen" Festival at the Girls' High School at Cork, and to this school also he presented a small collection of minerals. His list of it is also here printed (p. 530). In addition to the collection described by him, he gave to successive " Rose Queens," who deposited their gifts in the school, "a group of crystals; an exquisitely lovely, untouched crystal, exhibiting the most beautiful iridescent colours (combined with a perfect fairyland of form in its substance), as well as a nugget of pure gold, and a tress of native silver."[1] References to some of these specimens will be found in Ruskin's letters to the " Queens" which are printed in a later volume of this edition.

" *Grammar of Silica*."—The six principal collections, described above, formed in Ruskin's mind a connected scheme. He was endeavouring, as he explains (p. 458), " to organize a system of mineralogical instruction for schools, in which the accessible specimens to which it will refer in provincial towns, may be permanently connected by their numbers, both with each other, and with the great central examples of mineralogical structure" in the British Museum. In fulfilment of this purpose, Ruskin began (in MS.) the compilation of an Index to the specimens in the various catalogues. This work has here been completed (Index II., pp. 591 *seq.*).

It was his intention to supplement these various catalogues by writing an elementary handbook, which was to be called *Institutes of Mineralogy;* and another, dealing exclusively with siliceous minerals, to be called *The Grammar of Silica.* He did not finish this Grammar, but he had some sheets of it put into type; and as he refers to it in the published catalogues, these are here printed (pp. 533–541) from the proof. Its object was to systematise the various catalogues and to give " the definitions under which the specimens intended for parish schools may be arranged with intelligible conformity."[2]

Of the intended *Institutes of Mineralogy*, no portion was put into type, but twenty-one sheets exist in MS. These, however, are not sufficiently complete for publication. The sheets contain the beginnings of several chapters, the order of which he had not entirely fixed, thus: " Chap. I. The Three Weightless Elements"; " Chap. II. Of the Nature of Weight"; another sheet, also headed " Chap. II.," treats of " The Two Aerial Gases"; " Chap. III. The Two Terrestrial Gases";

---

[1] " Mr. Ruskin and the High School for Girls, Cork," in the *Ruskin Reading Guild Journal*, vol. i. p. 288.

[2] *The Guild of St. George: Master's Report, 1884,* § 6.

"Chap. IV. Interlude on Gases in General." At this point a second draft appears to begin with a chapter, "Of Metals Generally."

The following introductory passage will explain the standpoint from which Ruskin approached the subject:—

"1. The substances of which this earth and its atmosphere are composed have been examined by chemists in every state and relation in which they are never seen or found by people in general, and into which, till the world comes to an end, they are never likely to get of their own accord. But very few of them have been examined with even superficial attention in the conditions under which nature presents them and mankind uses them, so that the philosophers who squeeze the air into a blue liquid [1] remain unable to explain to us how the air squeezes itself into a breeze: those who compose the most violently corrosive acids out of fluorine cannot explain the building of a cube of fluor spar, nor those who inform us of the conditions of things at a temperature 200° below zero agree among themselves respecting the behaviour of ice at its ordinary point of congelation.

"2. The first object of this book is to interest young girls and boys in the aspects and natures of the things they can easily find, or which are familiar to them already; which they may safely and serenely examine without danger to themselves, the furniture, or the carpets; and of which, when examined, they can arrange examples which may be handled, and washed, and will neither explode, deliquesce, decompose, or evaporate. The various conditions of flint, rock-crystal, felspar, mica, and limestone, and of the four characteristic metals—gold, silver, copper, and iron—which are found in association with these minerals, furnish subjects of inexhaustible interest, and examples of them may be collected, many by a little care and patience, and the rest at very small expense, in quantities more than enough to occupy all the leisure which the modern life is ever likely to leave for their study.

"3. In the outset, however, it is necessary to state a few of the general conclusions at which experimental chemistry has arrived respecting the materials of the globe, and to explain the principal terms employed in defining their properties.

"The elements or simple substances out of which the world and the creatures that live in it are made, commonly appear to us distinguishable into solids, liquids, and vapours. But it is now ascertained that all vapours can be liquefied and most solids melted, so that we might, and logically should, speak of nearly everything in the world as the successive conditions of being frozen, melted, and evaporated.

[1] Compare *Queen of the Air*, Preface (Vol. XIX. p. 294).

But, practically, we can only describe things in the state they happen to be in, with certainty of knowing usually little enough of what they are like at other times. We must describe gold in its usually frozen state, without much troubling ourselves to think of it as melted into a bluish green liquid at a thousand degrees, and air as we breathe it and sail in it, undisturbed by the discovery that if it were two or three thousand times deeper than it is, the lower layer of it would smother the earth in an indigo-coloured deluge of ozone."

Another work which Ruskin planned was a *Grammar of Crystallography*.[1] In this he was to be helped by Mr. Collingwood :—

"It was to be a manual of the actual forms, the phenomenology, of native gold and silver and other minerals which crystallise into fronds and twigs and tangles, and pretty, plant-like shapes, unregarded by the mathematician, and quite unexplained by the elementary laws of crystallography. Illustrated from the beautiful specimens in his collection, with such exquisite drawings as he made of these tiny still-life subjects, it would have been a fairy-book of science." [2]

At one time he seems to have intended to give a wider scope to this work. "I believe it will be an extreme benefit to my younger readers," he said in 1884, "if I write for them a little *Grammar of Ice and Air*" (*The Storm-Cloud of the Nineteenth Century*, § 64). The volume of manuscripts relating to *Deucalion*, which is bound up at Brantwood, bears abundant testimony to the zeal with which Ruskin plunged into the study of crystallography. Mr. Fletcher purchased for him a set of models to show the normal forms of crystals, and he made page after page of diagrams in illustration of them; but the strength to bring all this material into form was in the end denied to him. Physical strength was failing, and Ruskin did not find himself equal to keeping abreast of new views and theories on the subjects of his inquiry. He had made a new friend, in connexion with these subjects, in Dr. George Harley, F.R.S., who was much interested in pearls.[3] Dr. Harley and his family had stayed with him at Brantwood,

---

[1] Ruskin was at work at this when interrupted by illness in February 1881: see *Master's Report on the St. George's Guild*, 1881, § 3.

[2] *Life and Work of John Ruskin*, 1900, p. 339.

[3] See his papers in the *Proceedings of the Royal Society*, vol. 43, 1888, pp. 461–465 ("The Chemical Composition of Pearls"); and vol. 45, 1889, pp. 612–614 ("The Structural Arrangement of the Mineral Matters in Sedimentary and Crystalline Pearls"). Pearls, according to Dr. Harley, are "diseased concretions." Ruskin, when he heard of this, begged Dr. Harley to speak of them as "deviations from the normal structure" (p. 300 of Mrs. Tweedie's *George Harley*.

and some subsequent correspondence [1] tells us how Ruskin had now (1887) laid his work on minerals aside:—

"BRANTWOOD, CONISTON, LANCASHIRE,
"*June* 15, 1887.

"DEAR DR. HARLEY,—It is entirely comforting and exhilarating to me to have your letter and Olga's,[2] and to hear of all those wonderful things in crystals, just when what I thought was clearest in *me* and hardest has become clouded and frangible. For every-thing I thought I *knew* of minerals (you know, I never *think* except what I thought I knew) has been made mere cloud and bewilder-ment by what I find in Judd's address at the Geological[3] of planes of internal motion, etc., and all my final purposes of writing elemen-tary descriptions of them—broken like reeds. I can only now study tadpoles with Olga. If, indeed, you could leave her here awhile on that Scottish run North, there will be an entirely kind and prudent chaperon staying in charge of a girl from Girton whom Olga would love, and who is going to undertake the elementary music at the village school. But I can't write more to-day. Olga's letter is beautifully written, but chills me with the fatal idea of her going to school! I think there should be *no* schools, nor pensions, nor academies, nor universities—that all children should be always at home with papa and mamma, seldom out.

"Ever gratefully yours,
"J. RUSKIN.

"Don't send me back the topaz. Let it stay with yours, which it properly companions. I have more here than I shall ever see."

"BRANTWOOD, CONISTON, LANCASHIRE,
"*June* 16, 1887.

"DEAR DR. HARLEY,—Indeed I *should* like to come, and have a little pink riband fastened by Olga to my breast-buttonhole, and be led wheresoever she chose; and, indeed, I should like not to talk, but to hear of the wonderful things you scientific people are

---

[1] Reprinted from pp. 234, 235 of *George Harley, F.R.S.: The Life of a London Physician.* Edited by his daughter, Mrs. Alec Tweedie, 1899. In line 7 of the first letter "Judd" is here a correction for "Field."

[2] Dr. Harley's younger daughter (then a little girl). When, says Mrs. Tweedie, Ruskin "discovered my sister Olga's love of animal life, he was quite enchanted, and entered into all her enthusiasm and pleasure, just as if he were himself a child; hand-in-hand the old gentleman and the little girl walked about for hours, he explaining the beauties of Nature and the habits of animal life to her interested little ears" (p. 234).

[3] The address was given on February 19, 1887: see *Quarterly Journal of the Geological Society,* vol. 43, p. 71 of the "Proceedings."

doing, for I am quite crushed now, and am resigned in a pulveru-lent state of mind to be radiated or coagulated into whatever new forms of belief or apprehension are possible to me in my old age.

"But, alas! I am not able any more but for the quiet of evening among the hills. And you know this is our loveliest time—the trees and grass at their greenest, and roses beginning to bud—it is the time we wait for all the year; and we are bound to do it reverence and be thankful.

"Please don't think me ungrateful—either Mrs. Harley or Olga or you; you have made me happy beyond my wont in thinking that you would care to have me, and I am, to all of you, a very faithful and grateful friend.

"J. RUSKIN.

"Your postscript is really tantalizing; I wait to hear what Olga says."

The more Ruskin studied, the more he found to study. The wonder and the beauty of the world grew upon him as he tried to fathom its mysteries and intricacies. "Any man's life," he says, "might be happily spent in merely describing and illustrating the various forms of calcite and galena" (p. 462, No. 27). "By the time the youngest pupil in the school is ninety," he says elsewhere, she may know something of "the infinitely multiplied interest" to be found in a piece of jasper (p. 530). Two years before he had written to Mr. Fletcher (October 13, 1885)—with an apology for his "vile always and of late brecciated or conglomerate writing": "This last illness has been a heavy warning to me; and I suppose my British Museum days are over, and that I must be content with quiet mineralogy by my lake shore." Ruskin had now reached the evening of his days; and he was content to sit still and wonder.

Account has now been given of all the pieces collected in this volume. It remains to bring together the summaries which Ruskin has given in various places of his work in the field of geology and mineralogy. One of these places is the last chapter of the first volume of *Deucalion*, where he explains that, though his chapters were divided, he was throughout making "advance in parallel columns"; on four subjects of geological theory he had "shown the necessity for revisal of evidence, and, in two cases, for reversal of judgment." These four subjects are "denudation, cleavage, crystallization and elevation, as causes of mountain form" (p. 274). Cleavage and elevation are presumably the subjects on which he claimed to have shown the necessity for "revisal of evidence." It is a constant theme with him to question the statements ordinarily given (see, for instance, pp. 30, 108, 112). The subjects of denudation (or erosion by ice and water and other natural causes) and of crystallisation are those on which he claimed to have proved the necessity for reversal of judgment.

Of Ruskin's contributions to the discussion of theories of glacier motion, account has already been given. A second question, with which many pages of this volume are occupied, is also concerned with glacier theories. "Do Glaciers Excavate?" The question has excited as much controversy as the methods and nature of glacier motion. On this subject Ruskin claimed, not without considerable justice, to have been somewhat of a pioneer. The theory of the excavating power of glaciers received a strong impetus from the publication in 1862 of Sir Andrew Ramsay's[1] paper "On the Glacial Origin of certain Lakes in Switzerland, the Black Forest, etc."[2] His hypothesis was that certain lake basins have been scooped out by glaciers now melted away; and few scientific papers have ever excited more interest or more controversy. The theory was opposed by Murchison;[3] but the idea that glaciers excavate received, and receives, much support, though, as Professor Bonney observes, "the hypothesis has not gathered its most ardent supporters from those with intimate personal knowledge of the Alps."[4] Among such persons was Ruskin. In 1863, in his lecture at the Royal Institution, he argued that the action of ice had been greatly overrated, and protested that glaciers had no capacity of scooping out lake basins (pp. 9 and n., 15, 16). In the following year he again strongly combated Ramsay's theory, in letters to the press, here reprinted in Appendix (II., p. 548). He lays it down that "ice has had small share in modifying even the higher ridges, and none in causing or forming the valleys" (p. 549); and that "the idea of the excavation of valleys by ice has become one of quite ludicrous untenableness" (p. 549). He then passed to explanations of glacial motion and action in connexion with the viscous theory. He does not question the glacier's power of abrasion, but confines its sculptural power within narrow limits. In 1865, in papers contributed

---

[1] Sir Andrew Crombie Ramsay (1814–1891), F.R.S., Lecturer in Geology at the Royal School of Mines, and Director-General of the Geological Survey.

[2] *Quarterly Journal of the Geological Society*, 1862, vol. xviii. p. 185.

[3] See below, pp. 23 and n., 117.

[4] *Geographical Journal*, June 1893, vol. i. p. 488.

to the *Geological Magazine*, he returned to the charge, in a passage of ironical agreement, which has been quoted in later discussions of the subject (see below, p. 23 *n.*). In *Deucalion*, he reiterates his conviction; for during his visit to Chamouni (1874) Ruskin found further confirmation of his views. "I was able," he says, "to cross the dry bed of a glacier, which I had seen flowing, two hundred feet deep, over the same spot, forty years ago; and there I saw, what before I had suspected, that modern glaciers, like modern rivers, were not cutting their beds deeper, but filling them up" (p. 126). Professor Bonney's observations a year later were, it is interesting to note, to the same general effect. "In 1875," he says, "at the foot both of the Glacier des Bois and of the Argentière glacier, was a stony plain. Both these proved to have been recently uncovered by the ice; in other words, the glacier had not been able to plough up a boulder-bed even at a place where, owing to the change of level, some erosive action might not unreasonably have been expected." [1]

As against those geologists who attributed sculptural force to erosion, Ruskin, then, constantly emphasised the importance of internal structure and original elevation (see, for instance, pp. 9, 365). And so, again, as against the mechanical theory of cleavage and jointing, he held that the phenomena were akin rather both in aspects and origin to crystalline cleavage.

This point brings us to Ruskin's mineralogy. His analysis of the structure of agates was, he claims, original work (p. 98); and he was led by it to suggestions which seemed to him of far-reaching importance. He held that agates and other siliceous substances have been formed by crystalline secretion of gelatinous silica, not by successive deposit of the material of the various layers; and that certain agates, "hitherto supposed to be formed by broken fragments of older agates, are in fact secretions out of a siliceous fluid" (p. 386). Finding a clue to large things in small, he suggested that contortions in gneiss on a small scale are similarly modes of crystallisation (p. 386); that the greater number of minor contortions in Alpine limestones were produced in like manner (pp. 214 *n.*, 386); and even that many of the faults and contortions in metamorphic rocks on a large scale are similarly caused. Just as he found the undulated structure in minerals to be produced by crystallisation, not by compression or violence, so he surmised that, on a larger scale, contortions "may be a crystalline arrangement assumed under pressure, but assuredly not a form assumed by ductile substance under mechanical force" (p. 259). Cleavage and

---

[1] *Geographical Journal*, June 1893, vol. i. p. 488.

jointing were, he held, often "a result of crystallization under polar forces" (p. 283).[1] The larger conclusions which Ruskin thus hinted are perhaps not likely to find favour with geologists. But in the matter of the formation of agates, it may well be considered an open question whether formations, commonly ascribed to successive deposits, were not rather the result of contemporaneous segregation.

Such were Ruskin's claims for his work in this field; the collection of his scattered contributions in this volume will enable them to be better considered.

Of the *illustrations* in this volume, the plates are as follows:—

(1) Two which were prepared from Ruskin's diagrams to illustrate the account in the *Geologist* of his lecture, "On the Forms of the Stratified Alps of Savoy." These are Plates I. and III. An additional plate (II.), illustrating the same lecture, is here given from one of Ruskin's drawings; the drawing (here reproduced full size) is at Brantwood (pencil and wash).

(2) Another new plate (prepared, but not published, in the author's lifetime) illustrates the papers, "On the Shape and Structure of some parts of the Alps, with reference to Denudation." This is Plate IV.

(3) The seven plates (V.–XI.) which Ruskin gave in the *Geological Magazine* to illustrate the papers on "Banded and Brecciated Concretions." Two of these were repeated by him in *Deucalion*, but are here given in their original places.

(4) Next, seven other plates which were added in *Deucalion*. These are here Plates XII., XIII.–XVI., XX., and XXI.

Two other plates which he issued with *Deucalion*, although intended for *Love's Meinie*, are in this edition given in the last-mentioned book (Nos. VII. and VIII. in Vol. XXV.). On the other hand, three new plates XVII.–XIX.) are here inserted in *Deucalion* to illustrate the lecture on Snakes (see further, p. 295 *n.*). The studies reproduced (full size) on Plate XIX. are in pencil and water-colours; the upper study on Plate XVIII. is in water-colours ($4\frac{3}{8} \times 5\frac{1}{2}$); these are in the possession

---

[1] Mr. Collingwood (*Life of Ruskin*, p. 249) cites in this connexion Professor Prestwich's *Geology*, 1886, vol. i. p. 283: "The system of joints seems to me to be not a simple mechanical action, but one combined with a condition of crystallisation; and though, from the influences of other mechanical forces to which the rocks have been exposed, and from the varying proportions of their constituent ingredients, we cannot expect the angles to present the exact definition which a crystal of the pure mineral would have, still there is every appearance of the plane-lines of shrinkage and jointing having been guided in many cases, if not in all, by planes of crystalline cleavage, in consequence of these being those of least resistance."

of Mr. William Ward. The upper study on Plate XVII.—in water-colours ($8\frac{1}{2} \times 6\frac{3}{4}$)—is in the possession of Mr. T. F. Taylor. Another new plate (XIIA.) is introduced to illustrate the chapter on "The Valley of Cluse"; this drawing, which is in pen and brown (9×18), is at Brantwood.

The *frontispiece* is reproduced, by the three-colour process, from a drawing by Ruskin of Australian opal. The drawing (here reproduced in its full size) is at Brantwood. Though the Plate loses something of the purity and sparkle of the original, it serves to give an idea of Ruskin's studies in this sort.

The plates (A and B) given in this Introduction have already been mentioned (p. xxiv.). The former (Mer de Glace) is in water-colours ($14 \times 21\frac{1}{2}$); the latter (Col de la Seigne) is also in water-colours (8×14). Both were given by Ruskin to Lady Simon, and by her to Mr. Herbert Severn.

The *woodcuts*, printed with the text, comprise all those originally given with Ruskin's papers in the *Geological Magazine*, and in *Deucalion*. Figs. 38-41 are new; they have been cut on wood by Mr. H. S. Uhlrich, from illustrations shown by Ruskin at his lecture on Snakes.

E. T. C.

I

ON THE FORMS OF THE STRATIFIED
ALPS OF SAVOY

(1863)

[*Bibliographical Note.*—This was a lecture given by Ruskin at the weekly evening meeting of the Royal Institution (Sir Henry Holland, Bart., in the Chair), June 5, 1863.

An abstract of the lecture, drawn up by Ruskin and here given (pp. 3–11), was published in the *Proceedings of the Royal Institution* (vol. iv. part ii., No. 38, pp. 142–146).

The abstract was reprinted separately with the following title :—

Royal Institution of Great Britain. | Weekly Evening Meeting, Friday, June 5, 1863. | Sir Henry Holland, Bart., M.D., D.C.L., F.R.S., Vice-President, | in the Chair. | John Ruskin, Esq. | On the Forms of the Stratified Alps of Savoy.

Octavo, pp. 4. The title occupies the upper portion of p. 1. Issued stitched and without wrappers; there is no imprint, and there are no headlines, the pages being numbered centrally.

The abstract was next reprinted in the *Geologist* for July 1863, vol. vi., No. 67, pp. 256–259.

It was included in *On the Old Road*, 1885, vol. i. pp. 721–727 (§§ 581–588) ; and again in the second edition of that book, 1899, vol. ii. pp. 359–366 (§§ 290–297). The sections are here renumbered.

A full report of the lecture, in the preparation of which the reporter clearly had access to Ruskin's MS., appeared in the *London Review* of June 13, 1863. As this report often reproduces Ruskin's own words, it is here appended (pp. 12–17).

An account of the lecture, for the most part translated from the *London Review*, appeared in French in the feuilleton of the *Journal de Genève* of September 2, 1863. The translation was made by Madame Adèle Roch, wife of a jeweller at Geneva.

In the *Geologist* for September 1863 (pp. 321–327), the editor (S. J. Mackie, F.G.S., F.S.A.) had an article on the lecture, with which were printed two woodcuts from diagrams shown by Ruskin ; these are here reproduced (Plates I. and III., pp. 6, 14). The article quotes several passages from the lecture ; these are here given in footnotes on pp. 3, 4, 5, 7, 9.]

ON THE FORMS OF THE STRATIFIED
ALPS OF SAVOY

1. THE purpose of the discourse was to trace some of the influences which have produced the present external forms of the stratified mountains of Savoy, and the probable extent and results of the future operation of such influences.[1]

The subject was arranged under three heads:—

    (I.) The Materials of the Savoy Alps.
    (II.) The Mode of their Formation.
    (III.) The Mode of their subsequent Sculpture.

2. (I.) *Their Materials.*—The investigation was limited to those Alps which consist, in whole or in part, either of Jura limestone, of Neocomian beds, or of the Hippurite limestone, and include no important masses of other formations. All these rocks are marine deposits; and the first question to be considered with respect to the development of mountains out of them is the kind of change they must undergo in being dried. Whether prolonged through vast periods of time, or hastened by heat and pressure, the drying and solidification of such rocks involved their contraction, and usually, in consequence, their being traversed throughout by minute fissures. Under certain conditions of pressure, these fissures take the aspect of slaty cleavage; under others, they become irregular cracks, dividing all the substance of the stone. If these are not

---

[1] ["'Geology,'" remarked Mr. Ruskin in his opening words, 'properly divides itself into two branches,—the study, first, of the materials and chronology of deposits; and, secondly, of their present forms'" (report in the *Geologist*, September 1863, p. 321).]

3

filled, the rock would become a mere heap of débris, and be incapable of establishing itself in any bold form. This is provided against by a metamorphic action, which either arranges the particles of the rock, throughout, in new and more crystalline conditions, or else causes some of them to separate from the rest, to traverse the body of the rock, and arrange themselves in its fissures; thus forming a cement, usually of finer and purer substance than the rest of the stone. In either case the action tends continually to the purification and segregation of the elements of the stone. The energy of such action depends on accidental circumstances: first, on the attractions of the component elements among themselves; secondly, on every change of external temperature and relation. So that mountains are at different periods in different stages of health (so to call it) or disease. We have mountains of a languid temperament, mountains with checked circulations, mountains in nervous fevers, mountains in atrophy and decline.

3. This change in the structure of existing rocks is traceable through continuous gradations, so that a black mud or calcareous slime is imperceptibly modified into a magnificently hard and crystalline substance, enclosing nests of beryl, topaz, and sapphire, and veined with gold. But it cannot be determined how far, or in what localities, these changes are yet arrested; in the plurality of instances they are evidently yet in progress.[1] It appears rational to suppose that as each rock approaches to its perfect type

---

[1] [The report in the *Geologist*, September 1863, p. 322, gives here some of Ruskin's words :—

"'Through the whole body of the mountain there runs, from moment to moment, year to year, age to age, a power which, as it were, makes its flesh to creep; which draws it together into narrower limits, and in the drawing, in the very act, supplies to every fissure its film, and to every pore its crystal.'

"And in this change the imaginative mind of Mr. Ruskin saw, perhaps with prophetic distinctness, how all terrestrial things were purifying themselves for some greater end, some more beautiful condition. All is advance, from disorder to system, from infection to purity; nor can any of us know at what point this ascent will cease. We can already trace the transformation from a grey flaky dust, which a rain-shower washes into black pollution, to a rock whose substance is of crystal, and which is

the change becomes slower; its perfection being continually neared, but never reached; its change being liable also
to interruption or reversal by new geological phenomena.
In the process of this change, rocks expand or contract;
and, in portions, their multitudinous fissures give them a
ductility or viscosity like that of glacier-ice on a larger
scale. So that many formations are best to be conceived
as glaciers, or frozen fields of crag, whose depth is to be
measured in miles instead of fathoms, whose crevasses are
filled with solvent flame, with vapour, with gelatinous flint,
or with crystallizing elements of mingled natures; the whole
mass changing its dimensions and flowing into new channels,
though by gradations which cannot be measured, and in
periods of time of which human life forms no appreciable
unit.

4. (II.) *Formation.*—Mountains are to be arranged, with
respect to their structure, under two great classes—those
which are cut out of the beds of which they are composed,
and those which are formed by the convolution or contortion of the beds themselves.[1] The Savoy mountains

starred with nests of beryl and sapphire. But we do not know if the
change is yet arrested, even in its apparently final results. We know in
its earlier stages it is yet in progress; but have we in any case seen its
end?"]
[1] [Here, again, the *Geologist*, pp. 323–324, gives some of the lecturer's words :—
"There is the mountain which is cut by streams or by more violent
forces out of a mass of elevated land, just as you cut a pattern in thick
velvet or cloth; and there is the mountain produced by the wrinkling
or folding of the land itself, as the more picturesque masses of drapery
are produced by its folds. Be clear in separating these two conditions.
There are two ways in which this folding of the hills may be effected. You
may have folds suspended or folds compressed. If underneath, a mass
comes up which sustains the folds,—a pendant wave; but if the force be
lateral, you have a compressed wave. And observe this further distinction :—if a portion be raised by a force from beneath, unless the beds
be as tenacious as they are ductile, they will be simply torn up and
dragged out of shape at that place, and on each side the country will be
undisturbed. But if they are pushed laterally into shape, the force of
the thrust must be communicated through them to beds beyond; nay,
the rock which immediately receives the shock may, if harder than those
beyond it, show little alteration of form, but pass on the force to weaker
beds at its side, and thus affect a much larger space of country than the
elevatory convulsion. Now the fact is that in the Alps both these actions
have taken place, and have taken place repeatedly, so that you have
evidence both of enormous lateral thrusts which have affected the country

are chiefly of this latter class. When stratified formations
are contorted, it is usually either by pressure from below,
which raises one part of the formation above the rest, or
by lateral pressure, which reduces the whole formation into
a series of waves. The ascending pressure may be limited
in its sphere of operation; the lateral one necessarily affects
extensive tracts of country, and the eminences it produces
vanish only by degrees, like the waves left in the wake of
a ship. The Savoy mountains have undergone both these
kinds of violence in very complex modes and at different
periods, so that it becomes almost impossible to trace separately and completely the operation of any given force at
a given point.

5. The speaker's intention was to have analysed, as far
as possible, the action of the forming forces in one wave
of simple elevation,[1] the Mont Salève, and in another of
lateral compression, the Mont Brezon: but the investigation
of the Mont Salève had presented unexpected difficulty.
Its façade had been always considered to be formed by
vertical beds, raised into that position during the tertiary
periods; the speaker's investigations had, on the contrary,
led him to conclude that the appearance of vertical beds
was owing to a peculiarly sharp and distinct cleavage, at
right angles with the beds, but nearly parallel to their
strike, elsewhere similarly manifested in the Jurassic series
of Savoy, and showing itself on the fronts of most of the

for hundreds of miles, and of local elevations independently operating
through them, and breaking their continuity of action. * * * The ripple of
a streamlet rises, glances, sighs, and is gone. An Atlantic wave advances
with the slow threatening of a cloud, and breaks with the prolonged
murmur of its thunder. Imagine that substance to be not of water, but
of ductile rock, and to nod towards its fall over a thousand vertical
fathoms instead of one, and you will see that we cannot assert, perhaps
cannot conceive, with what slowness of march or of decline the mountain-
wave may rise or rest. But whatever the slowness of process, the analogy
of action is the same. Only remember that this has taken place through
rocks of every various degree of consistence and elasticity, and as the
force thrills and swells from crag to crag, it is itself rent again and again
into variously recoiling, quenched, or contracted energy, and divides itself
against itself with destructive contradiction."
The asterisks are in the *Geologist*, apparently marking where the report omits a
passage in the lecturer's MS.]

[1] [For the Salève "wave," see also, below, p. 29.]

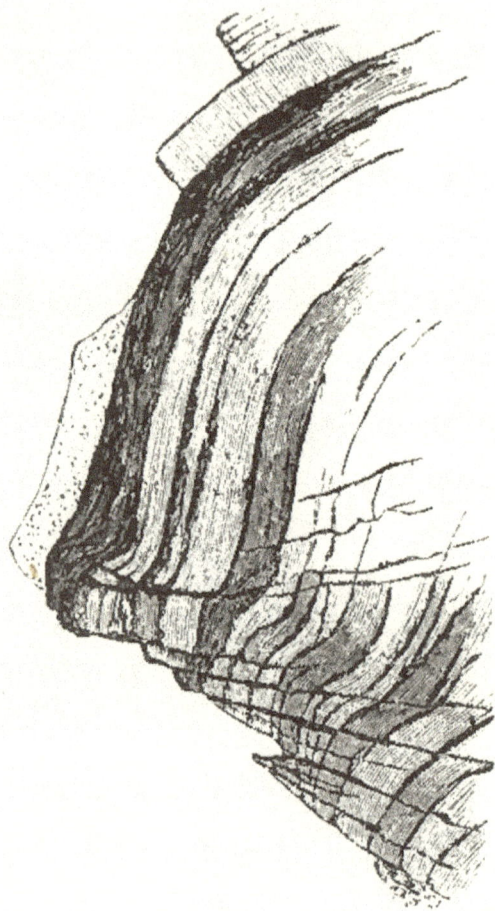

Ideal Section of the Salève

precipices formed of that rock. The attention of geologists was invited to the determination of this question.[1]

The compressed wave of the Brezon,[2] more complex in arrangement, was more clearly defined. A section of it was given[3] showing the reversed position of the Hippurite limestone in the summit and lower precipices. This limestone wave was shown to be one of a great series, running parallel with the Alps, and constituting an undulatory district, chiefly composed of chalk beds, separated from the higher limestone district of the Jura and lias by a long

[1] [The report of this part of the lecture is as follows in the *Geologist* (pp. 324–325) :—

"Saussure, Studer, and Favre, leading or copying each other perhaps, as geologists very often do, represent the face of the hill towards Geneva to be formed by vertical beds; but Mr. Ruskin's impression is

"'that these perpendicular plates of crag, clear and conspicuous though they are, are entirely owing to cleavage,—that is to say, to the splitting of the rock in consequence of the pressure undergone in its elevation; and that the true beds curve into the body of the hill. I dare not,' he adds, 'speak with any confidence in opposition to these great geologists, but I earnestly invite some renewed attention to the question, which is of no small importance in determining the nature of the shock which raised the walls of the Alps round the valley of Geneva.'"

The "ideal section of the Salève" given in Plate I. is from the diagram exhibited by Ruskin. Plate II. is engraved from a drawing by Ruskin at Brantwood, on which he has written :—

"Salève cliff, looking north, after climbing half-way up between Veyrier and the Grande Gorge; showing the likest part to true vertical stratification. But I believe it is *all* cleavage."

"He is fully aware," continues the *Geologist*, "of the difficulties which attend the verification of the section. All the lower part of the Salève is Jura limestone, as determined by Favre, and that this rises up in a nearly vertical sheet along the whole front, thrusting up the Neocomian and compressing it, Mr. Ruskin admits; 'but there is no doubt,' he contends, 'respecting these frontal clefts.' Neither does he deny that there are raised beds of Neocomian on parts of the mountain, as assigned by Favre; but [he contends] that at the Grande Gorge, where the natural section is clearest, there are the beds all following the curve of the summit, and that the vertical fissures are rather faults or cleavages, or partly both, the business being so complicated that one cannot tell which is which.' On this subject see further, below, p. 14; and compare W. G. Collingwood's *Limestone Alps of Savoy*, p. 100.]

[2] [The *Geologist* adds (p. 324): "notable in the dash and curve."]

[3] [That is, in the diagram here reproduced (Plate III., p. 14). In addition to this section, Ruskin must have shown a drawing of the mountain group. The report in the *Geologist* (p. 325) again gives the lecturer's words :—

"You see the group is composed of an isolated pyramidal mass, of a flat mass behind it which extends at both sides, and lastly, of a distant range of snowy summits, in which Mount Vergi and the Aiguille de Salouvre are conspicuous objects. Now these three masses are merely three parallel ridges of limestone-wave, formed mainly of originally horizontal beds of Rudisten-kalk, approaching you as you stand looking from

trench or moat, filled with members of the tertiary series—chiefly nummulite limestones and flysch. This trench might be followed from Faverges, at the head of the Lake of Annecy, across Savoy. It separated Mont Vergi from the Mont Dorons, and the Dent d'Oche from the Dent du Midi; then entered Switzerland, separating the Moleson from the Diablerets; passed on through the districts of Thun and Brientz, and, dividing itself into two, caused the zigzagged form of the Lake of Lucerne. The principal branch then passed between the High Sentis and the Glarnisch, and broke into confusion in the Tyrol. On the north side of this trench the chalk beds were often vertical, or cast into repeated folds, of which the escarpments were mostly turned away from the Alps; but on the south side of the trench, the Jurassic, Triassic, and Carboniferous beds, though much distorted, showed a prevailing tendency to lean towards the Alps, and turn their escarpments to the central chain.

6. Both these systems of mountains are intersected by transverse valleys, owing their origin, in the first instance, to a series of transverse curvilinear fractures, which affect the forms even of every minor ridge, and produce its principal ravines and boldest rocks, even where no distinctly excavated valleys exist. Thus, the Mont Vergi and the Aiguilles of Salouvre are only fragmentary remains of a range of horizontal beds, once continuous but broken by this transverse system of curvilinear cleavage, and worn or weathered into separate summits.

The means of this ultimate sculpture or weathering were lastly to be considered.

the Salève. Probably, I think, approaching at this moment, driven towards you by the force of the central Alps, the highest ridge broken into jags as it advances, which form the separate summits of Alpine fury and foam; the intermediate one joining both with a long flat swing and trough of sea, and the last, the Brezon, literally and truly breaking over and throwing its summit forward as if to fall upon the shore. There is the section of it (Plate III.); the height from base to summit is 4000 English feet,—the main mass of the façade, formed of vast sheets of Rudisten-kalk, 1000 feet thick,—plunging at last, as you see, in a rounded sweep to the plain."]

Salève cliff - looking north - after
climbing half way up between Mauris
Veirier and the grande gorge - showing
the likest part to true vertical stratification.
But I believe it is all cleavage.

7. (III.) *Sculpture.*—The final reductions of mountain form are owing either to disintegration, or to the action of water, in the condition of rain, rivers, or ice, aided by frost and other circumstances of temperature and atmosphere.

All important existing forms are owing to disintegration, or the action of water. That of ice had been curiously overrated.[1] As an instrument of sculpture, ice is much less powerful than water ; the apparently energetic effects of it being merely the exponents of disintegration. A glacier did not produce its moraine, but sustained and exposed the fragments which fell on its surface, pulverising these by keeping them in motion, but producing very unimportant effects on the rock below ; the roundings and striation produced by ice were superficial ; while a torrent penetrated into every angle and cranny, undermining and wearing continually, and carrying stones, at the lowest estimate, six hundred thousand times as fast as the glacier. Had the quantity of rain which has fallen on Mont Blanc in the form of snow (and descended in the ravines as ice) fallen as rain, and descended in torrents, the ravines would have been much deeper than they are now, and the glacier may so far be considered as exercising a protective influence. But its power of carriage is unlimited, and when masses of earth or rock are once loosened, the glacier carries them away, and exposes fresh surfaces. Generally, the work of water and ice is in mountain surgery like that of lancet and sponge—one for incision, the other for ablution.[2] No excavation by ice was possible on a large scale, any more

---

[1] [The *Geologist* (p. 326) reports the lecturer's words :—

"There have been suggestions made that the glaciers of the Alps may have scooped out the Lake of Geneva. You might as well think they had scooped out the sea. Once let a glacier meet with a hollow and it sinks into it, and becomes practically stagnant there, and can no more deepen or modify its receptacle than a custard can a pie-dish."

The suggestions referred to are those of Ramsay, whose famous paper on the subject had recently appeared : see Introduction, above, p. lxv.]

[2] [The *Geologist* (pp. 326–327) thus reports the lecturer's words :—

"The torrent cuts, the glacier cleanses ; one is for incision, the other for ablution and removal ; and so far as the present form is concerned, you may ignore the glacier altogether. It only helps the torrent here and there by exposing a surface and by carrying off the rubbish which

than by a stream of honey; and its various actions, with
their limitations, were only to be understood by keeping
always clearly in view the great law of its motion as a
viscous substance, determined by Professor James Forbes.[1]

8. The existing forms of the Alps are, therefore, traceable
chiefly to denudation as they rose from the sea, followed by
more or less violent aqueous action, partly arrested during
the glacial periods, while the produced diluvium was carried
away into the valley of the Rhine or into the North Sea.
One very important result of denudation had not yet been
sufficiently regarded; namely, that when portions of a thick
bed (as the Rudisten-kalk) had been entirely removed, the
weight of the remaining masses, pressing unequally on the
inferior beds, would, when these were soft (as the Neocomian
marls), press them up into arched conditions, like those of
the floors of coal-mines in what the miners called "creeps."[2]
Many anomalous positions of the beds of Spatangen-kalk
in the district of the Lake of Annecy were in all probability owing to this cause: they might be studied advanta-
geously in the sloping base of the great Rochers de Lanfon,
which, disintegrating in curved, nearly vertical flakes, each a
thousand feet in height, were nevertheless a mere outlying
remnant of the great horizontal formation of the Parmelan,
and formed, like it, of very thin horizontal beds of Rudisten-
kalk, imposed on shaly masses of Neocomian, modified by
their pressure. More complex forms of harder rock were
wrought by the streams and rains into fantastic outlines;
and the transverse gorges were cut deep where they had
been first traced by fault or distortion. The analysis of
this aqueous action would alone require a series of discourses; but the sum of the facts was that the best and
most interesting portions of the mountains were just those

the working water throws down; but the two sculptors are natural disintegration and the stream, and every existing form in the Alps is
distinctly traceable to one or other of these forces combined with the
internal geological structure."]

[1] [Here, again, see the Introduction, above, p. xxxv.]
[2] [See Lyell's *Elements of Geology*, ch. v. p. 50 (1865 edition).]

which were finally left, the centres and joints, as it were, of the Alpine anatomy. Immeasurable periods of time would be required to wear these away; and to all appearances, during the process of their destruction, others were rising to take their place, and forms of perhaps far more nobly organized mountain would witness the collateral progress of humanity.

J. R.

MR. RUSKIN ON THE ALPS OF SAVOY[1]

1. IT is not Mr. Ruskin as a word-painter, or Mr. Ruskin as an artist, that we speak of in this article, but Mr. Ruskin in an equally proper, though less familiar, capacity—Mr. Ruskin the geologist. On Friday last the Royal Institution contained one of those overflowing audiences such as Faraday used to draw; the lecturer was that brilliant word-painter, whose works some delight to abuse, thousands to admire, but all delight to read, and whom, it would seem, many wanted to see and hear.

2. Mr. Ruskin's geology was not only entertaining, but instructive and suggestive. The phases in which he viewed the Alps were just what we should have expected—eloquent and artistic pictures of their substance, formation, sculpture; what they were made of; how they were made; how their beautiful scenery was sculptured out by time, weather, wind and rains, contortions, dislocations, cracks, and fissures. Whenever the geologist stoops—or rather, it is seldom, rises—to description of structural phenomena, it is nearly always in districts which present them on the smallest and least complex scale. "We have never yet seen," Mr. Ruskin observed in his opening remarks, "a complete section of the Valley of the Rhone. I wonder at the indifference of travellers in such matters, and that it never occurs to them to ask how the scenery to which they owe so much enjoyment was first cast into its colossal shape; how the hills were withdrawn from the opening through which the Lake of Geneva expands between Clarens and Meillerie; or how that strange chasm, bent like forked lightning, was cleft through the rocks of Uri, to be filled with the waters of the four cantons."[2]

3. Mountain scenery is not the result of mere natural irregularities, mountains [are] no mere heaps of rock. They are truly, in Mr. Ruskin's own words, "true sculptured edifices." The hills of Savoy were selected because, though amongst the boldest, they were amongst the simplest, in displaying the main questions relating to structure. Then mentally the lecturer set forth to build a mountain, taking stone for brick and slime for mortar. The Savoy mountains are made of a great many things, chiefly from limestone; secondly, dark brown Neocomian sandstones; and, lastly, a hard grey rock, the Hippurite limestone of ages, equivalents of our Portland stone and lower chalk formations. All these rocks agree in this—they were formed at the bottom of the sea. Mr. Ruskin would not dwell on the difficulties of raising them. Sir Charles Lyell, he said, will allow

---

[1] [Report (under this heading) of the preceding lecture in the *London Review*: see above, p. 2. Sections 1–3 cover, it will be seen, the heading " I." in Ruskin's abstract ; §§ 4–7, his " II." ; and §§ 8–11, his " III." ; while § 12 is the reporter's summing-up.]

[2] [On this subject, compare *Deucalion*, below, p. 155.]

us to do that to any extent, if we only "take our time";[1] but he dwelt forcibly on the difficulty of *drying* them. Raised quickly or slowly—and the elevation of these mountains was, he believed, incalculably slow—the rocks must dry and settle. How long it takes to dry a wall, how long then to dry a mountain two miles deep! Consider, too, the consequences; by whatever means accomplished, drying is not merely a hardening, but a contraction of the whole body, and either the rock must become powdery like chalk, or it must crack and fall to pieces. In hard rocks these pores and cracks are filled with crystalline matter, matter moved and rearranged by that marvellous process, the finest particles taken out of the very flesh of the rock substance and carried in slow currents to be arranged with such strength and coherence that the very fissures which would have been sources of weakness become bonds of strength. The crack is not formed first, filled afterwards. It is filled as it is formed, or the rock would fall apart before it could be filled at all. The external aspect and hardness of a stone are no evidence of its real state. Though hard it may be getting harder; though soft, more tenacious; it may be contracting or expanding, in every case it is *changing*. Not one of the atoms of it is at rest. The particle of lime a thousand fathoms deep in rock that rings to the hammer's blow is no more at rest than the hardest worker in a railway tunnel. It is mining its way steadily as a mole through the mountain's heart. It is doing more than mining, it is purifying itself. All is advance from disorder to system, from infection to purity; and we can trace the transformation from grey flaky dust, which the rain washes into black pollution, to a rock whose substance is of crystal, starred with the beryl and sapphire. Nor do we know if the change is yet arrested. Is the imperfect granite to remain imperfect, or is it gathering itself still into better distinguishable crystals? Is the blotted marble to remain dull and indistinct, or is its purple glow to deepen and its variegation to involve itself in richer labyrinth?

4. It is sufficient for us at present to know that what we call hard and solid in rock is mobile and ductile. There are two great distinctive forms of elevated land—the mountain cut by streams out of a mass of strata, like a pattern out of thick velvet, and the mountain produced by the wrinkling and folding of the land itself. In England our valleys are cut by river, sea, or rain out of masses of raised land, and little interest attaches to them; but when mountains are the folds of stone-drapery gathered together and cast hither and thither under laws of complex harmony, the first and immediate interest that attaches to them is the wonderment what their consistence was when these folds were done. If from beneath a mass sustains the folds, we have a pendent wave. Lateral force causes a compressed wave. If a portion be raised, the beds above may be dragged or torn through, while the country on each side remains undisturbed; but lateral convulsions affect a much larger space of country than elevatory, so that we have evidence both of enormous lateral thrusts, affecting hundreds of miles of country, with local operations breaking through these and interrupting their continuity of action. Mr. Ruskin gave an apt simile of the Alps when he compared them to a shoal of rugged islands of igneous

---

[1] [See on this subject Lyell's *Principles of Geology*, vol. i. ch. v. : compare below, p. 117 *n.*]

rock emerging through a sea of limestone, tearing up fragments of rocks, and also, with more conspicuous power, wrinkling the edge of the limestone all round the wound, and sending off waves of lateral force to die away on the surrounding country—"waves of a slow Titanian storm which troubled the earth as the winds trouble the sea."

5. Imagine the mountain's substance not of water but of ductile rock, and to nod towards its fall over a thousand vertical fathoms. Perhaps we cannot conceive with what slowness of swell or of decline the mountain-wave may rise or rest; but as the force thrills from crag to crag, it recoils, divides against itself with destructive counteraction; and this has taken place not once, but many times—five or six periods of convulsion being marked at distant intervals. When one storm has been calmed, yet another stone-tempest from another point. So commingled are their actions in a complicated result, that it is not the work of one man, but of a multitude, not of one year, but of centuries, to decipher the flow and ebb of even a single mountain-tide. "I knew," said Mr. Ruskin, "of one marvellous outside breaker, not very high, but notable in the clash and curve—Mont Brezon." But as there were in it structures almost incredible and difficult to be explained, he determined to take an easy one, and selected Mont Salève, studied by geologists from De Saussure downwards. Even this mountain, one of mere elevation, which he supposed had been studied exhaustively, he found full of curious unrepresented structures; and the only distinct impression he could obtain was adverse to the three great observers, Saussure, Studer, and Favre, who all represent the face of the hill to be formed of vertical beds; while Ruskin sees plates of crag, entirely owing to cleavage—that is, to the splitting of the rock through the pressure it has undergone in its elevation; the true beds curving into the body of the hill, as were seen at the Grande Gorge.

6. Nothing daunted by the difficulties of the simpler mountain, Mr. Ruskin sketched out some of the prominent features of the more difficult Brezon, where the beds follow the curve of the summit, the vertical fissures being either faults or cleavages. The group mainly consists of a pyramidal mass, a flat mass behind it showing itself at both sides, terminating in two cliffs, and finally in the distant range of the snowy summits of Mont Vergy and the Aiguille de Salouvre. These are three parallel ridges of limestone, approaching probably at this moment, looking towards them from Le Salève, slowly driven by the force of the central Alps, "the highest range broken into jags, the separate summits of Alpine fury and foam"; the intermediate joining with a long flat swing and trough of sea; and, lastly, the Brezon breaking over and throwing its summit (4000 feet) forward as if to fall upon the shore. And really this wave-like action of the elevatory force, as shown in Mr. Ruskin's sections and views of the wonderfully contorted strata of these mighty hills, seems to have been caught and petrified in the very act, just as a sea-wave might be frozen into solid ice at the very point of curling. This Brezon and Vergy group are only a portion of the longitudinal waves which flow parallel with the Alps, and are cut across by those transverse precipitous valleys in which the grandest scenery occurs. From the Lake of Annecy to the Lake of Constance, along the north border of the chain, the mountains are divided into two belts. Outside of these runs a great continuous fault which separates the Tournette and

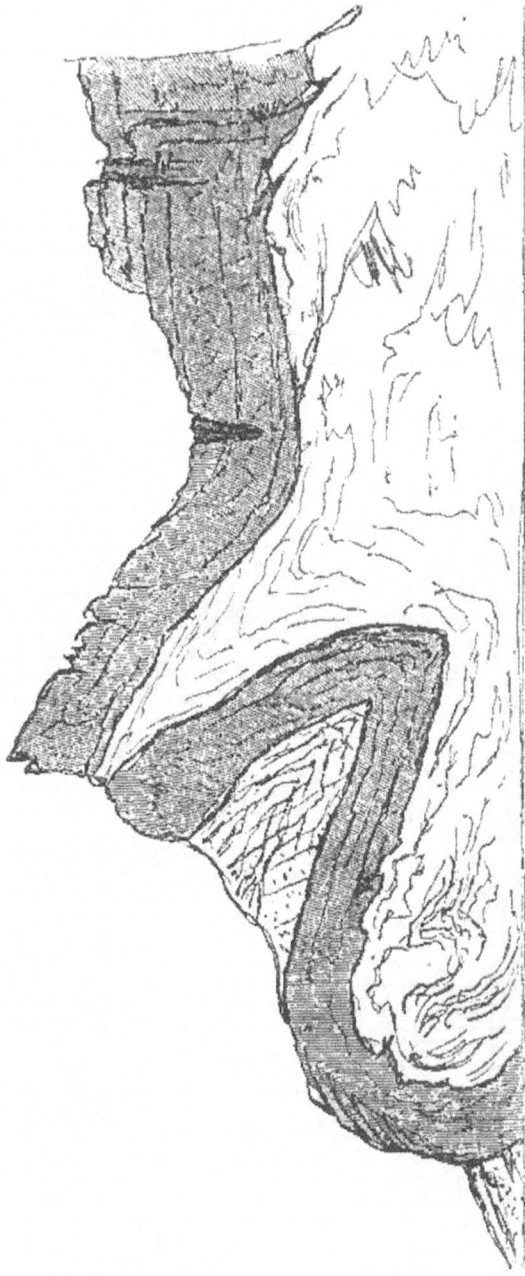

Section of the Brezon

Mont Vergy; and running between the Dent d'Oche and the Dent du Midi cuts the Moleson from the Diablerets, passes through the bottom of the Lake of Brientz, and splitting into two minor faults at Lake Lucerne, it goes into Canton Glarus and crosses the valley of the Rhine, ending in confusion in the Tyrol. On the hills within, the faults have their escarpments turned towards the Alps; but outside of it, in the broken and undulating ground, is the true wave-district of the Alps, where the hills are thrown vertically up, "as the timber of a wreck in a storm." Beyond this district the escarpments are turned away from the Alps. The valleys crossing these longitudinal ridges are neither gaps cut by river nor are they vertical faults. They are the expressions of disruptions in the unity of the long waves themselves, and they are accompanied by conditions of parallel fracture which mark a disposition throughout the entire body of the mountain to open in a similar manner.

7. In a novel and very able manner, too, Mr. Ruskin showed the action of the lateral pressure of rock masses in pressing up intermediate denuded areas into arches, as in the creep of the floors of coal-mines.

8. The sculpturing of mountains is either by disintegration, aided by chemical action, or by water acting as rain, as torrent, or as glacier; and Mr. Ruskin dwelt, we think rightly, on the more powerful agency of water in a sculpturing capacity than ice. In his own words, "The work of ice is so showy and superficial, and the artist's touch of water so cunning, quick, and tenderly fatal, that we are all apt to overrate the power of the one and underrate that of the other." Referring to Forbes' idea of the viscosity of ice, Mr. Ruskin contended against Professor Ramsay's doctrine of the scooping out of the great lake basins by the grinding of glaciers, comparing the viscous glacier-ice to honey or treacle, only less active. It is at no time, in his opinion, a very violent abrading agent, but wholly powerless when it falls into a pit. "There have been," he said, "suggestions made that the glaciers of the Alps may have scooped out the Lake of Geneva. You might as well think they had scooped out the sea. A glacier scoops out nothing; once let it meet with a hollow, and it spreads into it, and can no more deepen its receptacle than a custard can deepen a pie-dish."

9. That idea he considers the more singular, because, with its strongest and most concentrated force, the glacier of the Rhone has been unable to open for itself a passage between the two small contradictory rocks of the Gorge of St. Maurice. "So little effectual has it been in excavating them that the Rhone is still straightened for a passage, and a single town is fortalice enough to defend the pass where a key unlocks a kingdom; and yet we are asked to suppose that a glacier power which, concentrated, could not open a mountain gate, could dig out a sea-bottom when diffused." There is a more curious proof still of the excavating incapacity of ice. Full in the face of the deepest fall of this same Rhone glacier two impertinent little rocks stood up to challenge it. Don Quixote with his herd of bulls was rational in comparison.[1] But the glacier could make nothing of them. "It had to divide, slide, split, shiver itself over them, and ages afterwards, when it had vanished like an autumn vapour from the furrow

---

[1] [See *Don Quixote*, part ii. end of ch. lviii.]

of the Rhone, the little rocks still stood triumphant, and the Bishops of Sion built castles on their tops, and thence defied the torrent of the Reformation [1] coming up that valley as the rocks had done the passage of the glacier coming down it." From the shoulders of Mont Blanc the two great glaciers of Bossons and Taconay have each excavated for themselves a ravine in the shaly slates over which they descend, but the excavation is just as evident and as simple as a railroad trench. Down each gorge there falls an ice-stream a quarter of a mile wide, a hundred feet deep, falling at an average slope of 20° or 30°. They have gnawed away the rocks under them and beside them, and left between the sharp ridge of crumbling slate—Montagne de la Côte. If instead of ice-streams there had been waterfalls, cataracts four miles long, a hundred deep, and down a slope as steep as the roof of a gabled house, how long would De la Côte have stood? It would not, Mr. Ruskin thinks, have kept its present form a day. In a year it would have disappeared. Suppose on Mont Blanc, which rises 11,000 feet above Chamouni, instead of snow, the same quantity of rain fell and descended in the form of a torrent—the ravines of La Côte or Taconay would be far deeper than they are. The glaciers, so far from having a highly consuming, have a distinctly protective power. The water power of friction is diminished, not indeed in the rate of the diminished velocity, but in some large proportion to it. The swiftest glacier in summer does not move two feet a day; a torrent going down the same slope would run ten miles an hour at least—600,000 times as fast. With a certain weight of water, which, carrying stones, you have to grind rock with, will you have it in a vertical mass moved two feet a day, or will you have it in a horizontal sheet moved 1,200,000 feet a day? "Give me the level sheet and the fast pace," says Mr. Ruskin.

10. But, it will be said, under this weight of mural ice there are stones and sand like diamond dust in a lapidary's mill. There is hardly ever any such thing—a glacier does not like stones under it. They would make its life uncomfortable. Dirty and sandy above, it is clear as crystal below, and its action on the rocks beneath it is lambent, cleansing, silent, and soft. The glacier does not make the moraine, it only carries it. The moraine is only the sheddings of the rocks above. A glacier is a torrent turned on its back. Whatever is soft and decomposing the glaciers sponge away. What is hard and healthy they leave projecting and manifest. They are great carriers, a curious and effective parcels delivery company. Water in respect to them is as a flying lizard to a camel—it is all teeth and wings, but no back. For biting and carving, doing the sculptor's work, there is nothing the torrent cannot do. "Insidious, inevitable, patient, and passionate by turns; now hurling stones at its antagonist like a Titan, now sucking his strength like a vampire, piercing him with cavities like a pholas, sawing him in two like a toothed mill, and presently bedewing the remnants of him with hypocritical, or perhaps, repentant tears, and bringing handfuls of moss and wild flowers to heal the wounds it has made."

11. The present forms of the mountains Mr. Ruskin attributes not to

<hr>

[1] [For the bishopric of Sion as "the centre of Romanism in Switzerland," see *Modern Painters*, vol. iv. (Vol. VI. p. 411).]

the glacier, but to the two natural sculptors, flowing water and natural disintegration; every existing form in the Alps he believes to be distinctly traceable to one or other of these forces combined with internal geological structure.

12. Such were Mr. Ruskin's views, and very ably and eloquently they were put. We must not, however, forget that the lateral grinding force of ice is very different from the vertical scooping force. The ice of the glacier clings to the sides of the gorge, while its centre presses more quickly on or through. The under part of the glacier is always wet, it rests and slides on a sheet of water, and this water acting under pressure has increased chemical and mechanical action. The action under the glacier endures for ages, the action of the torrent is sudden and transient. The one is the tortoise, the other the hare, and the result in the fable is not given to the swift. Mr. Ruskin's attack, however, was very proper and highly philosophical, and one which it will require all the talent of the Jermyn Street Professor[1] to answer. He may not be ultimately conquered, but at any rate he has been driven back and has suffered a severe repulse, although the victory be not yet decisive.

[1] [For A. C. Ramsay, see the Introduction, above, p. lxv.]

II

NOTES ON THE SHAPE AND STRUCTURE OF SOME PARTS OF THE ALPS, WITH REFERENCE TO DENUDATION

(1865)

[*Bibliographical Note.*—These papers appeared in the *Geological Magazine* for February and May 1865, vol. ii., No. 8, pp. 49–54, No. 11, pp. 193–196.

They have not hitherto been reprinted. The sections are here numbered. In § 3, line 22, "Dovrefeldt" is here corrected to Dovre Fjeld. In the second article the illustrations were all included on a single plate. The first three of them (Figs. 1, 2, and 3) are here separately printed. The fourth was a rough woodcut of a subject which Ruskin afterwards engraved on a larger scale; this is here substituted, the lettering being added from the old woodcut. These changes have required slight alterations in the text; *e.g.*, in § 16 "Plate IV." for "Fig. 4." For a correction made in Fig. 1, see p. 29 *n.*]

NOTES ON THE SHAPE AND STRUCTURE OF SOME PARTS OF THE ALPS, WITH REFERENCE TO DENUDATION[1]

I

1. It is often said that controversies advance science. I believe, on the contrary, that they retard it—that they are wholly mischievous, and that all good scientific work is done in silence, till done completely. For party in politics, there are some conceivable, though no tenable, reasons; but scientific controversy in its origin must be always either an effort to obscure a discovery of which the fame is envied, or to claim credit for a discovery not yet distinctly established: and it seems to me there are but two courses for a man of sense respecting disputed statements;—if the matter of them be indeed doubtful, to work at it, and put questions about it, but not argue about it; so the thing will come out in its own time, or, if it stays in, will be no stumbling-block; but if the matter of them be not doubtful, to describe the facts which prove it, and leave them for what they are worth.

2. The subject of the existing glacial controversy between older and younger geologists seems to unite both characters. In some part, the facts are certain and need no discussion; in other points, uncertain, and incapable of being discussed. There are not yet data of measurement enough to enable us to calculate accurately the rate of diluvial or disintegrating action on mountains; there are

[1] [From the *Geological Magazine*, February 1865 (see above, p. 20).]

not data of experiment enough to enable us to reason respecting the chemical and mechanical development of mountains; but all geologists know that every one of these forces must have been concerned in the formation of every rock in existence: so that a hostile separation into two parties, severally maintaining a theory of Erosion, and a theory of Fracture, seems like dividing on the question whether a cracked walnut owes its present state to nature or the nutcrackers. In some respects, the dispute is even more curious; the Erosion party taking, in Geology, nearly the position which they would occupy zoologically, if they asserted that bears owed the sharpness of their claws to their mothers' licking, and chickens the shortness of their feathers to the friction of the falling bit of shell they had run away with on their heads. For indeed the Alps, and all other great mountains, have been tenderly softened into shape; and Nature still, though perhaps with somewhat molluscous tongue, flinty with incalculable teeth, watches over her craggy little Bruins—

> ". . . forms, with plastic care,
> Each growing lump, and brings it to a bear." [1]

Very assuredly, also, the Alps first saw the world with a great deal of shell on their heads, of which little now remains; and that little by no means so cunningly held together as the fragments of the Portland Vase. [2] No one will dispute that this shell has been deeply scratched, and clumsily patched; but the quite momentous part of the business is, that the creatures have been carefully Hatched! It is not the denudation of them, but the incubation, which is the main matter of interest concerning them. So that Professor Ramsay [3] may surely be permitted to enjoy his glacial theory without molestation—as long as it will

---

[1] [Pope: *Dunciad*, i. 102.]
[2] [For the story of this vase, which was broken into a hundred pieces, afterwards cunningly joined together, see E. T. Cook's *Popular Handbook to the Greek and Roman Antiquities in the British Museum*, pp. 656–657.]
[3] [See, again, the Introduction, above, p. lxv.]

last. Sir Roderick Murchison's temperate and exhaustive statement * seems to me enough for its extinction; but where would be the harm of granting it, for peace' sake, even in its complete expansion?

3. There were, we will suppose, rotatory glaciers—whirl-pools of ecstatic ice—like whirling Dervishes,¹ which exca-vated hollows in the Alps, as at the Baths of Leuk, or the plain of Sallenche, and passed afterwards out—"queue à queue"—through such narrow gates and ravines as those of Cluse. Gigantic glaciers in oscillation, like handsaws, severed the main ridge of the Alps, and hacked it away, for the most part, leaving only such heaps of sawdust as the chain of the Turin Superga; and here and there a fragment like the Viso and Cervin, to testify to the ancient height of the serrated ridge. Two vast longitudinal glaciers also split the spine of the Alps, east and west, like butcher's cleavers, each for sixty miles; then turned in accordance to the north ("Come si volge con le piante strette A terra, ed intra sè donna che balli" ²), cut down through the lateral limestones, and plunged, with the whole weight of their precipitate ice, into what are now the pools of Geneva and Constance. The lakes of Maggiore, of Como, and Garda, are similar excavations by minor fury of ice-foam;— the Adriatic was excavated by the great glacier of Lom-bardy;—the Black Sea, by the ice of Caucasus before Prometheus stole fire;—the Baltic, by that of the Dovre Fjeld, in the youth of Thor;—and Fleet Ditch in the days of the *Dunciad*³ by the snows of Snow Hill. Be it all so: but when all *is* so, there still was a Snow-hill for the snows to come down;—there still was a fixed arrangement

---

* Address at Anniversary Meeting of Royal Geographical Society, 1864.⁴

¹ [For a reference to this part of Ruskin's argument, in later discussions of the subject, see Sir Henry Howorth's *The Glacial Nightmare*, 1893, vol. ii. p. 621.]

² [*Purgatorio*, xxviii. 52, 53: "As when a lady, turning in the dance, doth foot it featly" (Cary).]

³ [The allusion here is to the third book of the *Dunciad*, where, from a mount of vision, the former state of Britain is disclosed.]

⁴ [See the *Journal of the Royal Geographical Society*, vol. 34, pp. clxiv.–clxxv.]

of native eminence, which determined the direction and concentrated the energies of the rotatory, precipitate, or oscillatory ice. If this original arrangement be once in-vestigated and thoroughly described, we may have some chance of ascertaining what has since happened to disturb it. But it is impossible to measure the disturbance before we understand the structure.

4. It is indeed true that the more we examine the Alps from sufficiently dominant elevations, the more the impres-sion gains upon us of their being rather one continuously raised tract, divided into ridges by torrent and decay, than a chain of independent peaks: but this raised tract differs wholly in aspect from groups of hills which owe their essential form to diluvial action. The outlying clusters of Apennine between Siena and Rome are as symmetrically trenched by their torrents as if they were mere heaps of sand; and monotonously veined to their summits with rami-fications of ravine; so that a large rhubarb-leaf, or thistle-leaf, cast in plaster, would give nearly a reduced model of any mass of them. But the circuit of the Alps, however sculptured by its rivers, is inherently fixed in a kind of organic form; its broad bar or islanded field of gneissitic rock, and the three vast wrinkled ridges of limestone which recoil northwards from it, like surges round a risen Kraken's[1] back, are clearly defined in all their actions and resistances: the chasms worn in them by existing streams are in due proportion to the masses they divide; the denudations which in English hill-country so often efface the external evidence of faults or fissures, among the Alps either follow their tracks, or expose them in sections; and the Tertiary beds, which bear testimony to the greater energy of ancient diluvial action, form now a part of the elevated masses, and are affected by their metamorphism: so that at the

----

[1] [A supposed sea-monster of vast size, said to have been seen off the coast of Norway and on the North American coasts; first mentioned in 1755 by Pontoppidan (*History of Norway*, ii. ch. vii. § 11). See Tennyson's poem (1830) "Kraken": "Far, far beneath in the abysmal sea, . . . The Kraken sleepeth."]

turn of every glen new structural problems present themselves, and new conditions of chemical change.

5. And over these I have now been meditating—or wondering—for some twenty years, expecting always that the advance of geology would interpret them for me: but time passes, and, while the aspect and anatomy of hills within five miles of Geneva remain yet unexplained, I find my brother-geologists disputing at the bottom of the lake. Will they pardon me if I at last take courage to ask them a few plain questions (respecting near and visible hills), for want of some answer to which I am sorely hindered in

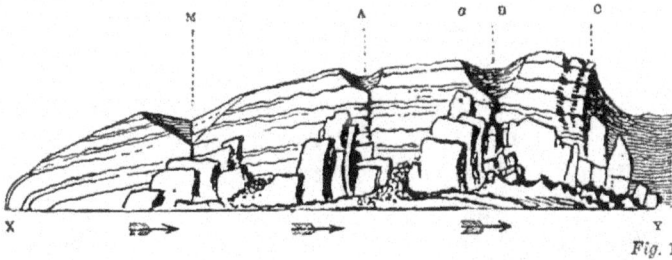

Fig. 1

NORTHERN PORTION OF THE RIDGE OF MONT SALÈVE

my endeavours to define the laws of mountain-form for purposes of art?

6. Fig. 1 is the front view, abstracted into the simplest terms, and laterally much shortened, of the northern portion of the ridge of the Mont Salève, five miles from Geneva.

It is distinguished from the rest of the ridge by the boldness of its precipices, which terminate violently at the angle C, just above the little village called, probably from this very angle, "Coin." The rest of the ridge falls back behind this advanced corner, and is softer in contour, though ultimately, in its southern mass, greater in elevation. Fig. 2 is the section, under *a*, as I suppose it to be; and Fig. 3, as it is given by Studer.[1] To my immediate purpose, it is of no consequence which is the true

[1] [See *Geologie der Schweiz*, 1851, vol. ii. p. 296.]

section; but the determination of the question, ultimately, is of importance in relation to many of the foliated precipices of the Alps, in which it is difficult to distinguish whether their vertical cleavage across the beds is owing merely to disintegration and expansion, or to faults. In

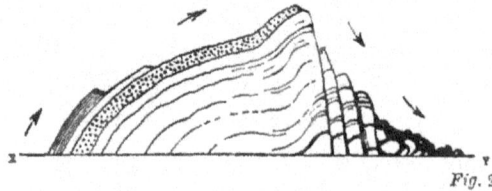

Fig. 2

SECTION OF MONT SALÈVE AT *a*, FIG. 1 (Ruskin)

all cases of strata arched by elevation, the flank of the arch (if not all of it) must be elongated, or divided by fissures. The condition, in abstract geometrical terms, is shown in Fig. 4. If A D was once a continuous bed, and the portion C D is raised to E F, any connecting portion, B C, will become of the form B E; and in doing this, either

Fig. 3

SECTION OF MONT SALÈVE AT *a*, FIG. 1 (Studer)

every particle of the rock must change its place, or fissures of some kind establish themselves. In the Alpine limestones, I think the operation is usually as at G H; but in the Salève the rock-structure is materially altered; so much so that I believe all appearance of fossils has been in portions obliterated. The Neocomian and the Coralline Jura of the body of the hill are highly fossiliferous; but I have scrambled among these vertical cleavages day after day in vain; and even Professor Favre renders no better account of them.*

7. The whole ridge of the mountain continues the curve of the eastern shore of the Lake of Geneva, and turns its rounded back to the chain of the Alps. The great Geneva glacier flowed by it, if ever, in the direction of the arrows from X to Y in Fig. 1; and, if it cut it into its present shape, turned very sharply round the corner at C! The

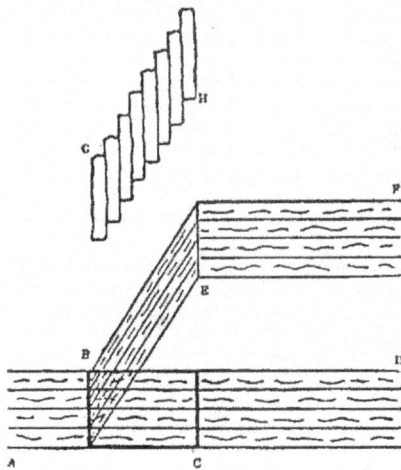

Fig. 4

DIAGRAM OF UPHEAVED BEDS

great Chamouni Glacier flowed over it, if ever, in the direction of the arrows from X to Y in Fig. 2. It probably never did, as there are no erratic blocks on the summits, though many are still left a little way down. But whatever these glaciers made of the mountain, or cut away from it, the existence of the ridge at all is originally owing to the elevation of its beds in a gentle arch longitudinally, and a steep semi-arch transversely; and the valleys or hollows by which this ridge is now traversed, or trenched

---

* *Considérations sur le Mont Salève*, Geneva, 1843, p. 12.[1]

[1] [For another reference to Favre, see Ruskin's Introduction (§ 10) to Collingwood's *Limestone Alps of Savoy* (below, p. 568).]

(M, the valley of Monnetier; A, the hollow called Petite Gorge; B, that called Grande Gorge; and C, the descent towards the Valley of Croisette), owe their origin to denudation, guided by curvilinear fissures, which affect and partly shape the summits of all the inner lateral limestone-ranges, as far as the Aiguille de Varens.

8. It is this guidance of the torrent-action by the fissures; the relation of the longitudinal fault to the great precipice; and the altered condition, not only of the beds on the cliff-side, but of the Molasse conglomerates on the eastern slope, to which I wish presently to direct attention: but I must give more drawings to explain the direction of these fissures than I have room for in this number of the Magazine; and also, before entering on the subject of the angular excavation of the valley at M, and curvilinear excavations at A and B, I want some answer to this question —one which has long embarrassed me:—The streams of the Alps are broadly divisible into three classes: 1st, those which fall over precipices in which they have cut no ravine whatever (as the Staubbach); 2nd, those which fall over precipices in which they have cut ravines a certain distance back (as the torrent descending from the Tournette to the Lake of Annecy); and, 3rd, those which have completed for themselves a sloping course through the entire mass of the beds they traverse (as the Eau Noire, and the stream of the Aletsch Glacier). The latter class—those which have completed their work—have often conquered the hardest rocks; the Eau Noire at Trient traverses as tough a gneiss as any in the Alps; while the Staubbach has not so much as cut back through the overhanging brow of its own cliff, though only of limestone! Are these three stages of work in anywise indicative of relative periods of time?—or do they mark different modes of the torrents' action on the rocks? I shall be very grateful for some definiteness of answer on this matter.[1]

---

[1] [In 1875 Ruskin was still asking the same question: see the letter to A. Tylor, above, p. xxv.]

II[1]

9. At the extremity of my sketch, Fig. 1, p. 25,[2] the beds appear to turn suddenly downwards. They are actually more inclined at this spot; but the principal cause of their apparent increase in steepness is a change in their strike. Generally parallel to the precipice, it here turns westwards (*i.e.*, towards the spectator); and, holding myself bound in candour to note, as I proceed, every circumstance appearing to make for the modern glacial theories, I must admit that, as the beds at this extremity of the cliff turn outwards from the Alps, it might not inaptly be concluded that the great Chamouni Glacier, which by its friction filed the mountain two thousand feet down at the top, by its pressure turned the end of it several points of the compass round at the bottom!

10. This change in the strike of beds, though over a very limited space, yet perfects the Salève as a typical example, entirely simple in its terms, of a wave of the undulatory district of the Savoy Alps. I call it the "undulatory" district, because, in common with a great belt of limestone ranges extending on the north side of the gneissose Alps as far as the Valley of the Rhine, it is composed of masses of rock which have bent like leather under the forces affecting them, instead of breaking like ice; and their planes of elevation are therefore all, more or less, curved.

11. This is one of the points on which I want help. I have hitherto met with no clear statement of the supposed or supposable differences between the mountains which rise bending, and those which rise rigid. The conglomerates of Central Switzerland, for instance, are raised always, I think, in rectilinear masses, league heaped on league of continuous slope, like tilted planks or tables. But the subjacent limestones of Altorf and Lauterbrunnen are thrown

---

[1] [From the *Geological Magazine*, vol. ii., May 1865 (see above, p. 20).]
[2] [The original paper here adds, "which by the printer's inadvertence is marked Y instead of X." The correction has been made in this edition.]

into fantastic curves. The gneissitic schists of Chamouni are all rectilinear; those of Val d'Aoste and the Black Forest are coiled like knots of passionate snakes. Is this difference caused by difference in the characters of the applied forces, or by differences in the state of the materials submitted to them? If by difference in the materials, it is not easy to understand how forces which could twist limestone-beds a thousand feet thick, as a laundress wrings linen, could have left conglomerates in any other state than that of unintelligible heaps of shingle and dust. But if the difference is in the manner of the agency, we have instantly a point of evidence of no small value respecting the date and sphere of any given elevatory force. But I have not yet seen any attempt to distinguish, among the several known periods of elevation in the Alps, those during which the action was accompanied by coiling pressures from those, if any, during which we have only evidence of direct heave. And the question is rendered both more intricate and interesting by the existence of the same structural distinction on a small scale. The metamorphic series, passing from gravel into gneiss, through the infinitely various "poudingues," is far more interesting than the transition from mud to gneiss through the schists, except only in this one particular, that the conglomerates, as far as I know them, do not distinctly coil. Their pebbles are wrenched, shattered, softened, pressed into each other; then veined and laminated; and at last they become crystalline with their paste: but they neither coil, nor wholly lose trace, under whatever pressure, of the consistent couching on their broad sides, which first directed Saussure's attention to the inclined position of their beds;[1] whereas limestones in the same transitional relation to the gneiss (those under the Castle of Martigny, for example) wrinkle themselves as if Falstaff's wit had vexed, or pleased, them, and made their faces "like a wet cloak ill laid up."[2]

---

[1] [On this subject, see below, p. 384.]
[2] [2 *Henry IV.*, Act v. sc. 1.]

12. It is true that where the conglomerates begin to take the aspect of shattered breccias, like those which accompany great faults, some aspect of coiling introduces itself also. I have not examined the conglomerate-junctions as I have the calcareous ones, because the mountain-forms of the breccias are so inferior (for my own special purposes of art) to those of the schists, that I never stay long in the breccia districts. But a few hours of study by the shores of the Reuss or Limmat are enough to show the general differences in aspect of compression between metamorphic schists and conglomerates; and the distinction on a large scale is everywhere notable; but complicated by this fact, which I have not until lately ascertained positively, that through even the most contorted beds of the limestones there run strange, long, rectilinear cleavages, extending in consistent slopes for leagues,—giving the mountain-mass, seen at right angles to their direction, an aspect of quite even stratification and elevation, with a strike entirely independent of its true beds and minor cleavages, and traversing them all.

13. Putting these gigantic cleavages (of the origin of which, even if I could guess any reason, I would still say nothing so long as I could do no more than guess) for the present out of the question, the mass of the Savoy limestones forms a series of surges, retreating from the Alps, undulatory in two directions at least, as at Fig. 5, A, and traversed by fissures usually at right angles to the strike of the longitudinal surge. When that strike varies consistently at the same time, we may get conditions of radiant curve and fracture, B; and the undulations themselves are seldom simple, as at *a*, but either complicated by successive emergence of beds, *b*, or more frequently by successive faults, *c*, farther modified by denudation of upper beds, *d*, and, locally, by reversals of their entire series, *e*.

14. All these complex phenomena will be produced by one consistent agency of elevatory or compressive force.

Any number of such tides of force may of course succeed each other at different epochs, each traversing the series of beds in new directions,—intersecting the forms already produced, and giving maxima and minima of elevation and depression where its own maxima and minima coincide with those resultant from previous forces.

15. Now, by every such passage of force, a new series of cleavages is produced in the rocks, which I shall for the present call " passive cleavages," as opposed to "native

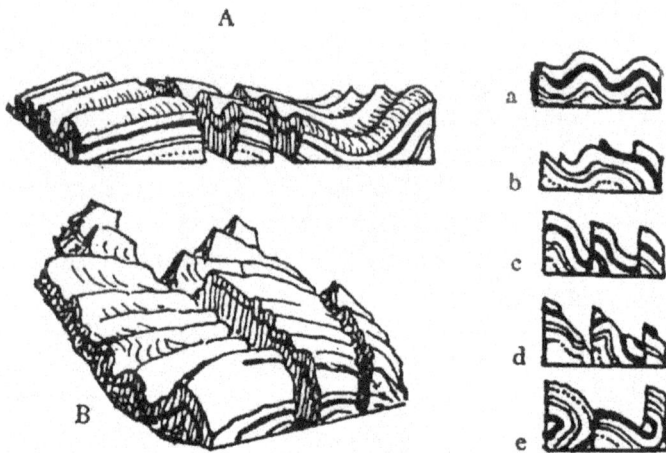

Fig. 5

cleavages." [1]   I do not care about the names ; anybody is welcome to give them what names they choose ; but it is necessary to understand and accept the distinction.   I call a cleavage "native" which is produced by changes in the relation of the constituent particles of a rock while the mass of it is in repose.   I call a cleavage "passive" which is produced by the motion of the entire mass under given pressures or strains.   Only I do not call the mere contraction and expansion of the rock *motion ;* though, in large formations, such changes in bulk may involve motion over leagues.   But I call every cleavage "native" which has

[1] [For Ruskin's later distinctions of various forms of cleavage, see *Deucalion,* below, pp. 289 *seq.*]

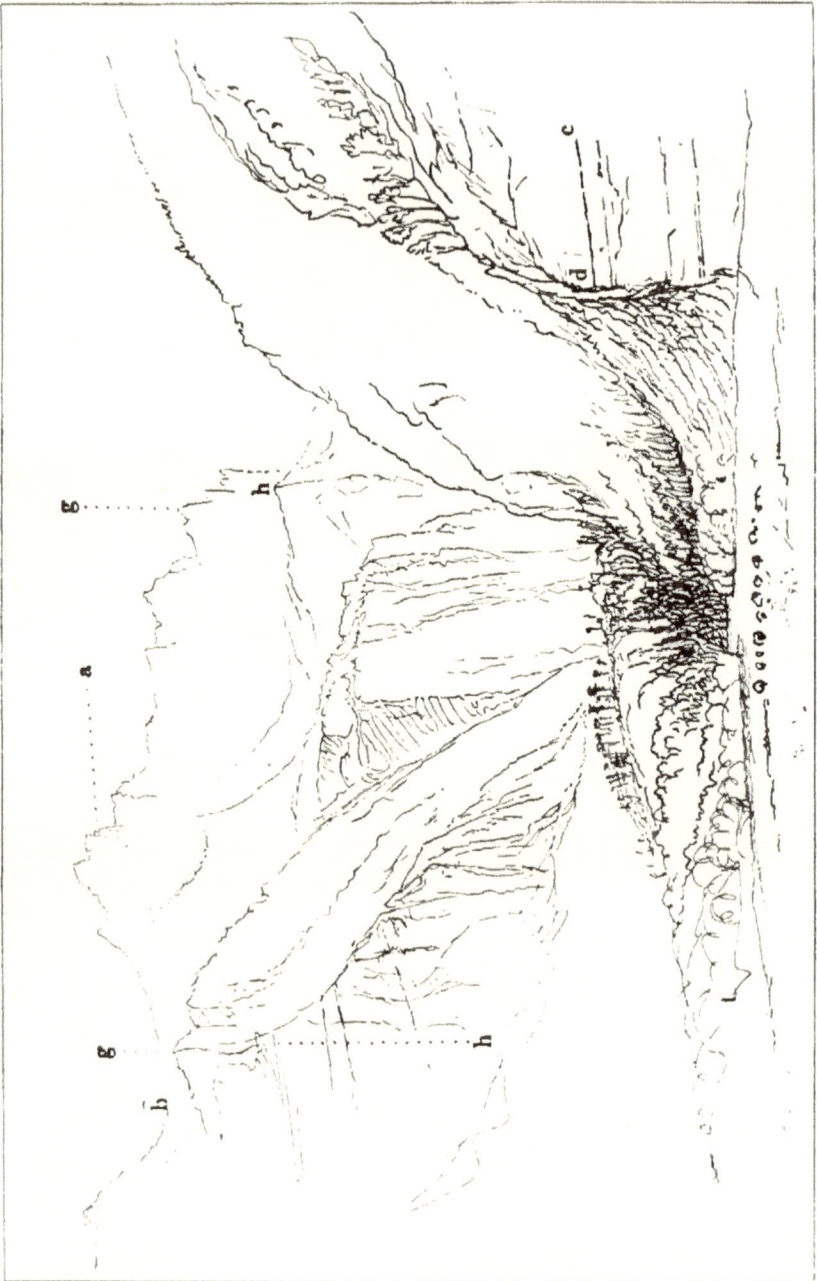

IV

been produced by contraction, expansion, segregation, or crystallization, whatever the space over which the rock may be moved by its structural change; and I only call a cleavage "passive" which has been caused by a strain on the rock under external force. Practically, the two cleavages, or rather the two groups of cleavages, mingle with and modify each other; but the "native" cleavage is universal;—the "passive" is local, and has more direct relations with the mountain-form.

16. In the range, for instance, of which the Aiguille de Varens forms the salient point, of which the rough outline is given in Fig. 6, seen in front, and in Plate IV. seen in profile, the real beds dip in the direction *a b*, Plate IV., being conspicuous in every aspect of the mountain in its profile: they are seen again on the opposite side of the valley (of Maglans)

Fig. 6

at *c d.* On the face of the mass, Fig. 6, they are seen to be contorted and wrinkled, on the left reversed in complete zigzags; * but through all these a native cleavage develops itself in the direction *e f*, accompanied by an elaborate network of diagonal veining with calcareous spar.

17. This cleavage directs the entire system of the descending streams, which, by help of it, cut the steps of precipice into oblique prisms, curved more and more steeply downwards, as the sweep of the torrent gains in power; so that, seen from a point a little farther to the right, the mountain seems composed of vast vertical beds, more or less curved in contour, Fig. 7. But the face of the precipice itself is hewn into steps and walls, with intermediate slopes, by a grand vertical passive cleavage, *g h*,

---

* This contortion is an important one, existing on both sides of the valley; but it is in reality farther to the left. I have crowded it in to complete the typical figure.

Plate IV., to which the direction and disposition of the entire Valley of Maglans are originally owing.

18. And thus in any given mountain-mass, before we can touch the question of denudation, we have to deter-

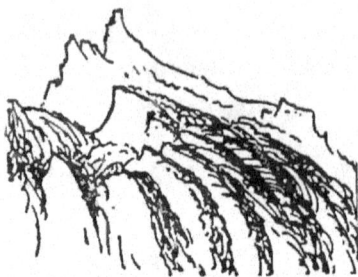

Fig. 7

mine the position it occupies in the wave-system of the country, —the connections of its cleavages with those of neighbouring masses,— and the probable points of maxima elevation which directed the original courses of glacier and stream. Then come the yet more intricate questions respecting the state of the materials at each successive elevation, and during the action of the successively destructive atmospheric influences.

I have no pretension to state more than a few of the main facts bearing on such questions in the Savoy districts of the Lower Chalk, which I will endeavour to do briefly in one or two following papers.[1]

[1] [No further papers, however, appeared.]

III

ON BANDED AND BRECCIATED
CONCRETIONS

(1867–1870)

[*Bibliographical Note.*—The papers "On Banded and Brecciated Concretions" first appeared in the *Geological Magazine* (1867-1870), as follows :—

    I. In vol. iv., No. 38, August   1867, pp. 337-339.
   II. ,, ,, iv., ,, 41, November   ,,   ,, 481-483.
  III. ,, ,, v., ,, 43, January 1868, ,, 12-13.
  IV. ,, ,, v., ,, 46, April   ,,   ,, 156-161.
   V. ,, ,, v., ,, 47, May   ,,   ,, 208-213.
  VI. ,, ,, vi., ,, 66, December 1869, ,, 529-534.
 VII. ,, ,, vii., ,, 67, January 1870, ,, 10-14.

One plate was given with each number of the Magazine. The plates were numbered according to their place in the Magazine ; thus the first plate was lettered at the top "*Geol. Mag. 1867. Vol. IX. Pl. XV.,*" and so on.

Two of the Plates were again used in *Deucalion ;* namely, Plate X. (in the numbering of the present volume), "Mural Agates" (which became Plate III. in *Deucalion*), and Plate VII., "Amethyst-quartz" (which became Plate IV.).

Each paper was next issued in pamphlet form, in brown paper wrappers. Paper I. is headed "[Extracted from the *Geological Magazine.* Vol. IV., No. 8.[1] August, 1867.]" Octavo, pp. 1-4. Imprint at the foot of the last page : "Stephen Austin, Printer, Hertford." Paper II. is headed similarly, with the same imprint, pp. 1-3. Paper III., pp. 1-7. Paper IV., pp. 1-5. Paper V. was presumably similarly issued, but the editors have not seen it.

Papers VI. and VII. were similarly issued, but in buff-coloured wrappers, lettered on the front (Paper VI.) "Banded & Brecciated Concretions. | [*Extracted from the* Geological Magazine, December, 1869.] | *Published by* Trübner & Co., 8 and 60, Paternoster Row, London, E.C.," pp. 3-6 (same imprint).

Paper VII., similarly lettered on the front wrapper ("January, 1870" for "December, 1869" being the only change), pp. 1-4 (same imprint).

These pamphlets are among the rarest of Ruskiniana.

Variæ Lectiones.—In this volume a few corrections have been made, in accordance with marks in Ruskin's copies of the Magazine (in Mr. Allen's possession). In Paper II., § 10, line 16, "concretions" for "secretions"; in Paper III., § 12 (III., line 24), see p. 48 *n.*, § 13, line 3, "cacholong" for "cacholony"; and in Paper VI., § 39, line 22, "acicular" for "or circular." References to figures and plates are altered throughout; and the usual numbering of sections is introduced.]

[1] *i.e.,* No. 8 of vol. iv., No. 38 from the start of the Magazine.

ON BANDED AND BRECCIATED
CONCRETIONS

I

[*August* 1867]

1. AMONG the metamorphic phenomena which seem to me
deserving of more attention than they have yet received,
I have been especially interested by those existing in the
brecciate formations. They are, of course, in the main,
twofold—namely, the changes of fragmentary or rolled-
pebble deposits into solid rocks, and of solid rocks, *vice
versâ*, into brecciate or gravel-like conditions. It is certainly difficult, in some cases, to discern by which of these
processes a given breccia has been produced; and it is
difficult, in many cases, to explain how certain conditions
of breccia can have been produced either way. Even the
pudding-stones of simplest aspect (as the common Molasse-
nagelfluhe of North Switzerland) present most singular conditions of cleavage and secretion, under metamorphic action;
the more altered transitional breccias, such as those of Valor-
sine, conceal their modes of change in a deep obscurity:
but the greatest mystery of all attaches to the alterations
of massive limestone which have produced the brecciated,
or apparently brecciated, marbles: and to the parallel
changes, on a smaller scale, exhibited by brecciated agate
and flint.

2. The transformations of solid into fragmentary rocks
may, in the main, be arranged under the five following
heads:—

(i.) Division into fragments by contraction or expansion,
and filling of the intervals with a secreted, injected, or

37

infused paste, the degree of change in the relative position of the fragments depending both on their own rate and degree of division, and on the manner of the introduction of the cement.

(ii.) Division into fragments by violence, with subsequent injection or secretion of cement. The walls of most veins supply notable instances of such action, modified by the influence of pure contraction or expansion.

(iii.) Homogeneous segregation, as in oolite and pisolite.

(iv.) Segregation of distinct substances from a homogeneous paste, as of chert out of calcareous beds. My impression is that many so-called siliceous " breccias" are segregations of knotted silex from a semi-siliceous paste ; and many so-called brecciated marbles are segregations of proportioned mixtures of iron, alumina, and lime, from an impure calcareous paste.

(v.) Segregation accompanied by crystalline action, passing into granitic and porphyritic formations.

3. Of these the fourth mode of change is one of peculiar and varied interest. I have endeavoured to represent three distinct and progressive conditions of it in the plate annexed [Plate V.]; but before describing these, let us observe the structure of a piece of common pisolite from the Carlsbad Springs.

It consists of a calcareous paste which arranges itself, as it dries, in imperfect spheres, formed of concentric coats which separate clearly from each other, exposing delicately smooth surfaces of contact : this deposit being formed in layers, alternating with others more or less amorphous. Now it is easy to put beside any specimen of this pisolite, a parallel example of stratified jasper, in which some of the beds arrange themselves in pisolitic concretions, while others remain amorphous. And I believe it will be found that the bands of agate, when most distinct and beautiful, are not successive coats, but pisolitic concretions of amorphous silica.

Of course, however, the two conditions must be often

united.   In all minerals of chalcedonic or reniform structure, stalactitic additions may be manifestly made at various periods to the original mass, while in the substance of the whole accumulation, a structural separation takes place, —separation (if the substance be siliceous) into bands, spots, dendritic nuclei, and flame-like tracts of colour.   But the separation into any of these states is not so simple a matter as might at first be supposed.

4. On looking more closely at the Carlsbad pisolite, we may discern here and there hemispherical concretions, of which the structure seems not easily to be accounted for; much less when it takes place to the extent shown in Fig. 1, Plate V., which represents, about one-third magnified, a piece of concretionary ferruginous limestone, in which I presume that the tendency of the iron oxide to form reniform concretions has acted in aid of the pisolitic disposition of the calcareous matter.   But there is now introduced a feature of notable difference.   In common piso-lite, the substance is homogeneous; here, every concretion is varied in substance from band to band, as in agates; and more varied still in degree of crystalline or radiant structure; while also sharp-angled fragments, traversed in one case by straight bands, are mingled among the spherical concretions: and series of brown bands, of varying thick-ness, connect, on the upper surfaces only, the irregular concretions together, in a manner not unusual in marbles, but nevertheless (to me) inexplicable.

5. Next to this specimen, let us take an example of what is usually called " brecciated " malachite (Fig. 2, in the same plate).   I think very little attention will show, in ordinary specimens of banded malachite, that the bands are concretionary, not successive; and in the specimen of which the section is represented in the plate, and in all like it, I believe the apparently brecciated structure is concretionary also.   This brecciation, it will be observed, results from two distinct processes: the rending asunder of the zoned concretions by unequal contraction which bends

the zones into conditions like the twisted fibres of a tree; and the filling up of the intervals with angular fragments, mixed with an ochreous dust (represented in the plate by the white ground), while the larger concretions of malachite are abruptly terminated only at right angles to the course of their zones, not broken raggedly across: a circumstance to be carefully noted as forbidding the idea of ordinary accidental fracture.

Whether concurrently with, or subsequent to, the brecciation (I believe concurrently), various series of narrow bands have been formed in some parts of the mass, binding the apparent fragments together, and connecting themselves strangely with the unruptured malachite, like the brown bands in example No. 1.

6. Now, if we compare this condition of the ore of copper with such a form of common brecciated agate as that represented in Plate V., Fig. 3, it will, I think, be manifest that the laws concerned in the production of this last—though more subtle and decisive in operation, are essentially the same as those under which the malachite breccia was formed,—complicated, however, by the energetically crystalline power of the (amethystine) quartz, which exerts itself concurrently with the force of segregation, and compels the zones developed by the latter to follow, through a great part of their course, the angular line of the extremities of the quartz-crystals co-temporaneously formed, while, in other parts of the stone, a brecciate segregation, exactly similar to that of the malachite, and only the fine ultimate perfectness of the condition of fragmentary separation which is seen incipiently in the pisolite (Fig. 1), interrupts the continuity both of the agate and quartz.

And finally, a narrow band, correspondent to the connecting zones of the malachite, surrounds the brecciated fragments in many places, while in others it loses itself in the general substance of the massive quartz.

7. I cannot, however, satisfy myself whether, in this last example, some conditions of violent rupture do not

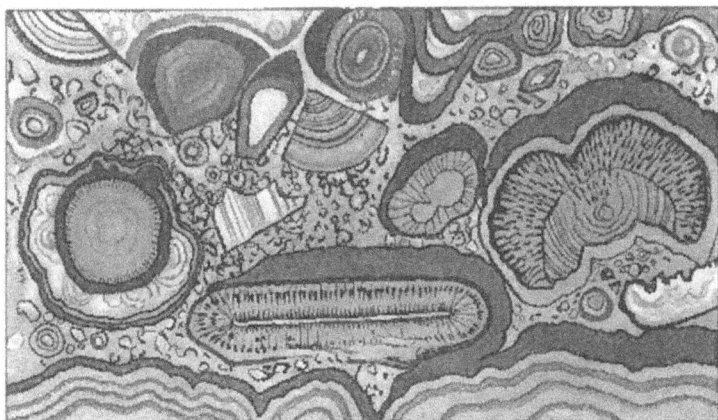

BANDED AND BRECCIATED CONCRETIONS

mingle with those of agatescent segregation; and I am sincerely desirous to know the opinions of better mineralogists than myself on these points of doubt: and this the more, because in proceeding to real and unquestionable states of brecciate rock, such as the fractured quartz and chalcedony of Cornwall, I cannot discern the line of separation, or fix upon any test by which a fragment truly broken and cemented by a siliceous paste which has modified or partly dissolved its edges, may be distinguished from a secretion contemporaneous with the paste, like the so frequent state of metalliferous ores dispersed in quartz.

Hoping for some help therefore, I will not add anything further in this paper; but if no one else will take up the subject, I shall proceed next month into some further particulars.

EXPLANATION OF PLATE V

Fig. 1. Section of a piece of concretionary ferruginous limestone, magnified about one-third.
Fig. 2. Section of a (so-called) "Brecciated" Malachite.
Fig. 3. Section of a Brecciated Agate.

II

[November 1867]

8. I wrote the first of these papers more with a view of obtaining some help in my own work than with any purpose of carrying forward the discussion of the subject myself. But no help having been given me, I must proceed cautiously alone, and arrange the order of my questions; since, when I have done my best as carefully as I can, the papers will be nothing but a series of suggestions for others to pursue at their pleasure.

Let me first give the sense in which I use some necessary words :—

(1.) Supposing cavities in rocks are produced by any accident, or by original structure (as hollows left by gas in lava), and afterwards filled by the slow introduction of a substance which forms an element of the rock in which the cavities are formed, and is finally present, in the cavities, in proportion to its greater or less abundance in the rock, I call the process " secretion."

(2.) But if the cavities are filled with a substance not present (or not in sufficient quantity present), in the surrounding rock, and therefore necessarily brought into them from a distance, I call the process, if slow, " infiltration "; if violent, " injection."

It is evident that water percolating a rock may carry a substance, present in the mass of it, by infiltration, into the cavities, and so imitate the process of secretion. But there are structural differences in the aspect of the two conditions hereafter to be noticed. The existence of permanent moisture is, however, to be admitted among conditions of secretion; but not of fluent moisture, introducing foreign elements.

(3.) If a crystalline or agatescent mass is formed by addition of successive coats, I call the process " accretion."

(4.) But if the crystalline or agatescent mass separates itself out of another solid mass, as an imbedded crystal, or nodule, and then, within its substance, divides itself into coats, I call the process " concretion." The orbicular granite of Elba is the simplest instance I can refer to of such manifest action: but all crystals, scattered equally through a solid enclosing paste, I shall call " concrete " crystals, as opposed to those which are constructed in freedom out of a liquid or vapour in cavities of rocks, and which I shall call " accrete."

The fluor nodules of Derbyshire, and amethystine nodules of some trap rocks, present, in their interiors, the most beautiful phenomena of concrete crystallization, of which I hope to give careful drawings.

9. It is true, as I said in the last paper, that these two processes are perpetually associated, and also that the difference between them is sometimes only between coats

attracted and coats imposed. A small portion of organic substance will, perhaps, attract silica to itself, out of a rock which contained little silica in proportion to its substance; and this first knot of silica will attract more, and, at last, a large mass of flint will be formed, which I should call "concrete"; but if a successive overflowing of a siliceous spring had deposited successive layers of silica upon it, I should call it "accrete." But the resemblance of the two processes in such instances need not interfere with the clearness of our first conception of them; nor with our sense of the firm distinction between the separation of a solid mass, already formed, into crystals or coats in its interior substance, and the increase of crystalline or coated masses by gradual imposition of new matter.

10. Now let me re-state the scope of the questions, for the following out of which I want to collect materials:—

(I.) I suspect that many so-called "conglomerates" are not conglomerates at all, but concretionary formations, capable, finally, of complete mechanical separation of parts; and therefore that even some states of apparently rolled gravel are only dissolutions of concretionary rock.

Of course, conglomerates, in which the pebbles are fragments of recognizable foreign rocks, are beyond all possibility of challenge; as also those in which the nodules could not, by any chemistry, have been secreted from the surrounding mass. But I have in my hand, as I write, a so-called "conglomerate" of red, rounded, flint "pebbles," much divided by interior cracks, enclosed by a finely crystallized quartz; and I am under the strongest impression that the enclosed pieces are not pebbles at all, but concretions—the spots on a colossal bloodstone. It is with a view to the solution of this large question that I am examining the minor structure of brecciated agates and flints.

(II.) It seems to me that some of the most singular conditions of crystalline metamorphic rocks are the result of the reduction of true conglomerates into a solid mass; and I want therefore to trace the changes in clearly recognizable conglomerates where they are affected by metamorphism, and arrange them in a consistent series.

(III.) I cannot, at present, distinguish in rocks the faults, veins, and brecciations caused by slow contraction from those occasioned by external pressure or violence. It seems to me now that many distortions and faults, which I have been in the habit of supposing the result of violence, are only colossal phenomena of retraction or contraction, and even that many apparent strata have been produced by segregation. A paper, on this subject, of Mr. George Maw's, put into my hands in May 1863, gave me the first suggestion of this possibility.[1]

11. I shall endeavour, as I have leisure, to present such facts to the readers of this Magazine as may bear on these three inquiries, and have first engraved the plate given in the present number in order to put clearly under their consideration the ordinary aspect of the veins in the first stage of metamorphism in the Alpine cherts and limestones. The three figures [Plate VI.] are portions of rolled fragments; it is impossible to break good specimens from the rock itself, for it always breaks through the veins, and it must be gradually ground down in order to get a good surface.

Fig. 1 is a portion of the surface of a black chertose mass, rent and filled by a fine quartzose deposit or secretion, softer than the black portions and yielding to the knife; neither black nor white parts effervesce with acids. It is as delicate an instance of a vein with rent fibrous walls as I could find (from the superficial gravel near Geneva).

Fig. 2 is from the bed of the stream descending from the Aiguille de Varens to St. Martin's. It represents the usual condition of rending and warping in the flanks of veins caused by slow contraction, the separated fragments showing their correspondence with the places they have

---

[1] [The editors have failed to trace this paper, and Mr. Maw cannot recall it. There are papers by him in the *Journal of the Geological Society* for 1867 and 1868, and in one of these (vol. xxiv. pp. 351–400) he discusses segregation as the chief agency in the variegation of ferruginous rocks; but though a short abstract of this paper appeared before Ruskin wrote (vol. xxiii. p. 114), the point is a different one.]

VI

Fig.L

Fig 2   Fig 3

ROCK VEINS.
UNDER CONTRACTION, AT TWO PERIODS.

seceded from; and it is evident that the secretion or injection of the filling white carbonate of lime must have been concurrent with the slow fracture, or else the pieces, unsupported, would have fallen asunder.

Fig. 3 is from the bed of the Arve at St. Martin's, and shows this condition still more delicately. The narrow black line traversing the white surface, near the top, is the edge of a film of slate, once attached to the dark broad vertical belt, and which has been slowly warped from it as the carbonate of lime was introduced. When the whole was partly consolidated, a second series of contractions has taken place, filled, not now by carbonate of lime, but by compact quartz, traversing in many fine branches the slate and calcite, nearly at right angles to their course.

I shall have more to say of the examples in this plate[1] in connection with others, of which engravings are in preparation.

III

[*January* 1868]

12. The states of semi-crystalline silica are so various, and so connected in their variety, that the best recent authorities have been content to group them all with quartz, giving to each only a few words of special notice; even the important chapters of Bischof[2] describe rather their states of decomposition and transition than the minerals themselves. Nevertheless, as central types, five conditions of silica are definable, structurally, if not chemically, distinct, and forming true species; and in entering on any detailed examination of agatescent arrangements, it is quite necessary to define with precision these typical substances, and their relation to crystalline quartz.

---

[1] [See below, § 12 (II.), p. 47.]

[2] [Carl Gustav Bischof: *Lehrbuch der chemischen und physikalischen Geologie*, 2 vols., Bonn, 1847, 1854. Translated (in the *Works of the Cavendish Society*) by B. H. Paul and J. Drummond, 1854. Chapter xlii. deals with "Quartz and other Siliceous Minerals." For other references to the book, see below, pp. 197, 207, 430 *n*.]

(1.) *Jasper.*—Opaque, with dull earthy fracture, and hard enough to take a perfect polish. When the fracture is conchoidal the mineral is not jasper, but stained flint. The transitional states are confused in fracture; but true jasper is absolutely separated from flint by two structural characters—on a small scale it is capable of the most delicate pisolitic [1] arrangement, and on a larger scale is continually found in flame-like concretions, beautifully involved and contorted. But flint is never pisolitic, and, in any fine manner, never coiled; nor do either of these structures take place in any transitional specimen, until the conchoidal fracture of the flint has given place to the dull earthy one of jasper; nor is even jasper itself pisolitic on the fracture, being too close-grained. The green base of heliotrope, with a perfectly even fracture, may be often seen, where it is speckled with white, to be arranged in exquisitely sharp and minute spherical concretions, cemented by a white paste, of which portions sometimes take a completely brecciated

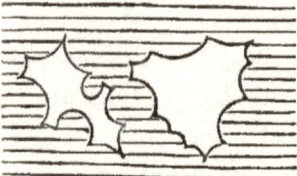

Fig. 8

aspect, each fragment being outlined by concave segments of circles (Fig. 8). Jasper is eminently retractile, like the clay in septaria,[2] and in agates often breaks into warped fragments, dragging the rest of the stone into distortion. In general, the imbedded fragments in any brecciated agate will be mainly of jasper; the cement, chalcedonic or quartzose.

(II.) *Flint.*—Amorphous silica, translucent on the edges, with fine conchoidal fracture. Opaque only when altered, nascent, or stained. Never coiled, never pisolitic, never reniform;[3] these essentially negative characters belonging to it as being usually formed by a slow accumulative secretion, and afterwards remaining unmodified (preserving therefore

---

[1] [For definition of this term, see § 14 (below, p. 49).]

[2] [Septaria are "concretions of considerable size and roughly spherical in shape, of which the parts nearest the centre have become cracked during the drying of the mass, the open spaces thus formed having been subsequently filled with some infiltrated mineral, usually calcite."]

[3] [For definition of this term, see, again, § 14 (below, p. 49).]

casts of organic forms with great precision). It is less retractile than jasper; its brecciate conditions being not so much produced by contraction or secession, as by true secretion, even when most irregular in shape (as a row of flints in chalk differ from the limestone fragments represented in Plate VI., Fig. 3, which might stand for a jasperine structure also). But there are innumerable transitions between these two states, affected also by external violence, which we shall have to examine carefully. Within these nodular concretions, flint is capable of a subsequently banded, though not pisolitic arrangement. (See Dr. S. P. Woodward's paper on banded flints, in this Magazine, vol. i., for October 1864, p. 145.[1])

(III.) *Chalcedony.*—Reniform silica, translucent when pure, opaque only when stained, nascent, or passing into quartz. The essential characteristic of chalcedony is its reniform structure, which in the pure mineral is as definite as in wavellite[2] or hæmatite, though when it is rapidly cooled or congealed from its nascent state of fluent jelly it may remain as a mere amorphous coating of other substances; very rarely, however, without some slight evidence of its own reniform crystallization. The study of its different degrees of congelation in agates is of extreme intricacy. As a free mineral in open cavities it is actively stalactitic, not merely pendent or accumulative, but animated by a kind of crystalline spinal energy, which gives to its processes something of the arbitrary arrangement of real crystals, modified always by cohesion, gravity, and (presumably) by fluid and gaseous currents.

There is no transition between chalcedony and flint. They may be intimately mixed at their edges, but the limit is definite. Impure brown and amber-coloured chalcedonies, and those charged with great quantities of foreign

---

[1] ["On the Nature and Origin of Banded Flints." Dr. Woodward's contention was that the origin of the bands is not organic but that they are produced by infiltration. For other references to this paper, see *Deucalion*, i. ch. ix. § 11 (below, p. 211), and pp. 399, 474. Samuel Pickworth Woodward (1821–1865) was an assistant in the department of geology and mineralogy, British Museum, 1848–1865.]

[2] [Ruskin calls special attention to this mineral in *Proserpina* (Vol. XXV. p. 215).]

matter, may closely resemble flint, but the two substances are entirely distinct. Between jasper and chalcedony the separation is still more definite in mass, jasper being never reniform,* and differing greatly in fracture; but the flame-like or spotted crimson stains of chalcedony often approach conditions of jasper; and there is, I suppose, no pisolitic formation of any substance without some inherent radiation, which associates it with reniform groups, so that pisolitic jasper must be considered as partly transitive to chalcedony. On the other hand, chalcedony seems to pass into common crystalline quartz through milky stellate quartz, associated in Auvergne with guttate and hemispherical forms.

(IV.) *Opal.*—Amorphous translucent silica, with resinous fracture, and essential water. Distinguished from chalcedony by three great structural characteristics—(*a*) its resinous fracture; (*b*) that it is never pisolitic or reniform; (*c*) that when zoned, in cavities or veins, its *zones are always rectilinear*, and transverse to the vein,[1] while those of chalcedony are usually undulating, and parallel to the sides of the vein; level only in lakes at the bottom of cavities.

(V.) *Hyalite.*—Amorphous transparent silica, with vitreous fracture, and essential water. Never reniform, nor pisolitic, nor banded; but composed of irregularly grouped bosses, generally elliptical or pear-shaped (only accidentally spherical), formed apparently by successive accretion of coats, but not showing banded structure internally (Fig. 9). Entirely transparent, with splendid smooth glassy fracture. Sometimes coating lava; sometimes in irregularly isolated patches upon it; apparently connected in structure with the roseate clusters of milky chalcedony of Auvergne. I shall keep the term "guttate" for this particular structure,

Fig. 9

* I have since found some reniform jasper and stained chalcedony. [Note in Ruskin's copy of the Magazine.]

---

[1] [For references to this point, see above, p. liii.; and below, p. 383.]

of which singular varieties also occur among the hornstones of Cornwall.

13. These five main groups are thus definable without embarrassment; two other conditions of silica, perhaps, ought to be separately named—namely, cacholong, which seems to take a place between chalcedony and opal, but which I have not yet been able satisfactorily to define; the other, the calcareous-looking, usually whitish agate, which often surrounds true translucent agate, as if derived from it by decomposition. I am under the impression that this is chalcedony, more or less charged with carbonate of lime, and that it might be arranged separately as lime-jasper, differing from aluminous jasper by being capable of reniform structure;[1] but it is certainly in some cases an altered state of chalcedony, which seems in its more opaque zones to get whiter by exposure to light. I shall therefore call it white agate, when it harmoniously follows the translucent zones, reserving the term jasper for granular aggregations. Perhaps ultimately it may be found that nascent chalcedony can take up either oxide of iron, or alumina, or lime, and might relatively be called iron-jasper, clay-jasper, and lime-jasper; but for any present descriptive purpose the simpler arrangement will suffice.

14. These, then, being the principal types of agatescent silica, it is of importance to define clearly the two structures I have severally called pisolitic and reniform.

A pisolitic mineral is one which has a tendency to separate by spherical fissures, or collect itself by spherical bands, round a central point.[2]

A reniform mineral is one which crystallizes in radiation from a central point, terminating all its crystals by an external spherical surface.[3] It is, however, difficult to define

---

[1] [But see Ruskin's note on p. 48.]

[2] [Pisolitic = having the structure indicated by the term pisolite which = limestone having a structure in which the individual grains or globules are as large as peas ($\pi i \sigma o s$).]

[3] [At a later date Ruskin urged the necessity of substituting the term "spheroidal" for "reniform," "every so-called 'kidney-shaped' mineral" being "an aggregate of spheroidal crystallization": see *Of the Distinctions of Form in Silica*, § 6 (below, p. 375).]

this character mathematically.   On the one hand, radiate crystals may be terminated by spherical curves, as in many zeolites, without being close set enough to constitute a reniform mass ; on the other, radiate crystals, set close, may be terminated so as to prevent smoothness of external spherical surface, and I am not sure whether this smoothness is a mere character of minute scale (so that chalcedony, seen delicately enough, might present pyramidal extremities of its fibres on the apparently smooth surface), or whether, in true reniform structure, the crystallization is actually arrested by a horizontal plane : I do not mean a crystalline plane, as in beryl, but one of imperfect crystallization, presenting itself only under a peculiar law of increase. Thus, in hæmatite, which is both reniform and pisolitic, the masses often divide in their interior by surfaces of jagged crystallization, while externally they are smooth and even lustrous ; but I put this point aside for future inquiry,[1] because it will require us to go into the methods of possible increment in quartz-crystals, and for our present purpose we need only a clear understanding of two plainly visible conditions of jasper and chalcedony, namely, that jasper will collect itself pisolitically, out of an amorphous mass, into concretion round central points, but not actively terminate its external surface by spherical curves ; while chalcedony will energetically so terminate itself externally, but will, in ordinary cases, only develop its pisolitic structure subordinately, by forming parallel bands round any rough surface it has to cover, without collecting into spheres, unless either provoked to do so by the introduction of a foreign substance, or encouraged to do so by accidentally favourable conditions of repose.

15. And here branch out for us two questions, both most intricate—first, as to the introduction of foreign bodies ; secondly, as to the crystalline disposition of chalcedony, under variable permission of repose.

First, as to foreign substances. I assume that in true

---

[1] [See below, pp. 56, 60, 71.]

pisolitic concretion, such as that of the jasper, roughly sketched in Fig. 10 (it is not a coral—the radiant lines are merely conventional indications of the grain of the jasper, so far as it is visible with a lens,*) no foreign body has provoked the orbicular arrangement. The jasper is red; the little dark circles are wells of pure chalcedony, each containing within it a white ball of crystallized quartz, forming a star on the section. The whole is magnified about three times in the drawing, being a portion of a horizontal layer, alternating with solid white jasper. It seems that the pisolitic structure is here truly native; but we must nevertheless grant the possibility that the balls of quartz may have had some organic atom for their nucleus. On the other hand, in the ordinary conditions of dendritic agate, in which stalactites of chalcedony surround branches of clearly visible chlorite,† or of oxide of iron or manganese, I assume that in the plurality of cases such sustaining substances have been first developed, and the chalcedonic stalactite afterwards superimposed, being, in the most literal sense of the word, "superfluous" silica; but I, nevertheless, see great reason for thinking that, in many cases, the core of the group is only a determination to its centre of elements which had been dispersed through the mass. In the generality of Mocha stones, the dendritic oxides, so far from being an original framework, are clearly of subsequent introduction, radically following the course of fissures from which they float partially into the body of the imperfectly congealed gelatinous mass; in other more rare, and singularly

Fig. 10

---

* In my woodcut diagrams I shall employ no fine execution; they will be merely illustrative, not imitative—diagrams, not drawings. In the plates, on the contrary, with Mr. Allen's good help, I shall do the best I can.

† Or green earth? I cannot find any good account of the green substance [1] which plays so important a part in the exterior coats of agates and Iceland chalcedonies.

---

[1] [See above, p. liii.; and below, p. 413.]

beautiful cases, the metallic oxides ramify in curves in the intervals of the pisolitic belts, and then there is nearly always a dark rod in the stalactitic centre, which may or may not be solid. In the finest Mocha stones, I think it is a black film round a chalcedonic nucleus; but in the associations of limonite with chalcedony, it is usually of solid radiate iron oxide, and doubtless of prior, though perhaps only of immediately prior, formation. A more complex state is presented by such stalactites, when enveloped in a chalcedonic solid paste, to which they do not communicate their own zoned structure. Ordinarily, the surrounding mass throws itself into zones parallel with those of the enclosed stalactite; but, in some cases, it is of quite adverse structure, perhaps laid level across the stalactitic fall.

The conditions admitting the interfusion of this solid paste are strangely connected with those which cause chalcedony to form true vertical stalactites and straight rods, instead of arborescent and twisted stalactites. I have never seen the twisted stalactite unless enveloping fibres of some foreign, perhaps organic, substance, enclosed in massive chalcedony; but the straight stalactite is perhaps oftener so than free (unless connected with limonite), and it would appear, therefore, as if the apparently interposed mass were really of contemporary formation, or else it would sometimes enclose the contorted stalactite. But this question respecting the causes of the vertical and twisted groups properly belongs to the second branch of our inquiry as to states of repose.

16. Second, conditions affecting mode of crystallization. It is evident that fluent deposits of silica contained in a rock-cavity must be affected, in course of their solidification, not only by every addition to their own mass, but by every change in the temperature or grain of the surrounding rock, so that we have innumerable modifications of state, dependent partly on accession and transmission of substance, partly on changes in external temperature and

pressure. And, under these influences, we perceive that the gelatinous silica occasionally obeys gravity,* and occasionally resists it, becoming sometimes pendent from the roof, and forming level lakes on the floor of cavities; at other times, throwing parallel bands on floor and roof alike, and in transitional periods, forming thick layers on the floor, and thin ones at the sides, the layers being liable, meantime, to different degrees of compression from their own modes of solidification, which give them, locally,

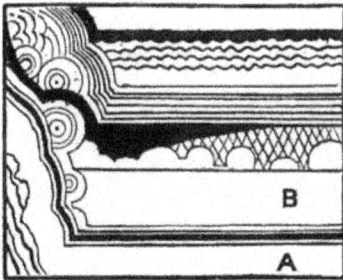

Fig. 11   *Fig.* 12

the appearance of an elastic compression and expansion: there seems no limit to the fineness of their lines at these compressed points, when their continuity is uninterrupted. Figures 11 and 12 illustrate, in two small pieces of agate, each here magnified about three times, most of the appearances which must be severally studied. In Fig. 11, the lowest band, A, level at the bottom, broken irregularly towards the rough side of the stone, is yet of nearly even thickness everywhere; above it, the one with a black central line encompasses the whole agate symmetrically. Then a white band, thin at the bottom, projects into concretions

---

* I use this word gravity in some doubt, not being quite sure that the straight beds are always horizontal, or always inferior to the rest deposited at the same time. I have one specimen in which, according to all analogies of structure, it would appear that the vacant space is *under* the level floor, between it and reniform chalcedony; and sometimes these floors cross pillars of stalactite like tiers of scaffolding.

on the flanks. Then, a thick white deposit, B, does not ascend at the flank at all; then a crystalline bed, with pisolitic concretions at the bottom of it, changes into dark chalcedony (drawn as black), which ascends at the flanks. Then another thin line at the bottom, in concretion at flanks, then one thick at the bottom, thin at the flanks, and so upwards. In Fig. 12, a level mass, itself composed of silica in two different states, one separating into flakes and the other even-laid, is surrounded by bands which melt into it with gradually diminishing thickness, these being evidently subordinate to an external formation of crystalline quartz, the whole terminated by a series of fine bands of graduated thickness and by clear chalcedony (drawn as black).

17. Now all these, and many more such variations, take place without any apparent disturbance of the gene-

Fig. 13

ral mass, each bed conforming itself perfectly to the caprice of its neighbour, and leaving no rents nor flaws. But an entirely different series of phenomena arise out of the fracture or distortion of one deposit by another, after the first has attained consistence. Thus, in Fig. 13, a yellow orbicular jasper is split into segments, singularly stellate or wheel-like, and then variously lifted and torn by superimposed chalcedony; and, in Fig. 14, a white and opaque agatescent mass is rent, while still ductile, the rents being filled with pure chalcedony: and from this state, in which the pieces are hardly separate, and almost hang together by connecting threads, we may pass on through every phase

Fig. 14

of dislocation to perfect breccia; but, all the while, we shall find the aspect of each formation modified by another kind of fault, which has no violent origin, and for illustration

FIG. 1

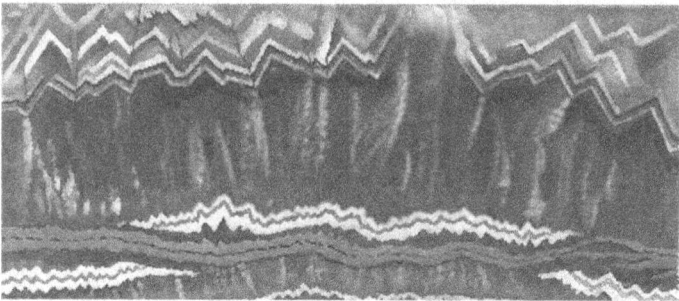

FIG. 2

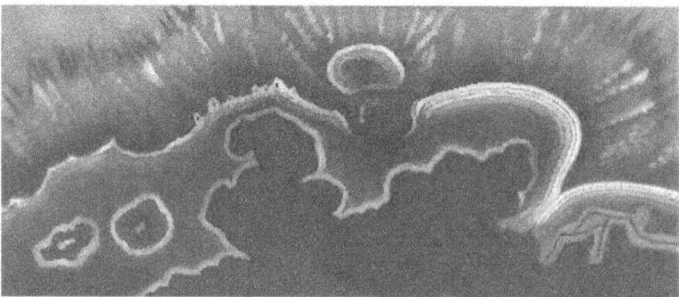

FIG. 3

AMETHYST-QUARTZ,
WITH WARPED FAULTS IN CONCRETION

of which I have prepared Plate VII. This plate represents (all the figures being of the natural size) three sections of amethystine agate, in which the principal material is amethyst quartz, and the white jasperine bands for the most part form between the points of the crystals.

All the three examples are types of pure concrete agatescence in repose, showing no trace whatever of external disturbance. The fault in the inclined bed at the base of the uppermost figure has some appearance of having been caused by a shock, but for that reason is all the more remarkable, the bed beneath it being wholly undisturbed, and its own fracture quite structural, and connected with the crystalline elevation and starry concretion above. I have no idea at present why the central portions of these concretions of dark amethyst are partly terminated by right lines, or what determines the greater number of bands on one side than on the other.

18. The second figure [Plate VII.] is of a less varied, but of still more curious interest.[1] There is no trace of violence or fracture in the stone, and the line of the crystallized amethystine mass is undisturbed at the summits, except by a partial dissolution in one part and mingling with the white bands above. But the white undulatory band at its base

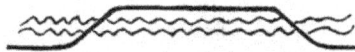

Fig. 15

is cut into three parts, and the intermediate portion lifted (or the flanks removed downwards) a quarter of an inch, by pure calm crystalline action, giving thus room for an interferent brown vein of less definite substance which proceeds without interruption, dividing the white band in a direction peculiarly difficult to explain, unless by supposing the interferent one to be the slow filling of a fissure originally opened in the direction of the black line in Fig. 15, and straightened in widening.

[1] [The specimens engraved in Figs. 1 and 2 on this plate are in the St. George's Museum, Sheffield: see below, p. 434 (M. 9, 10).]

19. But the third example [Plate VII.] is inexplicable, by any such supposition. It is the agatescent centre of a large amethyst nodule, in which a small portion, about the third of an inch long and a quarter of an inch thick, of its encompassing belt, is separated bodily from the rest, taken up into the surrounding concretion of quartz, and its place supplied by a confused segregation of chalcedony, with a sprinkled deposit of jasper spots on the surface exposed by this removal of its protecting coat; spots which, in the rest of the stone, form on the exterior of the coat itself, just under the quartz. There are many points in all these three examples which it is useless to take further note of at present, but to which I shall return[1] after collecting examples enough to form some basis of reasoning and comparison. I must apologize, as it is, for the length of this paper on a subject partly familiar, partly trivial, yet in which these definitions, not by skill of mine express- ible in less room, were necessary before I could proceed intelligibly.

IV

[April 1868]

20. I propose now to pursue my subject by describing in some detail a series of typical examples of the principal groups of agatescent minerals; noting, as we proceed, the circumstances in each which appear to afford proper ground for future general classification.

The upper figure in Plate VIII. represents, of its real size, the surface of a piece of jasperine agate in my own collection, belonging to the same general group as the speci- men, *a*, *b*, *d*, 5, in the British Museum.[2] This group con- sists, broadly, of irregular concretions of jasper affected by faults caused by contraction, having their interstices filled

---

[1] [This series of papers was never completed (see below, p. 84), and for the only further reference to Plate VII., see p. 61.]
[2] [The reference is to an arrangement of the minerals existing in 1868; the specimen cannot now be identified.]

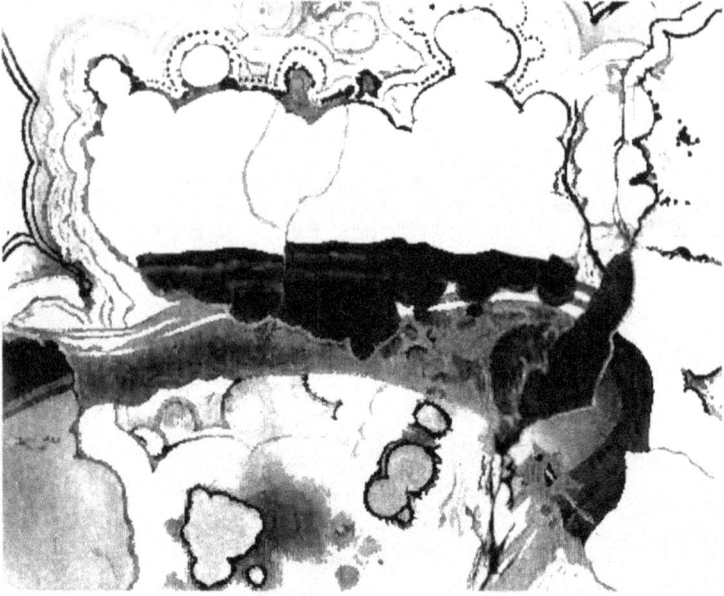

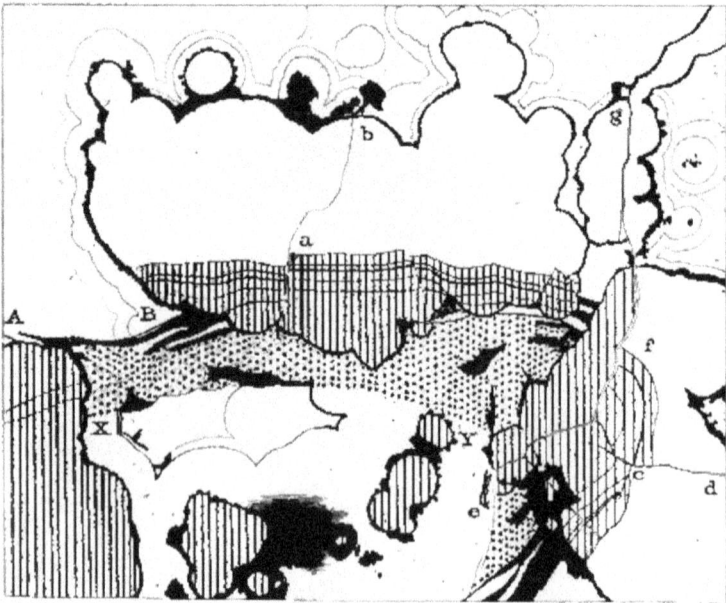

SECTION AND MAP OF A CONCRETE AGATE.

with chalcedony, and the whole enclosed by a quartzose crystalline mantle or crust.

The British Museum specimen (a, b, d, 5) is said to be Icelandic. Two others of the group are labelled "Oberstein"; one, of parallel construction, but slightly varied in character, is from Zweibrucken, in Bavaria. I do not know the locality of my own, but there is a community of feature in all the specimens, which assuredly indicates similarity of circumstance in their localities; and the more various the localities, the more interesting it will be eventually to determine their points of resemblance. I have not yet obtained an example of this group in the gangue,[1] but the crust of the stones themselves is in every case composed of quartz-crystals rudely formed, sometimes so minute as to look like a crumbling sandstone: in my own specimen they can only be seen with a lens, associated in filiform concretions like moss; within this crust two distinct formations have first taken place, and then a change of state is traceable affecting both in new directions. The map-diagram, Plate VIII., Fig. 2, is lettered, so as to permit accurate indication of the parts.*

The outer formation, next the crust, is composed of very pale whitish brown jasper. It is expressed by a shade of grey in the map, and is limited towards the interior of the stone by the strong line (with occasional projecting knots) thrown into curves, convex outwards.

The inner formation is of a finer jasper, with dark chalcedony in segregation. The vertical lines in the map indicate the chalcedony, and the pure white space, the inner jasper, terminated outwardly by convex curves. We are thus led at once to note the distinction between the two families of agates, formed from within outwards in

---

* I have carelessly worded the title of Plate VIII. as if the two figures were a vertical section and surface map; but the lower one is, of course, only explanatory of the upper.

---

[1] [i.e., the matrix in which an ore is found.]

knots, and from without inwards in nests.[1] The first group, to which our present example belongs, is usually agatescent in the interior, and crystallized on the surface; the second is agatescent in the coating, and crystallized in the interior.

21. Supposing the silica deposited under the same circumstances of solution, and the same time granted for solidification, the difference between these two structures would depend (and often does depend) only on the chance of the silica finding a hollow prepared for its reception, or a solid nucleus round which it can congeal; the ordinary deposits on the inner surface of a nest often become nodular or stalactitic as they project into its open space, and the greater part of the apparently independent concretions are probably mere fragments out of the hollows of larger ones. But there is, nevertheless, frequently a true distinction between the two modes of deposit. The agates formed on a central nucleus appear usually to have had a longer time for their construction than those which fill hollows, or, at least, they are the portion of the mass, in the hollow itself, which has crystallized most slowly; they are distinctly reniform in their chalcedony, and distinctly symmetrical in their crystals; while the nested agates run into level or irregularly continuous bands, and choke their cavities with confused network of quartz. I have difficulty in finding convenient names for these two families of agate, but merely for reference to them in these papers, I shall call those formed in knots, which are often conspicuously radiant in the lines of their crystals, " stellar " agates, and those evidently formed in cavities, " nested " agates.

22. I believe that the stellar forms, when independent, will be found most frequently under circumstances admitting the possibility of slow concretion at comparatively low temperatures, while the nested or bomb-like structure belongs characteristically to volcanic formations, in which the cavities might be filled by comparatively violent infusion,

---

[1] [For this distinction, see below, p. 378 n.]

and their contents in many cases quickly cooled. Both conditions, of course, sometimes agree in all their processes ; and we shall be able finally to classify these processes of deposit under description which will apply equally to the stellar and nested forms, marking afterwards the points of exceptional difference. Thus, for instance, the most frequent of all the forms of tranquil deposit, uninterrupted by flowing additions of material, is that in which a clear band of chalcedony, perfectly equal in breadth throughout, is first formed round the point (or branch) of nucleus, in stellar, or on the outer wall of the cavity in nested, agate. But after this has been formed in stellar agate, the succeeding belts will not usually show a minor pisolitic structure, whereas, in nested agates, marvellous groups of pisolitic hemispherical arches often rise from the inner surface of the clear external chalcedony, in section, like long bridges crossing a flat, and modify the whole series of bands above them, but, again, with this most important distinction between these and the bands of stellar agate— that stellar bands, the farther they retire from the nucleus, usually throw themselves with increasing precision into circular curves, till they sometimes terminate in perfect and exquisitely drawn segments of spheres; while in nested agate, the bands, if parallel, efface more and more the original minor curves as they approach the centre of the nest, and sweep over them in broad indeterminate lines, as successive coats of paint of equal thickness efface the projections and roughnesses of the surface they cover, or as successive falls of snow, undrifted, efface irregularities of ground.

23. And now, observe, we shall want a word expressive of an intermediate condition between the states above defined as pisolitic and reniform. A pisolitic mineral we define to be one which separates into more or less spherical layers by contraction ; and this kind of division takes place sometimes quite irrespectively of the crystalline structure, and on the grandest, as well as the most minute scale. In one of my

specimens of Indian Sard, there are multitudinous pisolitic
flaws, exquisitely perfect in spherical curvature, dividing the
parallel bands of the agate transversely in every direction,
looking like little paleæ of chaff in its clear substance; on
a large scale, the aiguilles of Chamouni are pisolitic, rending themselves into curved layers five or six hundred feet
in the sweep of their arcs, variously crossing their cleavage
(which is rectilinear), and often diametrically crossing their
beds. On the other hand, true reniform structure is perfectly compact, and dependent on minute radiating crystallization of substance. But between the two there is the fine
agatescent structure, in which bands of different materials,
jasperine and chalcedonic, are separated from each other
under a radiating law, and yet not divided by a mechanical
contraction; for though they are often so distinct as to
separate under the hammer-stroke, they never leave spaces
between, as true pisolitic beds do in ultimate separation.
For this intermediate action, the most frequent of all, I
shall keep the term "spheric"; and I was forced to admit
only a guarded use of the word "gravity" in last paper,[1]
because this spheric action is constant, as far as I know,
in all agatescent matter, so that I have never yet seen an
instance in level-laid agate of the transition from the lake
in the (lowest?) part of the cavity to the beds at the sides
being made under any subjection to the mechanical law of
gravity on fluent substance; but (as in the petrifaction of
the banks of Dante's Phlegethon: "lo fondo suo, ed *ambo
le pendici* fatt'eran pietra, e i margini da lato,"[2] "its bottom,
and both the slopes of its sides, and the margins at the
sides, were petrified") the flinty bands form in parallelism on
the slopes as well as the bottom, and retain this parallelism
undisturbed round the walls and vault of the agate. On
the other hand, I cannot but admit the idea that these
rectilinear tracts are formed under a modified influence
of gravity, because, first, I have never seen them laid in

---

[1] [See above, § 16, p. 53 *n.*]
[2] [*Inferno,* xiv. 82, 83.]

different directions in different parts of the same stone; and, secondly, whenever they are associated with pendent stalactites, they are at right angles to them. So that the aspect of one of these levelled agates in cavities may be approximately described as that of a polygonal crystal in which the position of one of its sides is determined by gravity; and the other sides modified into curves by radiating crystallization (of course the changes of form caused by gradual entrance or exit of material being at present withdrawn from consideration). In the example before us, which, though showing but feeble crystalline energy, belongs to the stellate group, the outer formation of rudely spheric white jasper withdraws itself confusedly from the sandy crust of quartz and becomes finer and finer towards the inner jasper, on the surface of which it throws down a coating of superb crimson (oxide of iron?), which is itself arranged every here and there in minute spherical concretions. The same formation exists in the same position under the quartzose outer bed and on the surface of the chalcedonic interior one, in the specimen figured in Plate VII., Fig. 3; and when we find it, as we often shall in future, under similar circumstances, I shall speak of it simply as the "medial oxide." In the map, this crimson deposit is throughout represented as black.

24. Proceeding next to examine the inner formation on the surface of which this medial oxide is deposited, we find it composed of two parts, sharply divided—a white jasper and dark grey translucent chalcedony. The white jasper has a spheric structure much more perfect than that of the outer coat, and so delicate as to be hardly visible without a lens (not that the spheres are small—they are on the average the third of an inch in diameter—but the lines of division are so subtle that the mass appears compact). In this character the inner deposit seems only a finer condition of the external one, but it differs *specifically* in being affected by sharp displacements apparently owing to contraction. To these faults, though minute, I would direct

the reader's special attention. They are by no means small in proportion to the extent of material affected by them; and they differ wholly from ordinary displacements, in this, that there is no trace whatever of movement at the limiting convex curves, but only at the edge of the chalcedony, so that the fault at *a* [Plate VIII.] seems owing to contraction within the space *a b*, and at *c*, to contraction within little more than the space *c d*; and farther, the fissures *a b*, *c d*, are not rugged or broken, as if caused by the displacement, but sinuously current, passing on through the chalcedony from *c* to *e*, *f*, and *g*; and, in fact, I am very certain that these veins are not caused by the contraction in question, but that the contraction takes place unequally on each side of the primarily formed vein. This kind of fault, of which we shall find frequent instances— the unequal contraction, namely, of beds on opposite sides of a vein or dyke—I shall call fault " by partition," and the violent fracture of beds at a point where no vein or dyke previously existed, I shall call fault " by divulsion." Deposits which fill compartments in fossil shells may often be seen, in a correspondent series of beds, to vary their proportionate thickness at each partition; the rectilinear bands of Labradorite may be found varying in thickness and position while they correspond in direction, in contiguous crystals; and I do not doubt but that even, on a great scale, displacements of beds, which at first sight might be supposed to have given rise to the fissures which divide them, will be found on examination to be the result of an unequal contractile action in the masses released by the fissure, protracted for long periods after it had given them their independence. Lastly, the separation of the chalcedony from the jasper does not take place only in the inner formation. It is an operation evidently subsequent to the deposition of both layers, and even in the outer one makes the entire dotted space, as far as the curved limit X Y, chalcedonic, and flushes it with a diffusion of the medial oxide from its edges; this medial oxide here drawing itself into bands, which being parallel with those of the grey chalcedony, are manifestly produced by a segregation which has taken place simultaneously in the two layers. This being clearly ascertained, the intensely sharp line, which separates the chalcedony from the white jasper, considered as a result of segregation, becomes highly remarkable, and a standard of possibility in sharpness of limit so produced.

25. The spots surrounded by dark lines in the lower part of the figure are portions of the inner formation cut off by the surface section. It is often difficult on a single plane to distinguish such spaces, the truncated summits of an inferior, or, as here, remnants of a superior, bed, from isolated concretions; and it is always necessary in examining agates to guard against mistaking variation of widths of belt caused by obliquity of section from true variations in vertical depth. All the difficulties of a geological survey sometimes meet in the space of a single flint. The gradated softness of edge in belts widened by oblique section is, however, usually an instant means of recognizing them; but in this stone the material is so fine that the oblique edges are as sharp as the vertical ones.

I could not without tediousness proceed farther in the description of this stone; it presents other phenomena peculiar to itself; but, resuming the points hitherto stated, we may define the family of agates, which it represents, as consisting of at least two formations enclosed by quartz; the inner formation being affected by dislocations which do not pass into the outer one. Generally their colour is brownish red and white, and their main material opaque and jasperine, their chalcedony developing itself subsequently and subordinately. The crimson veins and striæ, which in some examples traverse the inner formation, will furnish us with a study of separate interest after we have obtained determinate types of other large and typical groups; the minor details in each may then be examined with a better field for comparison. For convenience' sake I shall in future refer to the group described in this paper as

"Dipartite jaspers." Their division may, indeed, be into more than two coats or formations, but the operation of a contractile force in one, which does not affect another, sufficiently justifies the term for general purposes.

V

[*May* 1868]

26. The next group of agates which I have to describe belongs to the nested series, but is distinguished from all other varieties of that series by having a pure chalcedonic surface (unaffected, except in the form of it, by the material of its gangue), and by uniformity of colour, consisting only of white and transparent grey bands, wholly untinged by more splendid colours. But nearly all the agates of this group which now occur in the market have been dyed brown or black at Oberstein,[1] to the complete destruction of their loveliest phenomena.[2]

With the true agates of this group must be associated some transitional examples, in which the surface is more or less entangled with, and degraded by, the material of the gangue (the body of the stone then becoming susceptible of colouring by iron, or of chloritic arborescence from the exterior); and others, in which the mass is rudely egg-shaped, like a rolled pebble, and the crust is of a fine pale brown agatescent jasper in multitudinous concretions, plainly visible on the surface, like the convolutions of the brain of an animal. But in the typical examples of the whole series, no lines of concretion are visible on the surface; it is knotted and pitted, but not banded; it is of grey clear chalcedony, and the entire mass of the stone is often thrown into irregularly contorted folds, which are sometimes parallel to the interior bands, and from which I

---

[1] [In Oldenburg; the principal seat, since the sixteenth century, of the industry of cutting and setting agates. There is a "Memoir on the Oberstein Agate Quarries," by W. J. Hamilton, in the *Quarterly Journal of the Geological Society*, vol. iv, p. 214. For other references to Oberstein agates, see below, pp. 342, 382, 433.]

[2] [The MS. adds: ". . . phenomena, though the general arrangement of their beds may be thus more clearly seen."]

shall for convenience' sake give the name to the whole group of "Folded Agates."

27. I say "sometimes parallel," because the folds of the interior beds are much more complex than those of the surface, and often are most notable when the exterior is undisturbed; and they are specifically peculiar in two respects. First, they are formed out of beds which are in the greater part of their course accurately parallel, and arranged in gracefully sweeping continuous curves, while the bands of ordinary agates are broken into minor undulation, and run into irregular curves. Fig. 16 is the typical

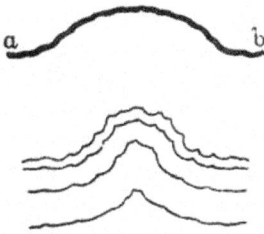
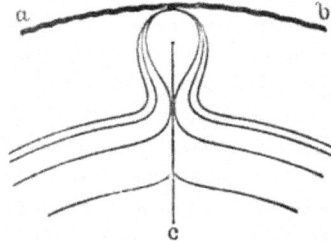

Fig. 16   *Fig.* 17

structure of common, and Fig. 17 of folded agate, the line *a b*, in each figure, representing the surface of the stone.

28. Secondly, these sweeping and beautifully parallel beds are at particular points of their course suddenly and systematically contracted, and bent outwards (outwards, that is to say, in nested agates—inwards in stellar agates, but the stellar formation is very rare in this group) like flowing drapery raised by a rod beneath it; and this ideal rod may either raise these sheets of drapery hanging over it, as clothes hang over a line, or on the end of it, as the sides of a tent hang from its pole,* with every variety of

---

* In Plate IX. Fig. 1 shows the clothes-line arrangement in pure surface-section, and Fig. 2 in perspective, seen through the transparent stone, the edges only of the pendent veils being at the surface. Of the tented arrangement I will give examples in succeeding plates,[1] but they are not specifically different arrangements; they are only accidental variations in the direction of the interrupting masses.

---

[1] [This was not done eighteen months elapsing before the publication of the next paper.]

beautiful curvature, intermediate between these two arrangements. The ideal rod is of course composed of the interior chalcedony or quartz; and I once supposed the entire range of these phenomena to be dependent on the former subtle influx of the dissolved silica at the points where the apparent rods or tubes reached the exterior of the stone; but I now believe rather that, taking Fig. 18 as a formal type of a perfect folded agate, the points *a*, *b*, *c*, etc., at the sides of the nest have been those of *impeded* secretion or deposit (if, which is not by any means clear to me, there has been successive deposit at all), and that the intermediate curved beds are the increasing stalactitic masses. The right lines indicating flaws at the intersection of these masses are essential in the typical structure. The two upper figures in Plate IX. will characteristically represent the phenomena principally resultant, though the complexity of these phenomena is so great that in detail they can only be followed out by the reader with good specimens of the stones in his hand.

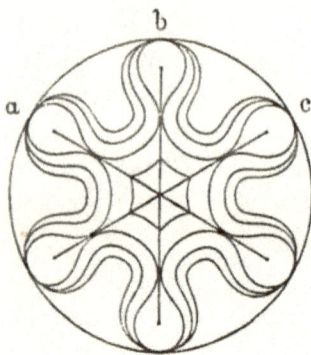

Fig. 18

29. Fig. 1 is from a very rare agate in my own collection, which unites the characters of the folded group with that of the nested agates which have level beds (the pure folded agates never, as far as I have seen, contain rectilinear tracts), and the folds, or tubes of arrest, in this stone are less regular in structure than in typical examples, and present somewhat the appearance of having been caused by contraction, the rent spaces being afterwards filled by the inner quartz. But I believe this appearance to be wholly deceptive. Whatever the cause of the interruptions may be, they are certainly not mere rents like those of septaria. The greater width of the white band at the top, which suggests the idea of large influx there, is a

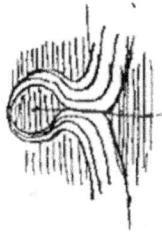
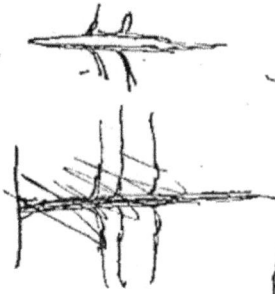
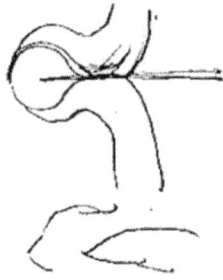
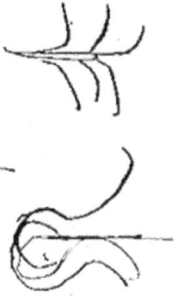

A Page of the MS. of the Papers "On Banded and Brecciated Concretions" (§ 29)

sectional deception; this white band is of equal thickness everywhere, and, with all the others, seems entirely concentric except when interrupted by the tubes, and by the changes in the direction of the films in its own substance which are connected with them. Fig. 2 is from a piece of perfect folded agate, showing the symmetrical arrangement of its successive beds round the tubes, and their lovely dependent curves as they detach themselves. In some cases, however, the tubes appear isolated in the mass of the stone, or interrupt the beds in their own thickness, but in whatever accidental relation to the secreted chalcedony, they assuredly indicate a peculiar state of its substance at the time of secretion; and their nature, and the conditions under which they develop themselves, must be understood before we can hope to explain the more complex tubular formation of dendritic chalcedonies.

30. And this investigation is rendered doubly difficult by the perpetual confusion in all agatescent bodies between the concretionary separation and successive deposit of their beds. If these folded agates were, indeed, formed in successive beds from without inwards, as it has been supposed, it should be possible sometimes to trace the point of influx of material and the sequence of the added bands from it, which I never yet have been able to do satisfactorily in a single instance in folded agates (and only with suspicion of the appearance of it, even in the brown coated and level bedded stones in which it seems to be of ordinary occurrence); and also, the beds ought to present some of the irregularly accumulate aspect of common calcareous stalactite, and in the interior we ought to find sometimes vacancies left by the failure of supply. But, on the contrary, folded agates are always *full*, so far as I have seen, except occasionally in the centres of their tubes or in hollows of outer folds, but they are always closed in their centres (differing, observe, again *essentially* from common agate in this circumstance), and their beds are not only parallel, instead of irregularly heaped, but involved in the strangest

way in reduplicate crystalline series. See the interior of the stone, Fig. 2 in Plate IX.

31. On the other hand, were they truly concrete, these beds ought to exhibit occasionally clear evidence of sub-ordinate concretion in their mass. Thus in the true con-crete jasperine agate, Fig. 19,* the beds which are simply concurrent on the right hand break up presently and separate into flamy and shell-like groups, transverse to the general bedding, and at last bend round a knotted nucleus; but nothing of this kind ever occurs in folded agates,

Fig. 19

though their veils of dependent film are sometimes covered with an exquisite dew of minute pisolitic concretions, making them look (under the lens) like a beautiful tissue of gossamer laden with dew, and connected with a peculiar complex basalt-like fracture; then finally, to finish the difficulty, these folded agates are connected by a series of scarcely distinguishable transitions with the group which we shall have next to examine, which seems to be in great part concretionary, but concretionary in right lines. The two lowest figures in Plate IX. are outlines of two of the most singular conditions of it. Fig. 3, Plate IX., is reduced in scale from a stone which I shall hereafter engrave of its real size,[1] as its mode of association of agatescent with crystalline structure is, as far as I know,

---

* Magnified about three times.

[1] [See the lower stone on Plate X. (p. 76).]

1

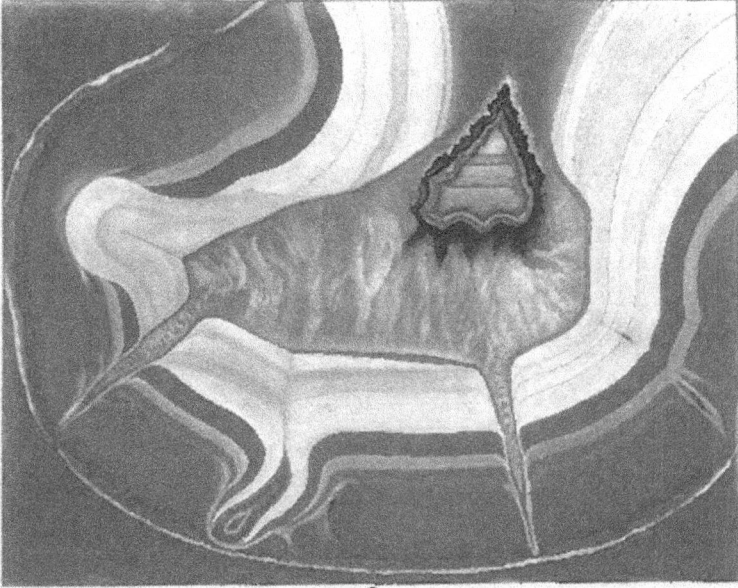

2

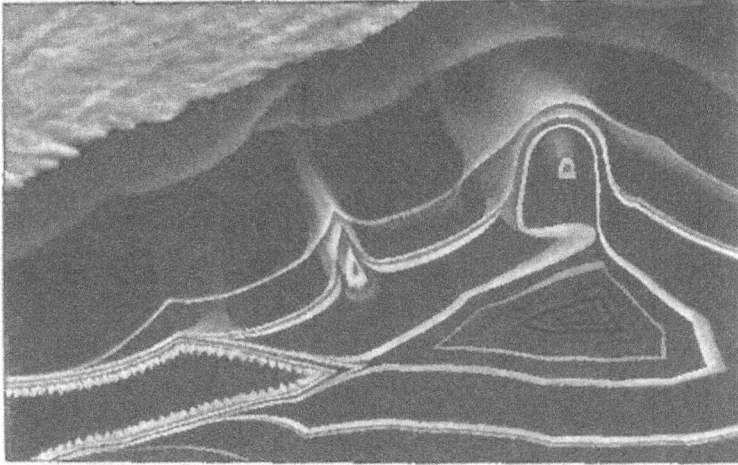

3

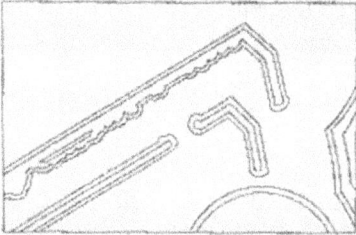

4

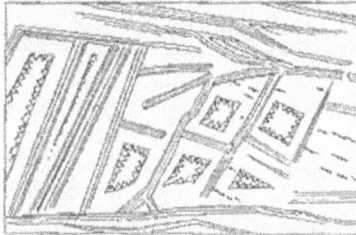

FOLDED AGATES AND MURAL AGATES

unique—and its proper discussion is connected with that of the modes of increase of crystals. Fig. 4, Plate IX., is from an agate of almost equal rarity, though I have seen other examples of its structure, but never so decisive in character. This figure is slightly enlarged, being of a portion of a mass which has crystallized out of a breccia in thin walls of linear brown agate enclosing opaque white agate, leaving internal spaces filled with quartz.

32. The entire group to which these examples belong, consisting of walls, or tabular crystallizations, of agate, I shall name Mural agates; and they are connected, on the one hand, with Folded agates, by a series in which tabular

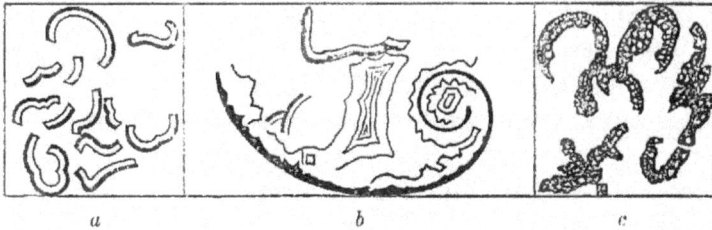

Fig. 20

portions of the external matrix are torn off like pieces of broken slate, lifted up into the agatescent mass, and then encrusted with folds of chalcedony; on the other hand, when the Mural fragments become curved, they are connected with a great jasperine group of the most curious interest, which I shall examine under the general term of Involute Agates, consisting of bands of a consistent structure broken up (or fragmentarily secreted), Fig. 20, _a_, in fine specimens disposed in curves resembling the contour of a haliotis shell, Fig. 20, _b_, but in less developed examples forming broken vermicular concretions in a jasperine paste, Fig. 20, _c_.

33. It is almost impossible without microscopic examination to distinguish some of these shell-like concretions (of which the most delicate are white, closely crowded, and surrounded by milky chalcedony) from true organic remains,

and to my mind perhaps the most singular fact, of all that are connected with minor physical phenomena, is this apparent effort of the occult natural powers to deceive their investigator, by making one thing resemble another. There seems to be a mocking spirit in Nature which sometimes plays with its creatures, as in the orchis tribe of plants, or the mantis group of insects; and sometimes deliberately connects two totally different systems of its work by deceptive resemblances,[1] causing prolonged difficulty or error in the attempt to discriminate them. In this subject before us, for instance, the inorganic secretions of chert and flint are connected, by the most subtle resemblances, with those which have organic nuclei; the filiform and foliated secretions of chlorite, and the flamelike and infinitely delicate mossy traceries of jasper, pass with the cunningest treason into the organisms of altered sponge and wood; the pisolitic and radiated-crystalline agates confuse themselves with true corals; the involute agates with shells; the rolled breccias with slowly knotted secretions; and all the phenomena of successive deposits, quite inextricably with those of segregation!

34. I imagine, however, that the reader must have had enough, for the present, of these mere statements of doubt, and as my next subject, mural agate, is a very difficult one, I shall delay the paper for some time, but meanwhile, if any good chemist would set briefly down for me what is now positively known of the fluent and gelatinous states of silica, and silicate of iron, with respect to their modes of separation, when undisturbed, from other substances, it would be of the greatest service to me (and not, I should imagine, irrelevant to the general purpose of this Magazine, for all inquiries respecting metamorphic rocks must rest on such chemical data primarily); and also I should be grateful to any mineralogist who would give me some tenable clue, or beginning of clue, to the laws which affect the

---

[1] [Compare below, pp. 72, 80.]

modes of crystalline increase; that is to say, which determine whether a prism of quartz or calcite shall increase at
the extremities or at the flanks, or consistently on both, or
inconsistently at different parts of the prism; and, especially,
by what law stellar or roseate aggregations take place, instead of confused ones, in groups of crystals; and by what
tendencies some minerals—fluor, for instance—are limited in
their expansions of the cubic or other common form, while
others, such as salt and the oxide of copper, are enabled
to shoot unlimitedly into prismatic needles; and others, like
sulphide of iron, will form in solid crystals on the outside
of calcite and in stellar acicular groups within it. If I can
get some help in this chemical and microscopic part of the
work, which I cannot do myself, I have hope of being
able to give something like a serviceable basis for future
description of the two great groups of calcite and silica,
and the modifications of iron which colour the concretions
of marble in the one case, and of agate in the other; and I
should do this piece of work with, perhaps, more zeal and
care than another person, owing to its connection with my
own speciality of subject, by the use of these two earth-
products in the arts, and the foundation of much of what
is most beautiful in architecture, and perfect in gem-engrav-
ing, on the accidents of congelation which have veined the
marble and the onyx.

DENMARK HILL, *22nd April,* 1868.

VI

[*December* 1869]

35. When we find at the sides of veins the veinstone
rent into laminæ, as I tried to represent in Plate VI.,
it is easy to think of the fracture as violent, and of the
disruption of the vein as sudden.

That, at least, this disruption must have been exceedingly slow, and that as it took place the rent must have

been filled by contemporary crystallization, is I think evident in the instances figured, and in the great number of cases which they represent.

And as I continue my inquiry, it becomes more and more questionable to me whether there has in such cases been disruption at all. For the more I endeavour to read Nature patiently, the more I find that she is always trying to deceive us while we are impatient, by pretending to do things in ways in which they never were done, and making things look like one another, which have no connection with each other.

36. For instance, in Fig. 21, which rudely sketches a piece of Cornish hornstone, it would seem at first sight

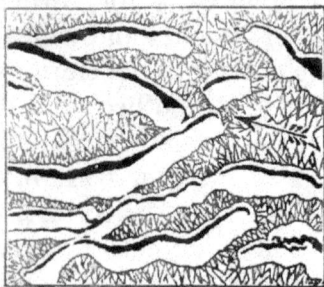

Fig. 21

that the detached black and white bands were pieces of a band once continuous, but which had been broken up, and re-cemented in disorder. And if, on a large scale, we had met with the fault in almost exactly coincident beds, to which the arrow points, we should have had little doubt of their former continuity. But in this stone they have never been in any other than their existing position, any more than the two upper beds on the left, of which one is an entirely undisturbed branch of the other, as much as any branch of stalactitic chalcedony is of the rest of the mass. Nor have any of these beds ever been broken at all. The whole is a tranquil determination of variously crystallizing substances, like that of the component minerals in granite. The white portions are hornstone; the black band in each is ferruginous, and the enclosing paste rudely crystalline quartz.

37. There is, however, one grave structural difference between this stone and common granite. The crystals in granite run in all directions. These zones of hornstone have a more or less parallel direction; and the black band,

with another narrow one succeeding it, is always at the same side of them.

I have placed the woodcut (Fig. 21) with the black beds uppermost, so that the resemblance may be seen between them, and the always uppermost grey beds in the highest division of Plate V. But in neither case can I say that their position has been influenced by gravity. For in Plate V. it will be observed that the elliptical bar of central calcite crystallizes in every direction, and in this piece of hornstone, very near the portion above figured, is a cavity, in which, while the bands whose separation forms it retain their relation un-

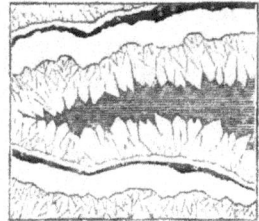

Fig. 22

changed, the quartz, having now room to crystallize, does so indifferently up and down, and from both sides, as in Fig. 22. I do not know the position of the stone *in situ*.

38. But though common granites show only arbitrary positions of crystals, in graphic granites we have a definitely parallel arrangement of them, somewhat resembling this of the hornstone, only more regular ; and in massive felspathic rock we get the same deceptive resemblance of faults exquisitely defined. Fig. 23 represents (of the real size, as are also Figs. 21 and 22) a portion of felspathic rock in which two crystals of labradorite are separated by apparent breccia, but really crystalline mass, of mixed labradorite and hyperstein. The oblique lines stand for this gangue (merely for a symbol—there are no lines nor cleavage in the gangue itself). The white spaces are pale blue labradorite, the horizontal lines indicate in each crystal a sharp, exquisitely defined, zone of vivid orange, and the vertical lines a zone of intense blue. There

Fig. 23

has evidently been no fracture in this case, any more than between the felspar crystals of common granite. And the —in this instance absolutely accurate—coincidence of direction in the zones of the detached pieces, with their fault-like variation in breadth and relative position, are both of them entirely crystalline phenomena.

39. Now we must always remember that in chalcedony and quartz we have two entirely distinct groups of crystalline forces—one radiant, endeavouring to throw the mass into spherical concretions; the other rectilinear, endeavouring to reduce it to hexagonal crystals; and that both of these are capable of producing phenomena of relative distortion.

Also, the group of the spheric forces associates itself delightedly with the spheric forces of hydrous oxide of iron, thus producing endlessly fantastic groups of mixed iron and chalcedony, while the rectilinear forces ally themselves in like manner to those of micaceous iron, bournonite, heavy spar, and calcite, producing tabular groups of crystals which present close analogies to the flat leaves of chalcedonies which have metallic or earthy laminæ for their support; while the iron-oxide, when it has no longer the power of modifying the shapes of the crystals, sets itself to imitate two other minerals frequently found in them. It mimics the globes of brown mica so exactly with its own bossy groups of clustered laminæ, that only a strong lens, or the knife, will distinguish them, and, in the interior of crystals, throws itself into golden-coloured radiant acicular[1] sheaves, which, when within amethyst, are the most beautiful things I know among minerals, but which it is a matter of great difficulty to distinguish in common quartz from minor forms of rutile. Finally, to crown the complexity of this iron and flint group, the sulphide of iron, varied beyond all minerals in the phantasies and grotesques which it can build out of its plastic and innumerable cubes, shoots its stellate crystals through the mass of the hydrous oxide

---

[1] [For "needle-crystals," see *Ethics of the Dust,* § 46 (Vol. XVIII. p. 258).]

and disputes with it the central position in stalactites of chalcedony.

40. But, through all this confusion, one generalization presents itself which is of great value. Whenever iron, whether oxide or sulphide, is associated with stalactitic chalcedony, it is always in the centre of the mass; but when iron, whether oxide or sulphide, is associated with quartz crystals, it is always (if determinately placed at all) either on the outside, or at a slight depth below the surface, under an external coat of clearer crystal. It may be indeterminately placed, in dispersed stars or cubes; but, if ordered at all, it is ordered so. Briefly, a crystal of quartz never has a *centre* of iron, and a crystal of chalcedony never a *coat* of it.*

41. And an important result seems to follow from this. If stalactites of chalcedony were formed by superfluent coats, some of these coats would have iron in solution at the outside as well as the interior, and would secrete it in successive films; whereas, on the contrary, the entire bulk of the iron, being always central, must surely have been secreted out of the entire mass; and, therefore, I believe that the true chalcedonic stalactite is indeed a long botryoidal crystal, like some of the forms of sulphide of iron, found in chalk, and not at all a drooping succession of fluent coats, except in cases of rapid deposit, which, as far as I remember, show no central iron.

42. Again, when iron is systematically associated with quartz, it is never in the centre of the crystal, but either on the surface, or under an externally imposed glaze. Hence it follows that the crystalline forces at work in forming quartz act nearly in the reverse of those that form chalcedony, as regards the direction of ferruginous elements, and that they have quite a peculiar power in finishing crystals,

* Of course I do not vouch for any so wide generalization as this absolutely. If ever one ventures to do such a thing, the next stone one takes up on a dealer's counter is sure to be an exception to the announced law; but I am confident that any mineralogist can fortify the statement from his own experience quite enough to justify our reasoning upon it.

which determines, at a given time, either a purer, or an amethystine, silica to the surface, often throwing down crystals of iron between the two.

43. I have already noticed the clear coat forming the exterior of many nested agates in basaltic cells, and the deposit of iron succeeding it, to which I gave the name of medial oxide. My impression is that the exterior of such agates, as relating to the crystalline power, may be considered identical with the centre of a stalactite, and I think it will be found that the iron in such stalactite centres, however delicate the fibre of it, is not solid, but tubular, leaving the absolute centre of clear silica correspondent to the surface of clear silica in a quartz crystal.

44. It is very strange that among these complicated forces certain conditions of chalcedony and quartz should be so constant, and the intermediate states, giving evidence of formation, so rare; but though the interior of almost every quartz crystal shows the forces of agatescence and straight crystallization in confused contest, I have only seven or eight specimens, out of a collection of some thousands, which clearly show the balance of the two powers in accomplished structure.

The uppermost figure in Plate X.[1] represents a portion of one of these, which is a stellar agate, formed of grey chalcedony, with white bands collected in a knot within radiant quartz. The precision of its lines is beyond all imitation, but Mr. Allen has succeeded in drawing and engraving it for us quite well enough to show the repeated efforts of the chalcedony to throw itself into straight crystalline planes, successful, tremulously, here and there for a quarter of an inch, and then thrust again into curvature by the lateral spheric force.

45. The second example, engraved in the lower figure in Plate X., shows the two forces reconciled in their reign:[2]

---

[1] [This plate was reproduced in *Deucalion*, and referred to in i. ch. ix. § 11 (below, p. 211).]

[2] [See above, p. 68.]

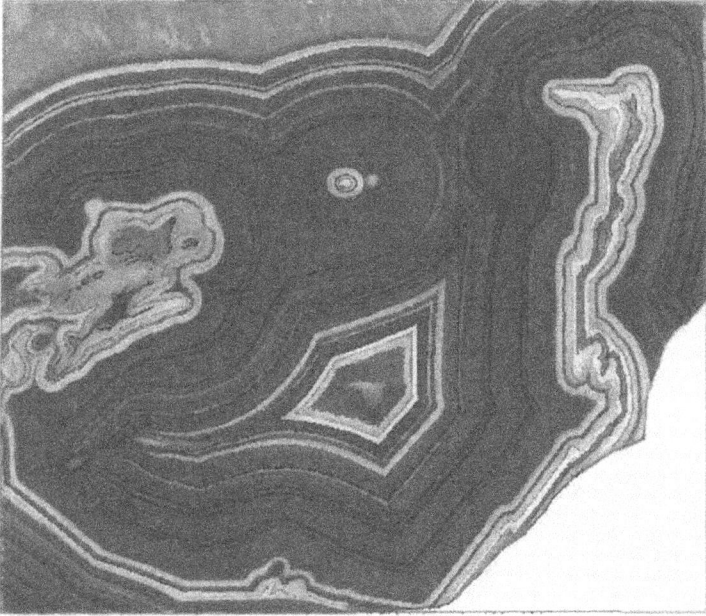

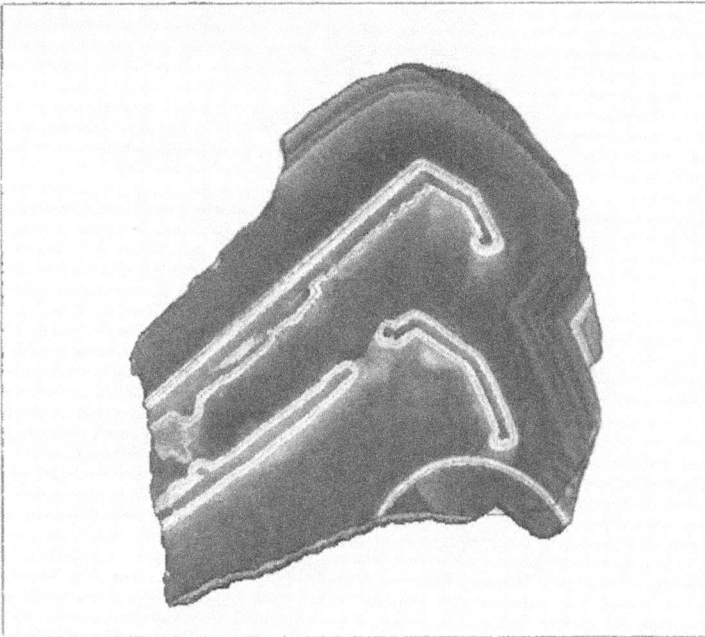

MURAL AGATES.

the crystalline or mural form is completely taken by the agatescent bands in one part of the stone and the spheric in another, while the bands themselves are arranged in double folds, turned at the extremities, like the back of a book.

46. Finally, the woodcut, Fig. 24, gives the rude outline of a stone in which the central nucleus of confused quartz has made vigorous, repeated, and, as far as I know quartz, I may even say super-quartzine efforts to gather itself into a single crystal, dragging the circumfluent agatescent lines

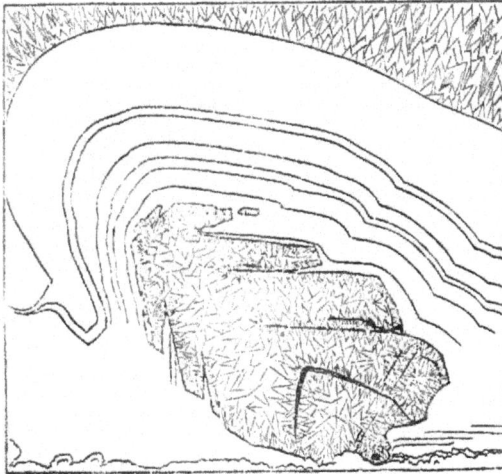

Fig. 24

one after another violently aside, to expire in the planes of its successive pyramids.

In all these instances the crystalline action is unmistakable, being at relative angles, of which only agatescent warping deranges the magnitude, but here (Fig. 25) is an example in which we have an apparently pendent stalactite (which is, however, the section of a vertical wall) without evidence of any relative planes, except the very short and secondary one on the left. Yet, between conditions of this kind and true stalactite agates, there is a gap which at present I cannot bridge. The mural agate consists of concretions in flat planes, formed irrespectively of gravity; the stalactite agate, of concretions on central rods, formed with reference to gravity.[1] I have, indeed, one example in which

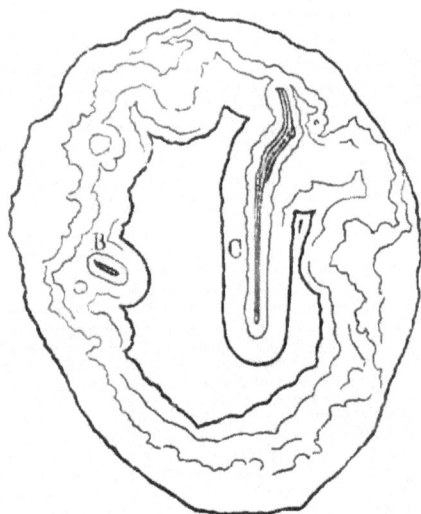

Fig. 25

these central rods are incipient in formation—are as fine as hairs, and are connected, as in Fig. 26, by drooping

Fig. 26

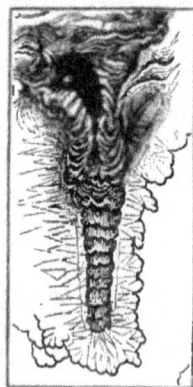

Fig. 27

branches concurrent with the successive outlines of the falling mass; and another, Fig. 27, in which the tubes of

[1] [Compare *Ethics of the Dust*, § 101, where Ruskin notes this distinction and threatens "a course of lectures" on the subject (Vol. XVIII. p. 332).]

a folded agate have become crystalline, and are clearly minded to determine themselves into straight lines. But these are both small, and of structures too unusual to found reasoning upon. I shall engrave them, however, hereafter,[1] but before examining these and the other structures illustrating the connection between mural and stalactite agates, it will be better to trace the closer connection on the other side between mural and conchoidal[2] agates. The states intermediate between these two will be the subject of my next paper.

VII

[*January* 1870]

47. We have now, I think, obtained sufficient evidence that the disposition of differently coloured or composed bands in agate is in most cases the result of crystalline segregation. We shall find, also, that the order of this segregation is constant under given conditions; and that, with fixed proportions of elements and fixed rate of cooling and drying, the agate will necessarily produce itself in a riband of a fixed succession or pattern of stripes: a spectrum of substances, which, if we had observed data enough, we might read like a spectrum of light; inferring, not the nature of the elements from its bars of colour, but the former conditions of solution from the bars of elements.

48. When the stone has been undisturbed, this riband or chord of its constituent elements will necessarily form quietly round it, either in its nest, or on its nucleus, with phases of level or vertical deposit under peculiar circumstances. But when the congelation has been disturbed, the chord of elements is broken up, and may then be traced here and there about the stone, forming where it may, and as it can. For instance, Fig. 28 represents rudely a

---

[1] [This, however, was not done.]
[2] [For "conchoidal" read "involute": see below, § 50, p. 81.]

quartzose band formed at a junction of fluor with siliceous sandstone. The dotted space is the grit, the undulating lines stand for a coarse mass of compact fluor spar, vaguely crystalline in that direction.

The faulting of the band is, I believe, entirely owing to fitfulness in the crystalline action; there is no trace of any kind of flaw or rent, either in the sandstone on one side, or fluor on the other.

The composition of the band here, as in the hornstone (Fig. 21, p. 72), is of one series of elements only; but

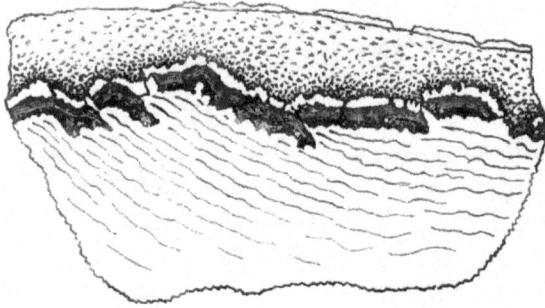

Fig. 28

very often the chord is composed of a central band, with corresponding opposite series on its sides. Here, in Fig. 29, is a very simple case, in which the chord has a thin white central line, with first a dark and then a broader white one on each side. The entire chord is flung irregularly about the stone, sometimes in continuity for a few folds, sometimes in broken segments; but the outer white band has the power of detaching itself from the chord occasionally, and of expanding here and there into wider spaces.

49. And, as in Fig. 28 we have a deceptive semblance of consecutive faults, so here we have an equally deceptive mimicry of brecciation by violence. But the two apparently broken portions of the band, in the centre of its own loop, are simply detached crystalline formations of it in those places. Here, Fig. 30, is a single example of such an one

from another stone, in which the enclosed banded segment is seen at once to be concurrent at its base with every undulation of the surrounding belt, though so trenchantly divided from it at the flanks.

50. And here we have to note a further separation of our subject into two branches, or, rather, into two threads of mesh (for its classification, like most true natural ones, is not branched, but reticulated). When the bands form in

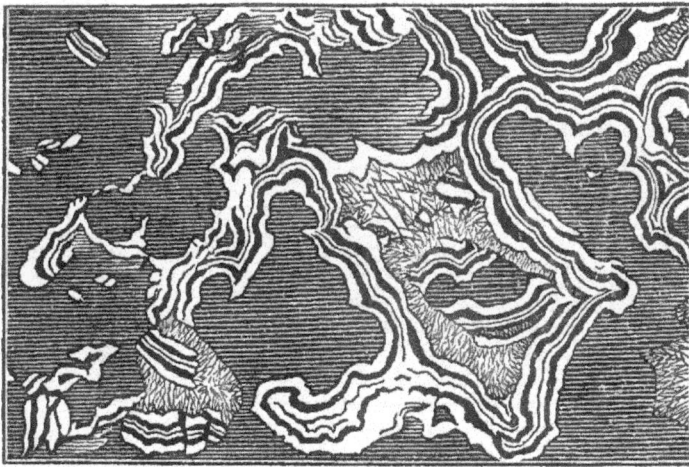

Fig. 29

several fragments in all directions, as in Fig. 29, we are conducted gradually to the most fantastic structures of abruptly brecciated agates. But when they are systematically affected by a consistent action of crystalline power, as in Fig. 28, we are conducted to the group I shall describe in the following paper [1] under the name of involute agates (I carelessly used the word "conchoidal" for involute, in § 46), which seems to me, as far as I have any clue to their mysterious structure, to be chiefly owing to the action, in a partially fluid substance, of the great diagonal—or spiral?—force of silica. This diagonal power of, or in, quartz, is to me one of the most interesting phenomena in mineral nature, both in itself, and as one of a group of powers like it—wholly

[1] [The present paper was, however, the last of the series.]

XXVI.   F

distinct from the crystalline ones, and acting with them, or dominant over them, at particular times and places, elsewhere and at other times remaining entirely passive.

51. Thus the growth of an ordinary quartz crystal depends on the regular imposition or secretion of parallel coats, which sometimes are capable afterwards of frank separation, forming "capped" quartz. But the flute-beak of Dauphiné is never capped. It is formed and wholly compacted under an oblique energy, which disciplines and guides together the hexagonal forces of the crystal. On the St. Gothard the same force, instead of terminating the

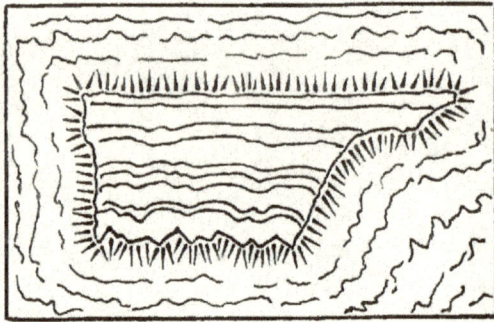

Fig. 30

crystals obliquely, unites them laterally, and leads them into long walls, warped into curves, sometimes like crowns or towers. Generally, when there is amianthus within crystals, the oblique force carries the filaments across the crystal diagonally; and it is very notable, as regards the time of secretion of these interior deposits, that while the iron oxides always arrange themselves in concurrence with the coats of the crystal, amianthus and rutile never do, but shoot clear through the whole body of it, if themselves long enough, and, if short, root themselves on an external plane, and shoot to the inside; while the iron oxides root themselves on internal planes and shoot to the outside.

52. Here (Fig. 31) is an example which will at once illustrate the power of the oblique force, and this relation of the oxides.

It is the section of a singly terminated, and apparently,

seen from the outside, an altogether single, crystal, one of a well-formed cluster, showing externally no signs of disturbance. They are all beautifully spotted with black iron oxide under a clear external coat, about one-seventh of an inch deep, which entirely covers them. These concretions of iron are represented in the woodcut accurately in section by the black spots; a minor series, not seen externally, is exposed by the section within the crystal, which is also shown by the section to be dual in the interior, separated into two parts by a perfectly straight line in the direction of its length, and nearly into two other parts by a jagged and broken one across it; all the interior beds being faulted by the oblique force, which acts,— in one direction softly, guiding,

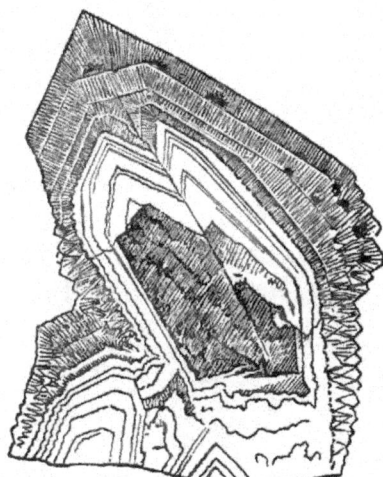

Fig. 31

without breaking, one part of the white beds (opaque white in the stone) into an angle beyond the other,— and in another direction violently, causing jagged flaws across the beds. Within the white beds, and under the great flaw, the quartz becomes again dark-clear.

53. Now, all these arrangements of substance take place under laws which surely need more investigation than they have yet received,* being quite distinct from those which

---

* I look with extreme interest to the result of the inquiries which Mr. W. Chandler Roberts has undertaken on the chemistry of silica.[1] I have to thank him already for some most valuable information communicated to me in the course of last year, of which, however, I will venture no statement until he has made public his discoveries in such form as he may think proper.

---

[1] [Sir William Chandler Roberts-Austen (1843–1902) assumed the latter name in 1885; F.R.S., K.C.B.; chemist and assayer to the Royal Mint, 1870. The editors do not find that he published, after the date of Ruskin's paper, inquiries into the chemistry of silica. He had already published papers "On Fungoid Growths in Aqueous Solutions of Silica and their Artificial Fossilization" (*Quarterly Journal of the Microscopical Society*, July 1868), and "On the Occurrence of Organic Appearances in Colloid Silica obtained by Dialysis" (*Journal of the Chemical Society*, 1868, N.S., vol. vi. p. 274).]

limit crystalline form, and bearing every semblance of a link between molecular and organic structure. For instance, pure crystalline force determines both gold and silver into cubes or octahedrons. So also it determines the diamond. But no force of aggregation supervenes to form branches or coils of diamonds; whereas an unexplained power, dominant over the crystalline one, extends the golden triangles into laminæ, and wreathes the cubes of silver into vermicular traceries. Agencies alike inexplicable twist the crystal of quartz like a piece of red-hot iron, and design the bands of agate into curves like those of a nautilus shell.

54. The transition from such coated crystals as that shown in Fig. 31, to these involute agates, may I think be traced without a break. The base of

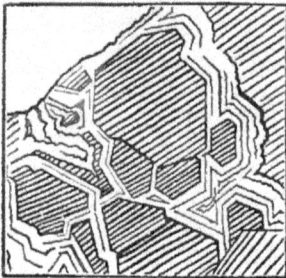

Fig. 32

this stone is formed of smaller and less perfect crystals, which, cut transversely, present themselves in honeycomb-like groups, Fig. 32. Each of these cells is a little mural agate, with no spherical force disturbing it. When quartz disposed to such formation gets mixed with jasper, or with any other uncrystallizable rock, the cells become shapeless, and we get results such as those represented in Plate XI. This stone there drawn shows the combination of angular cells with confusedly coiled ones, of which a close-set group is seen on the right, gathered together within broadly curved lines, which I think we shall be able to trace through succeeding examples, as they reduce themselves to the shell-like contours of true involute agate. On the other hand, in the centre of the stone, the less disciplined series of jasper veins, surrounding crystalline spaces, show the first origin of the groups of agate, which ultimately resemble a pebble breccia. I will endeavour in following papers to trace the two series through their gradual development.[1]

---

[1] [No more papers were, however, published.]

IV

DEUCALION

(1875–1883)

DEUCALION.

COLLECTED STUDIES

OF THE

LAPSE OF WAVES, AND LIFE OF STONES.

BY

JOHN RUSKIN, D.C.L., LL.D.,

HONORARY STUDENT OF CHRIST CHURCH, OXFORD; AND HONORARY FELLOW OF
CORPUS CHRISTI COLLEGE, OXFORD.

ἐπεὶ ἦ μάλα πολλὰ μεσηγύς,
οὔρεά τε σκιόεντα, θάλασσά τε ἠχήεσσα·

GEORGE ALLEN,
SUNNYSIDE, ORPINGTON, KENT.
1879.

[*Bibliographical Note.*—Portions of *Deucalion* were first delivered and (in some cases) printed as lectures. The book was first issued in Parts; and, in the case of the uncompleted volume ii. (Parts VII. and VIII.), it has hitherto been issued in no other form. This note, therefore, deals (1) with the lectures; (2) with the Parts; and (3) with editions of volume i. in collected form.

LECTURES

Some portions of *Deucalion* (viz., vol. i. ch. i., and partly ch. ii. and ch. iii.) were first delivered in a course of lectures delivered at Oxford in 1874. A preliminary announcement of "Three Lectures on the Relations of Rock and Snow Outline in the Alps" appeared in the *University Gazette* of March 3, 1874. "The lectures," it was added, "will be chiefly on geological or geometrical facts, and will be open only to Members of the University or strangers introduced by them." For the postponement of the course, see Vol. XXIII. p. xxx. In the *Gazette* of October 16, 1874, the course was announced under the heading "Mountain Form in the Higher Alps":—

Lecture I. The Alps and Jura (October 27).
  ,,   II. Alpine Forms produced by Snow (October 30).
  ,,   III. Alpine Forms produced by Ice (November 3).
  ,,   IV. Relations of Æsthetic to Mathematic Science of Form (November 6).

Other portions of *Deucalion* (viz., parts of Chapters II.–IV.) were delivered as a lecture at the London Institution on March 11, 1875, the subject being announced as "The Simple Dynamic Condition of Glacial Action among the Alps." This lecture was reported in the *Times* of March 15, 1875, as having been delivered "to a very crowded meeting." The lecturer, states the reporter, "said in his eloquent introduction, that the beauty of the Alps had first made him quit the study of geology for that of painting." The final passage of the report is given below, p. 163 *n.*

Chapter VII. of *Deucalion* was delivered as a lecture at the London Institution on February 17, 1876, the lecture being repeated there on March 28, 1876. The subject was announced as "And the Gold of that land is good; there is Bdellium and the Onyx Stone." The lecture was reported in the *Times*, February 21, 1876, and was the subject of a leading article in the *Daily Telegraph* (see below, p. 188).

The *Times* says that there was much inconvenience "owing to the overcrowding of the lecture-room and passages leading to it," in consequence of which Ruskin promised to repeat the lecture.

Chapter XII. was delivered as a lecture before the Literary and Scientific Institution, Kendal, on October 1, 1877 (under the title "Yewdale and its Streamlets"), and also at Eton College on December 8, 1877 (under the title "The Streams of Westmorland"). The Kendal lecture was reported in the *Kendal Times* of October 6, 1877, and also in the *Kendal Mercury* of the same date, and from the latter the report was reprinted in a pamphlet with the following title-page :—

"Yewdale and its Streamlets." | Report of a Lecture | by | Professor Ruskin, | delivered | In connection with the Kendal Literary and Scientific Institute, | At the Friends' Meeting House, | October 1st, 1877. | Reprinted from the "Kendal Mercury." | Price Threepence.

Octavo, pp. 18. Title-page (with blank reverse), pp. 1-2. On p. 3 is a statement explaining that the lecture "has been reprinted in the present form on account of its having been impossible to supply in full the demand for the issue" of the *Mercury*. Text of the report, pp. 5-18. At the end is a blank leaf, with an imprint on the reverse, "Printed at the 'Mercury' Office, Kendal." The pamphlet was issued sewn and without wrappers.

Some passages, omitted in *Deucalion* but contained in the report (and in one case in MS.), are in this edition printed in footnotes (see below, pp. 243, 260, 261, 266).

Chapter I. of volume ii. was delivered as a lecture at the London Institution on March 17, 1880, the lecture being repeated there on March 23. The title was "A Caution to Snakes." The lecture was briefly reported in the *Times*, *Daily News*, and *Daily Chronicle* of March 18, 1880. Some introductory remarks taken from these reports will be found on p. 295 *n*.

At this lecture a folded sheet (forming four quarto pages) was distributed, giving on pp. 1-2 a list of the "Diagrams" shown at it, and on p. 3 "Names of the Snake Tribe in the Great Languages." This is reprinted below, p. 330.

ISSUE IN PARTS

Each part had a title-page, which was the same in most particulars as shown here on the preceding leaf. At the foot of the reverse is the imprint, "Hazell, Watson, and Viney, Printers, London and Aylesbury." The title-pages of Parts I.-V. add to the author's description "and Slade Professor of Fine Art, Oxford"; the number of the part and the date differ, of course, in different parts. The parts were issued, octavo, in paper wrappers (pale grey or light buff), with the title-page (enclosed in a double-ruled frame) reproduced upon the front, with the addition of the rose above the publisher's imprint, and the words "Two Shillings and Sixpence" below the frame. The headlines were "Deucalion" on the left-hand pages, and on the right-hand pages the number and title of the several chapters.

PART I. *First Edition* (1875).—Issued in October 1875, containing pp. 1-48, thus: Introduction, pp. 1-7; Chapter I., pp. 8-25; Chapter II., pp. 26-39; part of Chapter III. (down to line 5 of § 12), pp. 40-48. With it was issued Plate I. (for the renumbering of the plates in the present volume, see below).

A *Second Edition* (with "Second Thousand" on the title-page and the date altered to "1883") was issued in that year.

PART II. *First Edition* (1875).—Issued in October 1875, containing pp. 49-96, thus: continuation of Chapter III., pp. 49-56; Chapter IV., pp. 57–72; Chapter V., pp. 73–83; Chapter VI., pp. 84–96. With it was issued Plate II.

A *Second Edition* was issued in 1883.

PART III. *First Edition* (1876).—Issued in November 1876, containing pp. 97–144, thus: Chapter VII., pp. 97–139; part of Chapter VIII. (down to line 16 of § 6), pp. 140–144. With it was issued Plate III.

A *Second Edition* was issued in 1883.

PART IV. *First Edition* (1876).—Issued in December 1876, containing pp. 145–192, thus: continuation of Chapter VIII., pp. 145–151; Chapter IX., pp. 152–170; Chapter X., pp. 171–189; part of Chapter XI. (down to line 10 of § 5), pp. 190–192. With it was issued Plate IV.

A *Second Edition* was issued in 1883.

PART V. *First Edition* (1878).—Issued in July 1878, containing pp. 193–240, thus: continuation of Chapter XI., pp. 193–203; Chapter XII., pp. 204–237; part of Chapter XIII. (down to the end of p. 268 here), pp. 238–240. With it were issued Plates V. and VI.

A *Second Edition* was issued in 1888.

PART VI. *First Edition* (1879).—Issued in October 1879, containing pp. 241–290, thus: continuation of Chapter XIII., pp. 241–246; Chapter XIV., pp. 247–276; Appendix, pp. 277–280; Index (by Ruskin), pp. 281–290. With this part the title-page and "Contents of Vol. I." were issued; also Plate VII.

PART VII. *First Edition* (1880).—The title-page of this part, commencing volume ii., followed that of Part VI., the number of the part and the date being changed; with Messrs. Hazell, Watson, and Viney's imprint at the foot of the reverse. Issued in July 1880, containing a leaf (with blank reverse) with the "Advice" (see here p. 295 n.), and pp. 1–48 (Chapter I.). With it were issued Plates VIII. and IX.

PART VIII. *First Edition* (1883). Title-page as in Parts I.–V. Issued in May 1883, containing pp. 49–88, thus: Chapter II., pp. 49–68; Chapter III., pp. 69–88. With it were issued Plates X. and XI.

No more of *Deucalion* was issued.

IN VOLUME FORM

The parts of the second volume (never having been completed) were also never collected. Those forming volume i. were collected in July 1882, and issued in mottled-grey paper boards, with a white paper label on the back, lettered "Ruskin. | Deucalion. | Vol. I." Subsequently (1894) issued in green cloth. Price 15s. Reduced in July 1900 to 10s.

There have been *unauthorised American editions* of all the parts of *Deucalion*.

Variæ Lectiones.—The second editions of various parts of *Deucalion* were not revised by Ruskin for press; but he noted some misprints, etc., in the index to the first volume, and in his copy for revision he noted some more and made several alterations. These are all incorporated in the text of the present volume.

The alterations made, in the present volume, in Ruskin's Index to vol. i. of *Deucalion* are of three kinds :—(1) Corrections of misprints, etc., which he noted in the Index have been made in the text, and the notes of them omitted from the Index. Thus the entry "BRECCIA (but for 'breccia' in these pages read 'conglomerate')," and a corresponding entry under "CON-GLOMERATE" are omitted. (2) Two notes given in the Index are transferred to the text. The note now given on p. 95 appeared in Ruskin's Index under "Passion"; and that on p. 96, under "Plans." (3) Additions, for which see p. 583 *n.*

The following is a list of all other variations in the text (other than minor matters of punctuation, spelling, and altered references). The more interesting of Ruskin's alterations are noted in footnotes to the text, and to these references only are here included :—

On the title-page the quotation from Homer (*Iliad*, i. 156, 157) is here corrected. The first line has hitherto been printed " ἐπεὶ μάλα πολλὰ μεταξὺ."

Volume i. : Ch. i. § 12, last line but one, and § 13, line 23, "breccia" is altered to "conglomerate" as noted by Ruskin in his Index; so also "buttresses" to "buttress" in § 21, line 12.

Ch. iv. § 5, line 2, "below" is here substituted for "opposite"; § 10, line 18, "serious" has hitherto been misprinted "series."

Ch. v. § 7, last lines, see p. 151 *n.*; § 12, line 23, the quotation from Keats has hitherto been misprinted "clasped like a missal shut where Paynims pray."

Ch. vi., § "15" has hitherto been misprinted for § "14."

Ch. vii. § 31, line 2, "below" is here substituted for "opposite"; § 32 (VI., line 9), "sanguine" is here a correction for "sable"; § 36, last line but one, "fourth" is a correction for "first."

Ch. ix. § 5, line 7, the words "in mountains" are added by Ruskin in his copy; § 12, Letter II., note (*d*), for "Plate X., lowest figure," the original edition reads, "the plate given in the last number" (namely, one of the plates from "Banded and Brecciated Concretions" which was given in *Deucalion*). Page 213, note (1), for the transference of this note to its present place, see p. 211 *n.*

Ch. x. § 2, third line from end, see p. 220 *n.*; § 8, line 16, "twenty-second" is here a correction for "twenty-third"; § 10, line 15, see p. 226 *n.*; line 35, "this" is Ruskin's correction in his copy for "it"; § 18, lines 2 and 3, similarly he corrected "he makes most dextrous use" to "he is most dextrous in the use," and italicises "it."

Ch. xii. § 8, line 5, Ruskin corrects in his copy the misprint "Weste-more-land" to "West-mere-land"; § 12, line 5, he corrects "that" to "the"; § 40, sixth line from end, the misprint of "science" for "morals" was corrected by Ruskin in his index.

Ch. xiv. § 31, line 6, "first" is here a correction for "second."

Volume ii. : Ch. i. § 33, line 15, "10" is here a correction for "6." Ch. ii. § 17, note, last line, "of the Alps" inserted by Ruskin in his copy.

Ch. iii. § 24, "spiculæ" here corrected to "spicula."

The numbering of the sections after § 18 is added in this edition. The following list shows how the original plates are numbered in this volume :—

Plate I.	is here Plate XII.	Plate VIII.	is Plate VII. in Vol. XXV.
" II.	" " XIII.		
" III.	" " X.	" IX.	is Plate VIII. in Vol. XXV.
" IV.	" " VII.		
" V.	" " XIV.	" X.	is here Plate XXII.
" VI.	" " XV.	" XI.	" " XXIII.
" VII.	" " XVI.		

CONTENTS OF VOLUME I

CONTENTS OF VOLUME II

DEUCALION

INTRODUCTION

Brantwood, 13th July, 1875.

1. I HAVE been glancing lately at many biographies, and have been much struck by the number of deaths which occur between the ages of fifty and sixty (and, for the most part, in the earlier half of the decade), in cases where the brain has been much used emotionally: or perhaps it would be more accurate to say, where the heart, and the faculties of perception connected with it, have stimulated the brain-action. Supposing such excitement to be temperate, equable, and joyful, I have no doubt the tendency of it would be to prolong, rather than depress, the vital energies. But the emotions of indignation, grief, controversial anxiety and vanity, or hopeless, and therefore uncontending, scorn, are all of them as deadly to the body as poisonous air or polluted water; * and when I reflect how much of the active part of my past life has been spent in these states,—and that what may remain to me of life can never more be in any other,—I begin to ask myself, with somewhat pressing arithmetic, how much time is likely to

---

* The reader would do well to study on this subject, with extreme care, the introductory clauses of Sir Henry Thompson's paper on Food, in the 28th number of *The Nineteenth Century*.[1]

---

[1] ["Food and Feeding," June 1879, vol. v. pp. 971 *seq.*: "The general outlines of a man's mental character and physical tendencies are doubtless largely determined by the impress of race and family. But to a large extent the materials and filling in of the framework depend upon his food and training. By the latter term may be understood all that relates to mental and moral and even to physical education."]

be left me, at the age of fifty-six, to complete the various designs for which, until past fifty, I was merely collecting materials.

2. Of these materials, I have now enough by me for a most interesting (in my own opinion) history of fifteenth-century Florentine art, in six octavo volumes; an analysis of the Attic art of the fifth century B.C., in three volumes; an exhaustive history of northern thirteenth-century art, in ten volumes; a life of Turner, with analysis of modern landscape art, in four volumes; a life of Walter Scott, with analysis of modern epic art, in seven volumes; a life of Xenophon, with analysis of the general principles of Education, in ten volumes; a commentary on Hesiod, with final analysis of the principles of Political Economy, in nine volumes; and a general description of the geology and botany of the Alps, in twenty-four volumes.*

3. Of these works, though all carefully projected, and some already in progress,[1]—yet, allowing for the duties of my Professorship, possibly continuing at Oxford, and for the increasing correspondence relating to *Fors Clavigera*,—it does not seem to me, even in my most sanguine moments, now probable that I shall live to effect such conclusion as would be satisfactory to me; and I think it will therefore be only prudent, however humiliating, to throw together at once, out of the heap of loose stones collected for this many-towered city which I am not able to finish,[2] such fragments of good marble as may perchance be useful to future builders; and to clear away, out of sight, the lime and other rubbish which I meant for mortar.

4. And because it is needful, for my health's sake, hence-forward to do as far as possible what I find pleasure, or

* I observe many readers have passed this sentence without recognizing its irony.

[1] [Still ironical; but Ruskin was at the time writing (on Florentine art) *The Laws of Fésole*, a translation of *The Economist* of Xenophon was in preparation (*Bibliotheca Pastorum*), and fragmentary notices of the Life of Scott were appearing in *Fors Clavigera*.]

[2] [See Luke xiv. 30.]

at least tranquillity, in doing, I am minded to collect first
what I have done in geology and botany; for indeed, had
it not been for grave mischance in earlier life (partly consisting in the unlucky gift, from an affectionate friend, of
Rogers's poems, as related in *Fors Clavigera* for August of
this year[1]), my natural disposition or these sciences would
certainly long ago have made me a leading member of the
British Association for the Advancement of Science; or
—who knows?—even raised me to the position which it
was always the summit of my earthly ambition to attain,
that of President of the Geological Society. For, indeed,
I began when I was only twelve years old, a *Mineralogical Dictionary*,[2] intended to supersede everything done
by Werner and Mohs[3] (and written in a shorthand composed of crystallographic signs now entirely unintelligible
to me),—and year by year have endeavoured, until very
lately, to keep abreast with the rising tide of geological
knowledge; sometimes even, I believe, pushing my way
into little creeks in advance of the general wave. I am
not careful to assert for myself the petty advantage of
priority in discovering what, some day or other, somebody
must certainly have discovered. But I think it due to my
readers, that they may receive what real good there may
be in these studies with franker confidence, to tell them[4]
that the first sun-portrait ever taken of the Matterhorn
(and as far as I know of any Swiss mountain whatever)
was taken by me in the year 1849;[5] that the outlines
(drawn by measurement of angle), given in *Modern Painters*,[6]

[1] [Letter 56, § 7 (reprinted in *Præterita*, i. ch. i. § 28).]
[2] [A page of this dictionary of minerals is exhibited in the Ruskin Museum at
the Coniston Institute. Ruskin refers to it again below, p. 553.]
[3] [Abraham Gottlob Werner (1750–1817), the father of German geology, author
of *Ueber die aüssern Kennzeichen der Fossilien* (1764) and numerous works on
geology and mineralogy. Friedrich Mohs, author of *Grund-Riss von Mineralogie*
(1822–1824).]
[4] [Here Ruskin wrote in his copy for revision, "Self-assertion : compare ch. xiv.
§ 2 (p. 273); and repeated too much: compare ch. x. § 5 (p. 222)." The chapters
of *Deucalion* were written and printed, it should be remembered, at considerable
intervals (see p. 90).]
[5] [See, again, below, p. 569.]
[6] [See, in this edition, Vol. VI. pp. 250 *seq.*, 284 *seq.*]

of the Cervin, and aiguilles of Chamouni, are at this day
demonstrable by photography as the trustworthiest then in
existence; that I was the first to point out, in my lecture
given in the Royal Institution,* the real relation of the
vertical cleavages to the stratification, in the limestone
ranges belonging to the chalk formation in Savoy; and
that my analysis of the structure of agates (*Geological
Magazine*[1]) remains, even to the present day, the only
one which has the slightest claim to accuracy of distinction,
or completeness of arrangement.   I propose therefore, if
time be spared me, to collect, of these detached studies, or
lectures, what seem to me deserving of preservation; together with the more carefully written chapters on geology
and botany in the latter volumes of *Modern Painters;*[2]
adding the memoranda I have still by me in manuscript,
and such further illustrations as may occur to me on revision.   Which fragmentary work,—trusting that among the
flowers or stones let fall by other hands it may yet find
service and life,—I have ventured to dedicate to Proserpina
and Deucalion.[3]

5. Why not rather to Eve, or at least to one of the
wives of Lamech, and to Noah? asks, perhaps, the pious
modern reader.

Because I think it well that the young student should
first learn the myths of the betrayal and redemption,[4] as
the Spirit which moved on the face of the wide first
waters,[5] taught them to the heathen world.   And because,
in this power, Proserpine and Deucalion are at least as
true as Eve or Noah; and all four together incomparably

* Reported in the *Journal de Genève,* date ascertainable, but of no
consequence.[6]

---

[1] [Part III. of this volume, above, pp. 37 *seq.*]
[2] [The reference is to the series of reprints commenced under the title of
In Montibus Sanctis (see Vol. III. p. lxii.), and to another projected work on botany
(see *ibid.,* p. xlix.).]
[3] [For these titles, see Vol. XXV. p. xlvii., and above, p. xlvi.]
[4] [For an explanation of this passage, see ii. ch. ii. § 6 (below, p. 335).]
[5] [See Genesis i. 2.]
[6] [In 1863. See now the first part of this volume, above, pp. 3 *seq.*]

truer than the Darwinian Theory. And in general, the reader may take it for a first principle, both in science and literature, that the feeblest myth is better than the strongest theory: the one recording a natural impression on the imaginations of great men, and of unpretending multitudes;[1] the other, an unnatural exertion of the wits of little men, and half-wits of impertinent multitudes.[2]

6. It chanced, this morning, as I sat down to finish my preface, that I had, for my introductory reading, the fifth chapter of the second book of Esdras; in which, though often read carefully before, I had never enough noticed the curious verse, "Blood shall drop out of wood, and the stone shall give his voice, and the people shall be troubled." Of which verse, so far as I can gather the meaning from the context, and from the rest of the chapter, the intent is, that in the time spoken of by the prophet, which, if not our own, is one exactly corresponding to it, the deadness of men to all noble things shall be so great, that the sap of trees shall be more truly blood, in God's sight, than their heart's blood; and the silence of men, in praise of all noble things, so great, that the stones shall cry out, in God's hearing, instead of their tongues; and the rattling of the shingle on the beach, and the roar of the rocks driven by the torrent, be truer Te Deum than the thunder of all their choirs. The writings of modern scientific prophets teach us to anticipate a day when even these lower voices shall be also silent; and leaf cease to wave, and stream to murmur, in the grasp of an eternal cold. But it may be, that rather out of the mouths of babes and sucklings[3] a better peace may be promised to the redeemed Jerusalem; and the strewn branches, and low-laid stones, remain at rest at the gates of the city, built in unity with herself, and saying with her human voice, "My King cometh."[4]

---

[1] [On the historical value of myths—on myths as facts—compare Vol. XXII. p. 444.]

[2] [Here, again, see ii. ch. ii. § 8 (below, p. 336).]

[3] [Psalms viii. 2.]

[4] [See Matthew xxi. 5.]

CHAPTER I

THE ALPS AND JURA

(Part of a Lecture given in the Museum of Oxford in October, 1874)

1. IT is often now a question with me whether the persons who appointed me to this Professorship have been disappointed, or pleased, by the little pains I have hitherto taken to advance the study of landscape. That it is my own favourite branch of painting seemed to me a reason for caution in pressing it on your attention;[1] and the range of art-practice which I have hitherto indicated for you, seems to me more properly connected with the higher branches of philosophical inquiry native to the University. But, as the second term of my Professorship will expire next year, and as I intend what remains of it to be chiefly employed in giving some account of the art of Florence and Umbria,[2] it seemed to me proper, before entering on that higher subject, to set before you some of the facts respecting the great elements of landscape, which I first stated thirty years ago; arranging them now in such form as my farther study enables me to give them. I shall not, indeed, be able to do this in a course of spoken lectures; nor do I wish to do so. Much of what I desire that you should notice is already stated, as well as I can do it, in

---

[1] [Compare *The Art of England,* § 156.]

[2] [The second period (of three years) expired in 1875. In the last term of 1874 Ruskin gave, first, the four lectures on "Mountain Form in the Higher Alps," which are partly embodied in *Deucalion;* and, secondly, the course on "The Æsthetic and Mathematic Schools of Art in Florence," now printed in Vol. XXIII.]

Modern Painters; and it would be waste of time to recast it in the form of address. But I should not feel justified in merely reading passages of my former writings to you from this chair;[1] and will only ask your audience, here, of some additional matters, as, for instance, to-day, of some observations I have been making recently, in order to complete the account given in *Modern Painters*, of the structure and aspect of the higher Alps.

2. Not that their structure—(let me repeat, once more, what I am well assured you will, in spite of my frequent assertion, find difficult to believe),—not that their structure is any business of yours or mine, as students of practical art.[2] All investigations of internal anatomy, whether in plants, rocks, or animals, are hurtful to the finest sensibilities and instincts of form. But very few of us have any such sensibilities to be injured ; and that we may distinguish the excellent art which they have produced, we must, by duller processes, become cognizant of the facts. The Torso of the Vatican[3] was not wrought by help from dissection ; yet all its supreme qualities could only be explained by an anatomical master. And these drawings of the Alps by Turner are in landscape, what the Elgin Marbles or the Torso are in sculpture. There is nothing else approaching them, or of their order. Turner made them before geology existed ;[4] but it is only by help of geology that I can prove their power.

3. I chanced, the other day, to take up a number of the *Alpine Journal* (May, 1871),[5] in which there was a review, by Mr. Leslie Stephen, of Mr. Whymper's *Scrambles among the Alps*, in which it is said that "if the Alpine

---

[1] [In 1877, however, Ruskin gave "Readings in *Modern Painters*"; but the readings were only a peg for discourses at large : see Vol. XXII. pp. xli., 508 seq.]

[2] [See *Eagle's Nest*, Vol. XXII. pp. 222 seq.]

[3] [See Vol. III. p. 608, and Vol. XXII. p. 95.]

[4] [See *Modern Painters*, vol. i. (Vol. III. p. 429 and n.).]

[5] [No. 33, vol. v. p. 235.]

Club has done nothing else, it has taught us for the first time really to see the mountains." I have not the least idea whom Mr. Stephen means by "us"; but I can assure him that mountains had been seen by several people before the nineteenth century; that both Hesiod and Pindar occasionally had eyes for Parnassus, Virgil for the Apennines, and Scott for the Grampians; and without speaking of Turner, or of any other accomplished artist, here is a little bit of old-fashioned Swiss drawing of the two Mythens, above the central town of Switzerland,* showing a degree of affection, intelligence, and tender observation, compared to which our modern enthusiasm is, at best, childish; and commonly also as shallow as it is vulgar.

4. Believe me, gentlemen, your power of seeing mountains cannot be developed either by your vanity, your curiosity, or your love of muscular exercise. It depends on the cultivation of the instrument of sight itself, and of the soul that uses it. As soon as you can see mountains rightly, you will see hills also, and valleys, with considerable interest; and a great many other things in Switzerland with which you are at present but poorly acquainted. The bluntness of your present capacity of ocular sensation is too surely proved by your being unable to enjoy any of the sweet lowland country, which is incomparably more beautiful than the summits of the central range, and which is meant to detain you, also, by displaying—if you have patience to observe them—the loveliest aspects of that central range itself, in its real majesty of proportion, and mystery of power.

5. For, gentlemen, little as you may think it, you can no more see the Alps from the Col du Géant, or the top of the Matterhorn, than the pastoral scenery of Switzerland

---

* In the Educational Series of my Oxford Schools [No. 286: see Vol. XXI. pp. 100, 129].

from the railroad carriage.[1] If you want to see the skeletons
of the Alps, you may go to Zermatt or Chamouni; but if
you want to see the body and soul of the Alps, you must
stay awhile among the Jura, and in the Bernese plain.
And, in general, the way to see mountains is to take a
knapsack and a walking-stick ; leave alpenstocks to be
flourished in each other's faces, and between one another's
legs, by Cook's tourists; and try to find some companionship in yourself with yourself; and not to be dependent for
your good cheer either on the gossip of the table-d'hôte, or
the hail-fellow and well met, hearty though it be, of even
the pleasantest of celebrated guides.

6. Whether, however, you think it necessary or not, for
true sight of the Alps, to stay awhile among the Jura or
in the Bernese fields, very certainly, for understanding, or
questioning of the Alps, it is wholly necessary to do so. If
you look back to the lecture, which I gave as the fourth
of my inaugural series,[2] on the Relation of Art to Use, you
will see it stated, as a grave matter of reproach to the
modern traveller, that, crossing the great plain of Switzerland nearly every summer, he never thinks of inquiring why
it is a plain, and why the mountains to the south of it are
mountains.

7. For solution of which, as it appears to me, not unnatural inquiry, all of you, who have taken any interest in
geology whatever, must recognize the importance of studying the calcareous ranges which form the outlying steps of
the Alps on the north ; and which, in the lecture just referred to, I requested you to examine for their crag scenery,
markedly developed in the Stockhorn, Pilate, and Sentis
of Appenzell. The arrangements of strata in that great
calcareous belt give the main clue to the mode of elevation of the central chain, the relations of the rocks over

---

[1] [Ruskin in his copy for revision here wrote, "Attach to this the preface of
Sesame;" that is, the preface to the second edition (only) of that book. See now
Vol. XVIII. p. 25.]
[2] [*Lectures on Art*, § 108 (Vol. XX. p. 102).]

the entire breadth of North Switzerland being, roughly, as in this first section :[1]—

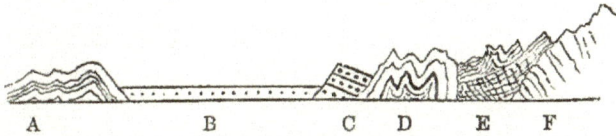

Fig. 33

A. Jura limestones, moderately undulating in the successive chains of Jura.
B. Sandstones of the great Swiss plain.
C. Pebble breccias of the first ranges of Alpine hills.
D. Chalk formations violently contorted, forming the rock scenery of which I have just spoken.
E. Metamorphic rocks lifted by the central Alps.
F. Central gneissic or granitic mass, narrow in Mont Blanc, but of enormous extent southwards from St. Gothard.

8. Now you may, for first grasp of our subject, imagine these several formations all fluted longitudinally, like a Gothic moulding, thus forming a series of ridges and valleys parallel to the Alps;—such as the Valley of Chamouni, the Simmenthal, and the great vale containing the lakes of Thun and Brientz; to which longitudinal valleys we now obtain access through gorges or defiles, for the most part cut across the formations, and giving geological sections all the way from the centres of the Alps to the plain.

9. Get this first notion very simply and massively set in your thoughts. Longitudinal valleys, parallel with the beds; more or less extended and soft in contour, and often occupied by lakes. Cross defiles like that of Lauterbrunnen, the Via Mala, and the defile of Gondo; cut down across the beds, and traversed by torrents, but rarely occupied by lakes. The Bay of Uri is the only perfect instance in Switzerland of a portion of lake in a diametrically cross valley; the crossing arms of the Lake Lucerne mark the exactly rectangular schism of the forces; the main direction being that of the lakes of Kussnacht and Alpnacht, carried

[1] [See also the upper figure in Plate XVI. (comparing ch. xiv. §§ 6, 12), pp. 275, 278, where Ruskin explains that the section is "arranged from Studer."]

on through those of Sarnen and Lungern, and across the low intervening ridge of the Brunig, joining the depressions of Brienz and Thun; of which last lake the lower reach, however, is obliquely transverse. Forty miles of the Lago Maggiore, or, including the portion of lake now filled by delta, fifty, from Baveno to Bellinzona, are in the longitudinal valley which continues to the St. Bernardino: and the entire length of the Lake of Como is the continuation of the great lateral Valtelline.

10. Now such structure of parallel valley and cross defile would be intelligible enough, if it were confined to the lateral stratified ranges. But, as you are well aware, the two most notable longitudinal valleys in the Alps are cut right along the heart of their central gneissic chain; how much by dividing forces in the rocks themselves, and how much by the sources of the two great rivers of France and Germany, there will yet be debate among geologists for many a day to come. For us, let the facts at least be clear; the questions definite; but all debate declined.

11. All lakes among the Alps, except the little green pool of Lungern, and a few small tarns on the cols, are quite at the bottom of the hills.[1] We are so accustomed to this condition, that we never think of it as singular. But in its unexceptional character, it is extremely singular. How comes it to pass, think you, that through all that wildness of mountain—raised, in the main mass of it, some six thousand feet above the sea, so that there is no col lower,—there is not a single hollow shut in so as to stay the streams of it; that no valley is ever barred across by a ridge which can keep so much as ten feet of water calm above it,—that every such ridge that once existed has been cut through, so as to let the stream escape?

I put this question in passing; we will return to it:[2] let me first ask you to examine the broad relations of the

---

[1] [Compare "Letters on the Conformation of the Alps," § 3 (below, p. 549).]
[2] [See § 19, p. 112.]

beds that are cut through. My typical section, Fig. 33, is stringently simple; it must be much enriched and modified to fit any locality; but in the main conditions it is applicable to the entire north side of the Alps, from Annecy to St. Gall.

12. You have first—(I read from left to right, or north to south, being obliged to do so because all Studer's sections are thus taken)—this mass of yellow limestone, called of the Jura, from its development in that chain; but forming an immense tract of the surface of France also; and, as you well know, this our city of Oxford stands on one of its softer beds, and is chiefly built of it. We may, I think, without entering any forbidden region of theory, assume that this Jura limestone extends under the plain of Switzerland, to reappear where we again find it on the flanks of the great range; where on the top of it the beds drawn with fine lines in my section correspond generally to the date of our English chalk, though they are far from white in the Alps. Curiously adjusted to the chalk beds, rather than superimposed, we have these notable masses of pebble conglomerate, which bound the sandstones of the great Swiss plain.

13. I have drawn that portion of the section a little more boldly in projection, to remind you of the great Rigi promontory; and of the main direction of the slope of these beds, with their backs to the Alps, and their escarpments to the plain. Both these points are of curious importance. Have you ever considered the reason of the fall of the Rossberg,[1] the most impressive physical catastrophe that has chanced in Europe in modern times? Few mountains in Switzerland looked safer. It was of inconsiderable height, of very moderate steepness; but its beds lay perfectly straight, and that over so large a space, that when the clay between two of them got softened by rain, one slipped off the other. Now this mathematical straightness is

---

[1] [See *Modern Painters*, vol. iv, (Vol. VI. pp. 195 n., 378-379), and Plate 50 (Turner's "Goldau").]

characteristic of these pebble beds,—not universal in them,
but characteristic of them, and of them only. The limestones underneath are usually, as you see in this section,
violently contorted; if not contorted, they are at least so
irregular in the bedding that you can't in general find a
surface of a furlong square which will not either by its
depression, or projection, catch and notch into the one
above it, so as to prevent its sliding. Also the limestones
are continually torn, or split, across the beds. But the conglomerates, though in many places they suffer decomposition, are curiously free from fissures and rents.[1] The hillside
remains unshattered unless it comes down in a mass. But
their straight bedding, as compared with the twisted limestone, is the notablest point in them; and see how very
many difficulties are gathered in the difference. The crushed
masses of limestone are supposed to have been wrinkled
together by the lateral thrust of the emerging protogines;
and these pebble beds to have been raised into a gable, or
broken into a series of colossal fragments set over each
other like tiles, all along the south shore of the Swiss
plain, by the same lateral thrust; nay, "though we may
leave in doubt," says Studer, "by what cause the folded
forms of the Jura may have been pushed back, there yet
remains to us, for the explanation of this gabled form of
the Nagelfluh, hardly any other choice than to adopt the
opinion of a lateral pressure communicated by the Alps to
the tertiary bottom. We have often found in the outer
limestone chains themselves clear evidence of a pressure
going out from the inner Alps; and the pushing of the
older over the younger formations along the flank of the
limestone hills, leaves hardly any other opinion possible."[2]

14. But if these pebble beds have been heaved up by
the same lateral thrust, how is it that a force which can
bend limestone like leather, cannot crush, everywhere, these
pebble beds into the least confusion? Consider the scale

---

[1] [Compare above, p. 30.]
[2] [B. Studer: *Geologie der Schweiz*, 2 vols., 1851-1853, vol. ii. p. 374.]

on which operations are carried on, and the forces of which
this sentence of Studer's so serenely assumes the action.
Here, A, Fig. 34, is his section[1] of the High Sentis of
Appenzell, of which the height is at least, in the parts thus
bent, 6000 feet. And here, B, Fig. 34, are some sheets of
paper, crushed together by my friend Mr. Henry Woodward,[2] from a length of four inches, into what you see;[3]
the High Sentis exactly resembles these, and seems to consist of four miles of limestone similarly crushed into one.
Seems, I say, remember: I never theorize, I give you the
facts only. The beds *do* go up and down like this: that
they have been crushed together, it is Mr. Studer who says

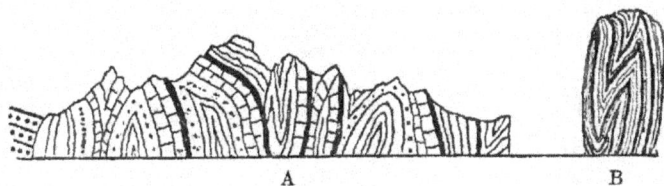

A   B

Fig. 34

or supposes; I can't go so far; nevertheless, I admit that
he appears to be right, and I believe he is right; only
don't be positive about it, and don't debate; but think of
it, and examine.

15. Suppose, then, you have a bed of rocks, four miles
long by a mile thick, to be crushed laterally into the space
of a mile. It may be done, supposing the mass not to be
reducible in bulk, in two ways: you may either crush it
up into folds, as I crush these pieces of cloth; or you may
break it into bits, and shuffle them over one another like
cards. Now, Mr. Studer, and our geologists in general,
believe the first of these operations to have taken place
with the limestones, and the second, with the breccias.

[1] [See *Geologie der Schweiz*, vol. ii. p. 193.]
[2] [Mr. Henry Woodward, LL.D., F.R.S., Keeper of the Geological Department,
British Museum; President of the Geological Society, 1894-1896; editor of *The
Geological Magazine*, 1864-1900.]
[3] [But see a correction of this passage below, ch. xiv. § 18 *n.*, p. 281.]

They are, as I say, very probably right: only just consider what is involved in the notion of shuffling up your breccias like a pack of cards, and folding up your limestones like a length of silk which a dexterous draper's shopman is persuading a young lady to put ten times as much of into her gown as is wanted for it! Think, I say, what is involved in the notion. That you may shuffle your pebble beds, you must have them strong and well knit. Then what sort of force must you have to break and to heave them? Do but try the force required to break so much as a captain's biscuit by slow push,—it is the illustration I gave long ago in *Modern Painters*,[1]—and then fancy the results of such fracturing power on a bed of conglomerate two thousand feet thick! And here is indeed a very charming bookbinder's pattern, produced by my friend in crushed paper, and the length of silk produces lovely results in these arrangements à la Paul Veronese. But when you have the cliffs of the Diablerets, or the Dent du Midi of Bex, to deal with; and have to fold *them* up similarly, do you mean to fold your two-thousand-feet-thick Jura limestone in a brittle state, or a ductile one? If brittle, won't it smash? If ductile, won't it squeeze?[2] Yet your whole mountain theory proceeds on the assumption that it has neither broken nor been compressed,—more than the folds of silk or coils of paper.

16. You most of you have been upon the Lake of Thun. You have been at least carried up and down it in a steamer; you smoked over it meanwhile, and countenanced the Frenchmen and Germans who were spitting into it. The steamer carried you all the length of it in half-an-hour; you looked at the Jungfrau and Blumlis Alp, probably, for five minutes, if it was a fine day; then took to your papers, and read the last news of the Tichborne case;[3] then you lounged about,—thought it a nuisance that the steamer couldn't take you up in twenty minutes, instead

---

[1] [See Vol. VI. p. 195.]
[2] [Ruskin in his copy here compares ch. xii. § 26 (below, pp. 256–257).]
[3] [For other contemptuous references to this case, see Vol. XXIII. p. 122.]

of half-an-hour; then you got into a row about your luggage at Neuhaus; and all that you recollect afterwards is that lunch where you met the So-and-sos at Interlaken.

17. Well, we used to do it differently in old times. Look here;—this * is the quay at Neuhaus, with its then travelling arrangements. A flat-bottomed boat, little better than a punt;—a fat Swiss girl with her schatz, or her father, to row it; oars made of a board tied to a pole: and so one paddled along over the clear water, in and out among the bays and villages, for half a day of pleasant life. And one knew something about the lake, ever after, if one had a head with eyes in it.

It is just possible, however, that some of you also who have been learning to see the Alps in your new fashion, may remember that the north side of the Lake of Thun consists, first, next Thun, of a series of low green hills, with brown cliffs here and there among the pines; and that above them, just after passing Oberhofen, rears up suddenly a great precipice, with its flank to the lake, and the winding wall of it prolonged upwards, far to the north, losing itself, if the day is fine, in faint tawny crests of rock among the distant blue; and if stormy, in wreaths of more than commonly torn and fantastic cloud.

18. To form the top of that peak on the north side of the Lake of Thun, you have to imagine forces which have taken—say, the whole of the North Foreland, with Dover Castle on it, and have folded it upside-down on the top of the parade at Margate,—then swept up Whitstable oyster-beds, and put them on the bottom of Dover Cliffs turned topsy-turvy,—and then wrung the whole round like a wet towel, till it is as close and hard as it will knit;—such is the beginning of the operations which have produced the lateral masses of the higher Alps.

19. Next to these, you have the great sculptural force,

---

* Turner's first study of the Lake of Thun, in 1803.[1]

---

[1] [The drawing was No. 7 in Ruskin's Exhibition of 1878 : see Vol. XIII. p. 417.]

which gave them, approximately, their present forms,—
which let out all the lake waters above a certain level,—
which cut the gorge of the Devil's Bridge—of the Via Mala
—of Gondo—of the Valley of Cluse ; which let out the
Rhone at St. Maurice, the Ticino at Faido, and shaped
all the vast ravines which make the flanks of the great
mountains awful.

20. Then, finally, you have the rain, torrent, and glacier,
of human days.

Of whose action, briefly, this is the sum.

Over all the high surfaces, disintegration—melting away
—diffusion—loss of height and terror.

In the ravines,—whether occupied by torrent or glacier,
—gradual incumbrance by materials falling from above ;
choking up of their beds by silt—by moraine—by continual
advances of washed slopes on their flanks ; here and there,
only, exceptional conditions occur in which a river is still
continuing feebly the ancient cleaving action, and cutting
its ravine deeper, or cutting it back.

Fix this idea thoroughly in your minds. Since the Valley
of Lauterbrunnen existed for human eyes,—or its pastures
for the food of flocks,—it has not been cut deeper, but
partially filled up by its torrents. The town of Interlachen
stands where there was once lake,—and the long slopes of
grassy sward on the north of it, stand where once was
precipice. Slowly,—almost with infinite slowness,—the declining and encumbering action takes place ;[1] but incessantly,
and,—as far as our experience reaches,—irredeemably.

21. Now I have touched in this lecture briefly on the
theories respecting the elevation of the Alps, because I
want to show you how uncertain and unsatisfactory they
still remain. For our own work,[2] we must waste no time
on them ; we must begin where all theory ceases ; and where
observation becomes possible,—that is to say, with the forms

---

[1] [Ruskin in his copy notes here, "Compare Lyell on slow raising of coal-mine
floor" : see ch. xxiv. of his *Elements of Geology*.]

[2] [Here, again, he explains : "'Our own work,' *i.e.*, at Oxford, is to see 'in what
strength and beauty of form,' etc." (p. 113).]

which the Alps have actually retained while men have dwelt among them, and on which we can trace the progress, or the power, of existing conditions of minor change. Such change has lately affected, and with grievous deterioration, the outline of the highest mountain of Europe, with that of its beautiful supporting buttress,—the Aiguille de Bionnassay. I do not care,—and I want you not to care,—how crest or aiguille was lifted, or where its materials came from, or how much bigger it was once.[1] I do care that you should know, and I will endeavour in these following pages securely to show you, in what strength and beauty of form it has actually stood since man was man, and what subtle modifications of aspect, or majesties of contour, it still suffers from the rains that beat upon it, or owes to the snows that rest.

[1] [Here Ruskin in his copy compares ch. xii. § 1 (below, p. 244).]

CHAPTER II

THE THREE ÆRAS

(Part of a Lecture given at the London Institution in March, 1875, with added pieces from Lectures in Oxford)

1. WE are now, so many of us, some restlessly and some wisely, in the habit of spending our evenings abroad, that I do not know if any book exists to occupy the place of one classical in my early days, called *Evenings at Home.*[1] It contained, among many well-written lessons, one, under the title of "Eyes and No Eyes," which some of my older hearers may remember, and which I should myself be sorry to forget. For if such a book were to be written in these days, I suppose the title and the moral of the story would both be changed; and, instead of "Eyes and No Eyes," the tale would be called "Microscopes and No Microscopes." For I observe that the prevailing habit of learned men is now to take interest only in objects which cannot be seen without the aid of instruments; and I believe many of my learned friends, if they were permitted to make themselves, to their own liking, instead of suffering the slow process of selective development, would give themselves heads like wasps', with three microscopic eyes in the middle of their foreheads, and two ears at the ends of their antennæ.

2. It is the fashion, in modern days, to say that Pope was no poet.[2] Probably our schoolboys, also, think Horace

---

[1] [*Evenings at Home; or, The Juvenile Budget Opened. Consisting of a Variety of Miscellaneous Pieces, for the Instruction and Amusement of Young Persons:* 6 vols., 1792–1796. First published anonymously; written by Dr. Aikin and his sister, Mrs. Barbauld. The chapter called "Eyes and No Eyes" was the "Nineteenth Evening" (vol. iv. pp. 93 *seq.*).]

[2] [For Ruskin's appreciation of Pope, see Vol. XVI. p. 446 *n.*]

none. They have each, nevertheless, built for themselves a monument of enduring wisdom ; and all the temptations and errors of our own day, in the narrow sphere of lenticular curiosity, were anticipated by Pope, and rebuked, in one couplet :

> " Why has not man a microscopic eye ?
> For this plain reason,—Man is not a fly."

While the nobler following lines,

> " Say, what avail, were finer optics given
> To inspect a mite, not comprehend the heaven ? " [1]

only fall short of the truth of our present dulness, in that we inspect heaven itself, without understanding it.

3. In old times, then, it was not thought necessary for human creatures to know either the infinitely little, or the infinitely distant ; nor either to see, or feel, by artificial help. Old English people used to say they perceived things with their five,—or it may be, in a hurry, they would say, their seven,—*senses ;* and that word " sense " became, and for ever must remain, classical English, derived from classical Latin, in both languages signifying, not only the bodily sense, but the moral one. If a man heard, saw, and tasted rightly, we used to say he had his bodily senses perfect. If he judged, wished, and felt rightly, we used to say he had his moral senses perfect, or was a man " in his senses." And we were then able to speak precise truth respecting both matter and morality ; and if we heard any one saying clearly absurd things,—as, for instance, that human creatures were automata,—we used to say they were out of their " senses," and were talking non-" sense."

Whereas, in modern days, by substituting analysis for sense in morals, and chemistry for sense in matter, we have literally blinded ourselves to the essential qualities of both

---

[1] [*Essay on Man*, Epistle I., lines 193–196. In the third of the lines quoted Pope wrote, "Say what the use, were," etc. Compare Vol. VI. p. 76 *n*.]

matter and morals; and are entirely incapable of understanding what is meant by the description given us, in a
book we once honoured, of men who "by reason of use,
have their *senses* exercised to discern both good and evil."[1]

4. And still, with increasingly evil results to all of us,
the separation is every day widening between the man of
science and the artist—in that, whether painter, sculptor,
or musician, the latter is pre-eminently a person who sees
with his Eyes, hears with his Ears, and labours with his
Body, as God constructed them; and who, in using instruments, limits himself to those which convey or communicate his human power, while he rejects all that increase it.
Titian would refuse to quicken his touch by electricity;
and Michael Angelo to substitute a steam-hammer for his
mallet. Such men not only do not desire, they imperatively
and scornfully refuse, either the force, or the information,
which are beyond the scope of the flesh and the senses of
humanity. And it is at once the wisdom, the honour, and
the peace, of the Masters both of painting and literature,
that they rejoice in the strength, and rest in the knowledge,
which are granted to active and disciplined life; and are
more and more sure, every day, of the wisdom of the
Maker in setting such measure to their being; and more
and more satisfied, in their sight and their audit of Nature,
that "the hearing ear, and the seeing eye,—the Lord hath
made even both of them."[2]

5. This evening, therefore, I venture to address you,
speaking limitedly as an artist; but, therefore, I think,
with a definite advantage in having been trained to the use
of my eyes and senses, as my chief means of observation:
and I shall try to show you things which with your own
eyes you may any day see, and with your own common-
sense, if it please you to trust it, account for.

Things which you may see, I repeat; not which you
might perhaps have seen, if you had been born when you

---

[1] [Hebrews v. 14.   Compare Vol. XXIII. p. 185.]
[2] [Proverbs xx. 12.]

were not born ; nor which you might perhaps in future see, if you were alive when you will be dead. But what, in the span of earth, and space of time, allotted to you, may be seen with your human eyes, if you learn to use them.

And this limitation has, with respect to our present subject, a particular significance, which I must explain to you before entering on the main matter of it.

6. No one more honours the past labour—no one more regrets the present rest—of the late Sir Charles Lyell,[1] than his scholar, who speaks to you. But his great theorem of the constancy and power of existing phenomena was only in measure proved,—in a larger measure disputable ; and in the broadest bearings of it, entirely false. Pardon me if I spend no time in qualifications, references, or apologies, but state clearly to you what Sir Charles Lyell's work itself enables us now to perceive of the truth. There are, broadly, three great demonstrable periods of the Earth's history. That in which it was crystallized ; that in which it was sculptured ; and that in which it is now being unsculptured, or deformed.[2] These three periods interlace with each other, and gradate into each other—as the periods of human life do. Something dies in the child on the day that it is born, —something is born in the man on the day that he dies : nevertheless, his life is broadly divided into youth, strength, and decrepitude. In such clear sense, the Earth has its three ages : of their length we know as yet nothing, except that it has been greater than any man had imagined.

---

[1] [Lyell (1797-1875) had died shortly before the delivery of this lecture. Lyell's standpoint, in approaching geological problems, was indicated in the subsidiary title of his great book, *The Principles of Geology*—"An Attempt to Explain the Former Changes of the Earth's Surface by Reference to Causes now in Operation"—the principle which he was the first geologist firmly to grasp being the power of gradual changes to produce great results if only time enough be allowed (see above, p. 13). The *Principles* was published in 1830-32-33, so that the revolution in geological ideas which it introduced, when it was flung into the strife of the schools severally known as Neptunists (see below, p. 556) and Vulcanists, was already in the air when Ruskin as a youth commenced the study of the subject. It is in this sense that he speaks of himself as "Lyell's scholar."]

[2] [With these three æras compare the division of the first paper in this volume into (1) materials, (2) formation, and (3) sculpture of the Alps (above, pp. 3 *seq.*).]

7. (THE FIRST PERIOD.)—But there was a period, or a succession of periods, during which the rocks which are now hard were soft; and in which, out of entirely different positions, and under entirely different conditions from any now existing or describable, the masses, of which the mountains you now see are made, were lifted and hardened, in the positions they now occupy, though in what forms we can now no more guess than we can the original outline of the block from the existing statue.

8. (THE SECOND PERIOD.)—Then, out of those raised masses, more or less in lines compliant with their crystalline structure, the mountains we now see were hewn, or worn, during the second period, by forces for the most part differing both in mode and violence from any now in operation, but the result of which was to bring the surface of the earth into a form approximately that which it has possessed as far as the records of human history extend. The Ararat of Moses's time, the Olympus and Ida of Homer's, are practically the same mountains now, that they were then.

9. (THE THIRD PERIOD.)—Not, however, without some calculable, though superficial, change; and that change, one of steady degradation. For in the third, or historical period, the valleys excavated in the second period are being filled up, and the mountains, hewn in the second period, worn or ruined down. In the second æra the valley of the Rhone was being cut deeper every day; now it is every day being filled up with gravel. In the second æra, the scars of Derbyshire and Yorkshire were cut white and steep; now they are being darkened by vegetation, and crumbled by frost. You cannot, I repeat, separate the periods with precision; but, in their characters, they are as distinct as youth from age.

10. The features of mountain form, to which during my own life I have exclusively directed my study, and which I endeavour to bring before the notice of my pupils in Oxford, are exclusively those produced by existing forces,

on mountains whose form and substance have not been materially changed during the historical period.

For familiar example, take the rocks of Edinburgh Castle, and Salisbury Craig. Of course we know that they are both basaltic, and must once have been hot. But I do not myself care in the least what happened to them till they were cold.* They have both been cold at least longer than young Harry Percy's spur;[1] and, since they were last brought out of the oven, in the shape which, approximately, they still retain, with a hollow beneath one of them, which, for aught I know, or care, may have been cut by a glacier out of white-hot lava, but assuredly at last got itself filled with pure, sweet, cold water, and called, in Lowland Scotch, the "Nor' Loch";—since the time, I say, when the basalt, above, became hard, and the lake beneath, drinkable, I am

---

* More curious persons, who *are* interested in their earlier condition, will find a valuable paper by Mr. J. W. Judd,[2] in the quarterly *Journal of the Geological Society*, May 1875; very successfully, it seems to me, demolishing all former theories on the subject, which the author thus sums, at p. 135.

"The series of events which we are thus required to believe took place in this district is therefore as follows:—

"A. At the point where the Arthur's Seat group of hills now rises, a series of volcanic eruptions occurred during the Lower Calciferous Sandstone period, commencing with the emission of basaltic lavas, and ending with that of porphyrites.

"B. An interval of such enormous duration supervened as to admit of—

　　a. The deposition of at least 3000 feet of Carboniferous strata.
　　b. The bending of all the rocks of the district into a series of great anticlinal and synclinal folds.
　　c. The removal of every vestige of the 3000 feet of strata by denudation.

"C. The outburst, after this vast interval, of a second series of volcanic eruptions upon the *identical site* of the former ones, presenting in its succession of events *precisely the same sequence,* and resulting in the production of rocks of *totally undistinguishable character.*

"Are we not entitled to regard the demand for the admission of such a series of extraordinary accidents as evidence of the *antecedent improbability* of the theory? And when we find that all attempts to suggest a period for the supposed second series of outbursts have successively failed, do not the difficulties of the hypothesis appear to be overwhelming?"

---

[1] [*2 Henry IV.*, Act i. sc. 1, line 42.]
[2] [Mr. John Wesley Judd, F.R.S.; on the Geological Survey, 1867–1870; President of the Geological Society, 1887–1888.]

desirous to examine with you what effect the winter's frost
and summer's rain have had on the crags and their hollows;
how far the "kittle nine steps"[1] under the castle-walls, or
the firm slope and cresting precipice above the dark ghost
of Holyrood, are enduring or departing forms; and how
long, unless the young engineers of New Edinburgh blast
the incumbrance away, the departing mists of dawn may
each day reveal the form, unchanged, of the Rock which
was the strength of their Fathers.

11. Unchanged, or so softly modified that eye can
scarcely trace, or memory measure, the work of time.
Have you ever practically endeavoured to estimate the
alterations of form in any hard rocks known to you, during
the course of your own lives? You have all heard, a thousand times over, the common statements of the school of
Sir Charles Lyell. You know all about alluviums and
gravels; and what torrents do, and what rivers do, and
what ocean currents do; and when you see a muddy stream
coming down in a flood, or even the yellow gutter more
than usually rampant by the roadside in a thunder-shower,
you think, of course, that all the forms of the Alps are to
be accounted for by aqueous erosion, and that it's a wonder
any Alps are still left. Well—any of you who have fished
the pools of a Scottish or Welsh stream,—have you ever
thought of asking an old keeper how much deeper they
had got to be, while his hairs were silvering? Do you
suppose he wouldn't laugh in your face?[2]

There are some sitting here, I think, who must have
themselves fished, for more than one summer, years ago, in
Dove or Derwent,—in Tweed or Teviot. Can any of you

---

[1] [Scott "records his pride in being found before he left the High School one
of the boldest and nimblest climbers of 'the kittle nine stanes,' a passage of difficulty
which might puzzle a chamois-hunter of the Alps, its steps, few and far between,
projected high in air from the precipitous black granite of the Castle rock" (Lock-
hart's *Life*, ch. iii.). See *Redgauntlet* as noted in Vol. XIII. p. 399 n.]

[2] [Mr. Wedderburn after the delivery of this lecture wrote to a Scotch farmer,
who had lived all his life under Ben Nevis, to ask about this, and he replied that
no pool had deepened in his lifetime. This reply was read out by Ruskin at the
next lecture.]

tell me a single pool, even in the limestone or sandstone,
where you could spear a salmon then, and can't reach one
now—(providing always the wretches of manufacturers have
left you one to be speared, or water that you can see
through)? Do you know so much as a single rivulet of
clear water which has cut away a visible half-inch of Highland rock, to your own knowledge, in your own day? You
have seen whole banks, whole fields washed away; and the
rocks exposed beneath? Yes, of course you have; and so
have I. The rains wash the loose earth about everywhere,
in any masses that they chance to catch—loose earth, or
loose rock. But yonder little rifted well in the native
whinstone by the sheepfold,—did the grey shepherd not put
his lips to the same ledge of it, to drink—when he and
you were boys together?

12. "But Niagara, and the Delta of the Ganges—and—
all the rest of it?" Well, of course a monstrous mass of
continental drainage, like Niagara, *will* wash down a piece
of crag once in fifty years (but only that, if it's rotten
below); and tropical rains will eat the end off a bank of
slime and alligators,—and spread it out lower down. But
does any Scotchman know a change in the Fall of Foyers?
—any Yorkshireman in the Force of Tees?

Except of choking up, it may be—not of cutting down.
It is true, at the side of every stream you see the places
in the rocks hollowed by the eddies. I suppose the eddies
go on at their own rate. But I simply ask, Has any
human being ever known a stream, in hard rock, cut its
bed an inch deeper down at a given spot?

13. I can look back, myself, now pretty nearly, I am
sorry to say, half a century, and recognize no change whatever in any of my old dabbling-places; but that some stones
are mossier, and the streams usually dirtier,—the Derwent
above Keswick, for example.

"But denudation does go on, somehow: one sees the
whole glen is shaped by it?" Yes, but not by the *stream*.
The stream only sweeps down the loose stones; frost and

chemical change are the powers that loosen them. I have indeed not known one of my dabbling-places changed in fifty years. But I have known the éboulement under the Rochers des Fyz, which filled the Lac de Chêde; I passed through the Valley of Cluse a night after some two or three thousand tons of limestone came off the cliffs of Maglans—burying the road and field beside it.[1] I have seen half a village buried by a landslip, and its people killed, under Monte St. Angelo, above Amalfi.[2] I have seen the lower lake of Llanberis destroyed, merely by artificial slate quarries; and the Waterhead of Coniston seriously diminished in purity and healthy flow of current by the débris of its copper mines. These are all cases, you will observe, of degradation ; diminishing majesty in the mountain, and diminishing depth in the valley, or pools of its waters. I cannot name a single spot in which, during my lifetime spent among the mountains, I have seen a peak made grander, a watercourse cut deeper, or a mountain pool made larger and purer.

14. I am almost surprised, myself, as I write these words, at the strength which, on reflection, I am able to give to my assertion. For, even till I began to write these very pages, and was forced to collect my thoughts, I remained under the easily adopted impression, that, at least among soft earthy eminences, the rivers were still cutting out their beds. And it is not so at all. There are indeed banks here and there which they visibly remove; but whatever they sweep down from one side, they sweep up on the other, and extend a promontory of land for every shelf they undermine: and as for those radiating fibrous valleys in the Apennines, and such other hills, which look symmetrically shaped by streams,—they are not lines of trench from below, but lines of wash or slip from above: they are the natural wear and tear of the surface, directed indeed in easiest descent by the bias of the stream, but not dragged

[1] [See Vol. III. p. 540 n.]
[2] [See Vol. I. p. 211.]

down by its grasp. In every one of those ravines the water is being choked up to a higher level; it is not gnawing down to a lower. So that, I repeat, earnestly, their chasms being choked below, and their precipices shattered above, all mountain forms are suffering a deliquescent and corroding change,—not a sculpturesque or anatomizing change. All character is being gradually effaced; all crooked places made straight,—all rough places, plain; and among these various agencies, not of *erosion*, but *corrosion*, none are so distinct as that of the glacier, in filling up, not cutting deeper, the channel it fills; and in rounding and smoothing, but never sculpturing, the rocks over which it passes.

In this fragmentary collection of former work, now patched and darned into serviceableness, I cannot finish my chapters with the ornamental fringes I used to twine for them; nor even say, by any means, all I have in my mind on the matters they treat of: in the present case, however, the reader will find an elucidatory postscript added at the close of the fourth chapter,[1] which he had perhaps better glance over before beginning the third.

[1] [See §§ 15-18, pp. 145-147.

CHAPTER III

OF ICE-CREAM

*(Continuation of Lecture delivered at London Institution, with added
Illustrations from Lectures at Oxford)*

1. THE statement at the close of the last chapter, doubtless
surprising and incredible to many of my readers, must,
before I reinforce it, be explained as referring only to
glaciers visible, at this day, in temperate regions. For of
formerly deep and continuous tropical ice, or of existing
Arctic ice, and their movements, or powers, I know, and
therefore say, nothing.* But of the visible glaciers couched

---

* The following passage, quoted in the *Geological Magazine* for June
of this year,[1] by Mr. Clifton Ward, of Keswick,[2] from a letter of Professor
Sedgwick's, dated May 24th, 1842, is of extreme value; and Mr. Ward's
following comments are most reasonable and just:—

"'No one will, I trust, be so bold as to affirm that an uninterrupted
glacier could ever have extended from Shap Fells to the coast of Holder-
ness, and borne along the blocks of granite through the whole distance,
without any help from the floating power of water. The supposition
involves difficulties tenfold greater than are implied in the phenomenon it
pretends to account for. The glaciers descending through the valleys of
the higher Alps have an enormous transporting power: but there is no
such power in a great sheet of ice expanded over a country without
mountains, and at a nearly dead level.'

"The difficulties involved in the theories of Messrs. Croll, Belt, Goodchild,
and others of the same extreme school,[3] certainly press upon me—and I

---

[1] [New Series, vol. ii. p. 285.]

[2] [The Rev. James Clifton Ward (1843–1880), curate at Keswick, and afterwards Vicar of Rydal; on the staff of the Geological Survey in Yorkshire 1865–1869,
and in the Lake District 1869–1877. For his contributions to *Deucalion*, see below,
pp. 267–271. For other references to Professor Adam Sedgwick (1785–1873), see
below, pp. 243 n., 284, 294, and Vol. VIII. p. xxv.]

[3] [That is, of the extreme glacial school. James Croll (1821–1890), F.R.S.,
Keeper of Maps and Correspondence of the Geological Survey of Scotland. Thomas
Belt (1832–1878), author of geological works on the glacial period. Mr. J. G.
Goodchild, author of numerous geological papers, etc., and Curator of the Collec-
tion of the Geological Survey of Scotland.]

124

upon the visible Alps, two great facts are very clearly ascertainable, which, in my lecture at the London Institution, I asserted in their simplicity, as follows :—

2. The first great fact to be recognized concerning them is that they are *Fluid* bodies. Sluggishly fluid, indeed, but definitely and completely so; and therefore, they do not scramble down, nor tumble down, nor crawl down, nor slip down; but *flow* down. They do not move like leeches, nor like caterpillars, nor like stones, but like, what they are made of, water.

That is the main fact in their state, and progress, on which all their great phenomena depend.

Fact first discovered and proved by Professor James Forbes, of Edinburgh, in the year 1842, to the astonishment of all the glacier theorists of his time ;—fact strenuously denied, disguised, or confusedly and partially apprehended, by all of the glacier theorists of subsequent times, down to our own day ;[1] else there had been no need for me to tell it you again to-night.

3. The second fact of which I have to assure you is partly, I believe, new to geologists, and therefore may be

think I may say also upon others of my colleagues—increasingly, as the country becomes more and more familiar in its features. It is indeed a most startling thought, as one stands upon the eastern borders of the Lake-mountains, to fancy the ice from the Scotch hills stalking boldly across the Solway, marching steadily up the Eden Valley, and persuading some of the ice from Shap to join it on an excursion over Stainmoor, and bring its boulders with it.

"The outlying northern parts of the Lake-district, and the flat country beyond, have indeed been ravaged in many a raid by our Scotch neighbours, but it is a question whether, in glacial times, the Cumbrian mountains and Pennine chain had not strength in their protruding icy arms to keep at a distance the ice proceeding from the district of the southern uplands, the mountains of which are not *superior* in elevation. Let us hope that the careful geological observations which will doubtless be made in the forthcoming *scientific* Arctic Expedition will throw much new light on our past glacial period.

"J. CLIFTON WARD.

" KESWICK, *April 26th*, 1875."

---

[1] [On this subject, see the Introduction, above, pp. xxxiii. *seq.*]

of some farther interest to you because of its novelty, though
I do not myself care a grain of moraine-dust for the new-
ness of things; but rather for their oldness; and wonder
more willingly at what my father and grandfather thought
wonderful (as, for instance, that the sun should rise, or a
seed grow), than at any newly-discovered marvel. Nor do
I know, any more than I care, whether this that I have
to tell you be new or not; but I did not absolutely *know*
it myself, until lately; for though I had ventured with
some boldness to assert it[1] as a consequence of other facts,
I had never been under the bottom of a glacier to look.
But, last summer, I was able to cross the dry bed of a
glacier,[2] which I had seen flowing, two hundred feet deep,
over the same spot, forty years ago. And there I saw,
what before I had suspected, that modern glaciers, like
modern rivers, were not cutting their beds deeper, but
filling them up.[3] These, then, are the two facts I wish
to lay distinctly before you this evening,—first, that glaciers
are fluent; and, secondly, that they are filling up their
beds, not cutting them deeper.[4]

4. (I.) Glaciers are fluent; slowly, like lava, but distinctly.

And now I must ask you not to disturb yourselves, as
I speak, with bye-thoughts about " the theory of regelation."[5]
It is very interesting to know that if you put two pieces
of ice together they will stick together; let good Professor

---

[1] [See above, p. 9, and below, pp. 550 *seq.*]

[2] [The Glacier des Bossons (as Ruskin notes in his copy).]

[3] [Compare W. G. Collingwood's *Limestone Alps of Savoy*, p. 174: "In our
district, quite independently of anything that has been written on the subject,
the fact cannot fail to strike a thoughtful observer, that even Alpine glaciers
during their greatest extension overflowed,—without ploughing away, as M. de
Mortillet thought,—the most friable Tertiary beds; as well as their own *moraine
profonde*. And even in modern small and steep Alpine glaciers, to which Professor
Geikie allows and attributes erosive power, what, as a matter of fact, do we find
underneath them? Where they have recently receded, is it bare, scraped rock?
Not at all; but a heaped-up causeway of dirt and stones. A visit to the Glacier
des Bossons, or that of the Bois, will show this."]

[4] [As Ruskin notes in his copy, "this second head is never entered upon here;
it comes into 'Yewdale and its Streamlets'" (see below, pp. 249 *seq.*); "but
compare § 17, and the whole of ch. ii. § 14."]

[5] [For Ruskin's dismissal of this theory, see below, ch. x. § 17, p. 230.]

Faraday have all the credit of showing us that;[1] and the human race in general, the discredit of not having known so much as that, about the substance they have skated upon, dropped through, and ate any quantity of tons of—these two or three thousand years.

It was left, nevertheless, for Mr. Faraday to show them that two pieces of ice will stick together when they touch —as two pieces of hot glass will. But the capacity of ice for sticking together no more accounts for the making of a glacier, than the capacity of glass for sticking together accounts for the making of a bottle. The mysteries of crystalline vitrification, indeed, present endless entertainment to the scientific inquirer; but by no theory of vitrification can he explain to us how the bottle was made narrow at the neck, or dishonestly vacant at the bottom. Those conditions of it are to be explained only by the study of the centrifugal and moral powers to which it has been submitted.

5. In like manner, I do not doubt but that wonderful phenomena of congelation, regelation, degelation, and gelation pure without preposition, take place whenever a schoolboy makes a snowball; and that miraculously rapid changes in the structure and temperature of the particles accompany the experiment of producing a star with it on an old gentleman's back. But the principal conditions of either operation are still entirely dynamic. To make your snowball hard, you must squeeze it hard; and its expansion on the recipient surface is owing to a lateral diversion of the impelling forces, and not to its regelatic properties.

6. Our first business, then, in studying a glacier, is to consider the mode of its original deposition, and the large forces of pressure and fusion brought to bear on it, with

---

[1] [Faraday had called attention, in a lecture at the Royal Institution in 1850, to the fact that if two pieces of ice, having throughout a temperature of 32° F. and each melting at its surface, are made to touch each other, they will freeze together at the points of contact. The lecture is reported in the *Athenæum*, June 15, 1850, and is referred to at p. xiii. of Forbes's *Occasional Papers*. Tyndall recalled the experiments in a lecture, January 23, 1857 : see, further, the Introduction, above, p. xxxvii.]

their necessary consequences on such a substance as we practically know snow to be,—a powder, ductile by wind, compressible by weight; diminishing by thaw, and hardening by time and frost; a thing which sticks to rough ground, and slips on smooth; which clings to the branch of a tree, and slides on a slated roof.

7. Let us suppose, then, to begin with, a volcanic cone in which the crater has been filled, and the temperature cooled, and which is now exposed to its first season of glacial agencies. Then let Plate XII.,[1] Fig. 1, represent this mountain, with part of the plains at its foot under an equally distributed depth of a first winter's snow, and place the level of perpetual snow at any point you like—for simplicity's sake, I put it half-way up the cone. Below this snow-line, all snow disappears in summer; but above it, the higher we ascend, the more of course we find remaining. It is quite wonderful how few feet in elevation make observable difference in the quantity of snow that will lie. This last winter, in crossing the moors of the peak of Derbyshire,[2] I found, on the higher masses of them, that ascents certainly not greater than that at Harrow from the bottom of the hill to the school-house, made all the difference between easy and difficult travelling, by the change in depth of snow.

8. At the close of the summer, we have then the remnant represented in Fig. 2, on which the snows of the ensuing winter take the form in Fig. 3; and from this greater heap we shall have remaining a greater remnant, which, supposing no wind or other disturbing force modified its form, would appear as at Fig. 4; and, under such necessary modification, together with its own deliquescence, would actually take some such figure as that shown at Fig. 5.

Now, what is there to hinder the continuance of accumulation? If we cover this heap with another layer of winter's snow (Fig. 6), we see at once that the ultimate

---

[1] [For references to Figs. 7–12 on this Plate, see below, pp. 158–160.]
[2] [In a driving tour in January 1875.]

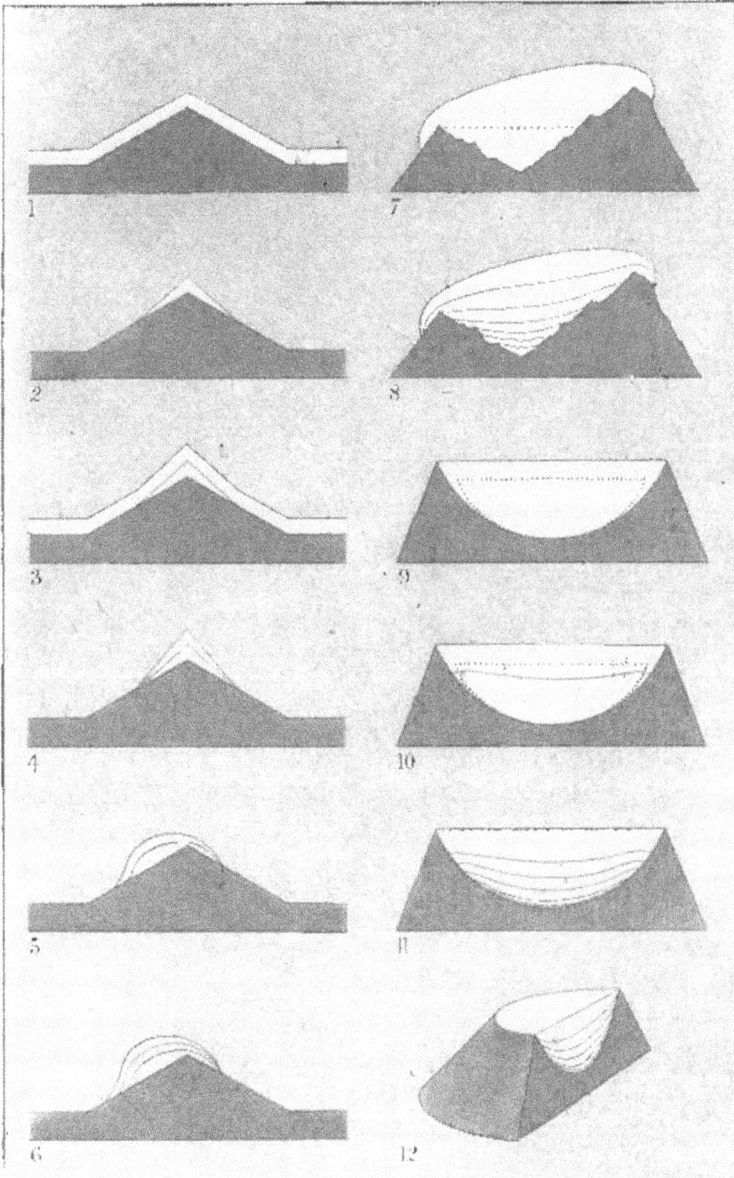

FIRST CONDITIONS OF ACCUMULATION AND FUSION
IN MOTIONLESS SNOW

condition would be, unless somehow prevented, one of enormous mass, superincumbent on the peak—like a colossal haystack, and extending far down its sides below the level of the snow-line.

You are, however, doubtless well aware that no such accumulation as this ever does take place on a mountain-top.

9. So far from it, the eternal snows do not so much as fill the basins between mountain-tops; but, even in these hollows, form depressed sheets at the bottom of them. The difference between the actual aspect of the Alps, and that which they would present if no arrest of the increasing accumulation on them took place, may be shown before you with the greatest ease; and in doing so I have, in all humility, to correct a grave error of my own, which, strangely enough, has remained undetected, or at least un-accused, in spite of all the animosity provoked by my earlier writings.

10. When I wrote the first volume of *Modern Painters*,[1] scarcely any single fact was rightly known by anybody, about either the snow or ice of the Alps. Chiefly the snows had been neglected: very few eyes had ever seen the higher snows near; no foot had trodden the greater number of Alpine summits; and I had to glean what I needed for my pictorial purposes as best I could,—and my best in this case was a blunder. The thing that struck me most, when I saw the Alps myself, was the enormous accumulation of snow on them; and the way it clung to their steep sides. Well, I said to myself, "of course it must be as thick as it can stand; because, as there is an excess which doesn't melt, it would go on building itself up like the Tower of Babel, unless it tumbled off. There must be always, at the end of winter, as much snow on every high summit as it can carry."[2]

There *must*, I said. That is the mathematical method

---

[1] [It was published in 1843.]
[2] [See Vol. III. p. 447.]

of science as opposed to the artistic. Thinking of a thing,
and demonstrating,—instead of looking at it. Very fine,
and very sure, if you happen to have before you all the
elements of thought ; but always very dangerously inferior
to the unpretending method of sight—for people who have
eyes, and can use them. If I had only *looked* at the
snow carefully, I should have seen that it wasn't anywhere as thick as it could stand or lie—or, at least, as a
hard substance, though deposited in powder, could stand.
And then I should have asked myself, with legitimate
rationalism, why it didn't; and if I had but asked——Well,
it's no matter what perhaps might have happened if I had.
I never did.

11. Let me now show you, practically, how great the
error was. Here is a little model of the upper summits
of the Bernese range. I shake over them as much flour
as they will carry ; now I brush it out of the valleys, to
represent the melting. Then you see what is left stands
in these domes and ridges, representing a mass of snow
about six miles deep. That is what the range would be
like, however, if the snow stood up as the flour does ; and
snow is at least, you will admit, as adhesive as flour.

12. But, you will say, the scale is so different, you can't
reason from the thing on that scale. A most true objection. You cannot ; and therefore I beg you, in like manner,
not to suppose that Professor Tyndall's experiments on "a
straight prism of ice, four inches long, an inch wide, and a
little more than an inch in depth,"* are conclusive as to
the modes of glacier motion.

In what respect then, we have to ask, would the difference in scale modify the result of the experiment made
here on the table, supposing this model was the Jungfrau
itself, and the flour supplied by a Cyclopean miller and his
men ?

* *Glaciers of the Alps,* p. 348.[1]

---

[1] [Ruskin's references to this book are to the first edition (1860).]

13. In the first place, the lower beds of a mass six miles deep would be much consolidated by pressure. But would they be *only* consolidated? Would they be in nowise squeezed out at the sides?

The answer depends of course on the nature of flour, and on its conditions of dryness. And you must feel in a moment that, to know what an Alpine range would look like, heaped with any substance whatever, as high as the substance would stand—you must first ascertain how high the given substance *will* stand—on level ground. You might perhaps heap your Alp high with wheat,—not so high with sand,—nothing like so high with dough; and a very thin coating indeed would be the utmost possible result of any quantity whatever of showers of manna, if it had the consistence, as well as the taste, of wafers made with honey.

14. It is evident, then, that our first of inquiries bearing on the matter before us must be, How high will snow stand on level ground, in a block or column? Suppose you were to plank in a square space, securely—twenty feet high—thirty—fifty; and to fill it with dry snow. How high could you get your pillar to stand, when you took away the wooden walls? and when you reached your limit, or approached it, what would happen?

Three more questions instantly propose themselves; namely, What happens to snow under given pressure? will it under some degrees of pressure change into anything else than snow? and what length of time will it take to effect the change?

Hitherto we have spoken of snow as dry only, and therefore as solid substance, permanent in quantity and quality. You know that it very often is not dry; and that, on the Alps, in vast masses, it is throughout great part of the year thawing, and therefore diminishing in quantity.

It matters not the least, to our general inquiry, how much of it is wet, or thawing, or at what times. I

merely at present have to introduce these two conditions
as elements in the business.   It is not dry snow always,
but often soppy snow—snow and water,—that you have to
squeeze.   And it is not freezing snow always, but very
often thawing snow,—diminishing therefore in bulk every
instant,—that you have to squeeze.

It does not matter, I repeat, to our immediate purpose,
when, or how far, these other conditions enter our ground;
but it is best, I think, to put the dots on the i's as we
go along. You have heard it stated, hinted, suggested,
implied, or whatever else you like to call it, again and again,
by the modern school of glacialists, that the discoveries of
James Forbes were anticipated by Rendu.[1]

15. I have myself more respect for Rendu than any
modern glacialist has. He was a man of De Saussure's
temper, and of more than De Saussure's intelligence; and
if he hadn't had the misfortune to be a bishop, would very
certainly have left James Forbes's work a great deal more
than cut out for him ;—stitched—and pretty tightly—in
most of the seams. But he was a bishop; and could only
examine the glaciers to an episcopic extent ;[2] and guess, the
best he could, after that. His guesses are nearly always
splendid; but he must needs sometimes reason as well as
guess; and he reasons himself with beautiful plausibility,
ingenuity, and learning, up to the conclusion—which he
announces as positive—that it always freezes on the Alps,
even in summer.[3] James Forbes was the first who ascertained the fallacy of this episcopal position; and who
announced[4]—to our no small astonishment—that it always
thawed on the Alps, even in winter.

16. Not superficially of course, nor in all places. But

---

[1] [On this subject, see the Introduction ; above, pp. xxxiii. *seq.*]

[2] [In his copy for a revision (which was, however, never carried out) Ruskin
here writes "Correct this sneer at bishops." "Episcopic," it may be noted, is
a coinage of Ruskin's, not included in Dr. Murray's *New English Dictionary.*]

[3] [See ch. ii. of Rendu's *Theory of the Glaciers of Savoy*, p. 24 of the English
translation.]

[4] [In the Sixteenth Letter on Glaciers : see *Occasional Papers on the Theory
of Glaciers, now first collected*, 1859, p. 226.]

internally, and in a great many places. And you will find
it is an ascertained fact—the first great one of which we
owe the discovery to him—that all the year round, you
must reason on the masses of aqueous deposit on the Alps
as, practically, in a state of squash. Not freezing ice or
snow, nor dry ice or snow, but in many places saturated
with,—everywhere affected by,—moisture ; and always subject, in enormous masses, to the conditions of change which
affect ice or snow at the freezing-point, and not below it.
Even James Forbes himself scarcely, I think, felt enough
the importance of this element of his own discoveries, in
all calculations of glacier motion. He sometimes speaks of
his glacier a little too simply, as if it were a stream of *un-
diminishing* substance, as of treacle or tar, moving under
the action of gravity only; and scarcely enough recognizes
the influence of the subsiding languor of its fainting mass,
as a constant source of motion ; though nothing can be
more accurate than his actual account of its results on
the surface of the Mer de Glace, in his fourth letter to
Professor Jameson.[1]

17. Let me drive the notion well home in your own
minds, therefore, before going farther. You may permanently secure it, by an experiment easily made by each
one of you for yourselves this evening, and that also on
the minute and easily tenable scale which is so approved
at the Royal Institution ;[2] for in this particular case the
material conditions may indeed all be represented in very
small compass. Pour a little hot water on a lump of sugar
in your tea-spoon. You will immediately see the mass
thaw, and subside by a series of, in miniature, magnificent and appalling catastrophes, into a miniature glacier,
which you can pour over the edge of your tea-spoon into
your saucer; and if you will then add a little of the
brown sugar of our modern commerce—of a slightly sandy

---

[1] [First published in the *Edinburgh New Philosophical Journal*, January 1843;
pp. 26 *seq.* in the *Occasional Papers.*]
[2] [The reference is to Tyndall's experiment as described above, p. 130.]

character,—you may watch the rate of the flinty erosion upon the soft silver of the tea-spoon at your ease, and with Professor Ramsay's help, calculate the period of time necessary to wear a hole through the bottom of it.

I think it would be only tiresome to you if I carried the inquiry farther by progressive analysis. You will, I believe, permit, or even wish me, rather to state summarily what the facts are:—their proof, and the process of their discovery, you will find incontrovertibly and finally given in this volume, classical, and immortal in scientific literature—which, twenty-five years ago, my good master Dr. Buckland [1] ordered me, in his lecture-room at the Ashmolean, to get,—as closing all question respecting the nature and cause of glacier movement,—James Forbes's *Travels in the Alps*.

18. The entire mass of snow and glacier (the one passing gradually and by infinite modes of transition into the other, over the whole surface of the Alps) is one great accumulation of ice-cream, poured upon the tops, and *flowing* to the bottoms, of the mountains, under precisely the same special condition of gravity and coherence as the melted sugar poured on the top of a bride-cake; but on a scale which induces forms and accidents of course peculiar to frozen water, as distinguished from frozen syrup, and to the scale of Mont Blanc and the Jungfrau, as compared to that of a bride-cake. Instead of an inch thick, the ice-cream of the Alps will stand two hundred feet thick,—no thicker, anywhere, if it can run off; but will lie in the hollows like lakes, and clot and cling about the less abrupt slopes in festooned wreaths of rich mass and sweeping flow, breaking away, where the steepness becomes intolerable, into crisp precipices and glittering cliffs.

19. Yet never for an instant motionless—never for an instant without internal change, through all the gigantic mass, of the relations to each other of every crystal grain.

---

[1] [See the Introduction; above, p. xx.]

That one which you break now from its wave-edge, and which melts in your hand, has had no rest, day nor night, since it faltered down from heaven when you were a babe at the breast; and the white cloud that scarcely veils yonder summit — seven-coloured in the morning sunshine — has strewed it with pearly hoar-frost, which will be on this spot, trodden by the feet of others, in the day when you also will be trodden under feet of men,[1] in your grave.

20. Of the infinite subtlety, the exquisite constancy of this fluid motion, it is nearly impossible to form an idea in the least distinct. We hear that the ice advances two feet in the day; and wonder how such a thing can be possible, unless the mass crushed and ground down everything before it. But think a little. Two feet in the day is a foot in twelve hours,—only an inch in an hour (or say a little more in the daytime, as less in the night),—and that is maximum motion in mid-glacier. If your Geneva watch is an inch across, it is three inches round, and the minute-hand of it moves three times faster than the fastest ice.[2] Fancy the motion of that hand so slow that it must take three hours to get round the little dial. Between the shores of the vast gulf of hills, the long wave of hastening ice only keeps pace with that lingering arrow, in its central crest; and that invisible motion fades away upwards through forty years of slackening stream, to the pure light of dawn on yonder stainless summit, on which this morning's snow lies—motionless.

21. And yet, slow as it is, this infinitesimal rate of current is enough to drain the vastest gorges of the Alps of their snow, as clearly as the sluice of a canal-gate empties a lock. The mountain basin included between the Aiguille Verte, the Grandes Jorasses, and the Mont Blanc, has an area of about thirty square miles, and only one outlet, little more than a quarter of a mile wide: yet, through this

----

[1] [Matthew v. 13.]

[2] [This passage, with ch. vi. § 15, formed originally the conclusion of the lecture at the London Institution: see below, p. 163 n.]

the contents of the entire basin are drained into the Valley
of Chamouni with perfect steadiness, and cannot possibly
fill the basin beyond a certain constant height above the
point of overflow.

Overflow, I say, deliberately; distinguishing always the
motion of this true fluid from that of the sand in an hour-
glass, or of stones slipping in a heap of shale. But that
the nature of this distinction may be entirely conceived by
you, I must ask you to pause with some attention at this
word, to "flow,"—which attention may perhaps be more
prudently asked in a separate chapter.

CHAPTER IV

LABITUR, ET LABETUR [1]

(Lecture given at London Institution, continued, with added Illustrations)

1. OF course—we all know what flowing means. Well, it is to be hoped so; but I'm not sure. Let us see. The sand of the hour-glass,—do you call the motion of that flowing?

No. It is only a consistent and measured fall of many unattached particles.

Or do you call the entrance of a gas through an aperture, out of a full vessel into an empty one, flowing?

No. That is expansion—not flux.

Or the draught through the keyhole? No—is your answer, still. Let us take instance in water itself. The *spring* of a fountain, or of a sea breaker into spray. You don't call that flowing?

No.

Nor the *fall* of a fountain, or of rain?

No.

Well, the *rising* of a breaker,—the current of water in the hollow shell of it,—is *that* flowing? No. After it has broken—rushing up over the shingle, or impatiently advancing on the sand! You begin to pause in your negative.

Drooping back from the shingle then, or ebbing from the sand? Yes; flowing, in some places, certainly, now.

You see how strict and distinct the idea is in our minds. Will you accept—I think you may—this definition of it? Flowing is "the motion of liquid or viscous matter over solid matter, under the action of gravity, without any other impelling force."

---

[1] [Horace: *Epistles*, i. 2, 43; quoted also in *Elements of Prosody*, § 12.]

2. Will you accuse me, in pressing this definition on you, of wasting time in mere philological nicety? Permit me, in the capacity which even the newspapers allow to me,—that of a teacher of expression,[1]—to answer you, as often before now, that philological nicety is philosophical nicety.[2] See the importance of it here. I said a glacier flowed. But it remains a question whether it does not also *spring*,—whether it can rise as a fountain, no less than descend as a stream.

For, broadly, there are two methods in which either a stream or glacier moves.

The first, by withdrawing a part of its mass in front, the vacancy left by which, another part supplies from behind.

That is the method of a continuous stream,—perpetual deduction,* by what precedes, of what follows.

The second method of motion is when the mass that is behind, presses, or is poured in upon, the masses before. That is the way in which a cataract falls into a pool, or a fountain into a basin.

Now, in the first case, you have catenary curves, or else curves of traction, going down the stream. In the second case, you have irregularly concentric curves, and ripples of impulse and compression, succeeding each other round the pool.

3. Now the Mer de Glace is deduced down its narrow channel, like a river; and the Glacier des Bossons is deduced down its deep ravine; and both were once injected into a pool of ice in the valley below, as the Glacier of the Rhone is still. Whereupon, observe, if a stream falls into a basin—level-lipped all round—you know when it runs over it must be pushed over—lifted over. But if ice is thrown into a heap in a plain, you can't tell, without the

---

* "Ex quo illa admirabilis a majoribus aquæ facta deductio est."—*Cic. de Div.*, 1. 44.

[1] [Compare Vol. XXV. p. 14.]
[2] [Compare *Lectures on Art*, § 68 (Vol. XX. p. 74); *Munera Pulveris*, § 101 (Vol. XVII. p. 225 n.), and *The Storm-Cloud of the Nineteenth Century*, § 66.]

closest observation, how violently it is pushed from behind,
or how softly it is diffusing itself in front; and I had never
set my eyes or wits to ascertain where compression in the
mass ceased, and diffusion began, because I thought Forbes
had done everything that had to be done in the matter.
But in going over his work again I find he has left just
one thing to be still explained; and that one chances to be
left to me to show you this evening, because, by a singular
and splendid Nemesis, in the obstinate rejection of Forbes's
former conclusively simple experiments, and in the endeavour
to substitute others of his own, Professor Tyndall has confused himself to the extreme point of not distinguishing
these two conditions of deductive and impulsive flux. His
incapacity of drawing, and ignorance of perspective, prevented him from constructing his diagrams either clearly
enough to show him his own mistakes, or prettily enough
to direct the attention of his friends to them;—and they
luckily remain to us, in their absurd immortality.

4. Forbes poured viscous substance in layers down a
trough; let the stream harden; cut it into as many sections
as were required; and showed, in permanence, the actual
conditions of such viscous motion.[1] Eager to efface the
memory of these conclusive experiments, Professor Tyndall
(*Glaciers of the Alps*, page 383) substituted this literally
"superficial" one of his own. He stamped circles on the
top of a viscous current; found, as it flowed, that they
were drawn into ovals; but had not wit to consider, or
sense to see, whether the area of the circle was enlarged
or diminished—or neither—during its change in shape. He
jumped, like the rawest schoolboy, to the conclusion that
a circle, becoming an oval, must necessarily be compressed!
You don't compress a globe of glass when you blow it into
a soda-water bottle, do you?

5. But to reduce Professor Tyndall's problem into terms.

---

[1] [Forbes described his experiments in the concluding chapter of his *Travels in the Alps of Savoy* (1843); and again in a paper read before the Royal Society (April 10, 1845), "Illustrations of the Viscous Theory of Glacier Motion," printed in the *Philosophical Transactions* for 1846, p. 143, and in the *Occasional Papers*, p. 77.]

Let A F, Fig. 35, below, be the side of a stream of any substance whatever, and $a\ f$ the middle of it; and let the particles at the middle move twice as fast as the particles at the sides. Now we cannot study all the phenomena of fluid motion in one diagram, nor any one phenomenon of fluid motion but by progressive diagrams; and this first one only shows the changes of form which would take place in a substance which moved with *uniform* increase of rapidity from side to centre. No fluid substance *would* so move; but you can only trace the geometrical facts step by step, from uniform increase to accelerated increase. Let the increase of rapidity, therefore, first be supposed uniform. Then, while the point A moves to B, the point a moves to c, and any points once intermediate in a right line between A and a, will now be intermediate in a right line between B and c, and their places determinable by verticals from each to each.

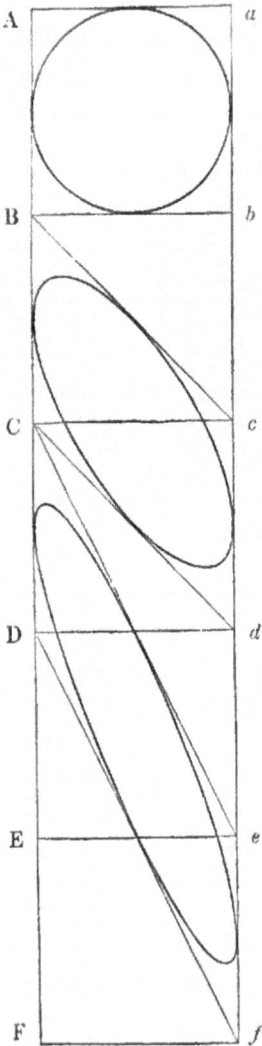

I need not be tedious in farther describing the figure. Suppose A b a square mile of the substance, and the origin of motion on the line A a. Then when the point A has arrived at B, the point B has arrived at C, the point a at c, and the point b at d, and the mile square, A b, has become the mile rhombic, B d, of the same area; and if there were a circle drawn in the square A b, it will become the fat ellipse in B d, and thin ellipse in C f, successively.

6. Compressed, thinks Professor Tyndall, one way, and stretched the other!

But the Professor has never so much as understood what "stretching" means. He thinks that ice won't stretch! Does he suppose treacle, or oil, *will?* The brilliant natural philosopher has actually, all through his two books on glaciers,[1] confused viscosity with elasticity! You can *stretch* a piece of Indian-rubber, but you can only *diffuse* treacle, or oil, or water.

"But you can draw these out into a narrow stream, whereas you cannot pull the ice?"

No; neither can you pull water, can you? In compressing any substance, you can apply any force you like; but in extending it, you can only apply force less than that with which its particles cohere. You can pull honey into a thin string, when it comes out of the comb; let it be candied, and you can't pull it into a thin string. Does that make it less a viscous substance? You can't stretch mortar either. It cracks even in the hod, as it is heaped. Is it, therefore, less fluent or manageable in the mass?

7. Whereas the curious fact of the matter is, that, in precise contrariety to Mr. Tyndall's idea, ice (glacier ice, that is to say) *will* stretch;[2] and that treacle or water won't! and that's just the plague of dealing with the whole glacier question—that the incomprehensible, untenable, indescribable ice will both squeeze and open; and is slipping through your fingers all the time besides, by melting away. You can't deal with it as a simple fluid; and still less as a simple solid. And instead of having less power to accommodate itself to the irregularities of its bed than water, it has much more;—a great deal more of it will subside into a deep place, and ever so much of it melt in passing over a shallow one; and the centre, at whatever rate it moves, will supply itself by the exhaustion of the sides, instead of raging round, like a stream in back-water.

8. However, somehow, I must contrive to deal at least

---

[1] [*Glaciers of the Alps* (1860) and *Forms of Water* (1872).]
[2] [Compare ch. vi. § 15 *n.*, p. 164.]

with the sure fact that the velocity of it is progressively
greater from the sides to the centre, and from the bottom
to the surface.

Now it is the last of these progressive increments which
is of chief importance to my present purpose.

For my own conviction on the matter; — mind, not
theory, for a man can always avoid constructing theories,
but cannot possibly help his conviction, and may sometimes feel it right to state them, — my own conviction is
that the ice, when it is of any considerable depth, no more
moves over the bottom than the lower particles of a running
stream of honey or treacle move over a plate; but that,
in entire rest at the bottom, except so far as it is moved
by dissolution, it increases in velocity to the surface in a
curve of the nature of a parabola, or of a logarithmic
curve, capable of being infinitely prolonged, on the supposition of the depth of the ice increasing to infinity.

9. But it is now my fixed principle not to care what
I think, when a fact can be ascertained by looking, or
measuring. So, not having any observations of my own
on this matter, I seek what help may be had elsewhere;
and find in the eleventh chapter of Professor Tyndall's
Glaciers of the Alps,[1] two most valuable observations, made
under circumstances of considerable danger, calmly encountered by the author, and grumblingly by his guide, —
danger consisting in the exposure to a somewhat close and
well-supported fire of round and grape from the glacier of
the Géant, which objected to having its velocity measured.
But I find the relation of these adventures so much dis-
tract me from the matter in hand, that I must digress

---

[1] [Chapter XI. of Part II. ; pp. 289 *seq*.: "The guide's attention had been divided
between his work and his safety, and he had to retreat more than a dozen times from
the falling boulders and débris. I, on the other hand . . . I took my axe, placed a
stake and an auger against my heart, buttoned my coat upon them, and cut an
oblique staircase up the wall of ice, until I reached a height of forty feet from the
bottom. Here the position of the stake being determined by Mr. Hirst, who was at
the theodolite, I pierced the ice with the auger, drove in the stake, and descended
without injury. During the whole operation, however, my guide growled audibly"
(p. 290).]

briefly into some notice of the general literary structure of this remarkable book.

10. Professor Tyndall never fails to observe with complacency, and to describe to his approving readers, how unclouded the luminous harmonies of his reason, imagination, and fancy remained, under conditions which, he rightly concludes, would have been disagreeably exciting, or even distinctly disturbing, to less courageous persons. And indeed I confess, for my own part, that my successfullest observations have always been made while lying all my length on the softest grass I could find;[1] and after assuring myself with extreme caution that if I chanced to go to sleep (which in the process of very profound observations I usually do, at least of an afternoon), I am in no conceivable peril beyond that of an ant-bite. Nevertheless, the heroic Professor does not, it seems to me, sufficiently recognize the universality of the power of English, French, German, and Italian gentlemen to retain their mental faculties under circumstances even of more serious danger than the crumbling of a glacier moraine; and to think with quickness and precision, when the chances of death preponderate considerably, or even conclusively, over those of life. Nor does Professor Tyndall seem to have observed that the gentlemen possessing this very admirable power in any high degree, do not usually think their own emotions, or absence of emotions, proper subjects of printed history, and public demonstration.

11. Nevertheless, when a national philosopher, under showers of granite grape, places a stake and auger against his heart, buttons his coat upon them, and cuts himself an oblique staircase up a wall of ice, nearly vertical, to a height of forty feet from the bottom; and there, unbuttoning his coat, pierces the ice with his auger, drives in his stake, and descends without injury, though during the whole operation his guide "growls audibly," we are bound to admit his claim to a scientific Victoria Cross—or at

---

[1] [Compare Vol. V. p. 164 and n.]

least crosslet, — and even his right to walk about in our London drawing-rooms in a gracefully cruciferous costume; while I have no doubt also that many of his friends will be interested in such metaphysical particulars and examples of serene mental analysis as he may choose to give them in the course of his autobiography.   But the Professor ought more clearly to understand that scientific writing is one thing, and pleasant autobiography another; and though an officer may not be able to give an account of a battle without involving some statement of his personal share in it, a scientific observer might with entire ease, and much convenience to the public, have published *The Glaciers of the Alps* in two coincident, but not coalescing, branches— like the glaciers of the Giant and Léchaud; and that out of the present inch and a half thickness of the volume, an inch and a quarter might at once have been dedicated to the Giant glacier of the autobiography, and the remaining quarter of an inch to the minor current of scientific ob- servation, which, like the Glacier de Léchaud, appears to be characterized by " the comparative shallowness of the upper portion," * and by its final reduction to " a driblet measur- ing about one-tenth of its former transverse dimensions."

12. It is true that the book is already divided into two portions,—the one described as " chiefly narrative," and the other as " chiefly scientific." The chiefly narrative portion is, indeed, full of very interesting matter fully justifying its title; as, for instance, " We tumbled so often in the soft snow, and our clothes and boots were so full of it, that we thought we might as well try the sitting posture in sliding down. We did so, and descended with extraordinary velocity " (p. 116). Or again: " We had some tea, which had been made at the Montanvert, and carried up to the Grand Mulets in a bottle. My memory of that tea is not pleasant " (p. 73). Or in higher strains of scientific wit and pathos: " As I looked at the objects which had now become

---

* *Glaciers of the Alps,* p. 288.

so familiar to me, I felt that, though not viscous, the ice did not lack the quality of adhesiveness, and I felt a little sad at the prospect of bidding it so soon farewell" [p. 88].

13. But the merely romantic readers of this section, rich though it be in sentiment and adventure, will find them-selves every now and then arrested by pools, as it were, of almost impassable scientific depth—such as the description of a rock "evidently to be regarded as an assemblage of magnets, or as a single magnet full of consequent points" (p. 140). While, on the other hand, when in the course of my own work, finding myself pressed for time, and eager to collect every scrap of ascertained data accessible to me, I turn hopefully to the eleventh chapter of the "chiefly scientific" section of the volume, I think it hard upon me that I must read through three pages of narrative describ-ing the Professor's dangers and address, before I can get at the two observations which are the sum of the scientific contents of the chapter, yet to the first of which "unfortu-nately some uncertainty attached itself" [p. 290], and the second of which is wanting in precisely the two points which would have made it serviceable. First, it does not give the rate of velocity at the base, but five feet above the base; and, secondly, it gives only three measurements of motion. Had it given four, we could have drawn the curve; but we can draw any curve we like through three points.

14. I will try the three points, however, with the most probable curve; but this being a tedious business, will re-serve it for a separate chapter,[1] which readers may skip if they choose: and insert, for the better satisfaction of any who may have been left too doubtful by the abrupt close of my second chapter, this postscript, written the other day after watching the streamlets on the outlying fells of Shap.

15. Think what would be the real result, if any stream among our British hills at this moment *were* cutting its bed deeper.

[1] [See ch. vi. § 1 (below, p. 156).]

In order to do so, it must of course annually be able
to remove the entire zone of débris moved down to its
bed from the hills on each side of it—and somewhat more.

Take any Yorkshire or Highland stream you happen to
know, for example; and think what quantity of débris must
be annually moved, on the hill surfaces which feed its
waters. Remember that a lamb cannot skip on their slopes,
but it stirs with its hoofs some stone or grain of dust which
will more or less roll or move downwards. That no shower
of rain can fall—no wreath of snow melt, without moving
some quantity of dust downwards. And that no frost can
break up, without materially loosening some vast ledges
of crag, and innumerable minor ones; nor without causing
the fall of others as vast, or as innumerable. Make now
some effort to conceive the quantity of rock and dust
moved annually, lower, past any given level traced on the
flanks of any considerable mountain stream, over the area
it drains—say, for example, in the basin of the Ken above
Kendal, or of the Wharfe above Bolton Abbey.

16. Then, if either of those streams were cutting their
beds deeper,—that quantity of rock, and something more,
must be annually carried down by their force, past Kendal
bridge, and Bolton stepping-stones. Which you will find
would occasion phenomena very astonishing indeed to the
good people of Kendal and Wharfedale.

17. "But it need not be carried down past the stepping-
stones," you say—"it may be deposited somewhere above."
Yes, that is precisely so ;—and wherever it *is* deposited, the
bed of the stream, or of some tributary streamlet, is being
raised. Nobody notices the raising of it ;—another stone
or two among the wide shingle—a tongue of sand an inch
or two broader at the burnside—who can notice that?
Four or five years pass ;—a flood comes ;—and Farmer So-
and-So's field is covered with slimy ruin. And Farmer
So-and-So's field is an inch higher than it was, for evermore—but who notices that? The shingly stream has gone
back into its bed : here and there a whiter stone or two

gleams among its pebbles, but next year the water stain
has darkened them like the rest, and the bed is just as far
below the level of the field as it was. And your careless
geologist says, "What a powerful stream it is, and how
deeply it is cutting its bed through the glen!"

18. Now, carry out this principle for existing glaciers.
If the glaciers of Chamouni were cutting their beds deeper,
either the annual line of débris of the Mont Blanc range
on its north side must be annually carried down past the
Pont Pelissier; or the Valley of Chamouni must be in
process of filling up, while the ravines at its sides are being
cut down deeper. Will any geologist, supporting the modern
glacial theories, venture to send me, for the next number
of *Deucalion*, his idea, on this latter, by him inevitable,
hypothesis, of the profile of the bottom of the Glacier
des Bossons, a thousand years ago; and a thousand years
hence ?

CHAPTER V

THE VALLEY OF CLUSE

1. WHAT strength of faith men have in each other; and how impossible it is for them to be independent in thought, however hard they try! Not that they ever ought to be; but they should know, better than they do, the incumbrance that the false notions of others is to them.

Touching this matter of glacial grinding action; you will find every recent writer taking up, without so much as a thought of questioning it, the notion adopted at first careless sight of a glacier stream by some dull predecessor of all practical investigation—that the milky colour of it is all produced by dust ground off the rocks at the bottom. And it never seems to occur to any one of the Alpine Club men, who are boasting perpetually of their dangers from falling stones; nor even to professors impeded in their most important observations by steady fire of granite grape,[1] that falling stones may probably knock their edges off when they strike; and that moving banks and fields of moraine, leagues long, and leagues square, of which every stone is shifted a foot forward every day on a surface melting beneath them, must in such shifting be liable to attrition enough to produce considerably more dust, and that of the finest kind, than any glacier stream carries down with it[2]— not to speak of processes of decomposition accelerated, on all services liable to them, by alternate action of frost and fierce sunshine.

2. But I have not, as yet, seen any attempts to determine

---

[1] [See above, p. 142.]

[2] [In his copy for revision Ruskin here notes the implied argument: "*therefore* the bed must be filling up."]

even the first data on which the question of attrition must be dealt with. I put it, in simplicity, at the close of last chapter. But, in its full extent, the inquiry ought not to be made merely of the bed of the Glacier des Bossons; but of the bed of the Arve, from the Col de Balme to Geneva; in which the really important points for study are the action of its waters at Pont Pelissier;—at the falls below Servoz;—at the portal of Cluse;—and at the northern end of the slope of the Salève.

3. For these four points are the places where, if at all, sculptural action is really going on upon its bed: at those points, if at all, the power of the Second Æra,[1] the æra of sculpture, is still prolonged into this human day of ours. As also it is at the rapids and falls of all swiftly descending rivers. The one vulgar and vast deception of Niagara[2] has blinded the entire race of modern geologists to the primal truth of mountain form, namely, that the rapids and cascades of their streams indicate, not points to which the falls have receded, but places where the remains of once colossal cataracts still exist, at the places eternally (in human experience) appointed for the formation of such cataracts, by the form and hardness of the local rocks. The rapids of the Amazon, the Nile, and the Rhine, obey precisely the same law as the little Wharfe at its Strid, or as the narrow "rivus aquæ"[3] which, under a bank of strawberries in my own tiny garden, has given me perpetual trouble to clear its channel of the stones brought down in flood, while, just above, its place of picturesque cascade is determined for it by a harder bed of Coniston flags, and the little pool, below that cascade, never encumbered with stones at all.

4. Now the bed of the Arve, from the crest of the Col de Balme to Geneva, has a fall of about 5000 feet; and if any young Oxford member of the Alpine Club is minded

---

[1] [See above, p. 118.]
[2] [Compare below, pp. 254, 370.]
[3] [From Horace's description of his little farm: *Odes*, iii. 16, 29.]

to do a piece of work this vacation, which in his old age, when he comes to take stock of himself, and edit the fragments of himself, as I am now sorrowfully doing, he will be glad to have done (even though he risked neither his own nor any one else's life to do it), let him survey that bed accurately, and give a profile of it, with the places and natures of emergent rocks, and the ascertainable depths and dates of alluvium cut through, or in course of deposition.

5. After doing this piece of work carefully, he will probably find some valuable ideas in his head concerning the proportion of the existing stream of the Arve to that which once flowed from the glacier which deposited the moraine of Les Tines; and again, of that torrent to the infinitely vaster one of the glacier that deposited the great moraine of St. Gervais; and finally of both, to the cliffs of Cluse, which have despised and resisted them. And ideas which, after good practical work, he finds in his head, are likely to be good for something : but he must not seek for them ; all thoughts worth having, come like sunshine, whether we will or no :[1] the thoughts not worth having are the little lucifer matches we strike ourselves.

6. And I hasten the publication of this number of *Deucalion*,[2] to advise any reader who cares for the dreary counsel of an old-fashioned Alpine traveller, to see the Valley of Cluse this autumn, if he may, rather than any other scene among the Alps ;—for if not already destroyed, it must be so, in a few months more, by the railway which is to be constructed through it, for the transport of European human diluvium. The following note of my last walk there,[3] written for my autumn lectures, may be worth preserving among the shingle of my scattered work.

7. I had been, for six months in Italy, never for a single moment quit of liability to interruption of thought.

---

[1] [Compare *Ethics of the Dust*, § 46 (Vol. XVIII. p. 259).]

[2] [Part II. ; issued in October 1875. For the construction of the railway, see Vol. XVIII. p. 25 *n*.]

[3] [In October 1874 : see the passages from Ruskin's diary given in Vol. XXIII. p. li.]

XIIA

THE VALLEY OF CLUSE

By day or night, whenever I was awake, in the streets of
every city, there were entirely monstrous and inhuman noises
in perpetual recurrence. The violent rattle of carriages,[1]
driven habitually in brutal and senseless haste, or creaking
and thundering under loads too great for their cattle, urged
on by perpetual roars and shouts: wild bellowing and howling of obscene wretches far into the night: clashing of
church bells, in the morning, dashed into reckless discord,
from twenty towers at once, as if rung by devils to defy
and destroy the quiet of God's sky, and mock the laws of
His harmony: filthy, stridulous shrieks and squeaks, reaching for miles into the quiet air, from the railroad stations
at every gate: and the vociferation, endless, and frantic, of
a passing populace whose every word was in mean passion,
or in unclean jest. Living in the midst of this, and of
vulgar sights more horrible than the sounds, for six months,
I found myself—suddenly, as in a dream—walking again
alone through the Valley of Cluse, unchanged since I knew
it first, when I was a boy of fifteen, quite forty years ago;
—and in perfect quiet, and with the priceless completion
of quiet, that I had no fear of any outcry or other base
disturbance of it.[2]

8. But presently, as I walked, the calm was deepened,
instead of interrupted, by a murmur—first low, as of bees,
and then rising into distinct harmonious chime of deep
bells, ringing in true cadences—but I could not tell where.
The cliffs on each side of the Valley of Cluse vary from
1500 to above 2000 feet in height; and, without absolutely echoing the chime, they so accepted, prolonged, and
diffused it, that at first I thought it came from a village
high up and far away among the hills; then presently it
came down to me as if from above the cliff under which I
was walking; then I turned about and stood still, wondering; for the whole valley was filled with the sweet sound,

---

[1] [Ruskin in his copy notes this passage as applying to Florence.]

[2] [The last words are here altered in accordance with Ruskin's marking in his copy.
This is a good instance of his care in revision, the earlier reading being, "that I
was without fear of any outcry or base disturbance of it."]

entirely without local or conceivable origin: and only after some twenty minutes' walk, the depth of tones, gradually increasing, showed me that they came from the tower of Maglans in front of me; but when I actually got into the village, the cliffs on the other side so took up the ringing, that I again thought for some moments I was wrong.

Perfectly beautiful, all the while, the sound, and exquisitely varied,—from ancient bells of perfect tone and series, rung with decent and joyful art.

"What are the bells ringing so to-day for,—it is no fête?" I asked of a woman who stood watching at a garden gate.

"For a baptism, sir."

And so I went on, and heard them fading back, and lost among the same bewildering answers of the mountain air.

9. Now that half-hour's walk was to me, and I think would have been to every man of ordinarily well-trained human and Christian feeling—I do not say merely worth the whole six months of my previous journey in Italy;— it was a reward for the endurance and horror of the six months' previous journey; but, as many here may not know what the place itself is like, and may think I am making too much of a little pleasant bell-ringing, I must tell you what the Valley of Cluse is in itself.

10. Of "Cluse," the closed valley,—not a ravine, but a winding plain, between very great mountains, rising for the most part in cliffs—but cliffs which retire one behind the other above slopes of pasture and forest. (Now as I am writing this passage in a country parsonage—of Cowley, near Uxbridge,[1]—I am first stopped by a railroad whistle two minutes and a half long,* and then by the rumble and grind of a slow train, which prevents me from hearing my own words, or being able to think, so that I must simply wait for ten minutes, till it is past.)

* Counted by watch, for I knew by its manner it would last, and measured it.

[1] [The diary shows a visit to his friends the Hilliards at Cowley in June 1875.]

It being past, I can go on. Slopes of pasture and forest, I said, mingled with arable land, in a way which you can only at present see in Savoy; that is to say, you have walnut and fruit trees of great age, mixed with oak, beech, and pine, as they all choose to grow—it seems as if the fruit trees planted themselves as freely as the pines. I imagine this to be the consequence of a cultivation of very ancient date under entirely natural laws; if a plum-tree or a walnut planted itself, it was allowed to grow; if it came in the way of anything or anybody, it would be cut down; but on the whole the trees grew as they liked; and the fields were cultivated round them in such spaces as the rocks left;—ploughed, where the level admitted, with a ploughshare lightly constructed, but so huge that it looks more like the beak of a trireme than a plough, two oxen forcing it to heave aside at least two feet depth of the light earth;—no fences anywhere; winding field walks, or rock paths, from cottage to cottage; these last not of the luxurious or trim Bernese type, nor yet comfortless châlets; but sufficient for orderly and virtuous life: in outer aspect, beautiful exceedingly, just because their steep roofs, white walls, and wandering vines had no pretence to perfectness, but were wild as their hills. All this pastoral country lapped into inlets among the cliffs, vast belts of larch and pine cresting or crowding the higher ranges, whose green meadows change as they rise, into mossy slopes, and fade away at last among the grey ridges of rock that are soonest silvered with autumnal snow.

11. The ten-miles' length of this valley, between Cluse and St. Martin's, include more scenes of pastoral beauty and mountain power than all the poets of the world have imagined;[1] and present more decisive and trenchant questions respecting mountain structure than all the philosophers of the world could answer: yet the only object which occupies

[1] [The valley, says Ruskin elsewhere, is "worth many Chamounis" (Vol. XXII. p. 69). For other references to it, see Vol. VI. pp. 183, 301; and *Præterita*, i. § 194, ii. § 214.]

the mind of the European travelling public respecting it, is to get through it, if possible, under the hour.

12. I spoke with sorrow, deeper than my words attempted to express, in my first Lecture,[1] of the blind rushing of our best youth through the noblest scenery of the Alps, without once glancing at it, that they might amuse, or kill, themselves on their snow. That the claims of all sweet pastoral beauty, of all pious domestic life, for a moment's pause of admiration or sympathy, should be unfelt, in the zest and sparkle of boy's vanity in summer play, may be natural[2] at all times; and inevitable while our youth remain ignorant of art, and defiant of religion; but that, in the present state of science, when every eye is busied with the fires in the Moon and the shadows in the Sun, no eye should occupy itself with the ravines of its own world, nor with the shadows which the sun casts on the cliffs of them; that the simplest,—I do not say problems, but bare facts, of structure,—should still be unrepresented, and the utmost difficulties of rock history untouched; while dispute, and babble, idler than the chafed pebbles of the wavering[3] beach, clink, jar, and jangle on from year to year in vain,—surely this, in our great University, I am bound to declare to be blameful, and to ask you, with more than an artist's wonder, why this fair Valley of Cluse is now closed indeed, and forsaken, "clasped like a missal where swart Paynims pray;"[4] and, with all an honest inquirer's indignation, to challenge— in the presence of our Master of Geology, happily one of its faithful and true teachers,* the Speakers concerning the

---

* Mr. Prestwich. I have to acknowledge, with too late and vain gratitude, the kindness and constancy of the assistance given me, on all occasions, when I asked it, by his lamented predecessor in the Oxford Professorship of Geology, Mr. Phillips.[5]

---

[1] [See above, p. 103.]
[2] [In his copy Ruskin here writes, "basely natural only."]
[3] [In his copy Ruskin thus explains the epithet : "beach advancing or receding with storm."]
[4] [Keats, *Eve of St. Agnes*; quoted also in Vol. XVII. p. 258.]
[5] [Sir Joseph Prestwich (1812–1896), Professor of Geology at Oxford, 1874–1888. For John Phillips (1800–1874), see Vol. XXII. p. 232; and below, pp. 275, 278, 286.]

Earth,—the geologists, not of England only, but of Europe and America,—either to explain to you the structure or sculpture of this * renownedest cliff in all the Alps, under which Tell leaped ashore;[1] or to assign valid reason for the veins in the pebbles which every Scotch lassie wears for her common jewellery.[2]

---

* The cliff between Fluelen and Brunnen, on the Lake of Uri, of which Turner's drawing[3] was exhibited at this lecture.

---

[1] [For an intended passage in which Ruskin began to explain his challenge here, see below, p. 368.]

[2] [Compare *Lectures on Art*, § 108 (Vol. XX. p. 102); and the closing words of the first volume of *Deucalion* (below, p. 292).]

[3] [No. 70 in Ruskin's Exhibition of 1878 : see Vol. XIII. p. 459, and Plate XXIV.]

CHAPTER VI

OF BUTTER AND HONEY [1]

1. THE last chapter, being properly only a continuation of the postscript to the fourth, has delayed me so long from my question as to ice-curves, that I cannot get room for the needful diagrams and text in this number: [2] which is perhaps fortunate, for I believe it will be better first to explain to the reader more fully why the ascertainment of this curve of vertical motion is so desirable.

To which explanation, very clear definition of some carelessly used terms will be essential.

2. The extremely scientific Professor Tyndall always uses the terms Plastic, and Viscous, as if they were synonymous. But they express entirely different conditions of matter. The first is the term proper to be used of the state of butter, on which you can stamp whatever you choose; and the stamp will stay; the second expresses that of honey, on which you can indeed stamp what you choose; but the stamp melts away forthwith.

And of viscosity itself there are two distinct varieties—one glutinous, or gelatinous, like that of treacle or tapioca soup; and the other simply adhesive, like that of mercury or melted lead.

And of both plasticity and viscosity there are infinitely various degrees in different substances, from the perfect and absolute plasticity of gold, to the fragile, and imperfect, but to man more precious than any quantity of gold,

---

[1] [This chapter also embodies passages from the lecture at the London Institution : see below, p. 163 n.]

[2] [Nor did he afterwards include them in any later number of *Deucalion*. There are diagrams and text, dealing with the subject, in the MS. material ; but they are not sufficiently in form for publication.]

plasticity of clay, and, most precious of all, the blunt and
dull plasticity of dough; and again, from the vigorous and
binding viscosity of stiff glue, to the softening viscosity of
oil, and tender viscosity of old wine. I am obliged therefore to ask my readers to learn, and observe very carefully
in our future work, these following definitions.

Plastic.—Capable of change of form under external force,
without any loss of continuity of substance; and of *retaining afterwards the form imposed on it.*

Gold is the most perfectly plastic substance we commonly know; clay, butter, etc., being more coarsely and
ruggedly plastic, and only in certain consistencies or at certain temperatures.

Viscous.—Capable of change of form under external
force, *but not of retaining the form imposed;* being languidly
obedient to the force of gravity, and necessarily declining
to the lowest possible level,—as lava, treacle, or honey.

Ductile.—Capable of being extended by traction without
loss of continuity of substance. Gold is both plastic and
ductile; but clay, plastic only, not ductile; while most
melted metals are ductile only, but not plastic.

Malleable.—Plastic only under considerable force.

3. We must never let any of these words entangle, as
necessary, the idea belonging to another.

A plastic substance is not necessarily ductile, though
gold is both; a viscous substance is not necessarily ductile,
though treacle is both; and the quality of elasticity, though
practically inconsistent with the character either of a plastic
body, or a viscous one, may enter both the one and the
other as a gradually superadded or interferent condition,
in certain states of congelation; as in Indian-rubber, glass,
sealing-wax, asphalt, or basalt.

I think the number of substances I have named in this
last sentence, and the number of entirely different states
which in an instant will suggest themselves to you, as
characteristic of each, at, and above, its freezing or solidify-
ing point, may show at once how careful we should be in

defining the notion attached to the words we use; and how inadequate, without specific limitation and qualification, *any* word must be, to express all the qualities of any given substance.

4. But, above all substances that can be proposed for definition of quality, glacier ice is the most defeating. For it is practically plastic; but *actually* viscous;—and that to the full extent. You can beat or hammer it, like gold; and it will stay in the form you have beaten it into, for a time;—and so long a time, that, on all instant occasions of plasticity, it is practically plastic. But only have patience to wait long enough, and it will run down out of the form you have stamped on it, as honey does, so that, actually and inherently, it is viscous, and not plastic.

5. Here then, at last, I have got Forbes's discovery and assertion put into accurately intelligible terms;—very incredible terms, I doubt not, to most readers.

There is not the smallest hurry, however, needful in believing them: only let us understand clearly what it is we either believe or deny; and in the meantime, return to our progressive conditions of snow on the simplest supposable terms, as shown in my first plate.

6. On a conical mountain, such as that represented in Fig. 6 (Plate XII.), we are embarrassed by having to calculate the subtraction by avalanche down the slopes. Let us therefore take rather, for examination, a place where the snow can lie quiet.

Let Fig. 7 (Plate XII.), represent a hollow in rocks at the summit of a mountain above the line of perpetual snow, the lowest watershed being at the level indicated by the dotted line. Then the snow, once fallen in this hollow, can't get out again; but a little of it is taken away every year, partly by the heat of the ground below, partly by surface sunshine and evaporation, partly by filtration of water from above, while it is also saturated with water in thaw-time, up to the level of watershed. Consequently it must subside every year in the middle; and, as the mass

remains unchanged, the same quantity must be added every
year at the top,—the excess being always, of course, blown
away, or dropped off, or thawed above, in the year it falls.

7. Hence the entire mass will be composed, at any
given time, of a series of beds somewhat in the arrangement given in Fig. 8 [Plate XII.]; more remaining of each
year's snow in proportion to its youth, and very little
indeed of the lowest and oldest bed.

It *must* subside, I say, every year;—but how much is
involved, of new condition, in saying this! Take the
question in the simplest possible terms; and let Fig. 9
[Plate XII.] represent a cup or crater full of snow, level in
its surface at the end of winter. During the summer, there
will be large superficial melting; considerable lateral melting
by reverberation from rock, and lateral drainage; bottom
melting from ground heat, not more than a quarter of an
inch,—(Forbes's *Travels*, page 364),[1]—a quantity which we
may practically ignore. Thus the mass, supposing the substance of it immovable in position, would be reduced by
superficial melting during the year to the form approximately
traced by the dotted line within it, in Fig. 9.

8. But how of the *interior* melting? Every interstice
and fissure in the snow, during summer, is filled either
with warm air, or warm water in circulation through it,
and every separate surface of crystal is undergoing its own
degree of diminution. And a constant change in the conditions of equilibrium results on every particle of the mass;
and a constant subsidence takes place, involving an entirely
different relative position of every portion of it at the end
of the year.

9. But I cannot, under any simple geometrical figure,
give an approximation to the resultant directions of change
in form; because the density of the snow must be in some
degree proportioned to the depth, and the melting less, in
proportion to the density.

---

[1] [Chapter xxi.; p. 365 in the reprint of 1900.]

Only at all events, towards the close of the year, the mass enclosed by the dotted line in Fig. 9 will have sunk into some accommodation of itself to the hollow bottom of the crater, as represented by the continuous line in Fig. 10. And, over that, the next winter will again heap the snow to the cup-brim, to be reduced in the following summer; but now through two different states of consistence, to the bulk limited by the dotted line in Fig. 10.

10. In a sequence of six years, therefore, we shall have a series of beds approximately such as in Fig. 11;—approximately observe, I say always, being myself wholly unable to deal with the complexities of the question, and only giving the diagram for simplest basis of future investigation, by the first man of mathematical knowledge and practical common-sense, who will leave off labouring for the contradiction of his neighbours, and apply himself to the hitherto despised toil of the ascertainment of facts. And when he has determined what the positions of the strata will be in a perfectly uniform cup, such as that of which the half is represented in perspective in Fig. 12, let him next inquire what would have happened to the mass, if, instead of being deposited in a cup enclosed on all sides, it had been deposited in an amphitheatre open on one, as in the section shown in Fig. 12. For that is indeed the first radical problem to be determined respecting glacier motion.

Difficult enough, if approached even with a clear head, and open heart; acceptant of all help from former observers, and of all hints from nature and heaven; but very totally insoluble, when approached by men whose poor capacities for original thought are unsteadied by conceit, and paralysed by envy.

11. In my next plate [XIII.], I have given, side by side, a reduction, to half-scale, of part of Forbes's exquisite chart of the Mer de Glace, published in 1845,[1] from his own

---

[1] [The chart was issued in a revised form in the second edition (1845) of Forbes's *Travels.*]

XIII

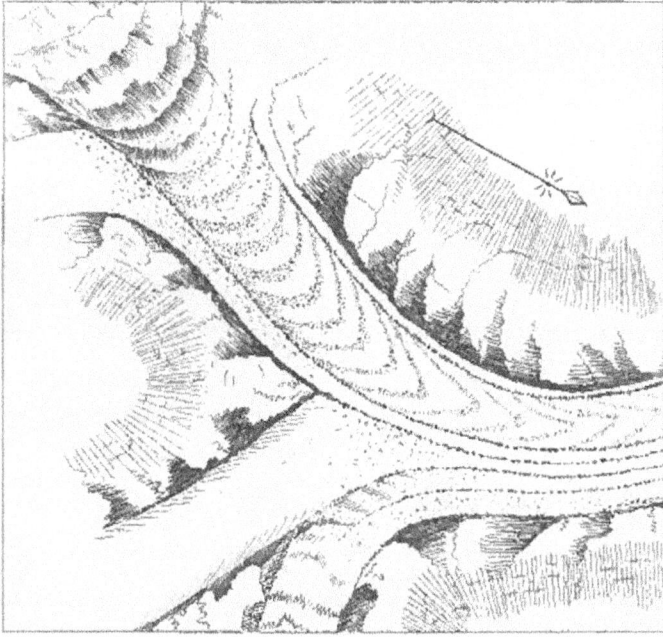

Tyndall 1860.

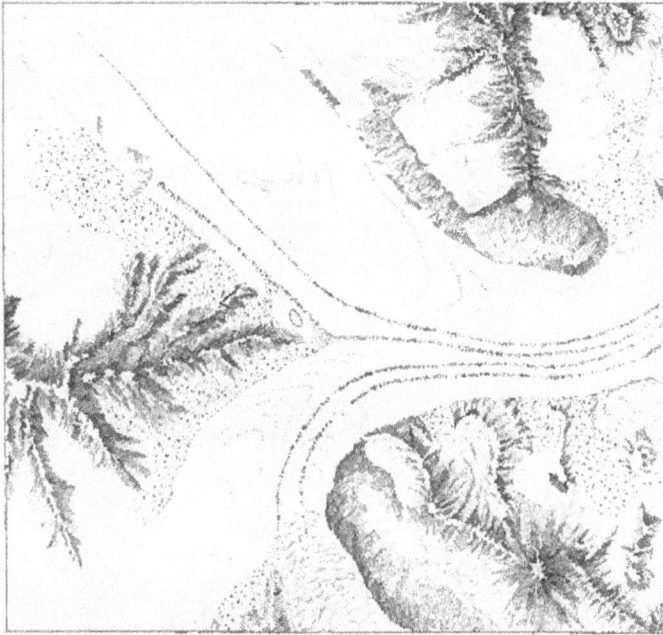

Forbes 1845.

THE PROGRESS OF MODERN SCIENCE
IN GLACIER SURVEY

survey made in 1842 ; and a reproduction, approximately in facsimile, of Professor Tyndall's woodcut, from his own "eye-sketch" of the same portion of the glacier "as seen from the cleft station, Trélaporte," published in 1860.*

That Professor Tyndall is unable to draw anything as seen from anywhere, I observe to be a matter of much self-congratulation to him; such inability serving farther to establish the sense of his proud position as a man of science, above us poor artists, who labour under the dis- advantage of being able with some accuracy to see, and with some fidelity to represent, what we wish to talk about. But when he found himself so resplendently in- artistic, in the eye-sketch in question, that the expression of his scientific vision became, for less scientific persons, only a very bad map, it was at least incumbent on his Royally- social Eminence to ascertain whether any better map of the same places had been published before. And it is indeed clear, in other places of his book, that he was conscious of the existence of Forbes's chart; but did not care to refer to it on this occasion, because it contained the correction of a mistake made by Forbes in 1842, which Professor Tyndall wanted, himself, to have the credit of correcting;[1] leaving the public at the same time to suppose it had never been corrected by its author.

12. This manner, and temper, of reticence, with its relative personal loquacity, is not one in which noble science can be advanced ; or in which even petty science can be increased. Had Professor Tyndall, instead of seek- ing renown by the exposition of Forbes's few and minute mistakes, availed himself modestly of Forbes's many and great discoveries, ten years of arrest by futile discussion and foolish speculation might have been avoided in the

---

* *Glaciers of the Alps*, p. 369. Observe also that my engraving, in consequence of the reduced scale, is grievously inferior to Forbes's work ; but quite effectually and satisfactorily reproduces Professor Tyndall's, of the same size as the original.

---

[1] [See *Glaciers of the Alps*, p. 370.]

annals of geology; and assuredly it would not have been left for a despised artist to point out to you, this evening, the one circumstance of importance in glacier structure which Forbes has not explained.

13. You may perhaps have heard I have been founding my artistic instructions lately on the delineation of a jam-pot.[1] Delighted by the appearance of that instructive object, in the Hôtel du Mont Blanc, at St. Martin's, full of Chamouni honey, of last year, stiff and white, I found it also gave me command of the best possible material for examination of glacial action on a small scale.[2]

Pouring a little of its candied contents out upon my plate, by various tilting of which I could obtain any rate of motion I wished to observe in the viscous stream; and encumbering the sides and centre of the said stream with magnificent moraines composed of crumbs of toast, I was able, looking alternately to table and window, to compare the visible motion of the mellifluous glacier, and its transported toast, with a less traceable, but equally constant, motion of the glacier of Bionnassay, and its transported granite. And I thus arrived at the perception of the condition of glacial structure, which though, as I told you just now,[3] not, I believe, hitherto illustrated, it is entirely in your power to illustrate for yourselves in the following manner.

If you will open a fresh pot of honey to-morrow at breakfast, and take out a good table-spoonful of it, you will see, of course, the surface generally ebb in the pot. Put the table-spoonful back in a lump at one side, and you will see the surface generally flow in the pot. The lump you have put on at the side does not diffuse itself over the rest; but it sinks into the rest, and the entire surface rises round it, to its former level.

---

[1] [See the "Instructions in Use of Rudimentary Series," Vol. XXI. pp. 257, 260.]

[2] [See Vol. XXIII. pp. lii., liii., for Ruskin's sojourn at St. Martin in October 1874. Ruskin, as appears from a note in his copy, repeated and confirmed this experiment at "Sallenches, 14th September 1882." He compares ch. xii. § 28 (below, p. 258).]

[3] [See above, p. 139.]

Precisely in like manner, every pound of snow you put on the top of Mont Blanc, eventually makes the surface of the glaciers rise at the bottom.*

14. That is not impulsive action, mind you. That is mere and pure viscous action—the communication of force equally in every direction among slowly moving particles. I once thought that this force might also be partially elastic, so that whereas, however vast a mass of honey you had to deal with,—a Niagara of honey,—you never could get it to leap like a sea-wave at rocks, ice might yet, in its fluency, retain this power of leaping; only slowly,— taking a long time to rise, yet obeying the same mathematic law of impulse as a sea-breaker; but ascending through æras of surge, and communicating, through æras, its recoil. The little ripple of the stream breaks on the shore, —quick, quick, quick. The Atlantic wave slowly uplifts itself to its plunge, and slowly appeases its thunder. The ice wave—if there be one—would be to the Atlantic wave, as the ocean is to the brook.[1]

If there be one! The question is of immense—of vital importance, to that of glacier action on crag: but, before attacking it, we need to know what the lines of motion are,—first, in a subsiding table-spoonful of honey; secondly,

---

* Practically hyperbolic expression, but mathematically true.

---

[1] [This passage, with ch. iii. § 20, formed originally the conclusion of the lecture at the London Institution. The report of it in the *Times* (March 15, 1875) is as follows:—

"It is worth making an effort to conceive the real periods of time involved in glacier motion. At the centre of any great glacier stream, low down where the ice is softest and moves fastest, its *maximum* average rate will not pass two feet in the twenty-four hours. The minute hand of a watch moves five or six times as fast. If therefore your ice-breaker rises at all, it rises through years of surge, returns through years of recoil. The little ripple of the stream breaks on the shore with quick repetition; the Atlantic wave slowly uplifts itself to plunge, and slowly appeases its thunder; the ice-wave, if there be one, would be to the Atlantic as the ocean's to the brook's. The question if there be such a wave is of vital importance to that of glacier action on crag. Mr. Ruskin added that he would give a definition of a glacier which he requested his audience to take note of: It is a tide which takes a year to rise, a cataract which takes fifty to fall, a torrent that is ribbed like a dragon, and a rock that is diffused like a lake."]

in an uprearing Atlantic wave; and, thirdly, in the pulsatory festoons of a descending cataract, obtained by the *relaxation* of its mass, while the same pulsatory action is displayed, as unaccountably, by a glacier cataract* in the *compression* of its mass.

And, on applying to learned men in Oxford and Cambridge † for elucidation of these modes of motion, I find that, while they can tell me everything I don't want to know, about the collision and destruction of planets, they are not entirely clear on the subject either of the diffusion of a drop of honey from its comb, or the confusion of a rivulet among its cresses. Of which difficult matters, I will therefore reserve inquiry to another chapter;[1] anticipating, however, its conclusions, for the reader's better convenience, by the brief statement, that glacier ice has no power of springing whatever;—that it cannot descend into a rock-hollow, and sweep out the bottom of it, as a cascade or a wave can; but must always sluggishly fill it to the brim before flowing over; and accumulate, beneath, under dead ice, quiet as the depths of a mountain tarn, the fallen ruins of its colossal shore.

---

* Or a stick of sealing-wax.[2] Warm one at the fire slowly through; and bend it into the form of a horseshoe. You will then see, through a lens of moderate power, the most exquisite facsimiles of glacier fissure produced by extension, on its convex surface, and as faithful image of glacier surge produced by compression, on its concave one.

In the course of such extension, the substance of the ice is actually expanded (see above, Chap. IV., § 7) by the widening of every minute fissure; and in the course of such compression, reduced to apparently solid ice, by their closing. The experiments both of Forbes and Agassiz appear to indicate that the original fissures are never wholly effaced by compression; but I do not myself know how far the supposed result of these experiments may be consistent with ascertained phenomena of regelation.

† I have received opportune and kind help, from the other side of the Atlantic waves, in a study of them by my friend Professor Rood.[3]

---

[1] [Ruskin here notes as a thing to be done on revision, "give more sure evidence" for the statement that "ice has *no* springing power." There are notes on the subject in the MS. material for *Deucalion*, but they are not sufficiently in form for publication.]

[2] [Compare below, p. 284.]

[3] [Ogden Nicholas Rood (1831–1902), for many years professor of physics in Columbia University. The study here referred to does not seem to have been published.]

CHAPTER VII

THE IRIS OF THE EARTH[1]

(Lecture given at the London Institution, February 17th and March 28th, 1876,
—the subject announced being, "AND THE GOLD OF THAT LAND IS GOOD:
THERE IS BDELLIUM AND THE ONYX STONE")*

1. THE subject which you permit me the pleasure of illustrating to you this evening, namely, the symbolic use of the colours of precious stones in heraldry, will, I trust, not interest you less because forming part both of the course of education in art which I have been permitted to found in Oxford; and of that in physical science, which I am about to introduce in the Museum for working men at Sheffield.

I say "to introduce," not as having anything novel to teach, or show; for in the present day I think novelty the worst enemy of knowledge, and my introductions are only of things forgotten. And I am compelled to be pertinaciously—it might even seem, insolently,—separate in effort from many who would help me, just because I am resolved that no pupil of mine shall see anything, or learn, but what the consent of the past has admitted to be beautiful, and the experience of the past has ascertained to be true. During the many thousand years of this world's existence, the persons living upon it have produced more lovely things than any of us can ever see; and have ascertained more profitable things than any of us can ever know. Of these

---

* The abrupt interpolation of this lecture in the text of *Deucalion* is explained in the next chapter [p. 197].

[1] [For this phrase, see *Stones of Venice*, vol. ii. (Vol. X. p. 187).]

infinitely existing beautiful things, I show to my pupils as many as they can thoroughly see,—not more; and of the natural facts which are positively known, I urge them to know as many as they can thoroughly know,—not more; and absolutely forbid all debate whatsoever. The time for debate is when we have become masters—not while we are students. And the wisest of masters are those who debate least.

2. For my own part—holding myself nothing better than an advanced student, guiding younger ones,—I never waste a moment of life in dispute, or discussion. It is at least ten years since I ceased to speak of anything but what I had ascertained; and thus becoming, as far as I know, the most practical and positive of men, left discourse of things doubtful to those whose pleasure is in quarrel;— content, for my pupils and myself, to range all matters under the broad heads of things certain, with which we are vitally concerned, and things uncertain, which don't in the least matter.

3. In the working men's museum at Sheffield, then, I mean to place illustrations of entirely fine metal-work, including niello and engraving;[1] and of the stones, and the Flora and Fauna, of Yorkshire, Derbyshire, Durham, and Westmoreland;* together with such foreign examples as may help to the better understanding of what we have at home. But in teaching metal-work, I am obliged to exhibit, not the uses of iron and steel only, but those also of the most precious metals, and their history; and for the understanding of any sort of stones, I must admit precious stones, and their history. The first elements of both these subjects, I hope it may not be uninteresting to you to follow out with me this evening.

---

* Properly, Westmoreland, the district of Western Meres.[2]

---

[1] [These intentions, however, were not carried out, except in the case of a mineral collection.]

[2] [Compare ch. xii. § 8 (below, p. 248); *Proserpina*, Vol. XXV. p. 431; and Preface to *Rock Honeycomb*, § 1.]

4. I have here, in my right hand, a little round thing, and in my left a little flat one, about which, and the like of them, it is my first business to explain, in Sheffield, what may *positively* be known. They have long been both, to me, subjects of extreme interest; and I do not hesitate to say that I know more about them than most people: but that, having learned what I can, the happy feeling of wonder is always increasing upon me—how little that is! What an utter mystery both the little things still are!

5. This first—in my right hand—is what we call a "pebble,"* or rolled flint, presumably out of Kensington gravel-pits. I picked it up in the Park,—the first that lay loose, inside the railings, at the little gate entering from Norfolk Street. I shall send it to Sheffield; knowing that, like the bit of lead picked up by Saadi in the *Arabian Nights*, it will make the fortune of Sheffield, scientifically, —if Sheffield makes the most of it, and thoroughly learns what it is.

6. What it *is*, I say,—you observe;—not merely, what it is *made of*. Anybody—the pitifullest apothecary round the corner, with a beggarly account of empty boxes[1]—can tell you that. It is made of brown stuff called silicon, and oxygen, and a little iron; and so any apothecary can tell what you all who are sitting there are made of:—you, and I, and all of us, are made of carbon, nitrogen, lime, and phosphorus, and seventy per cent. or rather more of water; but then, that doesn't tell us what we are,—what a child is, or what a boy is,—much less what a man is,—least of all, what supremely inexplicable woman is. And so, in knowing only what it is made of, we don't know what a flint is.

7. To know what it is, we must know what it can do, and suffer.

* 1. A. 1.[2] Sheffield Museum; see Chapter VIII. [p. 201].

---

[1] [*Romeo and Juliet* (in the description of the apothecary), Act v. sc. 1.]
[2] [F. 1 in the Catalogue: see below, p. 419.]

That it can strike steel into white-hot fire, but can itself be melted down like water, if mixed with ashes; that it is subject to laws of form one jot of which it cannot violate, and yet which it can continually evade, and apparently disobey; that in the fulfilment of these it becomes pure,—in rebellion against them, foul and base; that it is appointed on our island coast to endure for countless ages, fortifying the sea cliff; and on the brow of that very cliff, every spring, to be dissolved, that the green blades of corn may drink it with the dew;—that in its noblest forms it is still imperfect, and in the meanest, still honourable,—this, if we have rightly learned, we begin to know what a flint is.

8. And of this other thing, in my left hand,—this flat bit of yellow mineral matter,—commonly called a "sovereign," not indeed to be picked up so easily as the other— (though often, by rogues, with small pains);—yet familiar enough to the sight of most of us, and *too* familiar to our thought,—there perhaps are the like inquiries to be put. What *is* it? What can it do; and for whom? This shape given to it by men, bearing the image of a Cæsar;—how far does this make it a thing which is Cæsar's?[1] the opposed image of a saint, riding against a dragon—how far does this make it a thing which is of Saints? Is its testimony true, or conceivably true, on either side? Are there yet Cæsars ruling us, or saints saving us, to whom it does of right belong?

9. And the substance of it,—not separable, this into others, but a pure element,—what laws are over it, other than Cæsar's; what forms must it take, of its own, in eternal obedience to invisible power, if it escape our human hammer-stroke? How far, in its own shape, or in this, is it itself a Cæsar; inevitable in authority; secure of loyalty, lovable, and meritorious of love? For, reading its past history, we find it has been much beloved, righteously or iniquitously,—a thing to be known the grounds of, surely?

---

[1] [Matthew xxii. 21.]

10. Nay, also of this dark and despised thing in my right hand, we must ask that higher question, has it ever been beloved? And finding in its past history that in its pure and loyal forms, of amethyst, opal, crystal, jasper, and onyx, *it* also has been much beloved of men, shall we not ask farther whether it deserves to be beloved,—whether in wisdom or folly, equity or iniquity, we give our affections to glittering shapes of clay, and found our fortunes on fortitudes of stone; and carry down from lip to lip, and teach, the father to the child, as a sacred tradition, that the Power which made us, and preserves, gave also with the leaves of the earth for our food, and the streams of the earth for our thirst, so also the dust of the earth for our delight and possession: bidding the first of the Rivers of Paradise roll stainless waves over radiant sands, and writing, by the word of the Spirit, of the Rocks that it divided, "The gold of that land is good; there also is the crystal, and the onyx stone."

11. Before I go on, I must justify to you the familiar word I have used for the rare one in the text.[1]

If with mere curiosity, or ambitious scholarship, you were to read the commentators on the Pentateuch, you might spend, literally, many years of life, on the discussions as to the kinds of the gems named in it; and be no wiser at the end than you were at the beginning. But if, honestly and earnestly desiring to know the meaning of the book itself, you set yourself to read with such ordinary help as a good concordance and dictionary, and with fair knowledge of the two languages in which the Testaments have been clearly given to us, you may find out all you need know, in an hour.

12. The word "bdellium" occurs only twice in the Old Testament: here, and in the book of Numbers,[2] where you are told the manna was of the colour or look of bdellium.

[1] [Genesis ii. 12. See the title of this chapter; and compare *Fors Clavigera*, Letter 62, § 11.]

[2] [Numbers xi. 7. The other references in § 12 are to Exodus xvi. 14; Job xxxviii. 29; Genesis xxvii. 28; Deuteronomy xxxii. 2.]

There, the Septuagint uses for it the word κρύσταλλος, crystal, or more properly anything congealed by cold; and in the other account of the manna, in Exodus, you are told that, after the dew round the camp was gone up, "there lay a small round thing—as small as the *hoar-frost* upon the ground." Until I heard from my friend Mr. Tyrwhitt * of the cold felt at night in camping on Sinai, I could not understand how deep the feeling of the Arab, no less than the Greek,[1] must have been respecting the divine gift of the dew,—nor with what sense of thankfulness for miraculous blessing the question of Job would be uttered, "The hoary frost of heaven, who hath gendered it?" Then compare the first words of the blessing of Isaac: "God give thee of the dew of heaven, and of the fatness of earth;" and, again, the first words of the song of Moses: "Give ear, oh ye heavens,—for my speech shall distil as the dew;" and you will see at once why this heavenly food was made to shine clear in the desert, like an enduring of its dew;—Divine remaining for continual need. Frozen, as the Alpine snow—pure for ever.

13. Seize firmly that first idea of the manna, as the type of the bread which is the Word of God; † and then look on for the English word "crystal" in Job, of Wisdom, "It cannot be valued with the gold of Ophir, with the precious *onyx*, or the sapphire: the *gold and the crystal* shall not equal it, neither shall it be valued with pure

* See some admirable sketches of travelling in the Peninsula of Sinai, by this writer, in *Vacation Tourists*, Macmillan, 1864.[2] "I still remember," he adds in a private letter to me, "that the frozen towels stood on their edges as stiff as biscuits. By 11 A.M. the thermometer had risen to 85°, and was still rising."

† Sir Philip Sidney, in his translation of the ἄρτον οὐρανοῦ of the 105th Psalm [verse 40], completes the entire range of idea,

"Himself, from skies, their hunger to repel,
Candies the grasse with sweete congealed dew."

[1] [Compare *Queen of the Air*, §§ 38, 70 (Vol. XIX. pp. 334, 364).]
[2] ["Sinai," by the Rev. R. St. John Tyrwhitt, ch. vii. in *Vacation Tourists and Notes of Travel in 1862-1863*, edited by F. Galton. For the cold, see p. 339. For Mr. Tyrwhitt, see Vol. XV. pp. xxvi., xxx., 6.]

gold;"[1] in Ezekiel, "firmament of the terrible crystal," or in the Apocalypse, "A sea of glass, like unto crystal,—water of life, clear as crystal"—"light of the city like a stone most precious, even like a jasper stone, clear as crystal." Your understanding the true meaning of all these passages depends on your distinct conception of the permanent clearness and hardness of the Rock-crystal. You may trust me to tell you quickly, in this matter, what you may all for yourselves discover if you will read.

14. The three substances named here in the first account of Paradise, stand generally as types—the GOLD of all precious metals; the CRYSTAL of all clear precious stones prized for *lustre*;[2] the ONYX of all opaque precious stones prized for *colour*. And to mark this distinction as a vital one,—in each case when the stones to be set for the tabernacle-service are named, the onyx is named separately. The Jewish rulers brought "onyx stones, and stones to be set for the ephod, and for the breastplate."* And the onyx is used thrice, while every other stone is used only once, in the High Priest's robe; two onyxes on the shoulders bearing the twelve names of the tribes, six on each stone (Exod. xxviii. 9, 10), and one in the breastplate, with its separate name of one tribe (Exod. xxviii. 20).

15. A. Now note the importance of this grouping. The Gold, or precious metal, is significant of all that the power of the beautiful earth, gold, and of the strong earth, iron, has done for and against man. How much evil I need not say. How much good is a question I will endeavour to show some evidence on forthwith.

B. The Crystal is significant of all the power that jewels, from diamonds down through every Indian gem to the glass beads which we now make for ball-dresses, have had

---

* Exod. xxv. 7, xxxv. 27, comparing Job above quoted, and Ezekiel xxviii. 13.

---

[1] [Job xxviii. 16; Ezekiel i. 22; Revelation iv. 6, xxi. 11.]
[2] ["Lustre including light showing *through* colour" (Note in Ruskin's *copy*).]

over the imagination and economy of men and women—
from the day that Adam drank of the water of the crystal
river [1] to this hour.

How much evil that is, you partially know; how much
good, we have to consider.

c. The Onyx is the type of all stones arranged in
bands of different colours; it means primarily, nail-stone—
showing a separation like the white half-crescent at the
root of the finger-nail; not without some idea of its subjection to laws of life. Of these stones, part, which are
flinty, are the material used for cameos and all manner of
engraved work and pietra dura; but in the great idea of
banded or belted stones, they include the whole range of
marble, and especially alabaster, giving the name to the
alabastra, or vases used especially for the containing of
precious unguents, themselves more precious;* so that this
stone, as best representative of all others, is chosen to be
the last gift of men to Christ, as gold is their first;
incense with both : at His birth, gold and frankincense; at
His death, alabaster and spikenard.[2]

16. The two sources of the material wealth of all
nations were thus offered to the King of men in their
simplicity. But their power among civilized nations has
been owing to their workmanship. And if we are to ask
whether the gold and the stones are to be holy, much
more have we to ask if the worker in gold, and the
worker in stone, are to be conceived as exercising holy
function.

17. Now, as we ask of a stone, to know what it is,
what it can do, or suffer,[3] so of a human creature, to know
what it is, we ask what it can do, or suffer.

* Compare the "Nardi parvus onyx," which was to be Virgil's feast-
gift, in spring, to Horace.[4]

---

[1] [See Genesis ii. 10, and Revelation xxii. 1, 2.]
[2] [Matthew ii. 11; Mark xiv. 3. On the vase of the Magdalen, see below,
p. 183.]
[3] [See above, § 7.]
[4] [Odes, iv. 12, 17.]

So that we have two scientific questions put to us, in
this matter: how the stones came to be what *they* are—
or the law of Crystallization; and how the jewellers came
to be what *they* are—or the law of Inspiration. You see
how vital this question is to me, beginning now actually
to give my laws of Florentine art in English Schools![1]
How can artists be made artists,—in gold and in precious
stones? whether in the desert, or the city?—and if in the
city, whether, as at Jerusalem, so also in Florence, Paris,
or London?

Must we at this present time, think you, order the
jewellers, whom we wish to teach, merely to study and
copy the best results of past fashion? or are we to hope
that some day or other, if we behave rightly, and take
care of our jewels properly, we shall be shown also how to
set them; and that, merely substituting modern names for
ancient ones, some divine message will come to our crafts-
men, such as this: "See, I have called by name Messrs.
Hunt and Roskell, and Messrs. London and Ryder,[2] and
I have filled them with the Spirit of God, in wisdom and
in understanding, and in all manner of workmanship, to
work in gold, and in silver, and in brass, and in cutting
of stones"?[3]

18. This sentence, which, I suppose, becomes startling to
your ear in the substitution of modern for ancient names,
is the first, so far as I know, distinctly referring to the
ancient methods of instruction in the art of jewellery. So
also the words which I have chosen for the title (or, as
perhaps some of my audience may regretfully think it should
be called, the text) of my lecture, are the first I know
that give any account of the formation or existence of
jewels. So that the same tradition, whatever its value,
which gave us the commands we profess to obey for our

---

[1] [*The Laws of Fésole,* which Ruskin had in preparation at the time of this
lecture.]

[2] [For another reference to the latter of these two well-known Bond Street
jewellers, see *Fors Clavigera,* Letter 54, § 23.]

[3] [See Exodus xxxi. 2–5.]

moral law, implies also the necessity of inspired instruction for the proper practice of the art of jewellery; and connects the richness of the earth in gold and jewels with the pleasure of Heaven that we should use them under its direction. The scientific mind will of course draw back in scorn from the idea of such possibility; but then, the scientific mind can neither design, itself, nor perceive the power of design in others. And practically you will find that all noble design in jewellery whatsoever, from the beginning of the world till now, has been either instinctive, —done, that is to say, by tutorship of nature, with the innocent felicity and security of purely animal art,—Etruscan, Irish, Indian, or Peruvian gold being interwoven with a fine and unerring grace of industry, like the touch of the bee on its cell and of the bird on her nest,—or else, has been wrought into its finer forms, under the impulse of religion in sacred service, in crosier, chalice, and lamp; and that the best beauty of its profane service has been debased from these. And the three greatest masters of design in jewellery, the "facile principes" of the entire European School, are—centrally, the one who definitely worked always with appeal for inspiration—Angelico of Fésole;[1] and on each side of him, the two most earnest reformers of the morals of the Christian Church—Holbein, and Sandro Botticelli.[2]

19. I have first answered this, the most close home of the questions,—how men come to be jewellers. Next, how do stones come to be jewels? It seems that by all religious, no less than all profane, teaching or tradition, these substances are asserted to be precious,—useful to man, and sacred to God. Whether we have not made

---

[1] [For Angelico's jewel painting, see Vol. XXIII. p. 262. "It was the custom of Fra Angelico," says Vasari, "to abstain from retouching or improving any painting once finished. He altered nothing, but left all as it was done the first time, believing, as he said, that such was the will of God. It is also affirmed that he would never take pencil in hand until he had first offered a prayer" (vol. ii. p. 34, Bohn's edition).]

[2] [For these painters as reformers, see *Ariadne Florentina* (Vol. XXII. p. 328).]

them deadly instead of useful,—and sacrificed them to devils instead of God,—you may consider at another time.[1] To-night, I would examine only a little way the methods in which they are prepared by nature, for such service as they are capable of.

20. There are three great laws by which they, and the metals they are to be set in, are prepared for us; and at present all these are mysteries to us.

I. The first, the mystery by which " surely there is a vein for the silver, and a place for the gold whence * they fine it." No geologist, no scientific person whatsoever, can tell you how this gold under my hand was brought into this cleft in the bdellium; † no one knows where it was before, or how it got here: one thing only seems to be manifest—that it was not here always. This white bdellium itself closes rents, and fills hollows, in rocks which had to be rent before they could be rejoined, and hollowed before they could be refilled. But no one hitherto has been able to say where the gold first was, or by what process it came into this its resting-place. First mystery, then,— that there is a vein for the silver, and a place for the gold.

II. The second mystery is that of crystallization; by which, obeying laws no less arbitrary than those by which the bee builds her cell—the water produced by the sweet miracles of cloud and spring freezes into the hexagonal stars of the hoar-frost;—the flint, which can be melted and diffused like water, freezes also, like water, into *these* hexagonal towers of everlasting ice; ‡ and the clay, which can be dashed on the potter's wheel as it pleaseth the potter

---

* "Whence," not "where," they sift or wash it: ὅθεν διηθεῖται, LXX. [Job xxviii. 1].

† 20. A. 1. Sheffield Museum.[2]

‡ 1. Q. 11. Sheffield Museum.[3]

---

[1] [Ruskin considers, and answers, the question below, p. 193. See also Vol. X. p. 198, and Vol. XVIII. p. 217.]

[2] [See below, ch. viii. § 8, p. 201.]

[3] [See below, p. 432.]

to make it, can be frozen by the touch of Heaven into the hexagonal star of Heaven's own colour—the sapphire.[1]

III. The third mystery, the gathering of crystals themselves into ranks or bands, by which Scotch pebbles are made, not only is at present unpierced, but—which is a wonderful thing in the present century—it is even untalked about.[2] There has been much discussion as to the nature of metallic veins; and books have been written with indefatigable industry, and splendid accumulation of facts, on the limits, though never on the methods, of crystallization. But of the structure of banded stones not a word is ever said, and, popularly, less than nothing known; there being many very false notions current respecting them, in the minds even of good mineralogists.

And the basis of what I find to be ascertainable about them, may be told with small stress to your patience.

21. I have here in my hand,* a pebble which used to decorate the chimney-piece of the children's play-room in my aunt's house at Perth, when I was seven years old,[3] just half a century ago; which pebble having come out of the Hill of Kinnoull, on the other side of the Tay, I show you because I know so well where it came from, and can therefore answer for its originality and genuineness.

22. The Hill of Kinnoull, like all the characteristic crags or craigs of central Scotland, is of a basaltic lava—in which, however, more specially than in most others, these balls of pebble form themselves. And of these, in their first and simplest state, you may think as little pieces of flint jelly, filling the pores or cavities of the rock.

Without insisting too strictly on the analogy—for Nature is so various in her operations that you are sure to be deceived if you ever think one process has been in all

---

* 1. A. 8. Sheffield Museum.[4]

[1] [Compare the passage on the sapphire as "the consummation of the clay" in *Modern Painters*, vol. v. (Vol. VII. p. 208).]

[2] [Compare above, p. 155.]

[3] [See *Præterita*, i. ch. iii. ("The Banks of Tay"), § 74.]

[4] [See below, p. 435.]

respects like another—you may yet in most respects think of the whole substance of the rock as a kind of brown bread, volcanically baked, the pores and cavities of which, when it has risen, are filled with agate or onyx jelly, as the similar pores of a slice of quartern loaf are filled with butter, if the cook has spread it in a hurry.

23. I use this simile with more satisfaction, because, in the course of last autumn, I was making some practical experiments on glacial motion—the substances for experiment being supplied to me in any degree of congelation or regelation which might be required, by the perfectly angelic cook of a country friend,[1] who not only gave me the run of her kitchen, but allowed me to make domical mountains of her best dish-covers, and tortuous valleys of her finest napkins;—under which altogether favourable conditions, and being besides supplied with any quantity of ice-cream and blanc-mange, in every state of frost and thaw, I got more beautiful results, both respecting glacier motion, and interstratified rocks, than a year's work would have reached by unculinary analysis. Keeping, however—as I must to-night—to our present question, I have here a piece of this baked volcanic rock, which is as full of agate pebbles as a plum-pudding is of currants; each of these agate pebbles consisting of a clear green chalcedony, with balls of banded agate formed in the midst, or at the sides of them. This diagram * represents one enlarged.

And you have there one white ball of agate, floating apparently in the green pool, and a larger ball, which is cut through by the section of the stone, and shows you the banded structure in the most exquisite precision.

24. Now, there is no doubt as to the possible formation of these balls in melted vitreous substance as it cools, because we get them in glass itself, when gradually cooled

* This drawing is in Sheffield Museum.[2]

---

[1] [See Ruskin's letter from Broadlands in Vol. XXIV. p. xxi.]
[2] [In fact, however, Ruskin did not send it to Sheffield.]

in old glass-houses; and there is no more difficulty in accounting for the formation of round agate balls of this character than for that of common globular chalcedony. But the difficulty begins when the jelly is not allowed to remain quiet, but can run about while it is crystallizing. Then you get glutinous forms that choke cavities in the rock, in which the chalcedony slowly runs down the sides, and forms a level lake at the bottom; and sometimes you get the whole cavity filled with lake poured over lake, the liquid one over the frozen, floor and walls at last encrusted with onyx fit for kings' signets.*

25. Of the methods of engraving this stone, and of its general uses and values in ancient and modern days, you will find all that can interest you, admirably told by Mr. King, in his book on precious stones and gems,[1] to which I owe most of the little I know myself on this subject.

26. To-night, I would only once more direct your attention to that special use of it in the dress of the Jewish High Priest; that while, as one of the twelve stones of the breastplate, it was engraved like the rest with the name of a single tribe, two larger onyxes were used for the shoulder-studs of the ephod; and on these, the names of all the twelve tribes were engraved, six upon each.[2] I do not infer from this use of the onyx, however, any pre-eminence of value, or isolation of symbolism, in the stone; I suppose it to have been set apart for the more laborious piece of engraving, simply because larger surfaces of it were attainable than of true gems, and its substance was more easily cut. I suppose the bearing of the names on the shoulder to be symbolical of the priest's sacrificial office in bearing the guilt and pain of the people; while the bearing of them on the breast was symbolical of his

---

* I am obliged to omit here the part of the lecture referring to diagrams. It will be given in greater detail in the subsequent text.[3]

[1] [The Natural History, Ancient and Modern, of Precious Stones and Gems, and of the Precious Metals. By C. W. King, 1865. For the onyx, see pp. 254 seq.]

[2] [Exodus xxviii. 9–12, 20, xxxix. 6, 7, 13.]

[3] [See ch. xi. and Plate XIV. ("Structure of Lake Agate").]

pastoral office in teaching them: but, except in the broad distinction between gem and onyx, it is impossible now to state with any certainty the nature or meaning of the stones, confused as they have been by the most fantastic speculation of vain Jewish writers themselves.

There is no such difficulty when we pass to the inquiry as to the use of these stones in Christian Heraldry, on the breastplate and shield of the Knight; for that use is founded on natural relations of colour, which cannot be changed, and which will become of more and more importance to mankind in proportion to the degree in which Christian Knighthood, once proudly faithful to Death, in War, becomes humbly faithful to Life, in Peace.

27. To these natural relations of colour, the human sight, in health, is joyfully sensitive, as the ear is to the harmonies of sound; but what healthy sight is,[1] you may well suppose, I have not time to define to-night;—the nervous power of the eye, and its delight in the pure hues of colour presented either by the opal, or by wild flowers, being dependent on the perfect purity of the blood supplied to the brain, as well as on the entire soundness of the nervous tissue to which that blood is supplied. And how much is required, through the thoughts and conduct of generations, to make the new blood of our race of children pure—it is for your physicians to tell you, when they have themselves discovered this medicinal truth, that the divine laws of the life of Men cannot be learned in the pain and death of Brutes.[2]

28. The natural and unchangeable system of visible colour has been lately confused, in the minds of all students, partly by the pedantry of unnecessary science; partly by the formalism of illiberal art: for all practical service, it may be

---

[1] [The question is much discussed in *The Eagle's Nest*: see Vol. XXII. pp. 194 *seq.*; and on "the extreme rarity of finely-developed organic sight," see Vol. XVIII. p. 145, and Vol. XXV. p. 428 *n.*]

[2] [Ruskin's hostility to vivisection was the cause of his rupture with Oxford (see the Introduction to a later volume of this edition). For other passages in which he expresses the same view, see below, pp. 241, 344; and compare *Fors Clavigera*, Letters 64 (§ 8) and 75 (§ 10).]

stated in a very few words, and expressed in a very simple diagram.

There are three primary colours, Red, Blue, and Yellow; three secondary, formed by the union of any two of these; and one tertiary, formed by the union of all three.

If we admitted, as separate colours, the different tints produced by varying proportions of the composing tints, there would of course be an infinite number of secondaries, and a wider infinitude of tertiaries. But tints can be systematically arranged only by the elements of them, not the proportions of those elements. Green is only green, whether there be less or more of blue in it; purple only purple, whether there be less or more of red in it; scarlet only scarlet, whether there be less or more of yellow in it; and the tertiary grey only grey, in whatever proportions the three primaries are combined in it.

29. The diagram used in my drawing schools to express the system of these colours will be found coloured in the *Laws of Fésole*:[1]—this figure will serve our present purpose.*

The simple trefoil produced by segments of three circles in contact, is inscribed in a curvilinear equilateral triangle. Nine small circles are set,—three in the extremities of the foils, three on their cusps, three in the angles of the triangle.

The circles numbered 1 to 3 are coloured with the primitive colours; 4 to 6, with the secondaries; 7 with white; 8 with black; and the 9th, with the tertiary, grey.

---

* Readers interested in this subject are sure to be able to enlarge and colour it for themselves. I take no notice of the new scientific theories of primary colour:[2] because they are entirely false as applied to practical work, natural or artistic. Golden light in blue sky makes green sky; but green sky and red clouds can't make yellow sky.

---

[1] [See Vol. XV. p. 428; the diagram, however, was not in fact printed in colours.]

[2] [For a popular account of the theories in question, and of the application of them to the "three-colour process" of printing, the reader may be referred to the article and illustrative plate in the supplement to the *Encyclopædia Britannica*, 1902, vol. xxxii. p. 16. The principle on which the process is based is (in contradistinction to what Ruskin says in § 28) that "any possible tone of colour may be produced by blending the primary colours in the necessary proportions."]

30. All the primary and secondary colours are capable of infinitely various degrees of intensity or depression: they pass through every degree of increasing light, to perfect light, or white; and of increasing shade, to perfect absence of light, or black. And these are essential in the harmony required by sight; so that no group of colours can be perfect that has not white in it, nor any that has not

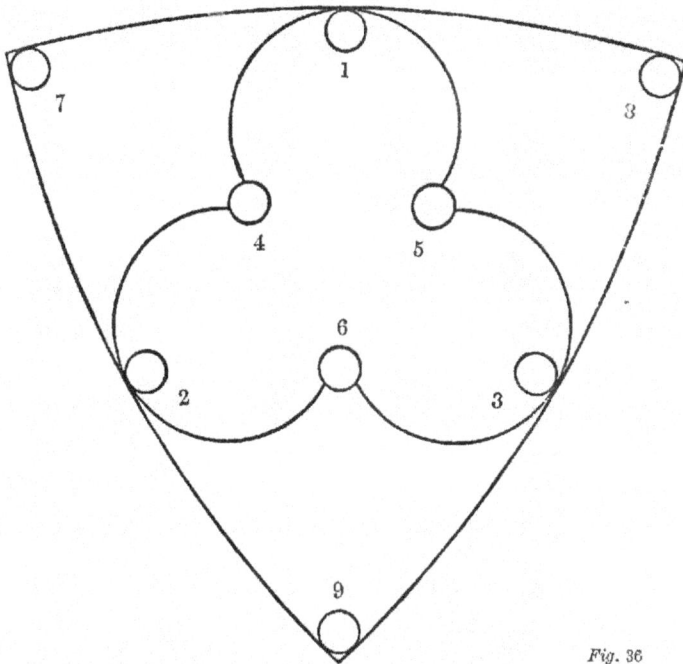

Fig. 36

black; or else the abatement or modesty of them, in the tertiary, grey. So that these three form the limiting angles of the field, or cloudy ground of the rainbow. " I do set my bow in the cloud."

And the nine colours of which you here see the essential group, have, as you know, been the messenger Iris; exponents of the highest purpose, and records of the perfect household purity and honour of men, from the days when Hesiod blazoned the shield of Heracles, to the day when the fighting *Téméraire* led the line at Trafalgar,—the

Victory following her, with three flags nailed to her masts, for fear one should be shot away.[1]

31. The names of these colours in ordinary shields of knighthood, are those given below, in the left-hand column. The names given them in blazoning the shields of nobles, are those of the correspondent gems: of heraldry by the planets, reserved for the shields of kings, I have no time to speak, to-night, except incidentally.

A. THE PRIMARY COLOURS.

1. Or.	Topaz.
2. Gules.	Ruby.
3. Azure.	Sapphire.

B. THE SECONDARY COLOURS.

4. Écarlate.	Jasper.
5. Vert.	Emerald.
6. Purpure.	Hyacinth.

C. THE TERTIARY COLOURS.

7. Argent.	Carbuncle.
8. Sable.	Diamond.
9. Colombin.	Pearl.

32. I. Or. Stands between the light and darkness; as the sun, who "rejoiceth as a strong man to run his course,"[2] between the morning and the evening. Its heraldic name, in the shields of kings, is Sol: the Sun, or Sun of Justice;[3] and it stands for the strength and honour of all men who

---

[1] [In this passage Ruskin means that the "colours" have come to be synonymous with "standards"—heraldic tinctures symbolising honour, and flag-signals standing for courage and patriotism. He refers to the rainbow, symbolised in Greek legend by Iris, the messenger of the Gods, and in the Bible set in the cloud for a token (Genesis ix. 13); to the "all-variegated" shield of the Greek hero (Hesiod, *Shield of Heracles*, 139 *seq.*); and to Nelson's flags and signals at Trafalgar—he "had, as usual, hoisted several flags, lest one should be shot away" (Southey's *Life of Nelson*, ch. ix.).]

[2] [Psalms xix. 5.]

[3] [See Vol. XVII. p. 59 *n.*]

run their race in noble work; whose path "is as the shining light, that shineth more and more unto the perfect day."[1]

For theirs are the works which are to shine before men, that they may glorify our Father. And they are also to shine before God, so that with respect to them, what was written of St. Bernard may be always true: "Opera sancti patris velut Sol in conspectu Dei."[2]

For indeed they are a true light of the world,[3] infinitely more good, in the sight of its Creator, than the dead flame of its sunshine; and the discovery of modern science, that all mortal strength is from the sun,[4] which has thrown irrational persons into stupid atheism, as if there were no God but the sun, is indeed the accurate physical expression of this truth, that men, rightly active, are living sunshine.

II. Gules (rose colour), from the Persian word "gul," for the rose.[5] It is the exactly central hue between the dark red, and pale red, or wild-rose. It is the colour of love, the fulfilment of the joy and of the love of life upon the earth. And it is doubly marked for this symbol. We saw earlier,[6] how the vase given by the Madeleine was precious in its material; but it was also to be indicated as precious in its form. It is not only the substance, but the form of the Greek urn, which gives it nobleness; and these vases for precious perfume were tall, and shaped like the bud of the rose. So that the rose-bud itself, being a vase filled with perfume, is called also "alabastron": and Pliny uses that word for it in describing the growth of the rose.[7]

The stone of it is the Ruby.

III. Azure. The colour of the blue sky in the height of it, at noon;—type of the fulfilment of all joy and love

---

[1] [Proverbs iv. 8; for the next quotation, see Matthew v. 16.]
[2] [The passage from "what was written of St. Bernard" down to "living sunshine," is repeated with some alterations in Fors Clavigera, Letter 63, § 4.]
[3] [John i. 9; viii. 12.]
[4] [See on this subject Eagle's Nest, § 100 (Vol. XXII. pp. 195, 196).]
[5] [See below, §§ 35 seq.]
[6] [See above, p. 172.]
[7] [Nat. Hist., xxi. 10.]

in heaven, as the rose-colour, of the fulfilment of all joy and love in earth. And the stone of this is the Sapphire; and because the loves of Earth and Heaven are in truth one, the ruby and sapphire are indeed the same stone; and they are coloured as if by enchantment,[1]—how, or with what, no chemist has yet shown,—the one azure, and the other rose.

And now you will understand why, in the vision of the Lord of Life to the Elders of Israel, of which it is written, "Also they saw God, and did eat and drink," you are told, "Under His feet was a plinth of sapphire, as it were, the body of Heaven in its clearness."[2]

IV. Écarlate (scarlet). I use the French word, because all other heraldic words for colours are Norman-French. The ordinary heraldic term here is "tenné" (tawny); for the later heralds confused scarlet with gules; but the colour first meant was the sacred hue of human flesh[3]—Carnation; —incarnation: the colour of the body of man in its beauty; of the maid's scarlet blush in noble love; of the youth's scarlet glow in noble war; the dye of the earth into which heaven has breathed its spirit: incarnate strength—incarnate modesty.

The stone of it is the Jasper, which, as we shall see,[4] is coloured with the same iron that colours the human blood; and thus you can understand why on the throne, in the vision of the returning Christ, "He that sat was to look upon like a jasper and a sardine stone."[5]

V. Vert (viridis), from the same root as the word "virtue" and "virgin,"[6]—the colour of the green rod in budding spring; the noble life of youth, born in the spirit,—as the scarlet means, the life of noble youth, in

---

[1] [Compare, again, the passage on the sapphire in *Modern Painters*, vol. v. (repeated in *Ethics of the Dust*): Vol. VII. p. 207, and Vol. XVIII. p. 359.]

[2] [Exodus xxiv. 10, 11 ("under his feet as it were a paved work of a sapphire stone," etc.]

[3] [Compare Vol. IV. p. 130.]

[4] [This was not done; but see *Two Paths*, § 155 (Vol. XVI. p. 384).]

[5] [Revelation iv. 3; quoted again below, p. 506.]

[6] [Compare *Val d'Arno*, § 64 (Vol. XXIII. p. 42).]

*flesh.** It is seen most perfectly in clear air after the sun has set,—the blue of the upper sky brightening down into it. It is the true colour of the eyes of Athena,—Athena Γλαυκῶπις,† looking from the west.[1]

The stone of it is the Emerald; and I must stay for a moment to tell you the derivation of that word.

Anciently, it did not mean our emerald, but a massive green marble, veined apparently by being rent asunder, and called, therefore, the Rent or Torn Rock.

Now, in the central war of Athena with the Giants, the sign of her victory was that the earth was rent, the power of it torn, and the graves of it opened. We know this is written for the sign of a greater victory than hers. And the word which Hesiod uses—the oldest describer of this battle—is twice over the same: the sea roared, the heavens thundered, the earth cried out in being rent, ἐσμαράγησε.[2] From that word you have "the rent rock,"—in Latin, smaragdus; in Latin dialect, smaraudus—softened into emeraudu, emeraude, emerald. And now you see why "there was a rainbow round about the throne in sight like unto an emerald."[3]

* Therefore, the Spirit of Beatrice is dressed in green, over *scarlet* (not rose;—observe this specially).

  "Sovra candido vel, cinta d' oliva,
      Donna m' apparve sotto verde manto,
      Vestita di color di *fiamma viva.*"[4]

† Accurately described by Pausanias, i. 14, as of the colour of a green lake, from the Tritonian pool;[5] compare again the eyes of Beatrice.[6]

---

[1] [Athena to Ruskin meant the clear air, and at sunset, "looking from the west," the clear air is bluish green—which colour he thus here reads into the eyes of the goddess; for other meanings which he found in the epithet. see *Queen of the Air* (Vol. XIX. pp. 306, 379); and compare *Laws of Fésole*, Vol. XV. p. 425.]

[2] [*Theogony*, 679, 693. Compare *Proserpina*, i. ch. x. § 2 (Vol. XXV. p. 334).]

[3] [Revelation iv. 3.]

[4] [*Purgatorio*, xxx. 30–33: "With white veil with olive wreathed, A virgin in my view appear'd, beneath Green mantle, robed in hue of living flame" (Cary).]

[5] ["Observing that Athena's image had blue eyes, I recognised the Libyan version of the myth. For the Libyans say that she is a daughter of Poseidon and the Tritonian lake, and that therefore she like Poseidon has blue eyes" (i. 14, 6).]

[6] [*Purgatorio*, xxxi. 116, where Dante speaks of the eyes of Beatrice as *smeraldi* (emeralds).]

VI. Purpure. The true purple of the Tabernacle, "blue, purple, and scarlet"[1]—the kingly colour, retained afterwards in all manuscripts of the Greek Gospels; therefore known to us absolutely by its constant use in illumination. It is rose colour darkened or saddened with blue; the colour of love in noble or divine sorrow; borne by the kings, whose witness is in heaven,[2] and their labour on the earth. Its stone is the Jacinth, Hyacinth, or Amethyst,— "like to that sanguine flower inscribed with woe."[3]

In these six colours, then, you have the rainbow, or angelic iris, of the light and covenant of life.

But the law of the covenant is, "I do set my bow in the cloud," on "the shadow of death"[4]—and the ordinance of it.

And as here, central, is the sun in its strength, so in the heraldry of our faith, the morning and the evening are the first day,—and the last.[5]

VII. Argent. Silver, or snow-colour; of the hoar-frost on the earth, or the star of the morning.

I was long hindered from understanding the entire group of heraldic colours, because of the mistake in our use of the word "carbuncle." It is not the garnet, but the same stone as the ruby and sapphire—only crystallized white, instead of red or blue. It is the white sapphire, showing the hexagonal star of its crystallization perfectly;[6] and therefore it becomes an heraldic bearing as a star.

And it is the personal bearing of that Geoffrey Plantagenet, who married Maud the Empress,[7] and became the sire of the lords of England, in her glorious time.

VIII. Sable (sable, sabulum), the colour of sand of the great hour-glass of the world, outshaken. Its stone is the

---

[1] [Exodus xxv. 4; compare *Giotto and his Works in Padua*, § 12 (Vol. XXIV. p. 25).]

[2] [See Job xvi. 19.]

[3] [*Lycidas*, 106. By a slip of the pen Ruskin here wrote "sable" for "sanguine." He quotes the line correctly in Vol. XVII. p. 405.]

[4] [Genesis ix. 3; Job iii. 5.]

[5] [Genesis i. 5.]

[6] [Compare *Ethics of the Dust*, Vol. XVIII. p. 257.]

[7] [See Vol. XVIII. p. 520 n.]

diamond—never yet, so far as I know, found but in the sand.* It is the symbol at once of dissolution, and of endurance: darkness changing into light—the adamant of the grave.

IX. Grey. (When deep, the second violet, giving Dante's full chord of the seven colours.)[1] The abatement of the light, the abatement of the darkness. Patience, between this which recedes and that which advances; the colour of the turtle-dove, with the message that the waters are abated;[2] the colour of the sacrifice of the poor,[3]—therefore of humility. Its stone is the Pearl; in Norman heraldry the Marguerite[4]—the lowest on the shield, yet of great price;[5] and because, through this virtue, open first the gates of Paradise,[6] you are told that while the building of the walls of it was of jasper, every several gate was of one pearl.[7]

33. You hear me tell you thus positively,—and without qualification or hesitation,—what these things mean. But mind, I tell you so, after thirty years' work, and that directed wholly to the one end of finding out the truth,[8] whether it was pretty or ugly to look in face of. During which labour I have found that the ultimate truth, the central truth, is always pretty; but there is a superficial truth, or half-way truth, which may be very ugly; and which the earnest and faithful worker has to face and fight, and pass over the body of,—feeling it to be his enemy; but which a careless seeker may be stopped by, and a misbelieving seeker will be delighted by, and stay with, gladly.

* Or in rock virtually composed of it.

---

1 [*Purgatorio*, xxix. 77-78: "seven bands, all in the colours of which the sun makes the rainbow, and the moon her halo."]

2 [Genesis viii. 3.]

3 [The dove was the sacrifice of the poor: see Leviticus xii. 8.]

4 [See the reference to *The Booke of St. Albans*, in Vol. XXV. p. 314; and compare the letter given in the Introduction, above, p. xliii.]

5 [Matthew xiii. 46.]

6 [Matthew v. 3.]

7 [Revelation xxi. 21.]

8 [Compare the motto from Wordsworth which Ruskin placed on the title-pages of *Modern Painters*, and see Vol. VII. p. 10.]

34. When I first gave this lecture, you will find the only reports of it in the papers, with which any pains had been taken, were endeavours to make you disbelieve it, or misbelieve it,—that is to say, to make "mescroyants" or "miscreants"[1] of you.

And among the most earnest of these, was a really industrious essay in the *Daily Telegraph*,[2]—showing evidence that the writer had perseveringly gone to the Heralds' Office and British Museum to read for the occasion; and, I think, deserving of serious notice because we really owe to the proprietors of that journal (who supplied the most earnest of our recent investigators with funds for his Assyrian excavations) the most important heraldic discoveries of the generations of Noah and Nimrod, that have been made since printing took the place of cuneiform inscription.[3]

I pay, therefore, so much respect to the archæologians of Fleet Street as to notice the results of their suddenly stimulated investigations in heraldry.

35. "The lecturer appeared to have forgotten," they said, "that every nation had its own code of symbols, and that gules, or red, is denominated by the French heralds gueules, and is derived by the best French philologers from the Latin 'gula,' the gullet of a beast of prey."

It is perfectly true that the best French philologists do give this derivation;[4] but it is also unfortunately true that the best French philologists are not heralds; and what is more, and worse, all modern heraldry whatsoever is, to the old science, just what the poor gipsy Hayraddin, in

---

[1] [On this word, see *Fors Clavigera*, Letter 25, §§ 19, 23.]

[2] [A leading article (doubtless by George Augustus Sala) in the *Daily Telegraph* of February 22, 1876.]

[3] [The reference is to George Smith (1840–1876), the Assyriologist. In 1867 he entered the service of the British Museum and set to work upon the cuneiform inscriptions; in 1872 he discovered and deciphered, among the tablets sent home by Layard, the Chaldean account of the Deluge. The *Daily Telegraph* then came forward and offered to defray the expenses of fresh researches at Nineveh under Smith. The results of these researches were published in 1875 under the title, *Assyrian Discoveries; an Account of Explorations and Discoveries on the Site of Nineveh during 1873 and 1874.*]

[4] [So Du Cange and Littré, though the latter, while giving the derivation from *gula* as "probable," mentions also the suggestion of derivation from the Persian *ghul*, rose.]

Quentin Durward,[1] is to Toison d'Or. But, so far from having "forgotten," as the writer for the press supposes I had, that there were knights of France, and Venice, and Florence, as well as England, it so happens that my first studies in heraldry were in *this* manuscript which is the lesson-book of heraldry written for the young Archduke Charles of Austria;[2] and in *this* one, which is a psalter written in the monastery of the Saint Chapelle for St. Louis, King of France;[3] and on the upper page of which, here framed,* you will see written, in letters of gold, the record of the death of his mother, Blanche of Castile, on the 27th of November, next after St. Geneviève's day; and on the under page, between the last lines of the Athanasian Creed, her bearing, the Castilian tower, alternating with the king's, —Azure, semé de France.

36. With this and other such surer authority than was open to the investigation of the press-writer, I will clear up for you his point about the word "gules." But I must go a long way back first. I do not know if, in reading the account of the pitching of the standards of the princes of Israel round the Tabernacle, you have ever been brought to pause by the singular covering given to the Tabernacle itself,—rams' skins dyed red, and *badgers'* skins.[4] Of rams' skins, of course, any quantity could be had from the flocks, but of badgers', the supply must have been difficult!

And you will find, on looking into the matter, that the so-called badgers' skins were indeed those which young ladies are very glad to dress in at the present day,—seal-skins; and that the meaning of their use in the Tabernacle

---

* The books referred to, in my rooms at Oxford, are always accessible for examination.

---

[1] [For Hayraddin Maugrabin, the Bohemian, as guide, see chaps. xv.–xx.; as false herald of William de la Marck, exposed by Toison d'Or, true herald of the Duke of Burgundy, see chap. xxxiii.]

[2] [Ruskin seems afterwards to have disposed of this MS. It was not at Brantwood at the time of his death.]

[3] ["Saint Louis's Psalter": see Vol. XXI. p. 15 *n*.]

[4] [Exodus xxxv. 7. Compare chap. x. ("The Heraldic Ordinaries") in *Eagle's Nest* (Vol. XXII. p. 276).]

was, that it might be adorned with the useful service of the *flocks* of the earth and sea: the multitude of the seals then in the Mediterranean being indicated to you both by the name and coinage of the city Phocæa; and by the attribution of them, to the God Proteus, in the fourth book of the *Odyssey*, under the precise term of flocks, to be counted by him as their shepherd.[1]

37. From the days of Moses and of Homer to our own, the traffic in these precious wools and furs, in the Cashmere wool, and the fur, after the seal disappeared, of the grey ermine (becoming white in the Siberian winter), has continued: and in the days of chivalry became of immense importance; because the mantle, and the collar fastening close about the neck, were at once the most useful and the most splendid piece of dress of the warrior nations, who rode and slept in roughest weather, and in open field. Now, these rams' skins, or fleeces, dyed of precious red, were continually called by their Eastern merchants "the red things," from the Zoroastrian word "gul,"—taking the place of the scarlet Chlamydes, which were among the richest wealth of old Rome.[2] The Latin knights could only render the eastern word "gul," by gula; and so in St. Bernard's red-hot denunciation of these proud red dresses, he numbers chiefly among them the little red-dyed skins,—pelliculas rubricatas,—which they called gulæ: "Quas gulas vocant."[3] These red furs, for wrist and neck, were afterwards supposed by bad Latinists to be called "gulæ," as *throat*-pieces. St. Bernard specifies them, also, in that office: "Even some of the clergy," he says, "have the red skins of weasels hanging from their necks—dependentes a collo";[4] this vulgar interpretation of gula became more commonly accepted, as

---

[1] [Compare *St. Mark's Rest*, § 72 (Vol. XXIV. p. 263).]

[2] [The scarlet chlamys or mantle, which came originally from Macedonia and Thessaly, was adopted by the Romans under the Emperors, and was often embroidered or interwoven with gold. Caligula, we are told, wore one so enriched (Suetonius, *Calig.* 19).]

[3] [*Epistolæ*, 42, c. 2 (the passage is quoted by Du Cange in his *Glossarium*, s. v. gula).]

[4] [*Declamatio De Vita et Moribus Clericorum*, c. 4: also quoted by Du Cange.]

intercourse with the East, and chivalric heraldry, diminished; and the modern philologist finally jumps fairly down the lion's throat, and supposes that the Tyrian purple, which had been the pride of all the Emperors of East and West, was named from a wild beast's gullet!

38. I do not hold for a mischance, or even for a chance at all, that this particular error should have been unearthed by the hasty studies of the *Daily Telegraph*. It is a mistake entirely characteristic of the results of vulgar modern analysis; and I have exposed it in detail, that I might very solemnly warn you of the impossibility of arriving at any just conclusions respecting ancient classical languages, of which this heraldry is among the noblest, unless we take pains first to render ourselves *capable of the ideas* which such languages convey. It is perfectly true that every great symbol, as it has, on one side, a meaning of comfort, has on the other one of terror; and if to noble persons it speaks of noble things, to ignoble persons it will as necessarily speak of ignoble ones.[1] Not under one only, but under all, of these heraldic symbols, as there is, for thoughtful and noble persons, the spiritual sense, so for thoughtless and sensual persons, there is the sensual one; and *can* be no other. Every word has only the meaning which its hearer can receive; you cannot express honour to the shameless, nor love to the unloving. Nay, gradually you may fall to the level of having words no more, either for honour or for love:

"There are whole nations," says Mr. Farrar, in his excellent little book on the families of speech, "people whom no nation now acknowledges as its kinsmen, whose languages, rich in words for all that can be eaten or handled, seem absolutely incapable of expressing the reflex conceptions of the intellect, or the higher forms of the consciousness; whose life seems confined to a gratification of animal wants, with no hope in the future, and no pride in the past. They are for the most part peoples without a literature, and without a history;—peoples whose tongues in some instances have twenty names for murder, but no name for love, no name for gratitude, no name for God."[2]

---

[1] [Compare *Queen of the Air*, § 4 (Vol. XIX. p. 299).]
[2] [*Families of Speech: Four Lectures.* By the Rev. F. W. Farrar, 1870, p. 156. Ruskin somewhat abbreviates the passage. For another reference to the book, see *Proserpina,* Vol. XXV. p. 334.]

39. The English nation, under the teaching of modern economists, is rapidly becoming one of this kind, which, deliberately living, not in love of God or man, but in defiance of God, and hatred of man, will no longer have in its heraldry, gules as the colour of love; but gules only as the colour of the throat of a wild beast. That will be the only part of the British lion symbolized by the British flag;—not the lion heart any more, but only the lion gullet.

And if you choose to interpret your heraldry in that modern fashion, there are volumes of instruction open for you everywhere. Yellow shall be to you the colour of treachery, instead of sunshine; green, the colour of putrefaction, instead of strength; blue, the colour of sulphurous hell-fire, instead of sunlit heaven; and scarlet, the colour of the harlot of Babylon,[1] instead of the Virgin of God. All these are legitimate readings,—nay, inevitable readings. I said wrongly just now that you might choose what the symbols shall be to you. Even if you would, you cannot choose. They can only reflect to you what you have made your own mind, and can only herald to you what you have determined for your own fate.

40. And now, with safe understanding of the meaning of purple, I can show you the purple and dove-colour of St. Mark's, once itself a sea-borne vase of alabaster full of incense of prayers; and a purple manuscript,—floor, walls, and roof blazoned with the scrolls of the gospel.[2]

They have been made a den of thieves,[3] and these stones of Venice here in my hand * are rags of the sacred

---

* Portions of the alabaster of St. Mark's torn away for recent restorations. The destruction of the floor of the church, to give work to modern mosaic-mongers, has been going on for years. I cannot bear the pain of describing the facts of it, and must leave the part of the lecture referring

[1] [Revelation xvii. 4.]
[2] [Compare the descriptions in *The Stones of Venice* (Vol. X. p. 112) and *St. Mark's Rest*, Vol. XXIV. pp. 204, 280.]
[3] [Matthew xxi. 13.]

robes of her Church, sold, and mocked like her Master.
They have parted her garments, and cast lots upon her
vesture.[1]

41. I return to our question at the beginning:[2] Are we
right in setting our hearts on these stones,—loving them,
holding them precious?

Yes, assuredly; provided it is the stone we love, and
the stone we think precious; and not ourselves we love, and
ourselves we think precious. To worship a black stone,
because it fell from heaven, may not be wholly wise, but
it is half-way to being wise; half-way to worship of heaven
itself. Or, to worship a white stone because it is dug
with difficulty out of the earth, and to put it into a log
of wood, and say the wood sees with it, may not be
wholly wise; but it is half-way to being wise; half-way to
believing that the God who makes earth so bright, may
also brighten the eyes of the blind. It is no true folly to
think that stones see, but it *is*, to think that eyes do not;
it is no true folly to think that stones live, but it *is*, to
think that souls die; it is no true folly to believe that, in
the day of the making up of jewels,[3] the palace walls shall
be compact of life above their corner-stone,—but it *is*, to
believe that in the day of dissolution the souls of the globe
shall be shattered with its emerald; and no spirit survive,
unterrified, above the ruin.

42. Yes, pretty ladies! love the stones, and take care
of them; but love your own souls better, and take care

to the colour of the marbles to be given farther on, in connection with
some extracts from my *Stones of Venice.* The superb drawing, by Mr.
Bunney, of the north portico, which illustrated them, together with the
alabasters themselves, will be placed in the Sheffield Museum.[4]

---

[1] [Psalms xxii. 18.]
[2] [See above, § 19, p. 175.]
[3] [Malachi iii. 17.]
[4] [See Vol. XXIV. pp. lviii. *seq.*, for account of the restorations in question.
Ruskin did not "farther on" give the extracts, etc., here promised. Bunney's water-
colour drawing of the north portico of St. Mark's, which remains at Sheffield, is shown
(by photogravure process) at p. 237 of William White's *Principles of Art as illustrated
in the Ruskin Museum.* The pieces of alabaster, which Ruskin brought from Venice,
remain in the Museum.]

of *them*, for the day when the Master shall make up
His jewels.   See that it be first the precious stones of
the breastplate of justice[1] you delight in, and are brave
in; not first the stones of your own diamond necklaces*
you delight in, and are fearful for, less perchance the lady's
maid miss that box at the station.   Get your breastplate
of truth first, and every earthly stone will shine in it.

Alas! most of you know no more what justice means,
than what jewels mean; but here is the pure practice of
it to be begun, if you will, to-morrow.

43. For literal truth of your jewels themselves, absolutely search out and cast away all manner of false, or
dyed, or altered stones.   And at present, to make quite
sure, wear your jewels uncut; they will be twenty times
more interesting to you, so.   The ruby in the British
crown is uncut; and is, as far as my knowledge extends,—
I have not had it to look at close,—the loveliest precious
stone in the world.   And, as a piece of true gentlewoman's
and true lady's knowledge, learn to know these stones
when you see them, uncut.   So much of mineralogy the
abundance of modern science may, I think, spare, as a
piece of required education for the upper classes.

44. Then, when you know them, and their shapes, get
your highest artists to design the setting of them.   Holbein,

---

* Do you think there was no meaning of fate in that omen of the
diamond necklace,[2] at the end of the days of queenly pride;—omen of
another line, of scarlet, on many a fair neck?   It was a foul story, you
say—slander of the innocent.   Yes, undoubtedly, fate meant it to be so.
Slander, and lying, and every form of loathsome shame, cast on the inno-
cently fading Royalty.   For the corruption of the best is the worst;[3] and
these gems, which are given by God to be on the breast of the pure priest,
and in the crown of the righteous king, sank into the black gravel of diluvium, under streams of innocent blood.

---

[1] [Ephesians vi. 14.]
[2] [The story of the necklace, presented through Mme. de Lamotte by Cardinal de
Rohan (as he supposed) to Marie Antoinette, is the subject, as the reader will remember, of one of Carlyle's Essays.   Recent discussion of the affair will be found
in F. Funck-Brentano's *The Diamond Necklace* (1901), and in Mr. Andrew Lang's
Historical Mysteries (1904).]
[3] [Compare Vol. XVII. p. 222.]

Botticelli, or Angelico, will always be ready to design a brooch for you. Then you will begin to think how to get your Holbein and Botticelli, which will lead to many other wholesome thoughts.

45. And lastly, as you are true in the choosing, be just in the sharing, of your jewels. They are but dross and dust, after all; and you, my sweet religious friends, who are so anxious to impart to the poor your pearls of great price, may surely also share with them your pearls of little price. Strangely (to my own mind at least), you are not so zealous in distributing your estimable rubies, as you are in communicating your *in*estimable wisdom.[1] Of the grace of God, which you can give away in the quantity you think others are in need of, without losing any yourselves, I observe you to be affectionately lavish; but of the jewels of God, if any suggestions be made by charity touching the distribution of *them*, you are apt, in your wisdom, to make answer like the wise virgins, " Not so, lest there be not enough for us and you."[2]

46. Now, my fair friends, doubtless, if the Tabernacle were to be erected again, in the middle of the Park, you would all be eager to stitch camels' hair for it;—some, to make presents of sealskins to it; and, perhaps, not a few fetch your jewel-cases, offering their contents to the selection of Bezaleel and Aholiab.[3]

But that cannot be, now, with so Crystal-Palace-like entertainment to you. The tabernacle of God is now with men;[4]—*in* men, and women, and sucklings also; which temple ye are, ye and your Christian sisters; of whom the poorest, here in London, are a very undecorated shrine indeed. *They* are the Tabernacle, fair friends, which you have got leave, and charge, to adorn. Not, in anywise, those charming churches and altars which you wreathe with garlands for

---

[1] [See Job xxviii. 18; Proverbs iii. 15.]
[2] [Matthew xxv. 9.]
[3] [See Exodus xxviii. 1 (Bezaleel and Aholiab), 14 (curtains of goat's hair), 19 (badgers' skins).]
[4] [Revelation xxi. 3.]

God's sake, and the eloquent clergyman's. You are quite
wrong, and barbarous in language, when you call *them*
" Churches " at all.[1] They are only Synagogues;—the very
same of which Christ spoke, with eternal meaning, as the
places that hypocrites would love to be seen in.[2] Here, in
St. Giles's, and the East, sister to that in St. George's and
the West, is the Church! raggedly enough curtained, surely!
Let those arches and pillars of Mr. Scott's[3] alone, young
ladies; it is *you* whom God likes to see well decorated, not
them. Keep your roses for your hair—your embroidery for
your petticoats. You are yourselves the Church, dears;
and see that you be finally adorned, as women professing
godliness,[4] with the precious stones of good works, which
may be quite briefly defined, for the present, as decorating
the entire Tabernacle; and clothing your poor sisters, with
yourselves.[5] Put roses also in *their* hair, put precious stones
also on *their* breasts; see that they also are clothed in your
purple and scarlet, with other delights;[6] that they also learn
to read the gilded heraldry of the sky; and, upon the earth,
be taught, not only the labours of it, but the loveliness.
For them, also, let the hereditary jewel recall their father's
pride, their mother's beauty; so shall your days, and theirs,
be long in the sweet and sacred land which the Lord your
God has given you:[7] so, truly, shall THE GOLD OF THAT
LAND BE GOOD, AND THERE, ALSO, THE CRYSTAL, AND THE
ONYX STONE.[8]

---

[1] [For Ruskin's discussion of the word "Church," see *Notes on the Construction
of Sheepfolds,* Vol. XII. pp. 524 *seq.*]

[2] [Matthew vi. 5.]

[3] [For Ruskin's numerous references to Sir Gilbert Scott, see General Index.]

[4] [1 Timothy ii. 10.]

[5] [Compare *Sesame and Lilies,* Vol. XVIII. pp. 176, 177.]

[6] [2 Samuel i. 24.]

[7] [Exodus xx. 12. Ruskin uses the same verse as a peroration to the chapter,
"The Heraldic Ordinaries," in *The Eagle's Nest:* see Vol. XXII. p. 287.]

[8] [See above, pp. 165, 169.]

CHAPTER VIII

THE ALPHABET

*(Chapter written to introduce the preceding Lecture ; but transposed, that
the Lecture might not be divided between two numbers)*

1. SINCE the last sentence of the preceding number of
Deucalion was written, I have been compelled, in preparing
for the arrangement of my Sheffield museum, to look with
nicety into the present relations of theory to knowledge in
geological science; and find, to my no small consternation,
that the assertions which I had supposed beyond dispute,
made by the geologists of forty years back, respecting the
igneous origin of the main crystalline masses of the primary
rocks, are now all brought again into question;[1] and that
the investigations of many of the most intelligent observers
render many former theories, in their generality, more than
doubtful. My own studies of rock structure, with reference
to landscape, have led me, also, to see the necessity of
retreating to, and securing, the very bases of knowledge in
this infinitely difficult science: and I am resolved, therefore, at once to make the series of *Deucalion* an absolutely
trustworthy foundation for the geological teaching in St.
George's schools;[2] by first sifting what is really known
from what is supposed; and then, out of things known,
sifting what may be usefully taught to young people, from
the perplexed vanity of prematurely systematic science.

2. I propose, also, in the St. George's Museum at Sheffield, and in any provincial museums hereafter connected
with it, to allow space for two arrangements of inorganic

---

[1] [The reference (as Ruskin explains in the next chapter, § 5) is to Gustaf Bischof's
Chemical Geology (for which, see p. 45 *n.*): see ch. ii. of the English translation.]
[2] [Compare *Proserpina*, Vol. XXV. p. 413.]

substances; one for mineralogists, properly so called, and the general public; the other for chemists, and advanced students in physical science. The mineralogical collection will be fully described and explained in its catalogue,[1] so that very young people may begin their study of it without difficulty, and so chosen and arranged as to be comprehensible by persons who have not the time to make themselves masters of the science of chemistry, but who may desire some accurate acquaintance with the aspect of the principal minerals which compose the world. And I trust, as I said in the preceding lecture,[2] that the day is near when the knowledge of the native forms and aspects of precious stones will be made a necessary part of a lady's education; and knowledge of the nature of the soils, and the building stones, of his native country, a necessary part of a gentleman's.

3. The arrangement of the chemical collection I shall leave to any good chemist who will undertake it: I suppose that now adopted by Mr. Maskelyne for the mineral collection in the British Museum may be considered as permanently authoritative.[3]

But the mineralogical collection I shall arrange myself, as aforesaid, in the manner which I think likely to be clearest for simple persons; omitting many of the rarer elements altogether, in the trust that they will be sufficiently illustrated by the chemical series; and placing the substances most commonly seen in the earth beneath our feet, in an order rather addressed to the convenience of memory than to the symmetries of classification.

4. In the outset, therefore, I shall divide our entire

---

[1] [See below, pp. 418 seq.]

[2] [See above, p. 194.]

[3] [Ruskin here notes in his copy for revision, "Correct"; meaning on revision to suggest some doubt whether Mr. Maskelyne's, or any other system of classification can be "permanently" authoritative (compare below, p. 418). The arrangement adopted by Professor Maskelyne at the British Museum "is virtually one published by Gustav Rose in 1852, as an improvement on the purely chemical system, and . . . is a mixed system, depending on two properties, chemical composition and crystalline form" (Introduction to the Study of Minerals, with a Guide to the Mineral Gallery, by L. Fletcher, F.R.S., 10th ed., 1906, pp. 59-69).]

collection into twenty groups, illustrated each by a separately bound portion of catalogue.[1]

These twenty groups will illustrate the native states, and ordinary combinations, of nine solid oxides, one gaseous element (fluorine), and ten solid elements, placed in the following order :—

1. Silica.	11. Carbon.
2. Oxide of Titanium.	12. Sulphur.
3. Oxide of Iron.	13. Phosphorus.
4. Alumina.	14. Tellurium.
5. Potassa.	15. Uranium.
6. Soda.	16. Tin.
7. Magnesia.	17. Lead.
8. Calcium.	18. Copper.
9. Glucina.	19. Silver.
10. Fluorine.	20. Gold.

5. A few words will show the objects proposed by this limited arrangement. The three first oxides are placed in one group, on account of the natural fellowship and constant association of their crystals.

Added to these, the next group of the alkaline earths will constitute one easily memorable group of nine oxides, out of which, broadly and practically, the solid globe of the earth is made, containing in the cracks, rents, or volcanic pits of it, the remaining eleven substances, variously prepared for man's use, torment, or temptation.

6. I put fluorine by itself, on account of its notable importance in natural mineralogy, and especially in that of Cornwall, Derbyshire, and Cumberland: what I have to say of chlorine and iodine will be arranged under the same head; then the triple group of anomalous substances created for ministry by fire, and the sevenfold group of the great metals, complete the list of substances which must be generally known to the pupils in St. George's schools.

---

[1] [The only portion, however, which was completed was No. 1, Silica.]

The phosphates, sulphates, and carbonates of the earths, will
be given with the earths ; and those of the metals, under
the metals.   The carburets, sulphurets, and phosphurets,*
under carbon, sulphur, and phosphorus.   Under glucina,
given representatively, on account of its importance in the
emerald, will be given what specimens may be desirable of
the minor or auxiliary earths—baryta, strontia, etc. ; and
under tellurium and uranium, the auxiliary metals—platinum, columbium, etc., naming them thus together, under
those themselves named from Tellus and Uranus.   With
uranium I shall place the cupreous micas, for their similarity of aspect.

7. The minerals referred to each of these twenty groups
will be further divided, under separate letters, into such
minor classes as may be convenient, not exceeding twenty :
the letters being initial, if possible, of the name of the
class ; but the letters I and J omitted, that they may not
be confused with numerals ; and any letter of important
sound in the mineral's name substituted for these, or for
any other that would come twice over.   Then any number
of specimens may be catalogued under each letter.

For instance, the siliceous minerals which are the
subject of study in the following lecture will be lettered
thus :—

A. Agate.   M. Amethyst.
C. Carnelian.   O. Opal.
H. Hyalite.   Q. Quartz.
L. Chalcedony.   S. Jasper.[1]

In which list, M is used that we may not have A repeated,
and will yet be sufficiently characteristic of Amethyst ; and

---

* I reject the modern term "sulphide" unhesitatingly.   It is as barbarous as "carbide."

---

[1] [Afterwards, Ruskin did not retain Carnelian as a separate division, classing it
instead with Jasper ; and his ultimate arrangement was : 1. Flint, "F" ; 2. Jasper,
"J" ; 3. Chalcedony, "C" ; 4. Opal, "O" ; 5. Hyalite, "H" ; 6. Quartz, "Q" ;
7. Amethyst, "M" ; and 8. Agate, "A." See below, p. 419.]

L, to avoid the repetition of C, may stand for Chalcedony; while S, being important in the sound of Jasper, will serve instead of excluded J, or pre-engaged A.

The complete label, then, on any (principally) siliceous mineral will be in such form as these following:—

1. A. 1,   meaning Silica, Agate, No. 1.
1. L. 40,   ,,   Silica, Chalcedony, No. 40.
1. Q. 520,   ,,   Silica, Quartz, No. 520.

8. In many of the classes, as in this first one of Silica, we shall not need all our twenty letters; but there will be a letter A to every class, which will contain the examples that explain the relation and connection of the rest. It happens that in Silica, the agates exactly serve this purpose; and therefore may have A for their proper initial letter. But in the case of other minerals, the letter A will not be the initial of the mineral's name, but the indication of its character, as explanatory of the succeeding series.

Thus the specimen of gold, referred to as 20. A. 1. in the preceding lecture,[1] is the first of the series exhibiting the general method of the occurrence of native gold in the rocks containing it; and the complete series in the catalogue will be—

A. Native Gold, in various geological formations.	G. Granulate Gold.
	K. Knitted Gold.
B. Branched Gold.	L. Leaf Gold.
C. Crystalline Gold.	M. Mossy Gold.
D. Dispersed Gold.	R. Rolled Gold.

9. It may be at once stated that I shall always retain the word "branched" for minerals taking either of the forms now called "arborescent" or "dendritic." The advance of education must soon make all students feel the absurdity of using the epithet "tree-like" in Latin, with a different meaning from the epithet "tree-like" in Greek. My general

---

[1] [See above, p. 175.]

word "branched" will include both the so-called "arborescent" forms (meaning those branched in straight crystals),
and the so-called "dendritic" (branched like the manganese
or oxide in Mocha stones); but with most accurate explanation of the difference; while the term "spun" will be reserved for the variously thread-like forms, inaccurately now
called dendritic, assumed characteristically by native silver
and copper.[1]

Of course, thread, branch, leaf, and grain, are all in most
cases crystalline, no less definitely than larger crystals; but
all my epithets are for practical service, not scientific definition; and I mean by "crystalline gold" a specimen which
distinctly shows octohedric or other specific form; and by
"branched gold" a specimen in which such crystalline forms
are either so indistinct or so minute as to be apparently
united into groups resembling branches of trees.

10. Every one of the specimens will be chosen for some
speciality of character; and the points characteristic of it
described in the catalogue; and whatever questions respecting its structure are yet unsolved, and significant, will be
submitted in succession, noted each by a Greek letter, so
that any given question may be at once referred to. Thus,
for instance: question *a* in example 20. G. 1. will be the
relation of the subdivided or granular condition of crystalline gold to porous states of the quartz matrix. As the

---

[1] [In the MS. of the first draft for ch. iii. in vol. ii. of *Deucalion* (see below,
p. 347 *n*.), there is the following note on this passage :—

"At page 147 [now 202] I promised an accurate account of the difference
between the forms commonly called arborescent, from the Latin word for
a tree, and dendritic, from the Greek one. When I wrote that sentence I
thought absolutely to reserve the word 'arborescent' for straight-branched
forms, but it must be extended at least so far as to include the definitely
systematized branching of ice and other truly crystalline substance on smooth
planes, as of glass,—though often more curved than straight. For the
irregular and unsystematized arborescence of manganese, and iron, I shall
reserve the term 'fucoid,' and the usually called 'dendritic' formations of
native silver, and any resembling them, I shall call 'filiform'; there being
nothing whatever resembling trees in them, but much that resembles threads.
As I find not a single word yet written by any mineralogist either on the
actual structure or the mode of production of such crystals, I may be
pardoned for deferring yet for a while the promised definition of them."
For the following passage of MS., see below, p. 211 *n*.]

average length of description required by any single specimen, chosen on such principle, ought to be at least half a
page of my usual type, the distribution of the catalogue
into volumes will not seem unnecessary; especially as in
due course of time, I hope that each volume will consist
of two parts, the first containing questions submitted, and
the second, solutions received.

The geological series will be distinguished by two letters
instead of one, the first indicating the principal locality of
the formation, or at least that whence it was first named.
And I shall distinguish *all* formations by their localities—
" M. L., Malham limestone"; " S. S., Skiddaw slate"; etc.,
—leaving the geologists to assign systematic or chronological names as they like. What is pliocene to-day may
be pleistocene to-morrow; and what is triassic in Mr. A.'s
system, tesserassic in Mr. B.'s; but Turin gravels and War-
wick sands remain where they used to be, for all that.

These particulars being understood, the lecture which
I gave this spring[1] on the general relations of precious
minerals to human interests, may most properly introduce
us to our detailed and progressive labour; and two paragraphs of it, incidentally touching upon methods of public
instruction, may fitly end the present chapter.[2]

11. In all museums intended for popular teaching, there
are two great evils to be avoided. The first is, super-
abundance; the second, disorder. The first is having too
much of everything. You will find in your own work that
the less you have to look at, the better you attend. You
can no more see twenty things worth seeing in an hour,
than you can read twenty books worth reading in a day.
Give little, but that little good and beautiful, and explain
it thoroughly. For instance, here in crystal, you may have
literally a thousand specimens, every one with something
new in it to a mineralogist; but what is the use of that to

---

[1] [The lecture printed in the preceding chapter.]
[2] [With §§ 11, 12 compare, in a later volume of this edition, the " Letters on a
Museum or Picture Gallery."]

a man who has only a quarter of an hour to spare in a week? Here are four pieces—showing it in perfect purity, —with the substances which it is fondest of working with, woven by it into tissues as fine as Penelope's; and one crystal of it stainless, with the favourite shape it has here in Europe—the so-called "flute-beak"[1] of Dauphiné,—let a man once understand that crystal, and study the polish of this plane surface, given to it by its own pure growth, and the word "crystal" will become a miracle to him, and a treasure in his heart for evermore.

12. Not too much, is the first law; not in disorder, is the second. Any order will do, if it is fixed and intelligible: no system is of use that is disturbed by additions, or difficult to follow; above all, let all things, for popular use, be *beautifully* exhibited. In our own houses, we may have our drawers and bookcases as rough as we please; but to teach our people rightly, we must make it a true joy to them to see the pretty things we have to show: and we must let them feel that, although, by poverty, they may be compelled to the pain of labour, they need not, by poverty, be debarred from the felicity and the brightness of rest; nor see the work of great artists, or of the great powers of nature, disgraced by commonness and vileness in the manner of setting them forth. Stateliness, splendour, and order are above all things needful in places dedicated to the highest labours of thought: what we willingly concede to the Graces of Society, we must reverently offer to the Muses of Seclusion; and out of the millions spent annually to give attractiveness to folly, may spare at least what is necessary to give honour to Instruction.

---

[1] [See Q. 12 in the Sheffield Catalogue, below, p. 432.]

CHAPTER IX

FIRE AND WATER

1. In examining any mineral, I wish my pupils first to be able to ascertain easily what it is; then to be accurately informed of what is *known* respecting the processes of its formation; lastly, to examine, with such precision as their time or instruments may permit, the effects of such formation on the substance. Thus, from almost any piece of rock, in Derbyshire, over which spring water has trickled or dashed for any length of time, they may break with a light blow a piece of brown incrustation, which, with little experience, they may ascertain to be carbonate of lime;—of which they may authoritatively be told that it was formed by slow deposition from the dripping water;—and in which, with little strain of sight, they may observe structural lines, vertical to the surface, which present many analogies with those which may be seen in coats of semi-crystalline quartz, or reniform chalcedony.

2. The more accurate the description they can give of the aspect of the stone, and the more authoritative and sifted the account they can render of the circumstances of its origin, the greater shall I consider their progress, and the more hopeful their scientific disposition.

But I absolutely forbid their proceeding to draw any logical inferences from what they know of stalagmite, to what they don't know of chalcedony. They are not to indulge either their reason or their imagination in the feeblest flight beyond the verge of actual experience; and they are to quench, as demoniacal temptation, any disposition they find in themselves to suppose that, because stalagmite and chalcedony both show lines of structure vertical to reniform surface, both have been deposited in a similar manner from a current solution. They are to address themselves to the investigation of the chalcedony precisely as if no stalagmite were in existence,—to inquire first what it is; secondly, when and how it is *known to be* formed; and, thirdly, what structure is discernible in it,—leaving to the close of their lives, and of other people's, the collection, from evidence thus securely accumulated, of such general conclusions as may then, without dispute, and without loss of time through prejudice in error, manifest themselves, not as "theories," but as demonstrable laws.

When, however, for the secure instruction of my thus restrained and patient pupils, I look, myself, for what is actually told me by eye-witnesses, of the formation of mineral bodies, I find the sources of information so few, the facts so scanty, and the connecting paste, or diluvial detritus, of past guesses, so cumbrously delaying the operation of rational diamond-washing, that I am fain, as the shortest way, to set such of my friends as are minded to help me, to begin again at the very beginning; and *r*eassert, for the general good, what their eyes can now see, in what their hands can now handle.

3. And as we have begun with a rolled flint, it seems by special guidance of Fors that the friend who has already first contributed to the art-wealth of the Sheffield Museum,[1] Mr. Henry Willett, is willing also to be the first contributor to its scientific treasuries of fact; and has set himself zealously to collect for us the phenomena observable in the chalk and flint of his neighbourhood.

Of which kindly industry, the following trustworthy notes [§ 12] have been already the result, which (whether the like observations have been made before or not being quite immaterial to the matter in hand) are assuredly themselves original and secure: not mere traditional gossip. Before giving them, however, I will briefly mark their relations to the entire subject of the structure of siliceous minerals.

4. There are a certain number of rocks in the world,

---

[1] [The reference is to some Japanese inlaid work presented to the Museum in 1876: see *Fors Clavigera*, Letter 64, § 20. For a letter to Mr. Willett about *Deucalion*, see the Introduction, above, p. xxxiii.]

which have been seen by human eyes, flowing, white-hot, and watched by human eyes as they cool down. The structure of these rocks is therefore absolutely known to have had something to do with fire.

There are a certain number of other rocks in the world which have been seen by human eyes in a state of wet sand or mud, and which have been watched, as they dried, into substances more or less resembling stone. The structure of these rocks is therefore known to have had something to do with water.

Between these two materials, whose nature is avouched by testimony, there occur an indefinite number of rocks, which no human eyes have ever seen, either hot or muddy; but which nevertheless show curious analogies to the ascertainably cooled substances on the one side, and to the ascertainably dried substances on the other. Respecting these medial formations, geologists have disputed in my ears during the half-century of my audient life (and had been disputing for about a century before I was born), without having yet arrived at any conclusion whatever; the book now held to be the principal authority on the subject, entirely contradicting, as aforesaid,[1] the conclusions which, until very lately, the geological world, if it had not accepted as incontrovertible, at least asserted as positive.

5. In the said book, however,—Gustaf Bischof's *Chemical Geology*,—there are, at last, collected a large number of important and secure facts, bearing on mineral formation: and principles of microscopic investigation have been established by Mr. Sorby, some years ago,[2] which have, I doubt not, laid the foundation, at last, of the sound knowledge of the conditions under which crystals are formed in mountains. Applying Mr. Sorby's method, with steady industry, to the rocks of Cumberland Mr. Clifton Ward has, so far as I can

---

[1] [See above, ch. viii. § 1, p. 197.]

[2] ["On the Microscopical Structure of Crystals, indicating the Origin of Minerals and Rocks," by H. C. Sorby, F.R.S., in the *Quarterly Journal of the Geological Society*, 1858, vol. xiv. pp. 453–500. On this paper as marking "one of the most prominent epochs of modern geology," see Geikie's *Founders of Geology*, 1897, p. 279. Ruskin was acquainted with Mr. Sorby, and refers to his experiments in another direction in Vol. XVII. p. 451.]

judge, placed the nature of *these*, at least, within the range of secure investigation.[1] Mr. Ward's kindness has induced him also to spare the time needful for the test of the primary phenomena of agatescent structure in a similar manner; and I am engraving the beautiful drawings he sent me, with extreme care, for our next number;[2] to be published with a letter from him, containing, I suppose, the first serviceable description of agatescent structure yet extant.*

6. Hitherto, however, notwithstanding all that has been accomplished, nobody can tell us how a common flint is made. Nobody ever made one; nobody has ever seen one naturally coagulate, or naturally dissolve; nobody has ever watched their increase, detected their diminution, or explained the exact share which organic bodies have in their formation. The splendid labours of Mr. Bowerbank have made us acquainted with myriads of organic bodies which have provoked siliceous concretion, or become entangled in it:[3] but the beautiful forms which these present have only increased the difficulty of determining the real crystalline modes of siliceous structure, unaffected by organic bodies.

7. Crystalline *modes*, I say, as distinguished from crystalline *laws*. It is of great importance to mineralogy that we

---

* I must, however, refer the reader to the valuable summary of work hitherto done on this subject by Professor Rupert Jones (*Proceedings of Geologists' Association*, vol. iv., No. 7) for examination of these questions of priority.[4]

---

[1] [*Memoirs of the Geological Survey: England and Wales. The Geology of the Northern Part of the English Lake District*, by J. Clifton Ward, 1876. The district was surveyed, as a preliminary "Notice" states, by Mr. W. T. Aveline and Mr. Ward.]

[2] [See below, p. 241, and Plate XIV.]

[3] [The reference is to successive Memoirs contributed in 1840, 1841, and 1849, by Dr. J. S. Bowerbank, F.R.S., on the silicification of sponges and other zoophytes, and on the formation of flint and other siliceous substances. See *Proceedings of the Geological Society*, vol. iii. pp. 278, 431; *Transactions of the Geological Society*, Series 2, vol. vi. p. 181; and *Quarterly Journal of the Geological Society*, vol. v. p. 319. Ruskin occasionally corresponded with Dr. Bowerbank: see below, p. 504.]

[4] ["On Quartz, Chalcedony, Agate, Flint, Chert, Jasper, and other Forms of Silica Geologically Considered," at pp. 439–458 of vol. iv. (1874–1876). Professor Rupert Jones's paper is for the most part a succinct account of contributions to the literature of the subject. Under the head of agates he refers (p. 445) to "The series of elaborate papers by Mr. Ruskin 'On Banded and Brecciated Concretions,' illustrated by some exquisite plates and many clear diagrams," as "a most valuable source of facts and suggestions to the student of agates."]

should carefully distinguish between the laws or limits which determine the possible angles in the form of a mineral, and the modes, or measures, in which, according to its peculiar nature or circumstances, it conducts itself under these restrictions.

Thus both cuprite and fluor are under laws which enforce cubic or octohedric angles in their crystals; but cuprite can arrange its cubes in fibres finer than those of the softest silk, while fluor spar only under rare conditions distinctly elongates its approximate cube into a parallelopiped.

Again, the prismatic crystals of wavellite arrange themselves invariably in spherical or reniform concretions;[1] but the rhombohedral crystals of quartz and hematite do so only under particular conditions, the study of which becomes a quite distinct part of their lithology.

8. This stellar or radiant arrangement is one essential condition in the forms and phenomena of agate and chalcedony; and Mr. Clifton Ward has shown in the paper to which I have just referred,[2] that it is exhibited under the microscope as a prevalent condition in their most translucent substance, and on the minutest scale.

Now all siliceous concretions, distinguishing themselves from the mass of the surrounding rocks, are to be arranged under two main classes; briefly memorable as knots, and nuts; the latter, from their commonly oval form, have been usually described by mineralogists as, more specially, "almonds."

"Knots" are concretions of silica round some central point or involved substance (often organic); such knots being usually harder and more solid in the centre than at the outside, and having their fibres of crystallization, if visible, shot outwards like the rays of a star, forming pyramidal crystals on the exterior of the knot.

9. "Almonds" are concretions of silica formed in cavities of rocks, or, in some cases, probably by their own energy producing the cavities they enclose; the fibres of crystallization,

---

[1] [Compare "Banded and Brecciated Concretions," § 12 (III.), above, p. 47.]
[2] [In the *Memoirs of the Geological Survey*, as noted above, p. 208 *n.*; see the earlier chapters, describing the rocks.]

if visible, being directed from the outside of the almond-shell towards its interior cavity.

10. These two precisely opposite conditions are severally represented best by a knot of sound black flint in chalk, and by a well-formed hollow agate in a volcanic rock.

I have placed in the Sheffield Museum a block of black flint, formed round a bit of Inoceramus shell; and an almond-shell of agate, about six times as big as a cocoa-nut, which will satisfactorily illustrate these two states.[1] But between the two, there are two others of distinctly gelatinous silica, and distinctly crystalline silica, filling pores, cavities, and veins, in rocks, by infiltration or secretion. And each of these states will be found passing through infinite gradations into some one of the three others, so that separate account has to be given of every step in the transitions before we can rightly understand the main types.

11. But at the base of the whole subject lies, first, the clear understanding of the way a knot of solid crystalline substance—say, a dodecahedral garnet—forms itself out of a rock-paste, say greenstone trap, without admitting a hair's-breadth of interstice between the formed knot and enclosing paste; and, secondly, clear separation in our thoughts, of the bands or layers which are produced by crystalline segregation, from those produced by successively accumulating substance. But the method of increase of crystals themselves, in an apparently undisturbed solution, has never yet been accurately described; how much less the phenomena resulting from influx of various elements, and changes of temperature and pressure. The frontispiece to the third number of *Deucalion*[2] gives typical examples of banded structure resulting from pure crystalline action; and the three specimens, 1. A. 21, 22, and 23, at Sheffield,[3] furnish parallel examples of extreme interest. But a particular

---

[1] [The "block of black flint" is F 9 in the Catalogue: see below, p. 422. The "almond-shell of agate" is not mentioned in the Catalogue. It is in a glass case by itself in the Museum; it is a large hollow geode, lined with quartz, measuring 15 in. × 8 × 9.]

[2] [As originally published. In this volume, Plate X. ("Mural Agates")—a plate which Ruskin repeated in *Deucalion* from the *Geological Magazine*.]

[3] [See below, p. 436.]

form of banding in flint, first noticed and described by
Mr. S. P. Woodward,* is of more interest than any other
in the total obscurity of its origin; and in the extreme
decision of the lines by which, in a plurality of specimens,
the banded spaces are separated from the homogeneous
ones, indicating the first approach to the conditions which
produce, in more perfect materials, the forms of, so-called,
"brecciated" agates. Together with these, a certain number
of flints are to be examined which present every appearance of having been violently fractured and re-cemented.
Whether fractured by mechanical violence, by the expansive
or decomponent forces of contained minerals, or by such slow
contraction and re-gelation as must have taken place in most
veins through masses of rock, we have to ascertain by the
continuance of such work as my friend has here begun.[1]

* *Geological Magazine,* 1864, vol. i., p. 145, Pl. VII. and VIII.[2]

---

[1] [The MS., cited above at p. 202 n., goes on to summarise the analysis given
in the present chapter :—

"There have been indicated, however, already in the course of the
preceding analysis, seven different 'modes' (compare Chap. IX. § 7*)
of crystallization, producing specifically different characters in groups of
crystals, namely :—

"1. Tendency to extended rather than massive forms; giving rise to
a generally tabular, rather than columnar structure (§ 5).

"2. Symmetrical attachment of arborescent groups of crystals to a
common centre (§ 6).

"3. Unsymmetrical attachment of arborescent groups of crystals variously
directed (§ 7 and § 10).

"4. Lateral aggregation of crystals without arborescence, and under no
uniform crystalline power (§ 11).

"5. Lateral aggregation of crystals without arborescence, but under
uniform crystalline power (§ 12).

"6. Lateral aggregation of crystals in fucoid instead of arborescent
groups (§ 13).

"7. Superimposed aggregation of crystals in irregularly elongated groups,
giving rise to a generally columnar or filiform, rather than tabular structure.

"And yet not one of these always wonderful and for the most part
beautiful conditions of mineral aggregation have even been described by
mineralogists, or even alluded to as implying any connected causes, or
any practically interesting effects, though every one of them is of course
to be traced to some perfectly distinct cause; and though the last of them
is among the most important practical powers in the Universe, as I hope
in some extent to show in the following chapter."

* See also my first statements on this subject at pages 12 and 13 of the *Geological
Magazine* for January 1870 [above, pp. 81-83].

The "following chapter" was, however, not written.]

[2] [Referred to also in the papers on "Banded and Brecciated Concretions":
see above, p. 47.]

Letter 1.*—*Introductory.*

12. " I am beginning to be perplexed about the number of flints, containing problems and illustrations, and wondering to what extent my inquiries will be of any use to you.

" I intended at first to collect only what was really beautiful in itself — " crystalline " ! but how the subject widens, and how the arbitrary divisions do run into one another ! What a paltry shifting thing our classification is ! One is sometimes tempted to give it all up in disgust, and I have a shrewd suspicion that all scientific classification (except for mutual aid to students) is absurd and pedantic : (*a*) varieties, species, genera, classes, orders, have most of them more in common than of divergence,—" a forming spirit" everywhere, for use and beauty.

" It is (to me) impossible to separate purely mineral and chemical siliceous bodies in chalk (*b*) from those which are partly formed by the silicate-collecting sponges, which seem to have given them their forms.

" Who is to say that the radiations and accretions of a crystal are not life, but that the same arrangements in a leaf or a tree are life ?—that the clouds which float in their balanced changeableness are not as much guided and defined as the clouds of the chalcedony, or the lenses of the human eye which perceives them ?

" I think the following facts are plain :

" 1. The chalk bands do go through the flint.

---

* I shall put my own notes on these and any future communications I may insert, in small print at the bottom of the pages ; and with letter-references—*a*, *b*, etc. ; but the notes of the authors themselves will be put at the end of their papers, in large print, and with number-references —1, 2, etc.[1]

(*a*) All, at least, is imperfect ; and most of it absurd in the attempt to be otherwise.

(*b*) It may be doubtful if any such exist in chalk ; but, if they exist, they will eventually be distinguishable.

---

[1] [Only one such note was, however, given ; and in this edition it is transferred from the end of the chapter to its place in the text (p. 213.]

"2. Fissures in flints are constantly repaired by fresh deposits of chalcedony and silex.

"3. Original sponge matter is preserved (*c*) and obliterated by siliceous deposit, in extent and degree varying infinitely, and apparently proportioned to the amount of iron present—*i.e.*, the iron preserves original form, unless when combined with sulphur enough to crystallize, when all the original structure disappears.

"4. Amygdaloids seem to be formed by a kind of independent or diverse arrangement of molecules, caused by slight admixture of foreign minerals."

LETTER II.—*Memoranda made at Mantell's Quarry, Cuckfield, on the banding noticed in the beds and nodules of the siliceous calciferous sandstone there, 31st May,* 1876.

Nos. I. and II. Ovate, concentric, ferruginous bandings; the centre apparently (1) free from banding.

III. Bands arranged at acute angles. These bands are not caused by fracture, but apparently by the intersection, at an acute angle, of the original lines of deposit. (*d*)

IV. In this specimen the newly fractured surfaces show no bandings, but the weathered surface develops the banding.

V. Ditto—*i.e.*, bands parallel; much more ferruginous, and consequently more friable when exposed to weathering.

May not something be learnt regarding the laws of banding in agates, flints, etc., from observing the arrangement of banding in rocks composed mainly of siliceous matter? (*e*)

May not some of the subtler influences which regulate the growth of trees in their lines of annual increase (magnetic probably) have some effect in the arrangement of minerals

(*c*) Q. The form or body of it only; is the matter itself ever preserved?

(*d*) These angular concretions require the closest study; see the segments of spheres in Plate X., lowest figure.

(*e*) More, I should say, from the agates, respecting the laws of banding in rocks: see the plate to the present number [Plate VII., p. 55]. When we can explain the interruptions of the bands on such scale as this, we may begin to understand some of those in larger strata.

in solution ?—nay, even of the higher vital processes, such as the deposition of osseous matter in teeth and bones ? (*f*)

(1) Probably the same arrangement exists (concentric), but has not been made visible because the iron has not been oxydized.

LETTER III.—*Memoranda respecting banded chalk.*

i. In the banded lines (ferruginous) noticed above and below the horizontal fissures beneath the cliff at the Hope Gap, Seaford, it is evident that these lines are not markings of original deposition, but are caused by successive infiltrations of water containing iron in solution. (*g*)

ii. Concentric markings of the same nature are observable in places where—

a. Iron pyrites are decomposing, and the iron in solution is being successively infiltrated into the surrounding chalk rock.

b. From dropping of ferruginous springs through crevices on horizontal surfaces.

c. This is observable also on surfaces of tabular flint.

iii. Very peculiar contorted bandings (similar to the so-called contorted rocks) are observable in certain places, notably in the face of the chalk-pit on the east side of Goldstone Bottom. This chalk-pit, or quarry, is remarkable—

1. *For the contorted bandings in the chalk rock, which are not markings of original deposition, being quite independent of original stratification.* (*h*)

(*f*) Yes, certainly; but in such case, the teeth and bones act by mineral law; not the minerals by teeth and bone law.

(*g*) Questionable. Bands are almost always caused by concretion, or, separation, not infiltration. However caused, the essential point, in the assertion of which this paper has so great value, is their distinction from strata.

(*h*) A most important point. It is a question with me whether the greater number of minor contortions in Alpine limestones may not have been produced in this manner.[1] When once the bands are arranged by segregation, chemical agencies will soon produce mechanical separation, as of original beds.

---

[1] [Compare " Distinctions of Form in Silica," § 29 (below, p. 386).]

2. For the excessive shattering and fissuring observable.

3. For the fact that these cracks and fissures have been refilled with distinctive and varying substances, as with flint, clay, Websterite, and intermediate admixtures of these substances.

4. For veins of flint, formerly horizontal, which show visible signs of displacement by subsidence.

5. For the numerous fissures in these veins of tabular flint being stained by iron, which apparently aids in the further process of splitting up and of widening the minute crevices in the flint. The iron also appears to be infiltrated at varying depths into the body of unfractured flint.

Qy. Has not ordinary flint the power or property of absorbing ferruginous fluid?

LETTER IV.—*Memoranda respecting brecciate flint.*

"*June* 7, 1876.

"I hasten to report the result of my fresh inquiry respecting the specimen I first sent to you as "breccia," but which you doubted.

"The site is the embouchure of the little tidal river Cuckmere, about two miles east of Seaford. I found a block at about the same spot (about three hundred yards east of the coastguard station, and about three-quarters of the distance west of the river's mouth).

"The rocks are here covered with sand, or with a bed of the old valley alluvium, not yet removed by wave action. Travelling westward, the transported blocks of breccia gradually increase in size (a pretty sure augury that they were derived from a western source). The whole coast is subject to a very rapid degradation and consequent encroachment of the sea, the average in some places being from twenty-five to thirty feet yearly. At a spot a hundred yards east of the coastguard station, blocks of one or two tons were visible. The denuded chalk rock is of chalk, seamed and fissured; the cliff of the same nature; but all the flints, and especially the tabular veins, are splintered and displaced to an unusual extent.

" Farther westward yet, the blocks of breccia weigh several tons, the cement being itself fissured, and in some places consisting of angular fragments stained with iron. From one mass I extracted a hollow circular flint split into four or five pieces, the fragments, although displaced, re-cemented in juxtaposition. (*i*)

" At the Hope Gap, the whole cliff becomes a fractured mass, the fissures being refilled, sometimes with calcareous cement, sometimes with clay, and in other places being hollow.

" From the sides of an oblique fissure filled with clay I extracted two pieces of a nodular flint, separated from each other by a two-inch seam of clay: when replaced (the clay having been removed) the two fitted exactly. An examination of the rocks shows that the fissures, which run in all directions, are largest when *nearly horizontal*, dipping slightly seawards.

" The upper and lower portions of some of these horizontal fissures are banded with iron stains, evidently derived from iron-water percolating the seams.

" If I am right, therefore, the mystery seems to be explained thus: (*k*)—

" I. Rain water, charged with carbonic acid, falling on the hills behind, trickles past the grass and humus beneath, through the cracks in the chalk, dissolving the carbonate of lime into a soluble bi-carbonate. Falling downwards, it escapes seawards through the horizontal fissures, widening them by its solvent power.

" II. The weight of the superincumbent mass by slow,

(*i*) I am not prepared to admit, yet, that any of these phenomena are owing to violence. We shall see.

(*k*) I think this statement of Mr. Willett's extremely valuable; and see no reason to doubt its truth, as an explanation of the subsidence of chalk and limestone in certain localities. I do not hitherto receive it as any explanation of fracture in flints. I believe Dover Cliffs might sink to Channel bottom without splitting a flint, unless bedded.

certain, irregular pressure, descends, maintaining the contact of surfaces, but still ever sinking at intervals, varied by the resisting forces of weight and pressure.

"III. This process is probably accelerated by the inflow and reflow of salt water at the ebb and flow of tide (into the fissures).

"IV. At certain periods, probably in the summer (as soluble bi-carbonate of lime becomes less soluble as temperature increases), a portion becomes redeposited as a hard semi-crystalline calcareous cement.

"V. This cement appears, in some instances, to be slightly siliceous, and may have a tendency, by the mutual attraction of siliceous matter, to form solid layers of tabular flint.

"VI. If these deductions be correct, it is probable that the great results involved in the sinking of limestone hills, and the consequent encroachment of the sea, may be traced (step by step) to the springs in valleys 'which run among the hills'; thence to the rain and dewdrops; higher up to the mists and clouds; and so onward, by solar heat, to the ocean, where at last again they find their rest."

LETTER V.—*Final Abstract.*

"*June* 13, 1876.

"In addition to the heat derived from summer and atmospheric changes, there will be a considerable amount of heat evolved from the friction produced between the sides of fissures when slipping and subsidence occur, and from the crushing down of flint supports when weight overcomes resistance.

"After heavy rainfall—

1. Fissures are filled.
2. Solution is rapid.
3. Hydraulic pressure increases.
4. Fissures are widened.

" After a period of dry weather—

1. Solution is diminished.
2. Hydraulic pressure relieved.
3. Subsidence and flint-crushing commence, or progress more rapidly.
4. Heat is evolved.
5. Carbonic acid discharged.
6. Semi-crystalline carbonate of lime is deposited around
    a. Fragments of crushed flint (at rest at intermitting intervals between motion of rocks).
    b. Angular fragments of original chalk rock.
    c. Angular fractured pieces of old cement.

" I have a dawning suspicion that siliceous deposits (as chalcedony, etc.) are made when the temperature falls, for reasons which I must postpone to a future paper."

CHAPTER X

"THIRTY YEARS SINCE"[1]

VILLAGE OF SIMPLON, 2nd September, 1876.

1. I AM writing in the little one-windowed room opening from the salle-à-manger of the Hôtel de la Poste; but under some little disadvantage, being disturbed partly by the invocation, as it might be fancied, of calamity on the heads of nations, by the howling of a frantic wind from the Col; and partly by the merry clattering of the knives and forks of a hungry party in the salon doing their best to breakfast adequately, while the diligence changes horses.

In that same room,—a little earlier in the year,—two-and-thirty years ago, my father and mother and I were sitting at one end of the long table in the evening: and at the other end of it, a quiet, somewhat severe-looking, and pale, English (as we supposed) traveller, with his wife; she, and my mother, working; her husband carefully completing some mountain outlines in his sketch-book.

2. Those days are become very dim to me; and I forget which of the groups spoke first. My father and mother were always as shy as children; and our busy fellow-traveller seemed to us taciturn, slightly inaccessible, and even Alpestre, and, as it were, hewn out of mountain flint, in his serene labour.

Whether some harmony of Scottish accent struck my father's ear, or the pride he took in his son's accomplishments prevailed over his own shyness, I think we first ventured word across the table, with view of informing the grave draughtsman that *we* also could draw. Whereupon my own sketch-book was brought out, the pale traveller

---

[1] [Adapted from Scott's title, "*Waverley; or, 'Tis Sixty Years Since.*"]

politely permissive.[1]   My good father and mother had
stopped at the Simplon for me (and now, feeling miserable
myself in the thin air, I know what it cost them), because
I wanted to climb the high point immediately west of the
Col, thinking thence to get a perspective of the chain
joining the Fletschhorn to the Monte Rosa.   I had been
drawing there the best part of the afternoon, and had
brought down with me careful studies of the Fletschhorn
itself, and of a great pyramid far westward,[2] whose name I
did not know, but, from its bearing, supposed it must be
the Matterhorn, which I had then never seen.

3. I have since lost both these drawings; and if they
were given away, in the old times when I despised the
best I did, because it was not like Turner, and any friend
has preserved them, I wish they might be returned to me;
for they would be of value in *Deucalion*, and of greater
value to myself; as having won for me, that evening, the
sympathy and help of James Forbes.   For his eye grew
keen, and his face attentive, as he examined the drawings;
and he turned instantly to me as to a recognized fellow-
workman,—though yet young, no less faithful than himself.

He heard kindly what I had to ask about the chain I
had been drawing; only saying, with a slightly proud
smile, of my peak supposed to be the Matterhorn,* "No,
—and when once you have seen the Matterhorn, you will
never take anything else for it!"

He told me as much as I was able to learn, at that
time, of the structures of the chain, and some pleasant
general talk followed; but I knew nothing of glaciers then,
and he had his evening's work to finish.   And I never
saw him again.

I wonder if he sees me now, or guided my hand as I

* It was the Weisshorn.

[1] [Compare the passage from Ruskin's diary given in the Introduction, above,
p. xxi.]

[2] ["Westward" is Ruskin's correction, in his copy, of "eastward" in the original
edition.]

cut the leaves of M. Viollet-le-Duc's *Massif du Mont Blanc*[1] this morning, till I came to page 58,—and stopped!

I must yet go back, for a little while, to those dead days.

4. Failing of Matterhorn on this side of the valley of the Rhone, I resolved to try for it from the other; and begged my father to wait yet a day for me at Brieg.

No one, then, had ever heard of the Bell Alp; and few English knew even of the Aletsch glacier. I laid my plans from the top of the Simplon Col; and was up at four, next day;[2] in a cloudless morning, climbing the little rock path which ascends directly to the left, after crossing the bridge over the Rhone, at Brieg; path which is quite as critical a little bit of walking as the Ponts of the Mer de Glace; and now, encumbered with the late fallen shatterings of a flake of gneiss of the shape of an artichoke leaf, and the size of the stern of an old ship of the line, which has rent itself away, and dashed down like a piece of the walls of Jericho, leaving exposed, underneath, the undulatory surfaces of pure rock, which, I am under a very strong impression, our young raw geologists take for real "muttoned"[3] glacier tracks.*

5. I took this path because I wanted first to climb the

---

* I saw this wisely suggested in a recent number of the *Alpine Journal*.[4]

---

[1] [*Le Massif du Mont Blanc. Étude sur sa constitution géodésique et géologique, sur les transformations et sur l'état ancien et moderne de ses glaciers*, par E. Viollet-le-Duc: Paris, 1876. An English translation of the book, by R. Bucknall, appeared in 1877. At p. 58 there are references to "les belles observations de M. Tyndall," and some rough woodcuts intended to illustrate the action of ice on the bed of a glacier.]

[2] [Ruskin in his copy writes "Correct," his ascent having in fact begun at ten in the morning of the day, on the evening of which he saw Forbes: see again the passage from Ruskin's diary of 1844 given in the Introduction, above, p. xxi.]

[3] ["The result of the attrition of fixed rocks attributed to glaciers is threefold. In the first place, the surface of rock, instead of being jagged, rugged, or worn into deep defiles, is even and rounded, often dome-shaped or spheroidal, showing the structure of the rock in section, and occasionally so smooth as to be difficultly accessible, as at the Höllenplatte near the Handeck. Such surfaces were called *roches moutonnées* by De Saussure" (Forbes: *Travels through the Alps of Savoy*, p. 52, ed. 1900).]

[4] [The reference is to a note headed "Roches Moutonnées," in the number for February 1876: vol. vii. p. 401.]

green wooded mass of the hill rising directly over the
valley, so as to enfilade the entire profiles of the opposite
chain, and length of the valley of the Rhone, from its
brow.

By midday I had mastered it, and got up half as high
again, on the barren ridge above it, commanding a little
tarn; whence, in one panorama are seen the Simplon and
Saas Alps on the south, with the Matterhorn closing the
avenue of the valley of St. Nicolas; and the Aletsch Alps
on the north, with all the lower reach of the Aletsch
glacier. This panorama I drew carefully; and slightly
coloured afterwards, in such crude way as I was then
able; and fortunately not having lost this, I place it in
the Sheffield Museum,[1] for a perfectly trustworthy witness
to the extent of snow on the Breithorn, Fletschhorn, and
Montagne de Saas, thirty years ago.

My drawing finished, I ran round and down obliquely
to the Bell Alp, and so returned above the gorge of the
Aletsch torrent—making some notes on it afterwards used
in *Modern Painters*,[2] many and many such a day of foot
and hand labour having been needed to build that book,
in which my friends nevertheless, I perceive, still regard
nothing but what they are pleased to call its elegant language, and are entirely indifferent, with respect to that and
all other books they read, whether the elegant language
tells them truths or lies.

That book contains, however (and to-day it is needful
that I should not be ashamed in this confidence of boasting), the first faithful drawings ever given of the Alps not
only in England, but in Europe; and the first definitions
of the manner in which their forms have been developed
out of their crystalline rocks.

6. "Definitions" only, observe, and descriptions; but no
"explanations." I knew, even at that time, far too much
of the Alps to theorize on them; and having learned,

---

[1] [Where it remains, in the Picture Gallery. The drawing is in five pieces.]
[2] [See vol. iv. ch. xvi. § 40 (Vol. VI. p. 315).]

in the thirty years since, a good deal more, with the only
consequence of finding the facts more inexplicable to me
than ever, laid M. Viollet-le-Duc's book on the seat of
the carriage the day before yesterday, among other stores
and preparations for passing the Simplon, contemplating on
its open first page the splendid dash of its first sentence
into space,—"La croûte terrestre, refroidie au moment du
plissement—qui a formé le massif du Mont Blanc,"—with
something of the same amazement, and same manner of
the praise, which our French allies are reported to have
rendered to our charge at Balaclava:—

> "C'est magnifique;—mais ce n'est pas"—la geologie.[1]

7. I soon had leisure enough to look farther, as the
steaming horses dragged me up slowly round the first ledges
of pines, under a drenching rain which left nothing but
their nearest branches visible. Usually, their nearest
branches, and the wreaths of white cloud braided among
them, would have been all the books I cared to read; but
both curiosity and vanity were piqued by the new utter-
ances, prophetic, apparently, in claimed authority on the
matters timidly debated by me in old time.

I soon saw that the book manifested, in spite of so
great false-confidence, powers of observation more true in
their scope and grasp than can be traced in any writer
on the Alps since De Saussure. But, alas, before we had
got up to Berisal, I had found also more fallacies than I
could count, in the author's first statements of physical
law; and seen, too surely, that the poor Frenchman's keen
natural faculty, and quite splendid zeal and industry, had
all been wasted, through the wretched national vanity
which made him interested in Mont Blanc only "since it
became a part of France,"[2] and had thrown him totally into

---

[1] ["C'est . . . mais ce n'est pas la guerre:" the saying attributed to General
Pélissier.]
[2] [See M. Viollet-le-Duc's introduction to Le Massif du Mont Blanc, p. vii.]

the clique of Agassiz and Desor;[1] with results in which neither the clique, nor M. Viollet, are likely, in the end, to find satisfaction.

8. Too sorrowfully weary of bearing with the provincial temper, and insolent errors, of this architectural restoration of the Gothic globe, I threw the book aside, and took up my Cary's *Dante*, which is always on the carriage seat, or in my pocket—not exactly for reading, but as an antidote to pestilent things and thoughts in general; and store, as it were, of mental quinine,—a few lines being usually enough to recover me out of any shivering marsh fever fit, brought on among foulness or stupidity.

It opened at a favourite old place, in the twenty-first canto of the *Paradise* (marked with an M. long ago, when I was reading Dante through to glean his mountain description [2]) :—

> " 'Twixt either shore
> Of Italy, nor distant from thy land," etc.; [3]

and I read on into the twenty-second canto, down to St. Benedict's

> " There, all things are, as they have ever been;
> Our ladder reaches even to that clime,
> Whither the patriarch Jacob saw it stretch
> Its topmost round, when it appeared to him
> With angels laden. But to mount it now
> None lifts his foot from earth; and hence my rule
> Is left a profitless stain upon the leaves.
> The walls, for abbey reared, turned into dens;
> The cowls, to sacks choked up with musty meal.
>
> .   .   .   .
>
> His convent, Peter founded without gold
> Or silver; I, with prayers and fasting, mine;
> And Francis, his, in meek humility.
> And if thou note the point whence each proceeds,
> Then look what it hath erred to, thou shalt find
> The white turned murky.
>   Jordan was turned back,
> And a less wonder than the refluent sea
> May, at God's pleasure, work amendment here."

---

[1] [See the Introduction, above, p. xxxvi.]
[2] [See *Modern Painters*, vol. iii. (Vol. V. pp. 303 *seq.*).]
[3] [Canto xxi., lines 95 *seq.* in Cary; the passage in Canto xxii. is lines 66–76, 86–93.

9. I stopped at this (holding myself a brother of the
third order of St. Francis[1]), and began thinking how long
it would take for any turn of tide by St. George's work,
when a ray of light came gleaming in at the carriage
window, and I saw, where the road turns into the high
ravine of the glacier galleries, a little piece of the Breithorn snowfield beyond.

Somehow, I think, as fires never burn, so skies never
clear, while they are watched; so I took up my Dante
again, though scarcely caring to read more; and it opened,
this time, not at an accustomed place at all, but at the
"I come to aid thy wish," of St. Bernard, in the thirty-
first canto. Not an accustomed place, because I always
think it very unkind of Beatrice to leave him to St.
Bernard; and seldom turn expressly to the passage: but it
has chanced lately to become of more significance to me,
and I read on eagerly, to the "So burned the peaceful
oriflamme,"[2] when the increasing light became so strong
that it awaked me, like a new morning; and I closed the
book again, and looked out.

We had just got up to the glacier galleries, and the
last films of rain were melting into a horizontal bar of
blue sky which had opened behind the Bernese Alps.

I watched it for a minute or two through the alternate
arch and pier of the glacier galleries, and then as we got
on the open hill flank again, called to Bernardo * to stop.

10. Of all views of the great mountains that I know
in Switzerland, I think this, of the southern side of the
Bernese range from the Simplon, in general the most dis-
appointing—for two reasons: the first, that the green mass
of their foundation slopes so softly to the valley that it
takes away half the look of their height; and the second,

---

* Bernardo Bergonza, of the Hôtel d'Italie, Arona, in whom any friend
of mine will find a glad charioteer; and they cannot anywhere find an
abler or honester one.

---

[1] [See the Introduction to Vol. XXIII. p. xlvii.]
[2] [Cary's Dante, *Paradiso*, xxxi. 61–119.]

that the greater peaks are confused among the crags immediately above the Aletsch glacier, and cannot, in quite clear weather, be recognized as more distant, or more vast. But at this moment, both these disadvantages were totally conquered. The whole valley was full of absolutely impenetrable wreathed cloud, nearly all pure white, only the palest grey rounding the changeful domes of it; and beyond these domes of heavenly marble, the great Alps stood up against the blue,—not wholly clear, but clasped by, and intertwined with, translucent folds of mist, traceable, but no more traceable, than the thinnest veil drawn over St. Catherine's or the Virgin's hair by Lippi or Luini; and rising as they [1] were withdrawn from such investiture, into faint oriflammes, as if borne by an angel host far distant; the peaks themselves strewn with strange light, by snow fallen but that moment,—the glory shed upon them as the veil fled;—the intermittent waves of still gaining seas of light increasing upon them, as if on the first day of creation.

"À present, vous pouvez voir l'hôtel sur le Bell Alp, bâti par Monsieur Tyndall."

The voice was the voice of the driver of the supplementary pair of horses from Brieg, who, just dismissed by Bernardo, had been for some minutes considering how he could best recommend himself to me for an extra franc.

I not instantly appearing favourably stirred by this information, he went on with increased emphasis, " Monsieur le *Professeur* Tyndall."

The poor fellow lost his bonnemain by this altogether—not out of any deliberate spite of mine; but because, at this second interruption, I looked at him, with an expression (as I suppose) so little calculated to encourage his hopes of my generosity that he gave the matter up in a moment, and turned away, with his horses, down the hill;

---

[1] [Ruskin, in his copy for revision, marks this sentence as " obscure," and explains that " they" are " the clouds." Three lines above, "clasped by and intertwined" is his correction for " clasped and entwined " in the original edition.]

—I partly not caring to be further disturbed, and being besides too slow—as I always am in cases where presence of mind is needful—in calling him back again.

11. For, indeed, the confusion into which he had thrown my thoughts was all the more perfect and diabolic, because it consisted mainly in the stirring up of every particle of personal vanity and mean spirit of contention which could be concentrated in one blot of pure black ink, to be dropped into the midst of my aerial vision.

Finding it totally impossible to look at the Alps any more, for the moment, I got out of the carriage, sent it on to the Simplon village; and began climbing, to recover my feelings and wits, among the mossy knolls above the convent.

They were drenched with the just past rain; glittering now in perfect sunshine, and themselves enriched by autumn into wreaths of responding gold.

The vast hospice stood desolate in the hollow behind them; the first time I had ever passed it with no welcome from either monk, or dog. Blank as the fields of snow above, stood now the useless walls; and for the first time, unredeemed by association; only the thin iron cross in the centre of the roof remaining to say that this had once been a house of Christian Hospitallers.

12. Desolate this, and dead the office of this,—for the present, it seems; and across the valley, instead, "l'hôtel sur le Bell Alp, bâti par Monsieur Tyndall," no nest of dreamy monks, but of philosophically peripatetic or perisaltatory "puces des glaces."

For, on the whole, that is indeed the dramatic aspect and relation of them to the glaciers; little jumping black things, who appear, under the photographic microscope, active on the ice-waves, or even inside of them;—giving to most of the great views of the Alps, in the windows at Geneva, a more or less animatedly punctuate and pulicarious character.

Such their dramatic and picturesque function, to any

one with clear eyes; their intellectual function, however, being more important, and comparable rather to a symmetrical succession of dirt-bands,—each making the ice more invisible than the last; for indeed, here, in 1876, are published, with great care and expense, such a quantity of accumulated rubbish of past dejection, and moraine of finely triturated mistake, clogging together gigantic heaped blocks of far-travelled blunder,—as it takes away one's breath to approach the shadow of.

13. The first in magnitude, as in origin, of these long-sustained stupidities,—the Pierre-à-Bot,[1] or Frog-stone, *par excellence*, of the Neuchâtel clique,—is Charpentier's Dilatation Theory, revived by M. Viollet-le-Duc, not now as a theory, but an assured principle!—without, however, naming Charpentier as the author of it; and of course without having read a word of Forbes's demolition of it.[2] The essential work of *Deucalion* is construction, not demolition; but when an avalanche of old rubbish is shot in our way, I must, whether I would or no, clear it aside before I can go on. I suppose myself speaking to my Sheffield men ; and shall put so much as they need know of these logs upon the line, as briefly as possible, before them.

14. There are three theories extant, concerning glacier-motion, among the gentlemen who live at the intellectual " Hôtel des Neuchâtelois."[3] These are specifically known as the Sliding,—Dilatation,—and Regelation, theories.

When snow lies deep on a sloping roof, and is not supported below by any cornice or gutter, you know that when it thaws, and the sun has warmed it to a certain extent, the whole mass slides off into the street.

That is the way the scientific persons who hold the " Sliding theory,"[4] suppose glaciers to move. They assume,

---

[1] [The name of a famous stone near Neuchâtel: see Vol. XVII. p. 476.]

[2] [The theory was propounded by Jean de Charpentier (1786–1855) in his *Essai sur les Glaciers* (1841). For Forbes's examination of it, see *Travels through the Alps of Savoy*, pp. 34 *seq.* (ed. 1900).]

[3] [See the Introduction ; above, p. xxxv.]

[4] [Forbes applied this term to Saussure's theory : see *ibid.*, p. 34.]

therefore, two things more; namely, first that all mountains are as smooth as house-roofs; and, secondly, that a piece of ice a mile long and three or four hundred feet deep will slide gently, though a piece a foot deep and a yard long slides fast,—in other words, that a paving-stone will slide fast on another paving-stone, but the Rossberg fall at the rate of eighteen inches a day.

There is another form of the sliding theory, which is that glaciers slide in little bits, one at a time; or, for example, that if you put a railway train on an incline, with loose fastening to the carriages, the first carriage will slide first, as far as it can go, and then stop; then the second start, and catch it up, and wait for the third; and so on, till when the last has come up, the first will start again.

Having once for all sufficiently explained the "Sliding theory" to you, I shall not trouble myself any more in *Deucalion* about it.

15. The next theory is the Dilatation theory. The scientific persons who hold *that* theory suppose that whenever a shower of rain falls on a glacier, the said rain freezes inside of it; and that the glacier being thereby made bigger, stretches itself uniformly in one direction, and never in any other; also that, although it can only be thus expanded in cold and wet weather, such expansion is the reason that it always goes fastest in hot and dry weather.

There is another form of the Dilatation theory, which is that the glacier expands by freezing its own meltings.

16. Having thus sufficiently explained the Dilatation theory to you, I shall not trouble myself in *Deucalion* farther about *it;* noticing only, in bidding it good-bye, the curious want of power in scientific men, when once they get hold of a false notion, to perceive the commonest analogies implying its correction. One would have thought that, with their thermometer in their hand to measure congelation with, and the idea of expansion in their head, the analogy between the tube of the thermometer, and a glacier channel, and the ball of the thermometer, and a

glacier reservoir, might, some sunshiny day, have climbed across the muddily-fissured glacier of their wits:—and all the quicker, that their much-studied Mer de Glace bears to the great reservoirs of ice above it precisely the relation of a very narrow tube to a very large ball. The vast "instrument" seems actually to have been constructed by Nature, to show to the dullest of savants the difference between the steady current of flux through a channel of drainage, and the oscillatory vivacity of expansion which they constructed their own tubular apparatus to obtain!

17. The last popular theory concerning glaciers is the Regelation theory.[1] The scientific persons who hold *that* theory, suppose that a glacier advances by breaking itself spontaneously into small pieces; and then spontaneously sticking the pieces together again;—that it becomes continually larger by a repetition of this operation, and that the enlargement (as assumed also by the gentlemen of the Dilatation party) can only take place downwards.

You may best conceive the gist of the Regelation theory by considering the parallel statement, which you may make to your scientific young people, that if they put a large piece of barley-sugar on the staircase landing, it will walk downstairs by alternately cracking and mending itself.

I shall not trouble myself farther, in *Deucalion*, about the Regelation theory.

18. M. Viollet-le-Duc, indeed, appears to have written his book without even having heard of *it*; but he is most dextrous in the use of the two others, fighting, as it were, at once with sword and dagger; and making his glaciers move on the Sliding theory when the ground is steep, and on the Dilatation theory when it is level. The woodcuts at pages 65, 66, in which a glacier is represented dilating itself up a number of hills and down again, and that at page 99, in which it defers a line of boulders, which by unexplained supernatural power have been deposited all across

---

[1] [See above, pp. 126–127.]

it, into moraines at its sides, cannot but remain triumphant among monuments of scientific error,—bestowing on their author a kind of St. Simeon-Stylitic pre-eminence of immortality in the Paradise of Fools.

19. Why I stopped first at page 58 of this singular volume, I see there is no room to tell in this number of *Deucalion;* still less to note the interesting repetitions by M. Viollet-le-Duc of the Tyndall-Agassiz demonstration that Forbes' assertion of the plasticity of ice in large pieces, is now untenable, by reason of the more recent discovery of its plasticity in little ones. I have just space, however, for a little woodcut from the *Glaciers of the Alps* (or *Forms of*

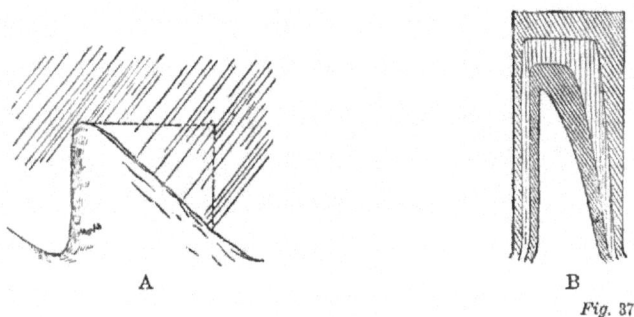

A   B

Fig. 37

Water, I forget which, and it is no matter),[1] in final illustration of the Tyndall-Agassiz quality of wit.

20. Fig. 37, A, is Professor Tyndall's illustration of the effect of sunshine on a piece of glacier, originally of the form shown by the dotted line, and reduced by solar power on the south side to the beautifully delineated wave in the shape of a wedge.

It never occurred to the scientific author that the sunshine would melt some of the top, as well as of the side, of his parallelopiped; nor that, during the process, even on the shady side of it, some melting would take place in the summer air. The figure at B[2] represents three stages

---

[1] [The woodcut is a portion of Fig. 5 in *Glaciers of the Alps* (p. 43).]

[2] [Ruskin in his copy notes, "My figure wants much more explanation." This he had intended to give; see vol. ii. ch. iii. § 23 *n.* (below, p. 357).]

of the diminution which would really take place, allowing
for these other somewhat important conditions of the question; and it shows, what may farther interest the ordinary
observer, how rectangular portions of ice, originally produced
merely by fissure in its horizontal mass, may be gradually
reduced into sharp, axe-edged ridges, having every appearance of splintery and vitreous fracture.   In next *Deucalion* [1]
I hope to give at last some account of my experiments on
gelatinous fracture, made in the delightful laboratory of
my friend's kitchen, with the aid of her infinitely conceding,
and patiently collaborating cook.

[1] [See below, p. 259.   For Ruskin's experiments in Lady Mount Temple's
kitchen, see above, p. 177.]

CHAPTER XI

OF SILICA IN LAVAS

1. THE rocks through whose vast range, as stated in the ninth chapter,[1] our at first well-founded knowledge of their igneous origin gradually becomes dim, and fades into theory, may be logically divided into these four following groups.

I. True lavas. Substances which have been rapidly cooled from fusion into homogeneous masses, showing no clear traces of crystallization.

II. Basalts.* Rocks in which, without distinct separation of their elements, a disposition towards crystalline structure manifests itself.

III. Porphyries. Rocks in which one or more mineral elements separate themselves in crystalline form from a homogeneous paste.

IV. Granites. Rocks in which all their elements have taken crystalline form.

2. These, I say, are logical divisions, very easily tenable. But Nature laughs at logic, and in her infinite imagination of rocks, defies all Kosmos, except the mighty one which we, her poor puppets, shall never discern. Our logic will help us but a little way ;—so far, however, we will take its help.

3. And first, therefore, let us ask what questions imperatively need answer, concerning indisputable lavas, seen by

---

* I use this word as on the whole the best for the vast class of rocks I wish to include ; but without any reference to columnar desiccation. I consider, in this arrangement, only internal structure.

---

[1] [See above, p. 207.]

living human eyes to flow incandescent out of the earth, and thereon to cool into ghastly slags.

On these I have practically burnt the soles of my boots, and in their hollows have practically roasted eggs; and in the lee of them, have been well-nigh choked with their stench; and can positively testify respecting them, that they were in many parts once fluid under power of fire, in a very fine and soft flux; and did congeal out of that state into ropy or cellular masses, variously tormented and kneaded by explosive gas; or pinched into tortuous tension, as by diabolic tongs; and are so finally left by the powers of Hell, to submit themselves to the powers of Heaven, in black or brown masses of adamantine sponge without water, and horrible honeycombs without honey, interlaid between drifted banks of earthy flood, poured down from merciless clouds whose rain was ashes.

The seas that now beat against these, have shores of black sand; the peasant, whose field is in these, ploughs with his foot, and the wind harrows.

4. Now of the outsides of these lava streams, and un-altered volcanic ashes, I know the look well enough; and could supply Sheffield with any quantity of characteristic specimens, if their policy and trade had not already pretty nearly buried them, and great part of England besides, under such devil's ware of their own production. But of the *insides* of these lava streams, and of the recognized alterations of volcanic tufa, I know nothing. And, accordingly, I want authentic answer to these following questions, with illustrative specimens.

5. (*a.*) In lavas which have been historically hot to perfect fusion, so as to be progressive, on steep slopes, in the manner of iron out of a furnace in its pig-furrows;— in such perfect lavas, I say—what kind of difference is there between the substance at the surface and at the ex-tremest known depths, after cooling? It is evident that such lavas can only accumulate to great depths in infer-nal pools or lakes. Of such lakes, which are the deepest

known? and of those known, where are the best sections? I want for Sheffield a series of specimens of any well-fused lava anywhere, showing the gradations of solidity or crystalline consolidation, from the outside to extreme depth.

(*b*.) On lavas which have not been historically hot, but of which there is no possible doubt that they were once fluent (in the air) to the above-stated degree, what changes are traceable, produced, irrespectively of atmospheric action, by lapse of time? What evidence is there that lavas, once cool to their centres, can sustain any farther crystalline change, or rearrangement of mineral structure?

(*c*.) In lavas either historically or indisputably once fluent, what forms of silica are found? I limit myself at present to the investigation of volcanic *silica:* other geologists will in time take up other minerals; but I find silica enough, and more than enough, for my life, or at least for what may be left of it.

Now I am myself rich in specimens of Hyalite, and Auvergne stellar and guttate chalcedonies; but I have no notion whatever how these, or the bitumen associated with them, have been developed; and I shall be most grateful for a clear account of their locality,—possible or probable mode of production in that locality,—and microscopic structure. Of pure quartz, of opal, or of agate, I have no specimen connected with what I should call a truly "living" lava; one, that is to say, which has simply cooled down to its existing form from the fluid state; but I have sent to the Sheffield Museum a piece of Hyalite, on a living lava,[1] so much like a living wasp's nest, and so incredible for a lava at all to the general observer, that I want forthwith some help from my mineralogical friends, in giving account of it.

6. And here I must, for a paragraph or two, pass from definition of flinty and molten minerals, to the more difficult definition of flinty and molten hearts; in order to explain why the Hyalite which I have just sent to the

---

[1] [H. 1: see below, p. 430.]

men of Sheffield, for their first type of volcanic silica,* is not at all the best Hyalite in my collection. This is because I practically find a certain quantity of selfishness necessary to live by ; and having no manner of saintly nature in me, but only that of ordinary men,—(which makes me all the hotter in temper when I can't get ordinary men either to see what I know they can see if they look, or do what I know they can do if they like),—I get sometimes weary of giving things away, letting my drawers get into disorder, and losing the powers of observation and thought which are connected with the complacency of possession, and the pleasantness of order. Whereupon I have resolved to bring my own collection within narrow limits ; but to constitute it resolutely and irrevocably of chosen and curious pieces, for my own pleasure ; trusting that they may be afterwards cared for by some of the persons who knew me, when I myself am troubled with care no more.†

7. This piece of Hyalite, however, just sent to Sheffield, though not my best, is the most curiously *definite* example I ever saw. It is on a bit of brown lava, which looks, as aforesaid, a little way off, exactly like a piece of a wasp's nest : seen closer, the cells are not hexagonal, but just like a cast of a spoonful of pease ; the spherical hollows having this of notable in them, that they are only as close to each other as they can be, to *admit of their being perfectly round :* therefore, necessarily, with little spaces of solid stone between them. I have not the slightest notion how such a lava can be produced. It is like an oolite with the yolks

---

* I give the description of these seven pieces of Hyalite at Sheffield,[1] in *Deucalion*, because their description is necessary to explain certain general principles of arrangement and nomenclature.

† By the way, this selfish collection is to be primarily of stones that will *wash*. Of petty troubles, none are more fretting than the effect of dust on minerals that can neither be washed nor brushed. Hence, my specialty of liking for silica, felspar, and the granite or gneissic rocks.

---

[1] [See, again, the Sheffield Catalogue, below, p. 430.]

of its eggs dropped out, and not in the least like a ductile substance churned into foam by expansive gas.

8. On this mysterious bit of gaseous wasp's nest, the Hyalite seems to have been dropped, like drops of glass from a melting glass rod. It seems to touch the lava just as little as it can; sticks at once on the edges of the cells, and laps over without running into, much less filling, them. There is not any appearance, and I think no possibility, of exudation having taken place; the silica cannot but, I think, have been deposited; and it is stuck together just as if it had fallen in drops, which is what I mean by calling Hyalite characteristically "guttate"; but it shows, nevertheless, a tendency to something like crystallization, in irregularities of surface like those of glacier ice, or the kind of old Venetian glass which is rough, and apparently of lumps coagulated. The fracture is splendidly vitreous,—the substance, mostly quite clear, but in parts white and opaque.

9. Now although no other specimen that I have yet seen is so manifestly guttate as this, all the hyalites I know agree in approximate conditions; and associate themselves with forms of chalcedony which exactly resemble the droppings from a fine wax candle. Such heated waxen efflu-ences,[1] as they congeal, will be found thrown into flattened coats; and the chalcedonies in question on the *under* surface precisely resemble them; while on the *upper* they become more or less crystalline, and, in some specimens, form lustrous stellar crystals in the centre.

10. Now, observe, this chalcedony, *capable of crystalliza-tion*, differs wholly from chalcedony properly so called, which may indeed be *covered* with crystals, but itself remains con-sistently smooth in surface, as true Hyalite does, also.

Not to be teazed with too many classes, however, I shall arrange these peculiar chalcedonies with Hyalite; and, accordingly, I send next to the Sheffield Museum, to follow this first Hyalite, an example of the transition from

---

[1] [On re-reading the book, Ruskin condemned this as an "affected term for wax droppings; but note importance of resemblance."]

Hyalite to dropped chalcedony (I. H. 2), being an Indian volcanic chalcedony, translucent, aggregated like Hyalite, and showing a *concave* fracture where a ball of it has been broken out.

11. Next (I. H. 3), pure dropped chalcedony. I do not like the word "dropped" in this use,—so that, instead, I shall call this in future *wax* chalcedony; then (I. H. 4) the same form, with crystalline surface,—this I shall henceforward call *sugar* chalcedony; and, lastly, the ordinary stellar form of Auvergne, *star* chalcedony (I. H. 5).

These five examples are typical, and perfect in their kind; next to them (I. H. 6) I place a wax chalcedony formed on a porous rock (volcanic ash?) which has at the surface of it small circular *concavities*, being also so irregularly coagulate throughout that it suggests no mode of deposition whatever, and is peculiar in this also, that it is thinner in the centre than at the edges, and that no vestige of its substance occurs in the pores of the rock it overlies.

Take a piece of porous broken brick, drop any tallowy composition over four or five inches square of its surface, to the depth of one-tenth of an inch; then drop more on the edges till you have a rampart round, the third of an inch thick; and you will have some likeness of this piece of stone: but how Nature held the composition in her fingers, or composed it to be held, I leave you to guess, for I cannot.

12. Next following, I place the most singular example of all (I. H. 7). The chalcedony in I. H. 6 is apparently dropped on the ashes, and of irregular thickness; it is difficult to understand *how* it was dropped, but once *get* Nature to hold the candle, and the thing is done.

But here, in I. H. 7, it is no longer apparently dropped, but apparently boiled! It rises like the bubbles of a strongly boiling liquid;—but not from a liquid mass; on the contrary (except in three places, presently to be described) it coats the volcanic ash in perfectly even thickness—a quarter of an inch, *and no more, nor less, everywhere*, over a space five inches square! and the ash, or lava, itself, instead of being porous throughout the mass, with the silica only on the surface, is filled with chalcedony in every cavity!

Now this specimen completes the transitional series from hyalite to perfect chalcedony; and with these seven specimens, in order, before us, we can define some things, and question of others, with great precision.

13. First, observe that all the first six pieces agree in two conditions,—*varying*, and *coagulated*, thickness of the deposit. But the seventh has the remarkable character of *equal*, and therefore probably crystalline, deposition everywhere.

Secondly. In the first six specimens, though the coagulations are more or less rounded, none of them are regularly spherical. But in the seventh, though the larger bubbles (so to call them) are subdivided into many small ones, every uninterrupted piece of the surface is *a portion of a sphere*, as in true bubbles.

Thirdly. The sugar chalcedony, I. H. 4, and stellar chalcedony, I. H. 5, show perfect power of assuming, under favourable conditions, prismatic crystalline form. But there is no trace of such tendency in the first three, or last two, of the seven examples. Nor has there ever, so far as I know, been found prismatic true hyalite, or prismatic true chalcedony.

Therefore we have here essentially three different minerals, passing into each other it is true; but, at a certain point, changing their natures definitely, so that *hyalite, becoming wax chalcedony, gains* the power of prismatic crystallization; and *wax chalcedony, becoming true chalcedony, loses* it again!

And now I must pause, to explain rightly this term "prismatic," and others which are now in use, or which are to be used, in St. George's Schools, in describing crystallization.

14. A prism (the *sawn* thing), in Newton's use of the word, is a triangular pillar with flat top and bottom.[1] Putting two or more of these together, we can make pillars of any number of plane sides, in any regular or irregular shape.

(1.) Crystals, therefore, which are columnar, and thick enough to be distinctly seen, are called "prismatic."

(2.) But crystals which are columnar, and so delicate that they look like needles, are called "acicular," from acus, a needle.

(3.) When such crystals become so fine that they look like hair or down, and lie in confused directions, the mineral composed of them is called "plumose."

(4.) And when they adhere together closely by their sides, the mineral is called "fibrous."

(5.) When a crystal is flattened by the extension of two of its planes, so as to look like a board, it is called "tabular"; but people don't call it a "tabula."

(6.) But when such a board becomes very thin, it *is* called a "lamina," and the mineral composed of many such plates, laminated.

(7.) When laminæ are so thin that, joining with others equally so, they form fine leaves, the mineral is "foliate."

(8.) And when these leaves are capable of perpetual subdivision, the mineral is "micaceous."

15. Now, so far as I know their works, mineralogists hitherto have never attempted to show cause why some minerals rejoice in longitude, others in latitude, and others in platitude. They indicate to their own satisfaction,—that

---

[1] [Prism, from the Greek πρίσμα (πρίζω = split). "A *prism* is a glass bounded with two equal and parallel triangular ends, and three plain and well polished sides, which meet in three parallel lines, running from the three angles of one end to the three angles of the other end" (Sir Isaac Newton : *On Opticks*).]

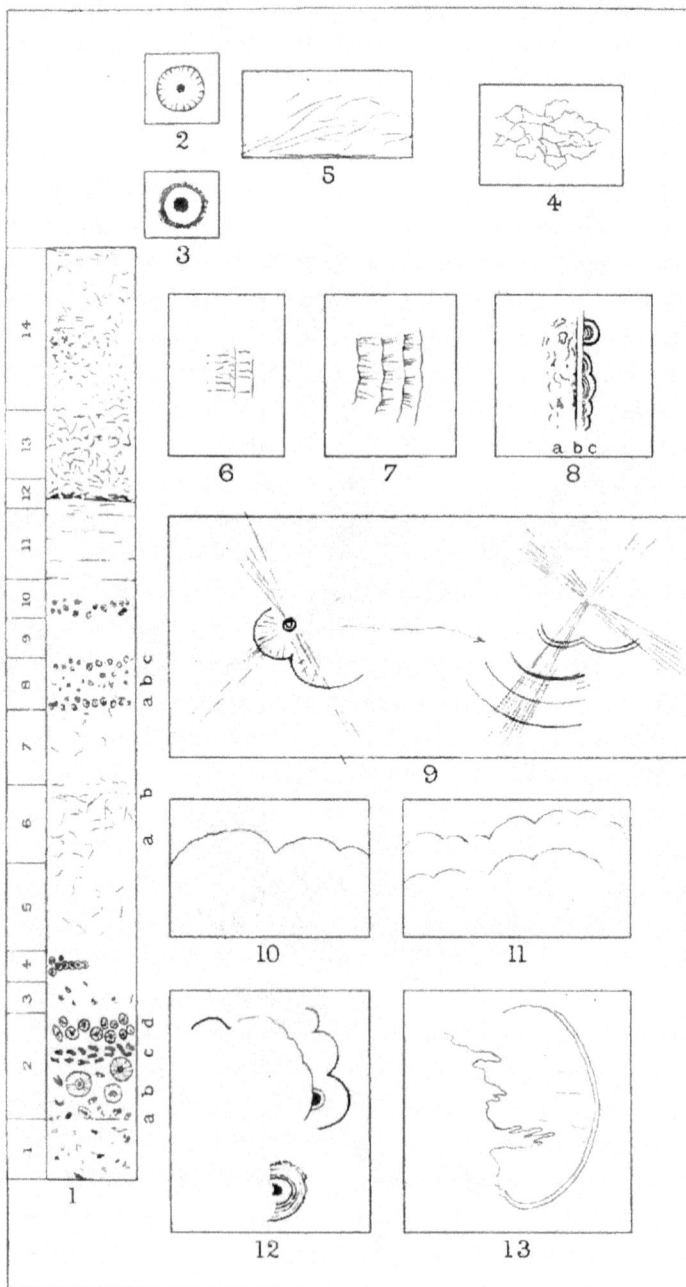

STRUCTURE OF LAKE AGATE

is to say, in a manner totally incomprehensible by the public,—all the modes of expatiation possible to the mineral, by cardinal points on a sphere: but why a crystal of ruby likes to be short and fat, and a crystal of rutile, long and lean; why amianth should bind itself into bundles of threads, cuprite weave itself into tissues, and silver braid itself into nests,—the use, in fact, that any mineral makes of its opportunities, and the cultivation which it gives to its faculties,—of all this, my mineralogical authorities tell me nothing. Industry, indeed, is theirs to a quite infinite degree, in pounding, decocting, weighing, measuring, but they have remained just as unconscious as vivisecting physicians that all this was only the anatomy of dust,—not its history.

But here at last, in Cumberland, I find a friend, Mr. Clifton Ward, able and willing to begin some true history of mineral substance, and far advanced already in preliminary discovery; and in answer to my request for help, taking up this first hyalitic problem, he has sent me the drawings —engraved, I regret to say, with little justice to their delicacy;*—in Plate XIV.

16. This plate represents, in Figure 1, the varieties of structure in an inch vertical section of a lake-agate; and in Figures 2, 3, 4, and 5, still farther magnified portions of the layers so numbered in Figure 1.

Figures 6 to 9 represent the structure and effect of polarized light in a lake-agate of more distinctly crystalline structure; and Figures 10 to 13, the orbicular concretions of volcanic Indian chalcedony. But before entering farther on the description of these definitely concretionary bands,[1] I think it will be desirable to take note of some facts regarding the larger bands of our Westmoreland mountains,

---

* But not by my fault, for I told the engraver to do his best; and took more trouble with the plate than with any of my own.

---

[1] [Ruskin in his copy refers to the resumption of the subject in ch. xiii.; below, pp. 267 *seq.*]

which become to me, the more I climb them, mysterious to a point scarcely tolerable; and only the more so, in consequence of their recent more accurate survey.

17. Leaving their pebbles, therefore, for a little while, I will ask my readers to think over some of the conditions of their crags and pools, explained as best I could, in the following lecture, to the Literary and Scientific Society of the town of Kendal. For indeed, beneath the evermore blessed Kendal-green of their sweet meadows and moors, the secrets of hill-structure remain, for all the work spent on them, in colourless darkness; and indeed, "So dark, Hal, that thou could'st not see thine hand."[1]

---

[1] [*1 Henry IV.*, Act ii. sc. 4.]

CHAPTER XII

YEWDALE AND ITS STREAMLETS

(Lecture delivered before the Members of the Literary and Scientific Institution, Kendal, 1st October, 1877) [1]

1. I FEAR that some of my hearers may think an apology due to them for having brought, on the first occasion of my being honoured by their audience, a subject before them which they may suppose unconnected with my own special work, past or present. But the truth is, I knew mountains long before I knew pictures; and these mountains of yours, before any other mountains. From this town, of Kendal, I went out, a child, to the first joyful excursions among the Cumberland lakes, which formed my love of landscape and of painting: and now, being an old man, I find myself more and more glad to return—and pray you to-night to return with me—from shadows [2] to the reality.

I do not, however, believe that one in a hundred of our youth, or of our educated classes, out of directly scientific circles, take any real interest in geology. [3] And for my own part, I do not wonder,—for it seems to me that geology tells us nothing really interesting. It tells us

---

[1] [For particulars of the lecture, see Bibliographical Note (above, p. 90). And on the general argument of this chapter see Vol. VI. p. 127 *n.*]

[2] [That is, from pictures; "the best in this kind are but shadows:" see Vol. XX. p. 300.]

[3] [The report of the lecture has this passage:—

"In a general audience, Mr. Ruskin said that he should not hope many would be interested in geological questions; even in this audience, inheriting as it did the light shed by its noble President, the great Sedgwick, whose name was at the foundation of the science of geology in England."]

much about a world that once was. But, for my part, a world that only was, is as little interesting as a world that only is to be. I no more care to hear of the forms of mountains that crumbled away a million of years ago to leave room for the town of Kendal, than of forms of mountains that some future day may swallow up the town of Kendal in the cracks of them. I am only interested— so ignoble and unspeculative is my disposition—in knowing how God made the Castle Hill of Kendal, for the Baron of it to build on, and how he brought the Kent through the dale of it, for its people and flocks to drink of.

2. And these things, if you think of them, you will find are precisely what the geologists cannot tell you. They never trouble themselves about matters so recent, or so visible; and while you may always obtain the most satisfactory information from them respecting the congelation of the whole globe out of gas, or the direction of it in space, there is really not one who can explain to you the making of a pebble, or the running of a rivulet.

May I, however, before pursuing my poor little inquiry into these trifling matters, congratulate those members of my audience who delight more in literature than science, on the possession, not only of dales in reality, but of dales in name? Consider, for an instant or two, how much is involved, how much indicated, by our possession in English of the six quite distinct words—vale, valley, dale, dell, glen, and dingle;—consider the gradations of character in scene, and fineness of observation in the inhabitants, implied by that six-foil cluster of words; as compared to the simple "thal" of the Germans, "valle" of the Italians, and "vallée" of the French, shortening into "val" merely for ease of pronunciation, but having no variety of sense whatever; so that, supposing I want to translate, for the benefit of an Italian friend, Wordsworth's "Reverie of Poor Susan," and come to "Green pastures she views in the midst of the dale," and look for "dale" in my Italian dictionary, I find "valle lunga e stretta tra poggi alti," and can only convey Mr. Wordsworth's meaning to my Italian listener by telling him that "la povera Susanna vede verdi prati, nel mezzo della valle lunga e stretta tra poggi alti"! It is worth while, both for geological and literary reasons, to trace the essential differences in the meaning and proper use of these words.

3. "Vale" signifies a large extent of level land, surrounded by hills, or nearly so; as the Vale of the White Horse, or Vale of Severn. The level extent is necessary to the idea; while the next word, "valley," means a large hollow among hills, in which there is little level ground, or none. Next comes "dale," which signifies properly a tract of level land on the borders of a stream, continued for so great a distance as to make it a district of importance as a part of the inhabited country; as Ennerdale, Langdale, Liddesdale. "Dell" is to dale, what valley is to vale; and implies that there is scarcely any level land beside the stream. "Dingle" is such a recess or dell clothed with wood;* and "glen" one varied with rocks. The term "ravine," a rent chasm among rocks, has its necessary parallel in other languages.

Our richness of expression in these particulars may be traced to the refinement of our country life, chiefly since the fifteenth century; and to the poetry founded on the ancient character of the Border peasantry; mingling agricultural with shepherd life in almost equal measure.

I am about to endeavour, then, to lay before you this evening the geological laws which have produced the "dale," properly so called, of which I take—for a sweet and near example—the green piece of meadow land through

---

* Connected partly, I doubt not, with Ingle, or Inglewood,—brushwood to burn (hence Justice Inglewood in *Rob Roy*).[1] I have still omitted "clough," or cleugh, given by Johnson in relation to "dingle," and constant in Scott, from "Gander-cleugh" to "Buc(k)-cleugh."

---

[1] [For Justice Inglewood, see ch. viii. For Gander-cleugh, see the Introduction to *Tales of my Landlord*; and for the Gander river, the Introductory Address to *Count Robert of Paris*.]

which flows, into Coniston Water, the brook that chiefly feeds it.

4. And now, before going farther, let me at once vindicate myself from the blame of not doing full justice to the earnest continuance of labour, and excellent subtlety of investigation, by which Mr. Aveline and Mr. Clifton Ward [1] have presented you with the marvellous maps and sections of this district, now in course of publication in the Geological Survey. Especially let me, in the strongest terms of grateful admiration, refer to the results which have been obtained by the microscopic observations of minerals instituted by Mr. Sorby,[2] and carried out indefatigably by Mr. Clifton Ward, forming the first sound foundations laid for the solution of the most secret problems of geology.

5. But while I make this most sincere acknowledgment of what has been done by these gentlemen, and by their brother geologists in the higher paths of science, I must yet in all humility lament that this vast fund of gathered knowledge is every bit of it, hitherto, beyond you and me. Dealing only with infinitude of space and remoteness of time, it leaves us as ignorant as ever we were, or perhaps, in fancying ourselves wiser, even more ignorant, of the things that are near us and around,—of the brooks that sing to us, the rocks that guard us, and the fields that feed.

6. To-night, therefore, I am here for no other purpose than to ask the simplest questions; and to win your interest, if it may be, in pleading with our geological teachers for the answers which as yet they disdain to give.

Here, in your long winding dale of the Kent,—and over the hills, in my little nested dale of the Yew,—will you ask the geologist, with me, to tell us how their pleasant depth was opened for us, and their lovely borders built? For, as yet, this is all that we are told concerning them, by accumulated evidence of geology, as collected in this

---

[1] [See above, p. 208 n.]
[2] [See above, p. 207 n.]

summary at the end of the first part of Mr. Clifton Ward's
volume on the geology of the lakes:—

"The most ancient geologic records in the district indicate marine conditions with a probable proximity of land. Submarine volcanoes broke out
during the close of this period, followed by an elevation of land, with continued volcanic eruptions, of which perhaps the present site of Keswick
was one of the chief centres. Depression of the volcanic district then
ensued beneath the sea, with the probable sensation of volcanic activity;
much denudation was effected; another slight volcanic outburst accompanied the formation of the Coniston Limestone, and then the old deposits
of Skiddaw Slate and volcanic material were buried thousands of feet
beneath strata formed in an upper Silurian sea. Next followed an
immensely long period of elevation, accompanied by disturbance and alteration of the rocks, and by a prodigious amount of marine and atmospheric
denudation. A subsequent depression, to a considerable extent, marked
the coming on of the Carboniferous epoch, heralded however, in all likeli-
hood, by a period of more or less intense cold. Then for succeeding ages,
the district elevated high above the surrounding seas of later times, under-
went that large amount of sub-aerial denudation which has resulted in the
formation of our beautiful English Lake-country."[1]

7. The only sentence in this passage of the smallest
service to us, at present, is that stating the large amount of
" sub-aerial denudation" which formed our beautiful country.

Putting the geological language into simple English, that
means that your dales and hills were produced by being
"rubbed down in the open air,"—rubbed down, that is to
say, in the manner in which people are rubbed down after
a Turkish bath, so as to have a good deal of their skin
taken off them. But observe, it would be just as rational to
say that the beauty of the human form was owing to the
immemorial and continual use of the flesh-brush, as that we
owe the beauty of our mountains to the mere fact of their
having been rubbed away. No quantity of stripping or denuding will give beauty when there is none to denude;—you
cannot rub a statue out of a sand-bank, or carve the Elgin
frieze with rottenstone for a chisel, and chance to drive it.

8. We have to ask then, first, what material there was
here to carve; and then what sort of chisels, and in what
workman's hand, were used to produce this large piece of

---

[1] [From p. 77 of the Memoir of the Geological Survey, as noted above, p. 208 n.]

precious chasing or embossed work, which we call Cumberland and West-mere-land.[1]

I think we shall get at our subject more clearly, however, by taking a somewhat wider view of it than our own dales permit, and considering what "sub-aerial denudation" means, on the surface of the world, instead of in Westmoreland only.

9. Broadly, therefore, we have, forming a great part of that surface, vast plains or steppes, like the levels of France, and lowlands of England, and prairies of America, composed mostly of horizontal beds of soft stone or gravel. Nobody in general talks of these having been rubbed down; so little, indeed, that I really do not myself know what the notions of geologists are on the matter. They tell me that some four-and-twenty thousand feet or so of slate—say, four miles thick of slate—must have been taken off the top of Skiddaw to grind that into what it is; but I don't know in the least how much chalk or freestone they think has been ground off the East Cliff at Brighton, to flatten that into what it is. They tell me that Mont Blanc must have been three times as high as he is now, when God, or the affinity of atoms, first made him; but give me no idea whatever how much higher the shore of the Adriatic was than it is now, before the lagoon of Venice was rubbed out of it.

10. Collecting and inferring as best I can, it seems to me they mean generally that all the mountains were much higher than they are now, and all the plains lower; and that what has been scraped off the one has been heaped on to the other: but that is by no means generally so; and in the degree in which it is so, hitherto has been unexplained, and has even the aspect of being inexplicable.

I don't know what sort of models of the district you have in the Museum, but the kind commonly sold represent the entire mountain surface merely as so much sand-heap washed into gutters. It is totally impossible for your

---

[1] [The land of western meres. Compare ch. vii. § 3 (above, p. 166).]

youth, while these false impressions are conveyed by the cheap tricks of geographical manufacture, to approach the problems of mountain form under any sense of their real conditions: while even advanced geologists are too much in the habit of thinking that every mountain mass may be considered as a heap of homogeneous clay, which some common plough has fretted into similar clods.

But even to account for the furrows of a field you must ask for plough and ploughman. How much more to account for the furrows of the adamantine rock. Shall one plough *there* with oxen?

I will ask you, therefore, to-night, to approach this question in its first and simplest terms, and to examine the edge of the weapon which is supposed to be still at work. The streamlets of the dale seem yet in many places to be excavating their glens as they dash down them,—or deepening the pools under their cascades. Let us in such simple and daily visible matters consider more carefully what are the facts.

11. Towards the end of July, this last summer, I was sauntering among the fern, beside the bed of the Yewdale stream, and stopped, as one does instinctively, at a place where the stream stopped also,—bending itself round in a quiet brown eddy under the root of an oak tree.

How many thousand thousand times have I not stopped to look down into the pools of a mountain stream,—and yet never till that day had it occurred to me to ask how the pools came there. As a matter of course, I had always said to myself, there must be deep places and shallow ones,—and where the water is deep there is an eddy, and where it is shallow there is a ripple,—and what more is there to say about it?

However, that day, having been of late in an interrogative humour about everything, it did suddenly occur to me to ask why the water should be deep there, more than anywhere else. This pool was at a bend of the stream, and rather a wide part of it; and it seemed to me that, for the most part, the deep pools I recollected *had* been at bends of streams, and in rather wide parts of them;—with the accompanying condition of slow circular motion in the water; and also, mostly under steep banks.

12. Gathering my fifty years' experience of brooks, this seemed to me a tenable generalization, that on the whole, where the bank was steepest, and one was most likely to tumble in, one was least likely to get out again.

And the gloomily slow and sullen motion on the surface, as if the bubbles were unwillingly going round in a mill,—this also I recollected as a usual condition of the deeper water,—*so* usual, indeed, that (as I say) I never once before had reflected upon it as the least odd. Whereas now, the thought struck me as I looked, and struck me harder as I looked longer, If the *bubbles* stay at the top, why don't the *stones* stay at the bottom? If, when I throw in a stick here in the back eddy, at the surface, it keeps spinning slowly round and round, and never goes down-stream—am I to expect that when I throw a stone into the same eddy, it will be immediately lifted by it out of the hole and carried away? And yet unless the water at the bottom of the hole has this power of lifting stones out of it, why is the hole not filled up?

13. Coming to this point of the question, I looked up the beck, and down. Up the beck, above the pool, there was a shallow rapid over innumerable stones of all sizes: and down the beck, just below the pool, there was a ledge of rock, against which the stream had deposited a heap of rolled shingle, and over the edges of which it flowed in glittering tricklets, so shallow that a child of four years old might have safely waded across; and between the loose stones above in the steep rapid, and the ledge of rock below—which seemed put there expressly for them to be lodged against—here was this deep, and wide, and quiet pool.

So I stared at it, and stared; and the more I stared, the less I understood it. And if you like, any of you may easily go and stare too, for the pool in question is visible enough from the coach-road, from Mr. Sly's Water-head Inn,[1] up to Tilberthwaite. You turn to the right from the bridge at Mr. Bowness's smithy, and then in a quarter of a mile you may look over the roadside wall into this quiet recess of the stream, and consider of many things. For, observe, if there were anything out of the way in the pool—I should not send you to look at it. I mark it only for one of myriads such in every mountain stream that ever trout leaped or ripple laughed in.

And beside it, as a type of all its brother deeps, these following questions may be wisely put to yourselves.

14. First—How are any of the pools kept clear in a stream that carries shingle? There is some power the water has got of lifting it out of the deeps hitherto un-explained—unthought of. Coming down the rapid in a rage, it drops the stones, and leaves them behind; coming to the deep hole, where it seems to have no motion, it picks them up and carries them away in its pocket. Explain that.

15. But, secondly, beside this pool let us listen to the wide murmuring geological voice, telling us—"To sub-aerial denudation you owe your beautiful lake scenery"!—Then, presumably, Yewdale among the rest?—Therefore we may look upon Yewdale as a dale sub-aerially denuded. That is to say, there was once a time when no dale was there, and the process of denudation has excavated it to the depth you see.

16. But now I can ask, more definitely and clearly, With what chisel has this hollow been hewn for us? Of course, the geologist replies, by the frost, and the rain, and the decomposition of its rocks. Good; but though frost may break up, and the rain wash down, there must have been somebody to cart away the rubbish, or still you would

---

[1] [The Waterhead Hotel at Coniston, long kept by Mr. Joseph Sly, and by his widow, after his death; now kept by Mr. Joseph Tyson. The old Coniston black-smith, Mr. Bowness, is long since dead.]

have had no Yewdale. Well, of course, again the geologist answers, the streamlets are the carters; and this stream past Mr. Bowness's smithy is carter-in-chief.

17. How many cartloads, then, may we suppose the stream has carried past Mr. Bowness's, before it carted away all Yewdale to this extent, and cut out all the northern side of Wetherlam, and all that precipice of Yewdale Crag, and carted all the rubbish first into Coniston Lake, and then out of it again, and so down the Crake into the sea? Oh, the geologists reply, we don't mean that the little Crake did all that. Of course it was a great river full of crocodiles a quarter of a mile long; or it was a glacier five miles thick, going ten miles an hour; or a sea of hot water fifty miles deep,—or,—something of that sort. Well, I have no interest, myself, in *any*thing of that sort: and I want to know, here, at the side of my little puzzler of a pool, whether there's any sub-aerial denudation going on still, and whether this visible Crake, though it can only do little, does *any*thing. Is it carrying stones at all, now, past Mr. Bowness's? Of course, reply the geologists; don't you see the stones all along it, and doesn't it bring down more every flood? Well, yes; the delta of Coniston Waterhead may, perhaps, within the memory of the oldest inhabitant, or within the last hundred years, have advanced a couple of yards or so. At that rate, those two streams, considered as navvies, are proceeding with the works in hand;—to that extent they are indeed filling up the lake, and to that extent sub-aerially denuding the mountains. But now, I must ask your attention very closely: for I have a strict bit of logic to put before you, which the best I can do will not make clear without some helpful effort on your part.

18. The streams, we say, by little and little, are filling up the lake. They did not cut out the basin of that. Something else must have cut out that, then, before the streams began their work. Could the lake, then, have been cut out all by itself, and none of the valleys that lead to

it ? Was it punched into the mass of elevated ground like a long grave, before the streams were set to work to cut Yewdale down to it ?

19. You don't for a moment imagine that. Well, then, the lake and the dales that descend with it, must have been cut out together. But if the lake not by the streamlets, then the dales not by the streamlets ? The streamlets are the consequence of the dales then,—not the causes; and the sub-aerial denudation to which you owe your beautiful lake scenery, must have been something, not only different from what is going on now, but, in one half of it at least, *contrary* to what is going on now. Then, the lakes which are now being filled up, were being cut down; and as probably, the mountains now being cut down, were being cast up.

20. Don't let us go too fast, however. The streamlets are now, we perceive, filling up the big lake. But are they not, then, also filling up the little ones ? If they don't cut Coniston Water deeper, do you think they are cutting Mr. Marshall's[1] tarns deeper ? If not Mr. Marshall's tarns deeper, are they cutting their own little pools deeper ? This pool by which we are standing—we have seen it is inconceivable how it is not filled up,—much more it is inconceivable that it should be cut deeper down. You can't suppose that the same stream which is filling up the Coniston Lake below Mr. Bowness's, is cutting out another Coniston Lake above Mr. Bowness's ? The truth is that, above the bridge as below it, and from their sources to the sea, the streamlets have the same function, and are filling, not deepening, alike tarn, pool, channel, and valley.

21. And that being so, think how you have been misled by seeking knowledge far afield, and for vanity's sake, instead of close at home, and for love's sake. You must go and see Niagara, must you ?—and you will brick up and make a foul drain of the sweet streamlet that ran past your doors. And all the knowledge of the waters and the earth that God meant for you, flowed with it, as water of life.

[1] [See Vol. XXIII. p. xxi.]

Understand, then, at least, and at last, to-day, Niagara is a vast Exception—and Deception.[1] The true cataracts and falls of the great mountains, as the dear little cascades and leaplets of your own rills, fall where they fell of old ;— that is to say, wherever there's a hard bed of rock for them to jump over. They don't cut it away—and they can't.[2] They do form pools *beneath* in a mystic way,—they excavate them to the depth which will break their fall's force —and then they excavate no more.*

We must look, then, for some other chisel than the streamlet ; and therefore, as we have hitherto interrogated the waters at their work, we will now interrogate the hills, in their patience.

22. The principal flank of Yewdale is formed by a steep range of crag, thrown out from the greater mass of Wetherlam, and known as Yewdale Crag.

It is almost entirely composed of basalt, or hard volcanic ash ; and is of supreme interest among the southern hills of the Lake District, as being practically the first rise of the great mountains of England, out of the lowlands of England.

And it chances that my own study window being just opposite this crag, and not more than a mile from it as the bird flies, I have it always staring me, as it were, in the face, and asking again and again, when I look up from writing any of my books,—"How did *I* come here ?"

I wrote that last sentence hurriedly, but leave it—as it was written ; for, indeed, however well I know the vanity of it, the question is still sometimes, in spite of my best effort, put to me in that old form by the mocking crags, as by a vast couchant Sphinx, tempting me to vain labour in the inscrutable abyss.

But as I regain my collected thought, the mocking

* Else, every pool would become a well, of continually increasing depth.

---

[1] [Compare above, p. 149, and below, p. 370.]
[2] [Compare above, p. 120.]

question ceases, and the divine one forms itself, in the voice of vale and streamlet, and in the shadowy lettering of the engraven rock.

"Where wast thou when I laid the foundation of the earth?—declare, if thou hast understanding."[1]

23. How Yewdale Crags came there, I, for one, will no more dream, therefore, of knowing, than the wild grass can know, that shelters in their clefts. I will only to-night ask you to consider one more mystery in the things they have suffered since they came.

You might naturally think, following out the idea of "sub-aerial denudation," that the sudden and steep rise of the crag above these softer strata was the natural consequence of its greater hardness; and that in general the district was only the remains of a hard knot or kernel in the substance of the island, from which the softer superincumbent or surrounding material had been more or less rubbed or washed away.*

24. But had that been so, one result of the process must have been certain—that the hard rocks would have resisted more than the soft; and that in some distinct proportion and connection, the hardness of a mountain would be conjecturable from its height, and the whole surface of the district more or less manifestly composed of hard bosses or ridges, with depressions between them in softer materials. Nothing is so common, nothing so clear, as this condition, on a small scale, in every weathered rock. Its quartz, or other hard knots and veins, stand out from the depressed surface in raised walls, like the divisions between the pits of Dante's eighth circle,[2]—and to a certain extent, Mr. Ward

* The most wonderful piece of weathering, in all my own district, is on a *projecting* mass of intensely hard rock on the eastern side of Goat's Water. It was discovered and shown to me by my friend the Rev. F. A. Malleson;[3] and exactly resembles deep ripple-marking, though nothing in the grain of the rock indicates its undulatory structure.

---

[1] [Job xxx. viii. 4.]
[2] [*Inferno*, viii. first lines.]
[3] [Vicar of Broughton-in-Furness. A series of letters from Ruskin to him is given in a later volume of this edition.]

tells us,[1] the lava dykes, either by their hardness or by their decomposition, produce walls and trenches in the existing surface of the hills. But these are on so small a scale, that on this map they cannot be discernedly indicated; and the quite amazing fact stands out here in unqualified and in-disputable decision, that by whatever force these forms of your mountain were hewn, it cut through the substance of them, as a sword-stroke through flesh, bone, and marrow, and swept away the masses to be removed, with as serene and indiscriminating power as one of the shot from the Devil's great guns at Shoeburyness goes through the oak and the iron of its target.

25. It is with renewed astonishment, whenever I take these sections into my hand, that I observe the phenomenon itself; and that I remember the persistent silence of geo-logical teachers on this matter, through the last forty years of their various discourse. In this shortened section, through Bowfell to Brantwood, you go through the summits of three first-rate mountains down to the lowland moors: you find them built, or heaped; barred, or bedded; here with forged basalt, harder than flint and tougher than iron,—there, with shivering shales that split themselves into flakes as fine as puff-paste, and as brittle as shortbread. And behold, the hewing tool of the Master Builder sweeps along the form-ing lines, and shapes the indented masses of them, as a draper's scissors shred a piece of striped sarcenet!

26. Now do but think a little of the wonderfulness in this. If the process of grinding was slow, why don't the hard rocks project? If swift, what kind of force must it have been? and why do the rocks it tore show no signs of rending? Nobody supposes it was indeed swift as a sword or a cannon-ball? but if not, why are the rocks not broken? Can you break an oak plank and leave no splinters, or cut a bed of basalt a thousand feet thick like cream-cheese?

---

[1] [See *Geology of the Northern Part of the English Lake District*, by J. Clifton Ward, pp. 13–29.]

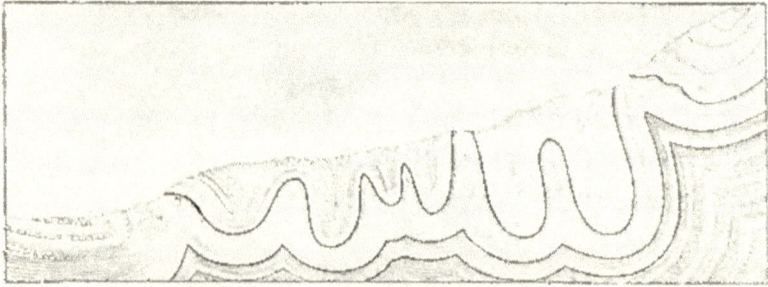

Fig. 1. Slates of Bull Crag and Maiden Moor (GEOL. SURVEY)

Fig. 2. Pie-Paste. Compression from the right, simple.

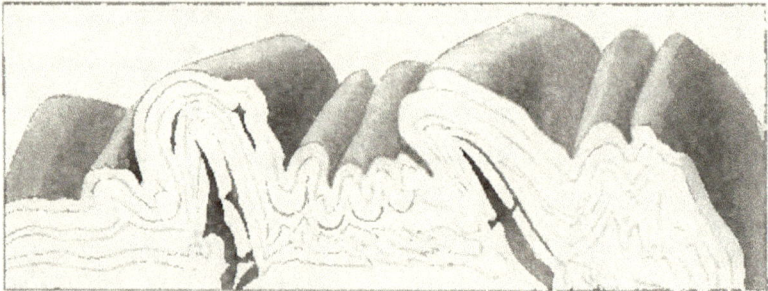

Fig. 3. Pie-Paste. Compression modified by elevatory forces.

Fig. 4. Pie-Paste. Compression restricted to the lower Strata
under a rigid upper one.

LATERAL COMPRESSION OF STRATA.
Fig. 1. Ideal. Figs. 2. 3. & 4. Practical.

But you suppose the rocks were soft when it was done. Why don't they squeeze then?

Make Dover Cliffs of baker's dough, and put St. Paul's on the top of them,—won't they give way somewhat, think you? and you will then make Causey Pike of clay, and heave Scawfell against the side of it; and yet shall it not so much as show a bruise?[1]

Yet your modern geologists placidly draw the folded beds of the Skiddaw and Causey Pike slate, *first*, without observing whether the folds they draw are *possible* folds in anything; and, *secondly*, without the slightest suggestion of sustained pressure, or bruise, in any part of them.

27. I have given in my diagram (Plate XV., Fig. 1) the section, attributed, in that last issued by the Geological Survey,[2] to the contorted slates of Maiden Moor, between Causey Pike and the erupted masses of the central mountains. Now, for aught I know, those contortions may be truly represented;—but if so, they are not contortions by lateral pressure. For, first, they are impossible forms in any substance whatever, capable of being contorted; and, secondly, they are doubly impossible in any substance capable of being squeezed.

Impossible, I say first, in any substance capable of being contorted. Fold paper, cloth, leather, sheets of iron,—what you will, and still you can't *have the folded bed at the top double the length of that at the bottom.* But here, I have measured the length of the upper bed, as compared with that of the lower, and it is twenty miles, to eight miles and a half.

Secondly, I say, these are impossible folds in any substance capable of being squeezed, for every such substance will change its form as well as its direction under pressure. And to show you how such a substance does actually behave and contort itself under lateral pressure, I have prepared the sections Figures 2, 3, and 4.

[1] [Compare ch. i. § 15 (above, p. 110.)]
[2] [*Memoirs of the Geological Survey: The Geology of the Northern Part of the English Lake District (quarter sheet 101 S.E.)*: 1876.]

28. I have just said you have no business to seek knowledge far afield, when you can get it at your doors. But
more than that, you have no business to go outside your
doors for it, when you can get it in your parlour. And it
so happens that the two substances which, while the foolish
little king was counting out his money, the wise little queen
was eating in the parlour, are precisely the two substances
beside which wise little queens, and kings, and everybody
else, may also think, in the parlour,—Bread, and honey.
For whatever bread, or at least dough, will do under pressure, ductile rocks, in their proportion, must also do under
pressure ; and in the manner that honey will move, poured
upon a slice of them,—in that manner, though in its own
measure, ice will move, poured upon a bed of them. Rocks,
no more than pie-crust, can be rolled out without squeezing
them thinner ; and flowing ice can no more excavate a
valley, than flowing treacle a teaspoon.[1]

29. I said just now, Will you dash Scawfell against
Causey Pike ?

I take, therefore, from the Geological Survey the section
of the Skiddaw slates, which continue the mass of Causey
Pike under the Vale of Newlands, to the point where the
volcanic mass of the Scawfell range thrusts itself up against
them, and laps over them. They are represented, in the
section, as you see (Plate XV., Fig. 1) ; and it has always
been calmly assumed by geologists that these contortions
were owing to lateral pressure.

But I must beg you to observe that since the uppermost of these beds, if it were straightened out, would be
more than twice the length of the lower ones, you could
only obtain that elongation by squeezing the upper bed
more than the lower, and making it narrower where it is
elongated. Now, if this were indeed at the surface of the
ground, the geologists might say the upper bed had been
thrown up because there was less weight on it. But, by

---

[1] [Compare ch. vi. § 13 (above, p. 162).]

their own accounts, there were five miles thick of rocks on the top of all this when it was bent. So you could not have made one bed tilt up, and another stay down; and the structure is evidently an impossible one.

30. Nay, answer the surveyors, impossible or not, it is there. I partly, in pausing, myself doubt its being there. This looks to me an ideal, as well as an impossible, undulation.

But if it is indeed truly surveyed, then assuredly whatever it may be owing to, it is not owing to lateral pressure.

That is to say, it may be a crystalline arrangement assumed under pressure, but it is assuredly *not* a form assumed by ductile substance under mechanical force. Order the cook to roll out half-a-dozen strips of dough, and to stain three of them with cochineal. Put red and white alternately one above the other. Then press them in any manner you like; after pressure, a wetted carving-knife will give you quite unquestionable sections, and you see the results of three such experiments in the lower figures of the plate.

31. Figure 2 represents the simplest possible case. Three white and three red dough-strips were taken, a red one uppermost (for the pleasure of painting it afterwards)! They were left free at the top, enclosed at the sides, and then reduced from a foot to six inches in length, by pressure from the right. The result, you see, is that the lower bed rises into sharpest gables; the upper ones are rounded softly. But in the geological section it is the upper bed that rises, the lower keeps down! The second case is much more interesting. The pastes were arranged in the same order, but bent up a little, to begin with, in two places, before applying the pressure. The result was, to my own great surprise, that at these points of previous elevation, the lower bed first became quite straight by tension as it rose, and then broke into transverse faults.

32. The third case is the most interesting of all. In this case, a roof of slate was put over the upper bed,

allowing it to rise to some extent only, and the pressure was applied to the two lower beds only.* The upper bed of course exuded backwards, giving these flame-like forms of which afterwards I got quite lovely complications by repeated pressures.[1] These I must reserve for future illustration,[2] concluding to-night, if you will permit me, with a few words of general advice to the younger members of this society, formed as it has been to trace for itself a straight path through the fields of literature, and over the rocks of science.

33. First.—Whenever you write or read English, write it pure, and make it pure if ill written, by avoiding all unnecessary foreign, especially Greek, forms of words yourself, and translating them when used by others. Above all, make this a practice in science. Great part of the

---

* Here I had to give the left-hand section, as it came more neatly. The wrinkled mass on the left coloured brown represents the pushing piece of wood, at the height to which it was applied.

---

[1] The report of the lecture has an additional passage here :—

"Having described some other experiments with the dough, Mr. Ruskin observed that there would have been the same results in the case of rocks under the influence of pressure in the same direction. But in these geological sections they would find no such action whatever indicated. Was there any other way in which the same results could be produced? He didn't know and he didn't care ; that was for geologists to determine. The present chief difficulty was to persuade geologists to look down at the things which might be seen and known thoroughly before they undertook observations on so large a scale. There were many contortions in small rocks which might be seen in process of formation ; contortions formed not by compression but by congelation. Mr. Ruskin then described some specimens of rock which he had on the table before him, one of which was a piece of quartz from the Coniston Rag, a bit of Yewdale Crag, containing a wonderful piece of "honeycomb," which, he said, he intended to leave at the Kendal Museum, and a piece of the actual basalt out of which these crags were cut. He could scarcely, to so large an audience, make clear the phenomena which required attention to minor details. It was no use enlarging on any characteristics of these rocks unless he could show them one by one all the features which mark particular cases, for he believed that a few words of general advice in whatever path of science they were pursuing—as he had found among his own pupils—were only useful to them in proportion to the degree in which they were earnest and faithful. He found that all the secrets of science were only to be approached with extreme humility, and with extreme care in framing the terms in which they were explained."]

[2] [This, however, was not done.]

supposed scientific knowledge of the day is simply bad English, and vanishes the moment you translate it.[1]

There is a farther very practical reason for avoiding all vulgar Greek-English. Greece is now a kingdom, and will I hope remain one, and its language is now living. The ship-chandler, within six doors of me on the quay at Venice, had indeed a small English sign—calling himself Ship-Chandler: but he had a large and practically more serviceable, Greek one, calling himself a "προμηθέττης τῶν πλοόων." Now when the Greeks want a little of your science, as in very few years they must, if this absurd practice of using foreign languages for the clarification of scientific principle still holds, what you, in compliment to Greece, call a "Dinotherium," Greece, in compliment to you, must call a "Nastybeastium,"[2]—and you know that interchange of compliments can't last long.[3]

34. (II.) Observe generally that all knowledge, little or much, is dangerous, in which your progress is likely to be broken short by any strict limit set to the powers of mortals: while it is precisely that kind of knowledge which provokes vulgar curiosity, because it seems so far away; an

---

[1] [Compare *Harbours of England*, § 16 (Vol. XIII. p. 28), and *Proserpina*, Vol. XXV. p. 200.]

[2] [Compare *Proserpina*, Vol. XXV. pp. 318, 321.]

[3] [The report of the lecture has an additional passage at this point which is here given, in Ruskin's own words, from a fragment of the MS. at Brantwood. He read most of it, with §§ 33–40 here, in the Oxford course of "Readings in *Modern Painters*" (see Vol. XXII. p. 519):—

"Secondly. In the very admirable and comprehensive address with which your session was opened by the chairman (Mr. Crewdson), there was one little rift within the lute which he will pardon me for at once riveting, that it may go no farther. It is the great mistake of our day to think ourselves wiser than the ancients; and I was much grieved to see this habit of thought had so far infected your teacher that he took upon himself to explain away one of the deepest and grandest truths ever written by one of the wisest Englishmen who ever lived.

"'A little knowledge is a dangerous thing.' The chairman rightly observed, though he might certainly have given Pope credit for having observed before him, that you must have a little knowledge before you can have more: and that if you refuse the little, you will never have much. But what Pope meant, and in his second line showed that he meant, was that a little knowledge was dangerous if you stopped at it. And the fact is . . . no step farther (§ 34).

"And observe generally that all knowledge . . ."

For the quotation from Pope (*Essay on Criticism*, ii. 15), compare Vol. XXII. p. 137.]

idle ambition, because it allows any quantity of speculation, without proof. And the fact is that the greater quantity of the knowledge which modern science is so saucy about, is only an asses' bridge, which the asses all stop at the top of, and which, moreover, they can't help stopping at the top of; for they have from the beginning taken the wrong road, and so come to a broken bridge—a Ponte Rotto[1] over the river of Death, by which the Pontifex Maximus allows them to pass no step farther.

35. For instance,—having invented telescopes and photography, you are all stuck up on your hobby-horses, because you know how big the moon is, and can get pictures of the volcanoes in it!

But you never can get any more than *pictures* of these, while in your own planet there are a thousand volcanoes which you may jump into, if you have a mind to; and may one day perhaps be blown sky-high by, whether you have a mind or not. The last time the great volcano in Java was in eruption,[2] it threw out a stream of hot water as big as Lancaster Bay, and boiled twelve thousand people. That's what I call a volcano to be interested about, if you want sensational science.

36. But if not, and you can be content in the wonder and the power of Nature, without her terror,—here is a little bit of a volcano, close at your very doors—Yewdale Crag, which I think will be quiet for our time,—and on which the anagallis tenella, and the golden potentilla, and the sundew,[3] grow together among the dewy moss in peace. And on the cellular surface of one of the blocks of it, you may find more beauty, and learn more precious things, than with telescope or photograph from all the moons in the milky-way, though every drop of it were another solar system.

I have a few more very serious words to say to the fathers, and mothers, and masters, who have honoured me

---

[1] [See Vol. XXIV. p. 177.]
[2] [June 10, 1877. The eruption of August 26, 1883, was yet more destructive.]
[3] [See Ruskin's description of these flowers in Vol. XXV. p. xli.]

with their presence this evening, with respect to the influence of these far-reaching sciences on the temper of children.

37. Those parents who love their children most tenderly, cannot but sometimes dwell on the old Christian fancy, that they have guardian angels. I call it an old fancy, in deference to your modern enlightenment in religion; but I assure you nevertheless, in spite of all that illumination, there remains yet some dark possibility that the old fancy may be true: and that, although the modern apothecary cannot exhibit to you either an angel, or an imp, in a bottle, the spiritual powers of heaven and hell are no less now, than heretofore, contending for the souls of your children; and contending with *you*—for the privilege of their tutorship.

38. Forgive me if I use, for the few minutes I have yet to speak to you, the ancient language,—metaphorical, if you will, of Luther and Fénelon, of Dante and Milton, of Goethe and Shakespeare, of St. John and St. Paul, rather than your modern metaphysical or scientific slang: and if I tell you, what in the issue of it you will find is either life-giving, or deadly, fact,—that the fiends and the angels contend with you daily for the spirits of your children: the devil using to you his old, his hitherto immortal, bribes, of lust and pride; and the angels pleading with you, still, that they may be allowed to lead your babes in the divine life of the pure and the lowly. To enrage their lusts, and chiefly the vilest lust of money, the devils would drag them to the classes that teach them how to get on in the world; and for the better pluming of their pride, provoke their zeal in the sciences which will assure them of there being no God in nature but the gas of their own graves.

And of these powers you may discern the one from the other by a vivid, instant, practical test. The devils always will exhibit to you what is loathsome, ugly, and, above all, dead ; and the angels, what is pure, beautiful, and, above all, living.

39. Take an actual literal instance. Of all known quadrupeds, the unhappiest and vilest, yet alive, is the sloth, having this farther strange devilry in him, that what activity he is capable of, is in storm, and in the night. Well, the devil takes up this creature, and makes a monster of it, —gives it legs as big as hogsheads, claws stretched like the roots of a tree, shoulders like a hump of crag, and a skull as thick as a paving-stone. From this nightmare monster he takes what poor faculty of motion the creature, though wretched, has in its minuter size ; and shows you, instead of the clinging climber that scratched and scrambled from branch to branch among the rattling trees as they bowed in storm, only a vast heap of stony bones and staggering clay, that drags its meat down to its mouth out of the forest ruin. This creature the fiends delight to exhibit to you,[1] but are permitted by the nobler powers only to exhibit to you in its death.*

40. On the other hand, as of all quadrupeds there is none so ugly or so miserable as the sloth, so, take him for all in all, there is none so beautiful, so happy, so wonderful, as the squirrel. Innocent in all his ways, harmless in his food, playful as a kitten, but without cruelty, and surpassing the fantastic dexterity of the monkey, with the grace

* The Mylodon.[2] An old sketch (I think, one of Leech's), in *Punch*, of Paterfamilias improving Master Tom's mind among the models on the mudbank of the lowest pond at Sydenham, went to the root of the matter. For the effect on Master Tom's mind of the living squirrel, compare the following account of the most approved modes of squirrel-hunting, by a clerical patron of the sport, extracted for me by a correspondent, from *Rabbits : How to Rear and Manage Them ; with Chapters on Hares, Squirrels, etc.* S. O. Beeton, 248, Strand, W.C.

"It may be easily imagined that a creature whose playground is the top twigs of tall trees, where no human climber dare venture, is by no means easy to capture—especially as its hearing is keen, and its vision remarkably acute. Still, among boys living in the vicinity of large woods and copses, squirrel-hunting is a favourite diversion, and none the less so because it is seldom attended by success. 'The only plan,' says the Rev. Mr. Wood, 'is to watch the animal until it has ascended an isolated tree, or, by a well-directed

---

[1] [For a note on this passage, see ii. ch. ii. § 23 (below, p. 346).]
[2] [For the Mylodon, see Vol. VIII. p. 72, and Vol. IX. p. 166.]

and the brightness of a bird, the little dark-eyed miracle of the forest glances from branch to branch more like a sun-beam than a living creature: it leaps, and darts, and twines, where it will;—a chamois is slow to it; and a panther, clumsy: grotesque as a gnome, gentle as a fairy, delicate as the silken plumes of the rush, beautiful and strong like the spiral of a fern,—it haunts you, listens for you, hides from you, looks for you, loves you, as if the angel that walks with your children had made it himself for their heavenly plaything.

And this is what *you* do, to thwart alike your child's angel, and his God,—you take him out of the woods into the town,—you send him from modest labour to competitive schooling,—you force him out of the fresh air into the dusty bone-house,—you show him the skeleton of the dead monster, and make him pore over its rotten cells and wire-stitched joints, and vile extinct capacities of destruction,— and when he is choked and sickened with useless horror and putrid air, you let him—regretting the waste of time —go out for once to play again by the woodside;—and the first squirrel he sees, he throws a stone at!

Carry, then, I beseech you, this assured truth away with you to-night. All true science begins in the love,

shower of missiles, to drive it into such a place of refuge, and then to form a ring round the tree so as to intercept the squirrel, should it try to escape by leaping to the ground and running to another tree. The best climber is then sent in chase of the squirrel, and endeavours, by violently shaking the branches, to force the little animal to loose its hold and fall to the earth. But it is by no means an easy matter to shake a squirrel from a branch, especially as the little creature takes refuge on the topmost and most slender boughs, which even bend under the weight of its own small body, and can in no way be trusted with the weight of a human being. By dint, however, of perseverance, the squirrel is at last dislodged, and comes to the ground as lightly as a snow-flake. Hats, caps, sticks, and all avail-able missiles are immediately flung at the luckless animal as soon as it touches the ground, and it is very probably struck and overwhelmed by a cap. The successful hurler flings himself upon the cap, and tries to seize the squirrel as it lies under his property. All his companions gather round him, and great is the disappointment to find the cap empty, and to see the squirrel triumphantly scampering up some tree where it would be useless to follow it.'"

not the dissection, of your fellow-creatures; and it ends in the love, not the analysis, of God. Your alphabet of science is in the nearest knowledge, as your alphabet of morals is in the nearest duty. "Behold, it is nigh thee, even at the doors."[1] The Spirit of God is around you in the air that you breathe,—His glory in the light that you see; and in the fruitfulness of the earth, and the joy of its creatures, He has written for you, day by day, His revelation, as He has granted you, day by day, your daily bread.[2]

---

[1] [See Mark xiii. 29.]

[2] [Matthew vi. 11. In responding to a vote of thanks for the lecture, Ruskin is thus reported (*Kendal Times*, October 6, 1877):—

"The lecturer in responding, quoting the words of Shakespeare's King Henry to the Bishop, 'Thou wert always good at sudden commendations, cousin of Winchester,' remarked that when as in this case the commendation had been so kindly given, he was helpless of sudden thanks. He felt that there was a debt on his side, for it was a kindness on their part to sit and listen to details which had appealed to them for sympathy in his embarrassment rather than conveyed instruction to an audience like the present. He had not known what a large audience he would meet, and had been prepared with a few details only with reference to his subject, and he now feared he had failed to afford that interest to his hearers that he could have wished. He did not, however, fail to appreciate the kindly hand that had been extended to him, the kindness and quiet courtesy he had received from high and low. Above all, he was delighted to live among Westmorland and Cumberland men, for he had always found those of the lower orders sagacious, modest, and gentle, willing to join him in his work and give him their assistance in the most simple and loving way. He thanked the Mayor for what he had said of his endeavour to teach kindly at Oxford and elsewhere. But it was not, as had been implied, to his instruction that he required obedience. He was not sure that he ever gave instruction. He taught them to look at something better than his teachings and writings, to look at the book of nature and learn from its laws. He wanted to see curiosity excited, and not vanity. He found it a great obstacle at the present day to the progress of the young that rewards were given for prominence in science. There was the reward of high praise, a man's name was taken up by the public, and it passed around the world at once, and there were few minds so tranquil that they could resist the temptation to labour for that magnificent prize rather than for the quiet possession of knowledge of the works of the Most High. He believed that the knowledge sought after for the sake of place was always more or less false, and he believed that nature deceived those who did not seek her honestly and heartily. Let them seek knowledge for its own sake and for the service of others, and it would then become delightful to them. This was all he had endeavoured to teach, and, in conclusion, he gave them the assurance that the way of nature and the way of duty were paths of pleasantness and peace."

For the quotation from Shakespeare at the beginning of this passage, see *Henry VIII.*, Act v. sc. 2; for the Bible reference at the end, Proverbs iii. 17 (compare Vol. XXIII. p. 259).]

CHAPTER XIII

OF STELLAR SILICA

1. THE issue of this number of *Deucalion* has been so long delayed, first by other work, and recently by my illness, that I think it best at once to begin Mr. Ward's notes on Plate XIV.: reserving their close, with full explanation of their importance and bearing, to the next following number.[1]

GRETA BANK COTTAGE, KESWICK,
June 13, 1876.

MY DEAR SIR,—I send you a few notes on the microscopic structure of the three specimens I have had cut. In them I have stated merely what I have seen. There has been much which I did not expect, and still more is there that I don't understand.

I am particularly sorry I have not the time to send a whole series of coloured drawings illustrating the various points; but this summer weather claims my time on the mountain-side, and I must give up microscopic work until winter comes round again.

The minute spherulitic structure—especially along the fine brown lines—was quite a surprise, and I shall hope on some future occasion to see more of this subject. Believe me, yours very truly,

J. CLIFTON WARD.

P.S.—There seems to be a great difference between the microscopic structure of the specimens now examined and that of the filled-up vesicles in many of my old lavas here, so far as my *limited* examination has gone.

---

[1] [Part IV. (chaps. viii. to xi.) appeared in December 1876; Part V. (containing chap. xii. and the first few pages of chap. xiii.), in July 1878. The remainder of chap. xiii. was given in Part VI., which, however, did not appear till October 1879. For Ruskin's illness early in 1878, see Vol. XXV. p. xxvi.]

SPECIMEN A

No. 1 commences at the end of the section farthest from A in specimen.

1. Transparent zone with irregular curious cavities (not liquid), and a few mossy-looking round spots (brownish).

Polarization. Indicating an indefinite semi - crystalline structure. (See note at page 269.)

2. Zone with minute seed-like bodies of various sizes (narrow brownish bands in the specimen of darker and lighter tints).

a. Many cavities, and of an indefinite oval form in general.

b. The large spherulites (2) are very beautiful, the outer zone (radiate) of a delicate greenish-yellow, the nucleus of a brownish-yellow, and the intermediate zone generally clear.

c. A layer of densely packed bodies, oblong or oval in form.

d. Spherulites generally similar to *b*, but smaller, much more stained of a brownish-yellow, and with more defined nuclei.

Polarization. The spherulites show a clearly radiate polarization, with rotation of a dark cross on turning either of the prisms; the intermediate ground shows the irregular semi-crystalline structure.

3. Clear zone, with little yellowish, dark, squarish specks.
Polarization. Irregular, semi-crystalline.

4. Row of closely touching spherulites with large nucleus and defined margin, rather furry in character (3). Margins and nuclei brown; intermediate space brownish-yellow.

Polarization. Radiate, as in the spherulites 2 *b*.

(This is a short brown band which does not extend down through the whole thickness of the specimen.)

5. Generally clear ground, with a brownish cloudy appearance in parts.
Polarization. Indefinite semi-crystalline.

6 *a*. On a hazy ground may be seen the cloudy margins of separately crystalline spaces.

Polarization. Definite semi-crystalline.*

6 *b*. A clear band with very indefinite polarization.

7. A clearish zone with somewhat of a brown mottled appearance (light clouds of brown colouring matter).

Polarization. Indefinite semi-crystalline.

8. Zone of brownish bodies (this is a fine brown line, about the middle of the section in the specimen).

a. Yellowish-brown nucleated disks.

b. Smaller, scattered, and *generally* non-nucleated disks.

c. Generally non-nucleated.

Polarization. The disks are too minute to show separate polarization effects, but the ground exhibits the indefinite semi-crystalline.

9. Ground showing indefinite semi-crystalline polarization.

10. Irregular line of furry-looking yellowish disks.

11. Zone traversed by a series of generally parallel and faint lines of a brownish-yellow. These are apparently lines produced by colouring matter alone,—at any rate, not by *visible* disks of any kind.

Polarization. Tolerably definite, and limited by the cross lines (6).

12. Dark-brown flocculent-looking matter, as if growing out from a well-defined line, looking like a moss-growth.

13. Defined crystalline interlocked spaces.

Polarization. Definite semi-crystalline.

14. Generally, not clearly defined spaces; central part rather a granular look (spaces very small).

Polarization. Under crossed prisms breaking up into tolerably definite semi-crystalline spaces.

---

* By "*indefinite* semi-crystalline" is meant the breaking up of the ground under crossed prisms with sheaves (5) of various colours not clearly margined.

By "*definite* semi-crystalline" is meant the breaking up of the ground under crossed prisms with a mosaic (4) of various colours clearly margined.

By "semi-crystalline" is meant the interference of crystalline spaces with one another, so as to prevent a perfect crystalline form being assumed.

SPECIMEN B

B 1. In the slice taken from this side there seems to be frequently a great tendency to spherulitic arrangement, as shown by the polarization phenomena. In parts of the white quartz, where the polarization appearance is like that of a mosaic pavement, there is even a semi-spherulitic structure. In other parts there are many spherulites on white and yellowish ground.

Between the many parallel lines of a yellowish colour the polarization (7) effect is that of fibrous coloured sheaves.

Here (8) there is a central clear band (*b*); between it and (*a*) a fine granular line with some larger granules (or very minute spherulites). The part (*a*) is carious, apparently with glass cavities. On the other side of the clear band, at *c*, are half-formed and adherent spherulites; the central (shaded) parts are yellow, and the outer coat, the intermediate portion clearish.

B 2. The slice from the end of the specimen shows the same general structure.

The general tendency to spherulitic arrangement is well seen in polarized light, dark crosses frequently traversing the curved structures.

Here (in Fig. 9) the portion represented on the left was situated close to the other portion, where the point of the arrow terminates, both crosses appearing together, and revolving in rotation of one of the prisms.

SPECIMEN C

The slice from this specimen presents far less variety than in the other cases. There are two sets of structural lines—those radiate (10), and those curved and circumferential (11). The latter structure is exceedingly fine and delicate, and not readily seen, even with a high power, owing to the fine radii not being marked out by any colour, the whole section being very clear and white.

A more decidedly nucleated structure is seen in part 12.

In (18) is a very curious example of a somewhat more glassy portion protruding in finger-like masses into a radiate clear, and largely spherical portion.

2. These notes of Mr. Clifton Ward's contain the first accurate statements yet laid before mineralogists respecting the stellar crystallization of silica, although that mode of its formation lies at the very root of the structure of the greater mass of amygdaloidal rocks, and of all the most beautiful phenomena of agates. And indeed I have no words to express the wonder with which I see work like that done by Cloizeaux in the measurement of quartz angles, conclude only in the construction of the marvellous diagram, as subtle in execution as amazing in its accumulated facts,* without the least reference to the conditions of varying energy which produce the spherical masses of chalcedony! He does not even use the classic name of the mineral, but coins the useless one, Geyserite, for the absolutely local condition of the Iceland sinter.

3. And although, in that formation, he went so near the edge of Mr. Clifton Ward's discovery as to announce that "leur masse se compose elle même de sphères enchâssées dans une sorte de pâte gélatineuse," he not only fails, on this suggestion, to examine chalcedonic structure generally, but arrested himself finally in the pursuit of his inquiry by quietly asserting, "ce genre de structure n'a jamais été rencontré jusqu'ici sur aucune autre variété de silice naturelle ou artificielle,"—the fact being that there is no chalcedonic mass whatever, which does *not* consist of spherical concretions more or less perfect, enclosed in a "pâte gélatineuse."

* Facing page 8 of the *Manuel de Minéralogie*.[1]

[1] [*Manuel de Minéralogie*, par A. Des Cloizeaux, 1st vol. Paris, 1862, 2nd vol., 1874–1893. Ruskin refers to this "marvellous diagram" again in ii. ch. ii., § 17 (below, p. 341). It shows in a single engraving all the angles in various forms of quartz. A similar diagram for calcite at p. 102 of vol. ii. is yet more elaborate and delicately engraved.]

4. In Professor Miller's manual, which was the basis of Cloizeaux's, chalcedony is stated to appear to be a mixture of amorphous with crystalline silica![1] and its form taken no account of. Malachite might just as well have been described as a mixture of amorphous with crystalline carbonate of copper!

5. I will not, however, attempt to proceed farther in this difficult subject until Mr. Clifton Ward has time to continue his own observations. Perhaps I may persuade him to let me have a connected series of figured examples, from pure stellar quartz down to entirely fluent chalcedony, to begin the next volume of *Deucalion* with;[2]—but I must endeavour, in closing the present one, to give some available summary of its contents, and clearer idea of its purpose; and will only trespass so far on my friend's province as to lay before him, together with my readers, some points noted lately on another kind of semi-crystallization, which bear not merely on the domes of delicate chalcedony, and pyramids of microscopic quartz, but on the far-seen chalcedony of the Dôme du Goûter, and the prismatic towers of the Cervin and dark peak of Aar.[3]

[1] [See p. 250 of the new edition (1852) by H. J. Brooke and W. H. Miller (Professor of Mineralogy in the University of Cambridge) of William Phillips's *Elementary Introduction to Mineralogy*. Des Cloizeaux states in the Preface to his *Manuel de Minéralogie* (1862) that his first intention had been to give "a simple translation of the remarkable work by Brooke and Miller."]

[2] [The next volume, however, took "a serpentine line of advance" to other subjects: see below, p. 295.]

[3] [For other references to the Finsteraarhorn, see Vol. III. p. 431, and Vol. VI. pp. 199 n., 205, 222.]

CHAPTER XIV

SCHISMA MONTIUM

1. THE index closing this volume of *Deucalion*,[1] drawn up by myself, is made as short as possible, and classifies the contents of the volume so as to enable the reader to collect all notices of importance relating to any one subject, and to collate them with those in my former writings. That they need such assemblage from their desultory occurrence in the previous pages, is matter of sincere regret to me, but inevitable, since the writing of a systematic treatise was incompatible with the more serious work I had in hand, on greater subjects. *The Laws of Fésole* alone might well occupy all the hours I can now permit myself in severe thought. But any student of intelligence may perceive that one inherent cause of the divided character of this book is its function of advance in parallel columns over a wide field; seeing that, on no fewer than four subjects, respecting which geological theories and assertions have long been alike fantastic and daring, it has shown at least the necessity for revisal of evidence, and, in two cases, for reversal of judgment.[2]

2. I say "it has shown," fearlessly; for at my time of life, every man of ordinary sense, and probity, knows what he has done securely, and what perishably. And during the last twenty years, none of my words have been set down untried;[3] nor has any opponent succeeded in overthrowing a single sentence of them.

3. But respecting the four subjects above alluded to

---

[1] [See now the end of this volume.]
[2] [On this passage, see the Introduction, above, p. lxiv.]
[3] [Compare *Ethics of the Dust*, Preface of 1877, § 2 (Vol. XVIII. p. 206).]

(denudation, cleavage, crystallization, and elevation, as causes of mountain form), proofs of the uncertainty, or even falseness, of current conceptions have been scattered at intervals through my writings, early and late, from *Modern Painters* to the *Ethics of the Dust*:[1] and, with gradually increasing wonder at the fury of so-called "scientific" speculation, I have insisted, year by year, on the undealt with, and usually undreamt of, difficulties which lay at the threshold of secure knowledge in such matters;—trusting always that some ingenuous young reader would take up the work I had no proper time for, and follow out the investigations of which the necessity had been indicated. But I waited in vain; and the rough experiments made at last by myself, a year ago, of which the results are represented in Plate XV. of this volume, are actually the first of which there is record in the annals of geology, made to ascertain the primary physical conditions regulating the forms of contorted strata. The leisure granted me, unhappily, by the illness which has closed my relations with the University of Oxford,[2] has permitted the pursuit of these experiments a little farther; but I must defer account of their results to the following volume,[3] contenting myself with indicating, for conclusion of the present one, to what points of doubt in existing theories they have been chiefly directed.

4. From the examination of all mountain ground hitherto well gone over, one general conclusion has been derived,— that wherever there are high mountains, there are hard rocks. Earth, at its strongest, has difficulty in sustaining itself above the clouds; and could not hold itself in any noble height, if knitted infirmly.

5. And it has farther followed, in evidence, that on the flanks of these harder rocks, there are yielding beds, which

---

[1] [See, *e.g.*, for cleavage, *Modern Painters*, vol. iv. (Vol. VI. pp. 475 *seq.*), and *Ethics of the Dust*, Vol. XVIII. p. 338; for elevation, *Modern Painters*, vol. iv. (Vol. VI. pp. 175 *seq.*), and *Ethics*, Vol. XVIII. p. 328; for denudation, *Modern Painters*, vol. iv. (Vol. VI. pp. 211 *seq.*); and for crystallization, *ibid.* pp. 130 *seq.*, and *Ethics*, Vol. XVIII. p. 330.]

[2] [He resigned the Professorship in 1878, but resumed it in 1883.]

[3] [This intention, however, was not carried out.]

appear to have been, in some places, compressed by them into wrinkles and undulations;—in others, shattered, and thrown up or down to different levels. My own interest was excited, very early in life,* by the forms and fractures in the mountain groups of Savoy; and it happens that the undulatory action of the limestone beds on each shore of the Lake of Annecy, and the final rupture of their outmost wave into the precipice of the Salève, present examples so clear, and so imposing, of each condition of form, that I have been led, without therefore laying claim to any special sagacity, at least into clearer power of putting essential questions respecting such phenomena than geologists of far wider experience, who have confused or amused themselves by collecting facts indiscriminately over vast spaces of ground. I am well convinced that the reader will find more profit in following my restricted steps; and satisfying or dissatisfying himself, with precision, respecting forms of mountains which he can repeatedly and exhaustively examine.

6. In the uppermost figure in Plate XVI., I have enlarged and coloured the general section given rudely above in Figure 33, page 105, of the Jura and Alps, with the intervening plain. The central figure is the southern, and the lowermost figure, which should be conceived as joining it on the right hand, the northern, series of the rocks composing our own Lake district, drawn for me with extreme care by the late Professor Phillips, of Oxford.[1]

---

* I well yet remember my father's rushing up to the drawing-room at Herne Hill, with wet and flashing eyes, with the proof in his hand of the first sentences of his son's writing ever set in type,[2]—"Inquiries on the Causes of the Colour of the Water of the Rhone" (*Magazine of Natural History*, September 1834; followed next month by "Facts and Considerations on the Strata of Mont Blanc, and on some Instances of Twisted Strata observable in Switzerland."[3] I was then fifteen). My mother and I eagerly questioning the cause of his excitement,—"It's—it's—only *print!*" said he. Alas! how much the "only" meant!

---

[1] [See above, p. 154.]
[2] [This is not strictly correct: see Vol. II. p. 265 n.]
[3] [For these papers, see Vol. I. pp. 191, 194.]

I compare, and oppose, these two sections, for the sake of fixing in the reader's mind one essential point of difference among many resemblances; but that they may not, in this comparison, induce any false impressions, the system of colour which I adopt in this plate, and henceforward shall observe, must be accurately understood.

7. At page 203 above, I gave my reasons for making no endeavour, at the Sheffield Museum, to certify the ages of rocks. For the same reason, in practical sections I concern myself only with their nature and position; and colour granite pink, slate purple, and sandstone red, without inquiring whether the granite is ancient or modern,—the sand trias or pliocene, and the slate Wenlock or Caradoc; but with this much only of necessary concession to recognized method, as to colour with the same tint all rocks which unquestionably belong to the same great geological formation, and vary their mineralogical characters within narrow limits. Thus, since, in characteristic English sections, chalk may most conveniently be expressed by leaving it white, and some of the upper beds of the Alps unquestionably are of the same period, I leave them white also, though their general colour may be brown or grey, so long as they retain cretaceous or marly consistence; but if they become metamorphic, and change into clay slate or gneiss, I colour them purple, whatever their historical relations may be.

8. And in all geological maps and sections given in *Deucalion*, I shall limit myself to the definition of the twelve following formations by the twelve following colours. It is enough for any young student at first to learn the relations of these great orders of rock and earth:—once master of these, in any locality, he may split his beds into any complexity of finely laminated chronology he likes;— and if I have occasion to split them for him myself, I can easily express their minor differences by methods of engraving.

9. But, primarily, let him be content in the recognition

of these twelve territories of Demeter, by this following colour heraldry :—

1. Granite will bear in the field,	Rose-red.
2. Gneiss and mica-slate	Rose-purple.
3. Clay-slate	Violet-purple.
4. Mountain limestone	Blue.
5. Coal measures and mill-stone grit	Grey.
6. Jura limestone	Yellow.
7. Chalk	White.
8. Tertiaries forming hard rock	Scarlet.
9. Tertiary sands and clays	Tawny.
10. Eruptive rocks not definitely volcanic	Green.
11. Eruptive rocks definitely volcanic, but at rest	Green, spotted red.
12. Volcanic rocks, active	Black, spotted red.

10. It will at once be seen, by readers of some geological experience, that approximately, and to the degree possible, these colours are really characteristic of the several formations ; and they may be rendered more so by a little care in modifying the tints. Thus the " scarlet " used for the tertiaries may be subdued as much as we please, to what will be as near a sober brown as we can venture without confusing it with the darker shades of yellow ; and it may be used more pure to represent definitely red sandstones or conglomerates ; while, again, the old red sands of the coal measures may be extricated from the general grey by a tint of vermilion which will associate them, as mineral substances, with more recent sand. Thus, in the midmost section of Plate XVI., this colour is used for the old red conglomerates of Kirby Lonsdale. And again, keeping pure light blue for the dated mountain limestones, which are indeed, in their emergence from the crisp turf of their pastures, grey, or even blue in shade, to the eye, a deeper blue may be kept for the dateless limestones which are associated with the metamorphic beds of the Alps ; as for my own Coniston Silurian limestone, which may be nearly as old as Skiddaw.

11. The colour called "tawny," I mean to be as nearly that of ripe wheat as may be, indicating arable land or hot prairie; while, in maps of northern countries, touched with points of green, it may pass for moorland and pasture; or, kept in the hue of pale vermilion, it may equally well represent desert alluvial sand. Finally, the avoidance of the large masses of fierce and frightful scarlet which render modern geological maps intolerable to a painter's sight (besides involving such geographical incongruities as the showing Iceland in the colour of a red-hot coal), and the substitution over all volcanic districts, of the colour of real greenstone, or serpentine, for one which resembles neither these, nor the general tones of dark colour either in lava or cinders, will certainly render all geological study less injurious to the eyesight and less harmful to the taste.

12. Of the two sections in Plate XVI., the upper one is arranged from Studer,[1] so as to exhibit in one view the principal phenomena of Alpine structure according to that geologist. The cleavages in the central granite mass are given, however, on my own responsibility, not his. The lower section was, as aforesaid, drawn for me by my kind old friend Professor Phillips, and is, I doubt not, entirely authoritative. In all great respects, the sections given by Studer, are no less so; but they are much ruder in drawing, and can be received only as imperfect summaries—perhaps, in their abstraction, occasionally involving some misrepresentation of the complex facts. For my present purposes, however, they give me all the data required.

13. It will instantly be seen, on comparing the two groups of rocks, that although nearly similar in succession, and both suggesting the eruptive and elevatory force of the granitic central masses, there is a wide difference in the manner of the action of these on the strata lifted by them. In the Swiss section, the softer rocks seem to have been crushed aside, like the ripples of water round any submersed

---

[1] [Arranged, that is, in accordance with Studer's views; not copied from any actual woodcut given in his book.]

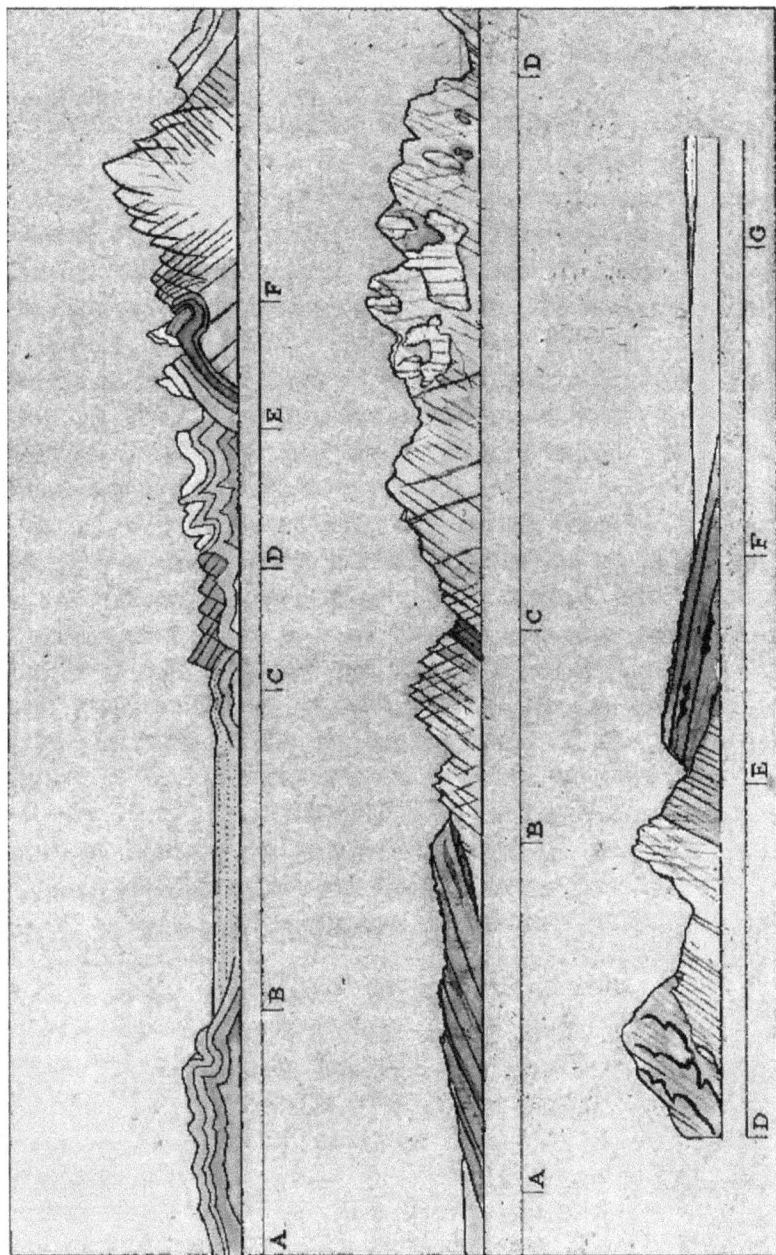

THE STRATA OF SWITZERLAND AND CUMBERLAND

object rising to the surface. In the English section, they seem to have undergone no such torsion, but to be lifted straight, as they lay, like the timbers of a gabled roof. It is true that, on the larger scale of the Geological Survey, contortions are shown at most of the faults in the Skiddaw slate; but, for the reasons already stated, I believe these contortions to be more or less conventionally represented; and until I have myself examined them, will not modify Professor Phillips' drawing by their introduction.

Some acknowledgment of such a structure is indeed given by him observably in the dark slates on the left in the lowermost section; but he has written under these undulatory lines "quartz veins," and certainly means them, so far as they are structural, to stand only for ordinary gneissitic contortion in the laminated mass, and not for undulating strata.

14. Farther. No authority is given me by Studer for dividing the undulatory masses of the outer Alps by any kind of cleavage-lines. Nor do I myself know examples of fissile structure in any of these mountain masses, unless where they are affected by distinctly metamorphic action, in the neighbourhood of the central gneiss or mica-schist. On the contrary, the entire courses of the Cumberland rock, from Kirby Lonsdale to Carlisle, are represented by Professor Phillips as traversed by a perfectly definite and consistent cleavage throughout, dipping steeply south, in accurately straight parallel lines, and modified only, in the eruptive masses, by a vertical cleavage, characterizing the pure granite centres.

15. I wish the reader to note this with especial care, because the cleavage of secondary rock has been lately attributed, with more appearance of reason than modern scientific theories usually possess, to lateral pressure, acting in a direction perpendicular to the lamination. It seems, however, little calculated to strengthen our confidence in such an explanation, to find the Swiss rocks, which appear to have been subjected to a force capable of doubling up

leagues of them backwards and forwards like a folded map, wholly without any resultant schistose structure; and the English rocks, which seem only to have been lifted as a raft is raised on a wave, split across, for fifty miles in succession, by foliate structures of the most perfect smoothness and precision.

16. It might indeed be alleged, in deprecation of this objection, that the dough or batter of which the Alps were composed, mostly calcareous, did not lend itself kindly to lamination, while the mud and volcanic ashes of Cumberland were, of a slippery and unctuous character, easily susceptible of rearrangement under pressure. And this view receives strong support from the dextrous experiment performed by Professor Tyndall in 1856,[1] and recorded, as conclusive, in 1872, wherein, first warming some wax, then pressing it between two pieces of glass, and finally freezing it, he finds the congealed mass delicately laminated; and attributes its lamination to the "lateral sliding of the particles over each other." *   But with his usual, and quite unrivalled, incapacity of following out any subject on the two sides of it, he never tells us, and never seems to have asked himself, how *far* the wax was flattened, and how far, therefore, its particles had been forced to slide; —nor, during the sixteen years between his first and final record of the experiment, does he seem ever to have used any means of ascertaining whether, under the observed conditions, real compression of the substance of the wax had taken place at all! For if not, and the form of the mass was only altered from a lump to a plate, without any increase of its density, a less period for reflection than sixteen years might surely have suggested to Professor Tyndall the necessity, in applying his result to geological matters, of providing mountains which were to be squeezed in one direction, with room for expansion in another.

* *Forms of Water* (King and Co.), 1872, p. 190 [§ 487].

[1] [At the Royal Institution on June 10, 1856: see *Glaciers of the Alps*, p. 6.]

17. For once, however, Professor Tyndall is not without fellowship in his hesitation to follow the full circumference of this question. Among the thousands of passages I have read in the works even of the most careful and logical geologists,—even such as Humboldt and De Saussure,—I remember *not one* distinct statement * of the degree in which they supposed the lamination of any given rock to imply real increase of its density, or only the lateral extension of its mass.

18. And the student must observe that in many cases lateral extension of mass is precisely avoided by the very positions of rocks which are supposed to indicate the pressure sustained. In Mr. Woodward's experiment with sheets of paper, for instance, (above quoted, p. 109 †) there is neither increase of density nor extension of mass, in the sheets of paper. They remain just as thick as they were,—just as long and broad as they were. They are only altered in direction, and no more compressed, as they bend, than a flag is compressed by the wind that waves it. In my own

---

* As these sheets are passing through the press, I receive the following most important note from Mr. Clifton Ward: "With regard to the question whether cleavage is necessarily followed by a reduction in bulk of the body cleaved, the following cases may help us to form an opinion. *Crystalline* volcanic rocks (commonly called trap), as a rule, are not cleaved, though the beds, uncrystalline in character, above and below them, may be. When, however, a trap is highly vesicular, it is sometimes well cleaved. May we not, therefore, suppose that in a rock, *wholly* crystalline, the particles are too much interlocked to take up new positions? In a purely fragmentary rock, however, the particles seem to have more freedom of motion; their motion under pressure leads to a new and more parallel arrangement of particles, each being slightly flattened or pulled out along the planes of new arrangement. This, then, points to a diminution of bulk at any rate in a direction at right angles to the planes of cleavage. The tendency to new arrangement of particles *under pressure points to accommodation under altered circumstances of space.* In rocks composed of fragments, the interspaces, being for the most part larger than the intercrystalline spaces of a trap rock, more freely allow of movement and new arrangement."

† There is a double mistake in the fifth line from the bottom in that page. I meant to have written, "from a length of four inches into the length of one inch,"—but I believe the real dimensions should have been "a foot crushed into three inches."

experiments with dough, of course the dough was no more compressible than so much water would have been. Yet the language of the geologists who attribute cleavage to pressure might usually leave their readers in the notion that clay can be reduced like steam ; and that we could squeeze the sea down to half its depth by first mixing mud with it ! Else, if they really comprehended the changes of form rendered necessary by proved directions of pressure, and did indeed mean that the paste of primitive slate had been " flattened out " (in Professor Tyndall's words [1]) as a cook flattens out her pastry-crust with a rolling-pin, they would surely sometimes have asked themselves,—and occasionally taken the pains to tell their scholars,—where the rocks in question had been flattened to. Yet in the entire series of Swiss sections (upwards of a hundred) given by Studer in his Alpine Geology, there is no hint of such a difficulty having occurred to him ;—none, of his having observed any actual balance between diminution of bulk and alteration of form in contorted beds ;—and none, show-ing any attempt to distinguish mechanical from crystalline foliation. The cleavages are given rarely in any section, and always imperfectly.

19. In the more limited, but steadier and closer, work of Professor Phillips on the geology of Yorkshire, the solitary notice of " that very obscure subject, the cleavage of slate " is contained in three pages (5 to 8 of the first chapter [2]), describing the structure of a single quarry, in which the author does not know, and cannot eventually discover, whether the rock is stratified or not ! I respect, and admire, the frankness of the confession ; but it is evident that before any affirmation of value, respecting cleavages, can be made by good geologists, they must both ascertain many laws of pressure in viscous substances at present unknown ; and describe a great many quarries with no

---

[1] [See Forms of Water, § 434.]
[2] [Of the second volume of Illustrations of the Geology of Yorkshire, by John Phillips, F.R.S., F.G.S., 1829, 1836.]

less attention than was given by Professor Phillips to this single one.

20. The experiment in wax, however, above referred to as ingeniously performed by Professor Tyndall, is not adduced in the *Forms of Water* for elucidation of cleavage in rocks, but of riband structure in ice—(of which more presently[1]). His first display of it, however, was, I believe, in the lecture delivered in 1856 at the Royal Institution, —this, and the other similar experiments recorded in the Appendix to the *Glaciers of the Alps*, being then directed mainly to the confusion of Professor Sedgwick, in that the Cambridge geologist had—with caution—expressed an opinion that cleavage was a result of crystallization under polar forces.

21. Of that suggestion Professor Tyndall complimentarily observed that "it was a bold stretch of analogies," and condescendingly—that "it had its value—it has drawn attention to the subject." Presently, translating this too vulgarly intelligible statement into his own sublime language, he declares of the theory in debate that it, and the like of it, are "a dynamic power which operates against intellectual stagnation."[2] How a dynamic power differs from an undynamic one—(and, presumably, also, a potestatic dynamis from an unpotestatic one); and how much more scientific it is to say, instead of—that our spoon stirs our porridge,—that it "operates against the stagnation" of our porridge, Professor Tyndall trusts the reader to recognize with admiration. But if any stirring, or skimming, or other operation of a duly dynamic character, could have clarified from its scum of vanity the pease-porridge of his own wits, Professor Tyndall would have felt that men like the Cambridge veteran,—one of the very few modern men of science who possessed real genius,—stretch no analogies farther than they will hold; and, in this particular case, there were two facts, familiar to Sedgwick, and with which

[1] [See below, § 25, p. 285.]
[2] [*Glaciers of the Alps*, Appendix ("Comparative View of the Cleavage of Crystals and Slate-Rocks"), p. 431.]

Professor Tyndall manifests no acquaintance, materially affecting every question relating to cleavage structure.

22. The first, that all slates whatever, among the older rocks, are more or less metamorphic; and that all metamorphism implies the development of crystalline force. Neither the chiastolite in the slate of Skiddaw, nor the kyanite in that of St. Gothard, could have been formed without the exertion, through the whole body of the rock, of crystalline force, which, extracting some of its elements, necessarily modifies the structure of the rest. The second, that slate-quarries of commercial value, fortunately rare among beautiful mountains, owe their utility to the unusual circumstance of cleaving, over the quarryable space, practically in one direction only. But such quarryable spaces extend only across a few fathoms of crag, and the entire mass of the slate mountains of the world is cloven, not in one, but in half-a-dozen directions, each separate and explicit; and requiring, for their production on the pressure theory, the application of half-a-dozen distinct pressures, of which none shall neutralize the effect of any other! That six applications of various pressures, at various epochs, might produce six cross cleavages, may be conceived without unpardonable rashness, and conceded without perilous courtesy; but before pursuing the investigation of this hex-foiled subject, it would be well to ascertain whether the cleavage of any rock whatever does indeed accommodate itself to the calculable variations of a single pressure, applied at a single time.

23. Whenever a bed of rock is bent, the substance of it on the concave side must be compressed, and the substance of it on the convex side, expanded. The degree in which such change of structure must take place may be studied at ease in one's arm-chair, with no more apparatus than a stick of sealing-wax and a candle;[1] and as soon as I am shown a bent bed of any rock with distinct lamination on its concave side, traceably gradated into distinct

---

[1] [Compare above, p. 164 n.]

crevassing on its convex one, I will admit without farther debate the connection of foliation with pressure.

24. In the meantime, the delicate experiments by the conduct of which Professor Tyndall brought his audiences into what he is pleased to call "contact with facts"[1] (in older times we used to say "grasp of facts"; modern science for its own part prefers, not unreasonably, the term "contact," expressive merely of occasional collision with them), must remain inconclusive. But if in the course of his own various "contact with facts" Professor Tyndall has ever come across a bed of slate squeezed between two pieces of glass—or anything like them—I will thank him for a description of the locality. All metamorphic slates have been subjected assuredly to heat—probably to pressure; but (unless they were merely the shaly portions of a stratified group) the pressure to which they have been subjected was that of an irregular mass of rock ejected in the midst of them, or driven fiercely against them; and their cleavage—so far as it is indeed produced by that pressure, must be such as the iron of a target shows round a shell;—and not at all representable by a film of candle-droppings.

25. It is further to be observed,—and not without increasing surprise and increasing doubt,—that the experiment was shown, on the first occasion, to explain the lamination of slate, and on the second, to explain the riband structure of ice. But there are no ribands in slate, and there is no lamination in ice. There are no regulated alternations of porous with solid substance in the one; and there are no constancies of fracture by plane surfaces in the other; moreover—and this is to be chiefly noted,—slate lamination is always straight; glacier banding always bent. The structure of the pressed wax might possibly explain one or other of these phenomena; but could not possibly explain both, and does actually explain neither.

26. That the arrangement of rock substance into fissile

---

[1] ["The first step is to put oneself into contact with nature, to seek facts" (*Glaciers of the Alps*, p. 431).]

folia does indeed take place in metamorphic aluminous
masses under some manner of pressure, has, I believe, been
established by the investigations both of Mr. Sorby and
of Mr. Clifton Ward. But the reasons for continuity of
parallel cleavage through great extents of variously contorted
beds ;—for its almost uniform assumption of a high angle ;
—for its as uniform non-occurrence in horizontal laminæ
under vertical pressure, however vast ;—for its total disre-
gard of the forces causing upheaval of the beds ;—and its
mysteriously deceptive harmonies with the stratification, if
only steep enough, of neighbouring sedimentary rock,—
remain to this hour, not only unassigned, but unsought.

27. And it is difficult for me to understand either the
contentment of geologists with this state of things, or the
results on the mind of ingenuous learners, of the partial
and more or less contradictory information hitherto obtainable on the subject. The section given in the two lower
figures of Plate XVI. was drawn for me, as I have already
said, by my most affectionately and reverently remembered
friend, Professor Phillips, of Oxford. It goes through the
entire crest of the Lake district from Lancaster to Carlisle,
the first emergent rock-beds being those of mountain limestone, A to B, not steeply inclined, but lying unconform-
ably on the steeply inclined flags and grit of Furness Fells,
B to C. In the depression at C lies Coniston Lake ; then
follow the masses of Coniston Old Man and Scawfell,
C to D, sinking to the basin of Derwentwater just after
the junction, at Grange, of their volcanic ashes with the
Skiddaw slate. Skiddaw himself, and Carrock Fell, rise
between D and E ; and above E, at Caldbeck, again
the mountain limestone appears in unconformable bedding,
declining under the Trias of the plain of Carlisle, at
the northern extremity of which a few rippled lines do
service for the waves of Solway.

28. The entire ranges of the greater mountains, it will
be seen, are thus represented by Professor Phillips as consisting of more or less steeply inclined beds, parallel to

those of the Furness shales; and traversed by occasional cleavages at an opposite angle. But in the section of the Geological Survey, already referred to,[1] the beds parallel to the Furness shales reach only as far as Wetherlam, and the central mountains are represented as laid in horizontal or slightly basin-shaped swirls of ashes, traversed by ejected trap, and divided by no cleavages at all, except a few vertical ones indicative of the Tilberthwaite slate quarries.

29. I think it somewhat hard upon me, now that I am sixty years old, and short of breath in going up hills, to have to compare, verify for myself, and reconcile as I may, these entirely adverse representations of the classical mountains of England:—no less than that I am left to carry forward, in my broken leisure, the experiments on viscous motion instituted by James Forbes thirty years ago.[2] For the present, however, I choose Professor Phillips' section as far the most accurately representative of the general aspect of matters, to my present judgment; and hope, with Mr. Clifton Ward's good help, to give more detailed drawings of separate parts in the next volume of *Deucalion*.[3]

30. I am prepared also to find Professor Phillips' drawing in many respects justifiable, by my own former studies of the cleavage structure of the central Alps, which, in all the cases I have examined, I found to be a distinctly crystalline lamination, sometimes contorted according to the rock's own humour, fantastically as Damascus steel; but presently afterwards assuming inconceivable consistency with the untroubled repose of the sedimentary masses into whose company it had been thrust. The junction of the contorted gneiss through which the gorge of Trient is cleft, with the micaceous marble on which the tower of Martigny is built, is a transition of this kind within reach of the least adventurous traveller; and the junction of the gneiss of the Montanvert with the porous limestone which underlies it, is

---

[1] [See p. 257.]
[2] [See above, pp. 133, 139.]
[3] [This, however, was not done. Mr. Clifton Ward died in 1881 (see below, p. 570).]

certainly the most interesting, and the most easily explored, piece of rock-fellowship in Europe. Yet the gneissitic lamination of the Montanvert has been attributed to stratification by one group of geologists, and to cleavage by another, ever since the valley of Chamouni was first heard of: and the only accurate drawings of the beds hitherto given are those published thirty years ago in *Modern Painters.*[1] I had hoped at the same time to contribute some mite of direct evidence to their elucidation, by sinking a gallery in the soft limestone under the gneiss, supposing the upper rock hard enough to form a safe roof; but a decomposing fragment fell, and so nearly ended the troubles, with the toil, of the old miner who was driving the tunnel, that I attempted no farther inquiries in that practical manner.[2]

31. The narrow bed, curved like a sickle, and coloured vermilion, among the purple slate, in the uppermost section of Plate XVI., is intended to represent the position of the singular band of quartzite and mottled schists ("bunte Schiefer"), which, on the authority of Studer's section at page 178 of his first volume, underlies, at least for some thousands of feet, the granite of the Jungfrau; and corresponds, in its relation to the uppermost cliff of that mountain, with the subjacence of the limestone of Les Tines to the aiguilles of Chamouni. I have coloured it vermilion in order to connect it in the student's mind with the notable conglomerates of the Black Forest, through which their underlying granites pass into the Trias; but the reversed position which it here assumes, and the relative dominance of the central mass of the Bernese Alps, if given by Studer with fidelity, are certainly the first structural phenomena which the geologists of Germany should benevolently qualify themselves to explain to the summer society of Interlachen. The view of the Jungfrau from the Castle of Manfred[3] is probably the most beautiful natural vision in Europe; but, for all that modern science can hitherto tell us, the

---

[1] [See above, p. 97.]
[2] [For these excavations, see pp. xxvii., 545–547.]
[3] [For a note on this castle in the valley of Lauterbrunnen, see Vol. V. p. lxiii.]

construction of it is supernatural, and explicable only by the Witch of the Alps.

32. In the meantime I close this volume of *Deucalion* by noting firmly one or two letters of the cuneiform language in which the history of that scene has been written.

There are five conditions of rock cleavage which the student must accustom himself to recognize, and hold apart in his mind with perfect clearness, in all study of mountain form.

I. The Wave cleavage: that is to say, the condition of structure on a vast scale which has regulated the succession of summits. In almost all chains of mountains not volcanic, if seen from a rightly chosen point, some law of sequence will manifest itself in the arrangement of their eminences. On a small scale, the declining surges of pastoral mountain, from the summit of Helvellyn to the hills above Kendal, seen from any point giving a clear profile of them, on Wetherlam or the Old Man of Coniston, show a quite rhythmic, almost formal, order of ridged waves, with their steepest sides to the lowlands; for which the cause must be sought in some internal structure of the rocks, utterly untraceable in close section. On vaster scale, the succession of the aiguilles of Chamouni, and of the great central aiguilles themselves, from the dome of Mont Blanc through the Jorasses, to the low peak of the Aiguille de Trient, is again regulated by a harmonious law of alternate cleft and crest, which can be studied rightly only from the far-distant Jura.

The main directions of this vast mountain tendency might always be shown in a moderately good model of any given district, by merely colouring all slopes of ground inclined at a greater angle than thirty degrees, of some darker colour than the rest. No slope of talus can maintain itself at a higher angle than this (compare *Modern Painters*, vol. iv., p. 318[1]); and therefore, while the mathematical laws

---

[1] [Ruskin's reference is to the first edition: see now Vol. VI. p. 376.]

of curvature by aqueous denudation, which were first ascertained and systematized by Mr. Alfred Tylor,[1] will be found assuredly to regulate or modify the disposition of masses reaching no steeper angle, the cliffs and banks which exceed it, brought into one abstracted group, will always display the action of the wave cleavage on the body of the yet resisting rocks.

33. II. The Structural cleavage.

This is essentially determined by the arrangement of the plates of mica in crystalline rocks, or—where the mica is obscurely formed, or replaced by other minerals—by the sinuosities of their quartz veins. Next to the actual bedding, it is the most important element of form in minor masses of crag; but in its influence on large contours, subordinate always to the two next following orders of cleavage.

34. III. The Asphodeline cleavage;—the detachment, that is to say, of curved masses of crag more or less concentric, like the coats of an onion. It is for the most part transverse to the structural cleavage, and forms rounded domes and bending billows of smooth contour, on the flanks of the great foliated mountains, which look exactly as if they had been worn for ages under some river of colossal strength. It is far and away the most important element of mountain form in granitic and metamorphic districts.

35. IV. The Frontal cleavage. This shows itself only on the steep escarpments of sedimentary rock, when the cliff has been produced in all probability by rending elevatory force. It occurs on the faces of nearly all the great precipices in Savoy, formed of Jura limestone, and has been in many cases mistaken for real bedding. I hold it one of the most fortunate chances attending the acquisition of Brantwood, that I have within three hundred yards of me, as I

---

[1] [See his paper "On the Action and Formation of Rivers, Lakes, and Streams, with Remarks on Denudation and the Causes of the Great Changes of Climate which occurred just prior to the Historical Period," *Geological Magazine*, new series, decade ii., vol. ii. (1875), pp. 433–473 (on pp. 474–476 is a list of Tylor's scientific papers). For other references to this article, see above, p. xxv., and below, p. 368 n.]

write, jutting from beneath my garden wall, a piece of crag knit out of the Furness shales, showing frontal cleavage of the most definite kind, and enabling me to examine the conditions of it as perfectly as I could at Bonneville or Annecy.

36. V. The Atomic cleavage.

This is the mechanical fracture of the rock under the hammer, indicating the mode of coherence between its particles, irrespectively of their crystalline arrangements. The conchoidal fractures of flint and calcite, the raggedly vitreous fractures of quartz and corundum, and the earthy transverse fracture of clay slate, come under this general head. And supposing it proved that slaty lamination is indeed owing either to the lateral expansion of the mass under pressure, or to the filling of vacant pores in it by the flattening of particles, such a formation ought to be considered, logically, as the ultimate degree of fineness in the coherence of crushed substance; and not properly a "structure." I should call this, therefore, also an "atomic" cleavage.

37. The more or less rectilinear divisions, known as "joints," and apparently owing merely to the desiccation or contraction of the rock, are not included in the above list of cleavages, which is limited strictly to the characters of separation induced either by arrangements of the crystalline elements, or by violence in the methods of rock elevation or sculpture.

38. If my life is spared, and my purposes hold, the second volume of *Deucalion* will contain such an account of the hills surrounding me in this district, as shall be, so far as it is carried, trustworthy down to the minutest details in the exposition of their first elements of mountain form.[1] And I am even fond enough to hope that some of the youths of Oxford, educated in its now established schools of Natural History and Art, may so securely and consistently follow out such a piece of home study by the delineation

---

[1] [This purpose, however, was not fulfilled.]

of the greater mountains they are proud to climb, as to
redeem, at last, the ingenious nineteenth century from the
reproach of having fostered a mountaineering club, which
was content to approve itself in competitive agilities, without
knowing either how an aiguille stood, or how a glacier
flowed; and a Geological Society, which discoursed with
confidence on the catastrophes of chaos, and the processes
of creation, without being able to tell a builder how a slate
split, or a lapidary how a pebble was coloured.[1]

[1] [See the reference to this passage, below, p. 374.]

APPENDIX

When I began *Deucalion,* one of the hopes chiefly connected with it was that of giving some account of the work done by the real masters and fathers of Geology. I must not conclude this first volume without making some reference (more especially in relation to the subjects of inquiry touched upon in its last chapter) to the modest life and intelligent labour of a most true pioneer in geological science, Jonathan Otley. Mr. Clifton Ward's sketch of the good guide's life,[1] drawn up in 1877 for the Cumberland Association for the Advancement of Literature and Science, supplies me with the following particulars of it, deeply—as it seems to me—instructive and impressive.

He was born near Ambleside, at Nook House, in Loughrigg, January 19th, 1766. His father was a basket-maker; and it is especially interesting to me, in connection with the resolved retention of Latin as one of the chief elements of education in the system I am arranging for St. George's schools,[2] to find that the Westmoreland basket-maker was a good Latin scholar; and united Oxford and Cambridge discipline for his son with one nobler than either, by making him study Latin and mathematics, while, till he was twenty-five, he worked as his father's journeyman at his father's handicraft. "He also cleaned all the clocks and watches in the neighbourhood, and showed himself very skilful in engraving upon copper-plates, seals and coin." In 1791 he moved to Keswick, and there lived sixty-five years, and died, ninety years old and upwards.

I find no notice in Mr. Ward's paper of the death of the father, to whose good sense and firmness the boy owed so much. There was yet a more woful reason for his leaving his birthplace. He was in love with a young woman named Anne Youdale, and had engraved their names together on a silver coin. But the village blacksmith, Mr. Bowness, was also a suitor for the maiden's hand; and some years after, Jonathan's niece, Mrs. Wilson, asking him how it was that his name and Anne Youdale's were engraved together on the same coin, he replied, "Oh, the blacksmith beat me." * He never married, but took to mineralogy, watchmaking, and other

---

* I doubt the orthography of the fickle maid's name, but all authority of antiquaries obliges me to distinguish it from that of the valley. I do so, however, still under protest—as if I were compelled to write Lord Lonsdale, "Lownsdale," or the Marquis of Tweeddale, "Twaddle," or the victorious blacksmith, "Beauness." The latter's family still retain the forge by Elter Water—an entirely distinct branch, I am told, from our blacksmiths of the Dale: see above, pp. 251, 252.

---

[1] [See "Jonathan Otley, the Geologist and Guide," by J. Clifton Ward, at pp. 125–169 of Part II., 1876–1877, of *Transactions of the Cumberland Association for the Advancement of Literature and Science.* Otley died in 1857, aged ninety.]

[2] [See *Fors Clavigera,* Letter 8, § 10.]

consolatory pursuits, with mountain rambling—alike discursive and attentive. Let me not omit what thanks for friendly help and healthy stimulus to the earnest youth may be due to another honest Cumberland soul,—Mr. Crosthwaite.[1] Otley was standing one day (before he removed to Keswick) outside the Crosthwaite Museum,* when he was accosted by its founder, and asked if he would sell a curious stick he held in his hand. Otley asked a shilling for it, the proprietor of the Museum stipulating to show him the collection over the bargain. From this time congenial tastes drew the two together as firm and staunch friends.

He lived all his life at Keswick, in lodgings,—recognized as " Jonathan Otley's, up the steps,"—paying from five shillings a week at first, to ten, in uttermost luxury; and being able to give account of his keep to a guinea, up to October 18, 1852,—namely, board and lodging for sixty-one years and one week, £1325; rent of room extra, fifty-six years, £164, 10s. Total keep and roof overhead, for the sixty usefullest of his ninety years, £1489, 10s.

Thus housed and fed, he became the friend, and often the teacher, of the leading scientific men of his day,—Dr. Dalton the chemist, Dr. Henry the chemist, Mr. Farey the engineer, Airy the Astronomer Royal, Professor Phillips of Oxford, and Professor Sedgwick of Cambridge. He was the first accurate describer and accurate map-maker of the Lake District; the founder of the geological divisions of its rocks,—which were accepted from him by Sedgwick, and are now finally confirmed;—and the first who clearly defined the separation between bedding, cleavage, and joint in rock,—hence my enforced notice of him in this place. Mr. Ward's memoir gives examples of his correspondence with the men of science above named; both Phillips and Sedgwick referring always to him in any question touching Cumberland rocks, and becoming gradually his sincere and affectionate friends. Sedgwick sate by his death-bed.

I shall have frequent occasion to refer to his letters, and to avail myself of his work.[2] But that work was chiefly crowned in the example he left —not of what is vulgarly praised as self-*help* (for every noble spirit's watchword is " God us ayde ")—but of the rarest of moral virtues, self-*possession*. " In your patience, possess ye your souls."[3]

I should have dwelt at greater length on the worthiness both of the tenure and the treasure, but for the bitterness of my conviction that the rage of modern vanity must destroy, in our scientific schoolmen, alike the casket, and the possession.

---

* In that same museum, my first collection of minerals—fifty specimens—total price, if I remember rightly, five shillings—was bought for me, by my father, of Mr. Crosthwaite. No subsequent possession has had so much influence on my life. I studied Turner at his own gallery, and in Mr. Windus's portfolios;[4] but the little yellow bit of "copper ore from Coniston," and the "Garnets" (I never could see more than one!) from Borrowdale, were the beginning of science to me which never could have been otherwise acquired.

---

[1] [See Vol. II. p. 296 n.]
[2] [This, however, was not done.]
[3] [Luke xxi. 19.]
[4] [See Vol. III. p. 235.]

DEUCALION

VOL. II

CHAPTER I

LIVING WAVES[1]

1. THE opening of the second volume of *Deucalion* with a lecture on Serpents[2] may seem at first a curiously serpentine mode of advance towards the fulfilment of my promise that the said volume should contain an account of the hills surrounding me at Coniston (above, vol. i., p. 291, § 38).

---

[1] [To this chapter, as originally published, the following "Advice" was prefixed :—

> "Photographs from the diagrams used in illustration of this lecture are in preparation : and may be ordered of MR. WARD, 2, Church Terrace, Richmond. The two plates now given with it were engraved for illustration of beak-structure in 'Love's Meinie'; but may be of some present use here : and are better printed than lying by to rust."

Six sheets of photographs for binding up with *Deucalion* were issued. Two of these gave the snakes here shown on Plate XVIII. ; a third, those on Plate XIX. ; a fourth, the heads of duck and crocodiles on Plate XVII. ; a fifth contained what are here Figures 38 and 39; a sixth, what are here Figures 40 and 41. The "two plates" in illustration of beak-structure are now Plates VII. and VIII. in Vol. XXV.]

[2] [The lecture was first delivered on March 17, 1880, and repeated on March 23 (see above, p. 90). The reports in the *Times* and *Daily News* here collated contain the following introductory remarks :—

> "Yesterday evening the theatre of the London Institution was densely crowded with members and visitors assembled to listen to a lecture by Mr. J. Ruskin entitled 'A Caution to Snakes.' Dr. Warren De la Rue, F.R.S., President of the Institution, was in the chair. Alluding to the fact that Professor Huxley had opened the lectures for the season in that theatre with one on Snakes, Mr. Ruskin said he trusted the seeming antagonism between himself and that distinguished man, to whose genius as well as to that of Mr. Darwin he paid a high tribute of respect, would not be misunderstood. The subject was appropriate to St. Patrick's Day. He had intended to invoke the grace and protection of the Saint, but would not do so for want of time. All he would say in this conjuncture of politics was that whatever was good for England was good for Ireland, and whatever was bad for Ireland was bad for England; and nothing would go thoroughly well in either country until everybody who had got a house and land lived in the one and stayed on the other."]

But I am obliged now in all things to follow in great part the leadings of circumstance: and although it was only the fortuitous hearing of a lecture by Professor Huxley[1] which induced me to take up at present the materials I had by me respecting snake motion, I believe my readers will find their study of undulatory forces dealt through the shattered vertebræ of rocks, very materially enlivened, if not aided, by first observing the transitions of it through the adjusted vertebræ of the serpent. I would rather indeed have made this the matter of a detached essay, but my distinct books are far too numerous already; and, if I could only complete them to my mind, would in the end rather see all of them fitted into one colubrine chain of consistent strength, than allowed to stand in any broken or diverse relations.

There are, however, no indications in the text of the lecture itself of its possible use in my geological work. It was written as briefly and clearly as I could, for its own immediate purpose: and is given here, as it was delivered, with only the insertion of the passages I was forced to omit for want of time.

2. The lecture, as it stands, was, as I have just said, thrown together out of the materials I had by me; most of them for a considerable time; and with the help of such books as I chanced to possess,—chiefly, the last French edition of Cuvier,— Dr. Russell's *Indian Serpents*, — and Bell's *British Reptiles*.[2] Not until after the delivery of the lecture for the second time, was I aware of the splendid work done recently by Dr. Günther, nor had I ever seen drawings of serpents for a moment comparable, both in action and in detail of scale, to those by Mr. Ford which

---

[1] [Huxley's lecture on Snakes was delivered at the London Institution on December 1, 1879. It was not published ; a short notice of it may be seen in the *Times*, December 2, 1879. Ruskin indicates the general thesis of the lecture, below, p. 343.]

[2] [The volume "Les Reptiles" in Cuvier's *Le Règne Animal* (for the English translation, see p. 317). *History of British Reptiles*, by Thomas Bell, Professor of Zoology in King's College, London, 1839 (2nd edition 1849). For Russell's *Indian Serpents*, see p. 321 *n*.]

illustrate Dr. Günther's descriptions; or, in colour, and refinement of occasional action, to those given in Dr. Fayrer's *Thanatophidia of India*.[1] The reader must therefore understand that anything generally said, in the following lecture, of modern scientific shortcoming, or error, is not to be understood as applying to any publication by either of these two authors, who have, I believe, been the first naturalists to adopt the artistically and mathematically sound method of delineation by plan and profile; and the first to represent serpent action in true lines, whether of actual curve, or induced perspective.

What follows, then, is the text of what I read, or, to the best of my memory, spoke, at the London Institution.

3. In all my lectures on Natural History at Oxford[2] I virtually divided my subject always into three parts, and asked my pupils, first, to consider what had been beautifully thought about the creature; secondly, what was accurately known of it; thirdly, what was to be wisely asked about it.

First, you observe, what was, or had been, beautifully thought about it; the effect of the creature, that is to say, during past ages, on the greatest human minds. *This*, it is especially the business of a gentleman and a scholar to know. It is a king's business, for instance, to know the meaning of the legend of the basilisk, the King of Serpents, who killed with a look, in order that he may not himself become like a basilisk. But that kind of knowledge would be of small use to a viper-catcher.

Then the second part of the animal's history is—what is truly known of it, which one usually finds to be extremely little.

And the third part of its history will be—what remains

---

[1] [*The Thanatophidia of India, being a Description of the Venomous Snakes of the Indian Peninsula, with an Account of the Influence of their Poison upon Life*, by J. Fayrer, C.S.I., M.D. (2nd edition 1874). The illustrations were executed by students at the School of Art in Calcutta. *The Reptiles of British India*, by Albert C. L. G. Günther, published for the Ray Society, 1864 (Plates by G. H. Ford).]

[2] [See *Eagle's Nest*, § 180 (Vol. XXII. p. 245); and *Love's Meinie*, § 35 (Vol. XXV. p. 40).]

to be asked about it—what it now behoves us, or will be profitable to us, to discover.

4. It will perhaps be a weight off your minds to be assured that I shall waive to-night the first part of the subject altogether;—except so far as thoughts of it may be suggested to you by Mr. Severn's beautiful introductory diagram,* and by the references I have to make to it,[1] though shown for the sake of the ivy, not the Eve,—its subject being already explained in my Florentine Guide to the Shepherd's Tower. But I will venture to detain you a few moments while I point out how, in one great department of modern science, past traditions may be used to facilitate, where at present they do but encumber, even the materialistic teaching of our own day.

5. When I was furnishing Brantwood, a few years ago, I indulged myself with two bran-new globes, brought up to all the modern fine discoveries. I find, however, that there's so much in them that I can see nothing. The names are too many on the earth, and the stars too crowded in the heaven. And I am going to have made for my Coniston parish school a series of drawings in dark blue, with golden stars, of one constellation at a time, such as my diagram No. 2[2] with no names written to the stars at all. For if the children don't know their names without print on their diagram, they won't know them without print on the sky. Then there must be a school-manual of the constellations, which will have the legend of each told as simply as a fairy tale; and the names of the chief stars given on a map of them, corresponding to the blue diagram, —both of course drawn as the stars are placed in the sky;

---

* The Creation of Eve, bas-relief from the tower of Giotto. The photograph may be obtained from Mr. Ward.

---

[1] [No. 1 in the printed list which was distributed at the lecture: see below, p. 330. The bas-relief is shown on Plate 43 of Vol. XXIV. For the explanation, see *Mornings in Florence*, § 130 (*ibid.*, pp. 421–422).]

[2] [See No. 2 in the printed list, below, p. 330.]

or as they would be seen on a concave celestial globe, from the centre of it. The having to look down on the stars from outside of them is a difficult position for children to comprehend, and not a very scientific one, even when comprehended.

6. But to do all this rightly, I must have better outlines than those at present extant. The red diagram, No. 3,[1] which has I hope a little amused you, more than frightened, is an enlargement of the outline given on my new celestial globe, to the head of the constellation Draco. I need not tell you that it is as false to nature as it is foolish in art; and I want you to compare it with the uppermost snake head in No. 4,[2] because the two together will show you in a moment what long chapters of *Modern Painters* were written to explain,—how the real faculty of imagination is always true, and goes straight to its mark:[3] but people with no imagination are always false, and blunder or drivel about their mark. That red head was drawn by a man who didn't know a snake from a sausage, and had no more imagination in him than the chopped pork of which it is made. Of course he didn't know that, and with a scrabble of lines this way and the other, gets together what he thinks an invention—a knot of gratuitous lies, which you contentedly see portrayed as an instrument of your children's daily education. While—two thousand and more years ago—the people who had imagination

Fig. 36

---

[1] [See, again, the printed list, below, p. 330.]

[2] [The Hydra of Lerna: see list, below, p. 330. A similar design on a coin of Phæstus may be seen in the exhibition of electrotypes at the British Museum, II. B. 38.]

[3] [See Vol. IV. p. 250.]

enough to believe in Gods, saw also faithfully what was to be seen in snakes; and the Greek workman gives, as you see in this enlargement of the silver drachma of Phæstus, with a group of some six or seven sharp incisions, the half-dead and yet dreadful eye, the flat brow, the yawning jaw, and the forked tongue, which are an abstract of the serpent tribe for ever and ever.

Fig. 29

And I certify you that all the exhibitions they could see in all London would not teach your children so much of art as a celestial globe in the nursery, designed with the force and the simplicity of a Greek vase.

7. Now, I have done alike with myths and traditions;[1] and perhaps I had better forewarn you, in order, what I am next coming to. For, after my first delivery of this lecture, one of my most attentive hearers, and best accustomed pupils, told me that he had felt it to be painfully unconnected,—with much resultant difficulty to the hearer in following its intention. This is partly inevitable when one endeavours to get over a great deal of ground in an hour; and indeed I have been obliged, as I fastened the leaves together, to cut out sundry sentences of adaptation or transition—and run my bits of train all into one, without buffers. But the actual divisions of what I have to say are clearly jointed for all that; and if you like to jot them down from the leaf I

[1] [Other references to the snake in art and mythology will be found in Vol. VII. p. 184; Vol. XI. p. 166; Vol. XIX. pp. 363–367.]

have put here at my side for my own guidance, these are
the heads of them:[1]—

I. Introduction—Imaginary Serpents, §§ 3–6.
II. The Names of Serpents, §§ 8–15.
III. The Classification of Serpents, §§ 16–32.
IV. The Patterns of Serpents, §§ 33–34.
V. The Motion of Serpents, §§ 35–39.
VI. The Poison of Serpents, §§ 40–47.
VII. Caution, concerning their Poison, §§ 48–51.
VIII. The Wisdom of Serpents, §§ 52–53.
IX. Caution, concerning their Wisdom, §§ 54–56.

It is not quite so bad as the sixteenthly, seventeenthly,
and to conclude, of the Duke's chaplain, to Major Dalgetty;[2]
but you see we have no time to round the corners, and
must get through our work as straightly as we may.

We have got done already with our first article, and
begin now with the names of serpents; of which those
used in the great languages, ancient and modern, are all
significant, and therefore instructive, in the highest degree.[3]

8. (i.) The first and most important is the Greek "ophis,"
from which you know the whole race are called, by scientific
people, ophidia. It means the thing that sees all round;
and Milton is thinking of it when he makes the serpent,
looking to see if Eve be assailable, say of himself, " Her
husband, *for I view far round*, not near."[4] Satan says that,
mind you, in the person of the Serpent, to whose faculties,
in its form, he has reduced himself. As an angel, he would
have *known* whether Adam was near or not: in the serpent,
he has to look and see. This, mind you further, however,
is Miltonic fancy, not Mosaic theology;—it is a poet and
a scholar who speaks here,—by no means a prophet.

---

[1] [The numbered sections, dealing with the several subjects, are here added.]
[2] [See Scott's *Legend of Montrose*, ch. vi.]
[3] [See, again, the printed list, which was distributed at the lecture, below, p. 332.]
[4] [*Paradise Lost*, ix. 482.]

9. Practically, it has never seemed to me that a snake *could* see far round, out of the slit in his eye, which is drawn large for you in my diagram[1] of the rattlesnake;* but either he or the puff-adder, I have observed, seem to see with the backs of their heads as well as the fronts, whenever I am drawing them. You will find the question entered into at some length in my sixth lecture in the *Eagle's Nest;*[2] and I endeavoured to find out some particulars of which I might have given you assurance to-night, in my scientific books; but though I found pages upon pages of description of the scales and wrinkles about snakes' eyes, I could come at no account whatever of the probable range or distinctness in the sight of them; and though extreme pains had been taken to exhibit, in sundry delicate engravings, their lachrymatory glands and ducts, I could neither discover the occasions on which rattlesnakes wept, nor under what consolations they dried their eyes.

10. Next (ii.) for the word dracon, or dragon. We are accustomed to think of a dragon as a winged and clawed creature; but the real Greek dragon, Cadmus's or Jason's, was simply a serpent, only a serpent of more determined vigilance than the ophis, and guardian therefore of fruit, fountain, or fleece. In that sense of guardianship, not as a protector, but as a sentinel, the name is to be remembered as well fitted for the great Greek lawgiver.

The dragon of Christian legend is more definitely malignant, and no less vigilant. You will find in Mr. Anderson's supplement to my *St. Mark's Rest,* "The Place of Dragons," a perfect analysis of the translation of classic into Christian tradition in this respect.[3]

* See the careful drawing of the eye of Daboia Russellii, *Thanatophidia*, p. 14 *n.*

---

[1] [No. 7 in the list of diagrams: see below, p. 331.]
[2] [See *Eagle's Nest,* §§ 101 *seq.*, and especially § 109 (Vol. XXII. p. 200).]
[3] [See Vol. XXIV. pp. 370 *seq.*]

11. (iii.) Anguis. The strangling thing, passing into the French "angoisse" and English "anguish"; but we have never taken this Latin word for our serpents, because we have none of the strangling or constrictor kind in Europe. It is always used in Latin for the most terrible forms of snake, and has been, with peculiar infelicity, given by scientific people to the most innocent, and especially to those which can't strangle anything. The "Anguis fragilis" breaks like a tobacco-pipe; but imagine how disconcerting such an accident would be to a constrictor!

12. (iv.) Coluber, passing into the French "couleuvre," a grandly expressive word. The derivation of the Latin one is uncertain,[1] but it will be wise and convenient to reserve it for the expression of coiling. Our word "coil," as the French "cueillir," is from the Latin "colligere," to collect; and we shall presently see[2] that the way in which a snake "collects" itself is no less characteristic than the way in which it diffuses itself.

13. (v.) Serpens. The winding thing. This is the great word which expresses the progressive action of a snake, distinguishing it from all other animals; or, so far as modifying the motion of others, making them in that degree serpents also, as the elongated species of fish and lizard. It is the principal object of my lecture this evening to lay before you the law of this action, although the interest attaching to other parts of my subject has tempted me to enlarge on them so as to give them undue prominence.

14. (vi.) Adder. This Saxon word, the same as nieder or nether, "the grovelling thing," was at first general for all serpents, as an epithet of degradation, "the deaf adder that stoppeth her ears."[3] Afterwards it became provincial, and has never been accepted as a term of science. In the most scholarly late English it is nearly a synonym with "viper," but that word, said to be a contraction for vivipara,

---

[1] [According to some etymologists, the word is akin to "celer."]
[2] [See below, § 37, p. 318.]
[3] [Psalms lviii. 4.]

bringing forth the young alive, is especially used in the New Testament of the Pharisees, who compass heaven and earth to make one proselyte.[1] The Greek word used in the same place, echidna, is of doubtful origin, but always expresses treachery joined with malice.

15. (vii.) Snake. German, "schlange," the crawling thing; and with some involved idea of sliminess, as in a snail. Of late it has become partly habitual, in ordinary English, to use it for innocent species of serpents, as opposed to venomous; but it is the strongest and best general term for the entire race; which race, in order to define clearly, I must now enter into some particulars respecting classification, which I find little announced in scientific books.

16. And here I enter on the third division of my lecture, which must be a disproportionately long one, because it involves the statement of matters important in a far wider scope than any others I have to dwell on this evening. For although it is not necessary for any young persons, nor for many old ones, to know, even if they *can* know, anything about the origin or development of species, it is vitally necessary that they should know what a species *is*, and much more what a genus or (a better word) *gens*, a race, of animals is.[2]

17. A gens, race, or kinship, of animals, means, in the truth of it, a group which can do some special thing nobly and well. And there are always varieties of the race which do it in different styles,—an eagle flies in one style, a windhover in another, but both gloriously,—they are "Gentiles" —gentlemen creatures, well born and bred. So a trout belongs to the true race, or gens, of fish: he can swim perfectly; so can a dolphin, so can a mackerel: they swim in different styles indeed, but they belong to the true kinship of swimming creatures.

18. Now between the gentes, or races, and between the

---

[1] [Matthew xxiii. 15, 33.]
[2] [See Vol. IV. p. 164 (note of 1883), where Ruskin refers to the definitions here given.]

species, or families, there are invariably links—mongrel creatures, neither one thing nor another,—but clumsy, blundering, hobbling, misshapen things. You are always thankful
when you see one that you are not *it*. They are, according
to old philosophy, in no process of development up or down,
but are necessary, though much pitiable, where they are.
Thus between the eagle and the trout, the mongrel or
needful link is the penguin. Well, if ever you saw an eagle
or a windhover flying,[1] I am sure you must have sometimes
wished to be a windhover; and if ever you saw a trout or
a dolphin swimming, I am sure, if it was a hot day, you
wished you could be a trout. But did ever anybody wish
to be a penguin?

So, again, a swallow is a perfect creature of a true gens;
and a field-mouse is a perfect creature of a true gens; and
between the two you have an accurate mongrel—the bat.
Well, surely some of you have wished, as you saw them
glancing and dipping over lake or stream, that you could
for half-an-hour be a swallow: there have been humble
times with myself when I could have envied a field-mouse.
But did ever anybody wish to be a bat?

19. And don't suppose that you can invert the places of
the creatures, and make the gentleman of the penguin, and
the mongrel of the windhover,—the gentleman of the bat,
and mongrel of the swallow. All these living forms, and
the laws that rule them, are parables, when once you can
read; but you can only read them through love, and the
sense of beauty; and some day I hope to plead with you
a little, of the value of that sense, and the way you have
been lately losing it.[2] But as things are, often the best way
of explaining the nature of any one creature is to point out
the other creatures with whom it is connected, through
some intermediate form of degradation. There are almost
always two or three, or more, connected gentes, and between

---

[1] [For a description of a windhover flying, see a passage from Ruskin's diary
given in Vol. XXIV. p. xxix.]

[2] [Ruskin touched partly on this subject in his lectures at the London Institution in 1884: see *The Storm-Cloud of the Nineteenth Century*, § 24.]

each, some peculiar manner of decline and of reascent. Thus, you heard Professor Huxley explain to you that the true snakes were connected with the lizards through helpless snakes, that break like withered branches; and sightless lizards, that have no need for eyes or legs. But there are three other great races of life, with which snakes are connected in other and in yet more marvellous ways. And I do not doubt being able to show you, this afternoon, the four quarters, or, as astrologers would say, the four houses, of the horizon of serpent development, in the modern view, or serpent relation, in the ancient one. In the first quarter, or house, of his nativity, a serpent is, as Professor Huxley showed you, a lizard that has dropped his legs off. But in the second quarter, or house, of his nativity, I shall show you that he is also a duck that has dropped her wings off. In the third quarter, I shall show you that he is a fish that has dropped his fins off. And in the fourth quarter of ascent, or descent, whichever you esteem it, that a serpent is a honeysuckle, with a head put on.

20. The lacertine relations having been explained to you in the preceding lecture by Professor Huxley, I begin this evening with the Duck. I might more easily, and yet more surprisingly, begin with the Dove; but for time-saving must leave your own imaginations to trace the transition, easy as you may think it, from the coo to the quack, and from the walk to the waddle. Yet that is very nearly one-half the journey. The bird is essentially a singing creature, as a serpent is a mute one; the bird is essentially a creature singing for love, as a puff-adder is one puffing for anger; and in the descent from the sound which fills that verse of Solomon's Song, "The time of the singing of birds is come, and the voice of the turtle is heard in our land,"[1] to the recollection of the last flock of ducks which you saw disturbed in a ditch, expressing their dissatisfaction in that peculiar monosyllable which from its senselessness

---

[1] [Song of Solomon ii. 2.]

has become the English expression for foolish talk,* you have actually got down half-way; and in the next flock of geese whom you discompose, might imagine at first you had got the whole way, from the lark's song to the serpent's hiss.

21. But observe, there is a variety of instrumentation in hisses. Most people fancy the goose, the snake, and we ourselves, are alike in the manner of that peculiar expression of opinion. But not at all. Our own hiss, whether the useful and practical ostler's in rubbing down his horse, or that omnipotent one which, please, do not try on me just now!—are produced by the pressure of our soft round tongues against our teeth. But neither the goose nor snake can hiss that way, for a goose has got no teeth, to speak of, and a serpent no tongue, to speak of. The sound which imitates so closely our lingual hiss is with them only a vicious and vindictive sigh,—the general disgust which the creature feels at the sight of us expressed in a gasp. Why do you suppose the puff-adder is called puffy?† Simply because he swells himself up to hiss, just as Sir Gorgius Midas might do to scold his footmen, and then actually and literally "expires" with rage, sending all the air in his body out at you in a hiss. In a quieter way, the drake and gander do the same thing; and we ourselves do the same thing under nobler conditions, of which presently.

22. But now, here's the first thing, it seems to me, we've got to ask of the scientific people, what use a serpent has for his tongue, since it neither wants it to talk with, to taste with, to hiss with, nor, so far as I know, to lick with,‡ and least of all to sting with,—and yet, for people who do not know the creature, the little vibrating forked

* The substantive "quack" in its origin means a person who quacks,— i.e., talks senselessly; see Johnson.

† In more graceful Indian metaphor, the "Father of Tumefaction."— (Note from a friend.)

‡ I will not take on me to contradict, but I don't in the least believe, any of the statements about serpents licking their prey before they swallow it.

thread, flashed out of its mouth, and back again, as quick as lightning, is the most threatening part of the beast; but what is the use of it? Nearly every other creature but a snake can do all sorts of mischief with its tongue. A woman worries with it, a chameleon catches flies with it, a snail files away fruit with it, a humming-bird steals honey with it, a cat steals milk with it, a pholas digs holes in rocks with it, and a gnat digs holes in *us* with it; but the poor snake cannot do any manner of harm with it whatsoever; and what is *his* tongue forked for?

23. I must leave you to find out that at your leisure; and to enter at your pleasure into the relative anatomical questions respecting forms of palate, larynx, and lung, in the dove, the swan, the goose, and the adder,—not unaccompanied by serpentive extension and action in the necks of the hissing birds, which show you what, so to speak, Nature is thinking of. These mechanical questions are all —leather and prunella,[1] or leather and catgut;—the *moral* descent of the temper and meaning in the sound, from a murmur of affection to a gasp of fury, is the real transition of the creature's being. You will find in Kinglake's account of the charge of the Greys in the battle of Balaclava, accurate record of the human murmur of long-restrained rage, at last let loose;[2] and may reflect, also at your leisure, on the modes of political development which change a kindly Scot into a fiery dragon.

24. So far of the fall of the bird-angels from song to hiss: next consider for a minute or two the second phase of catastrophe—from walk to waddle. Walk,—or, in prettier creatures still, the run. Think what a descent it is, from the pace of the lapwing, like a pretty lady's,—"Look, where Beatrice, like a lapwing, runs;"[3] or of the cream-coloured

---

[1] [Pope: *Essay on Man*, iv. 204.]

[2] ["The Scots Greys gave no utterance except to a low, eager, fierce moan of rapture—the moan of outbursting desire" (A. W. Kinglake's *Invasion of the Crimea*, vol. v. p. 118, 6th edition).]

[3] [*Much Ado about Nothing*, Act iv. sc. 1.]

courser * of the African desert, whom you might yourselves
see run, on your own downs, like a little racehorse, if you
didn't shoot it the moment it alighted there,—to the re-
spectable, but, to say the least, unimpressive, gait from
which we have coined the useful word to "waddle." Can
you remember exactly how a duck does walk? You can
best fancy it by conceiving the body of a large barrel carried
forward on two short legs, and rolling alternately to each
side at every step. Once watch this method of motion
attentively, and you will soon feel how near you are to
dispensing with legs altogether, and getting the barrel to
roll along by itself in a succession of zigzags.

25. Now, put the duck well under water, and he *does*
dispense with his legs altogether.

There is a bird who—my good friend, and boat-builder,
Mr. Bell,[1] tells me—once lived on Coniston Water, and
sometimes visits it yet, called the saw-bill duck, who is
the link, on the ducky side, between the ducks and divers:
his shape on the whole is a duck's, but his habits are a
diver's,—that is to say, he lives on fish, and he catches
them deep under water—swimming, under the surface, a
hundred yards at a time.

26. We do not at all enough dwell upon this faculty in
aquatic birds. Their feet are only for rowing—not for
diving. Those little membranous paddles are no use whatever, once under water. The bird's full strength must be
used in diving: he dives with his wings—literally flies under

---

* Cursorius isabellinus (Meyer), Gallicus (Gould).[2]

[1] ["Mr. William Bell was one of the celebrities of this date. In his youth he
had been a sort of right-hand mate of John Beever of the Thwaite, brother to
the ladies of *Hortus Inclusus*, and author of *Practical Fly-Fishing*. On the death of
his father, William Bell became the leading carpenter of the place, and the leading Liberal. . . . Ruskin sent word that he would like to come and have a talk
about politics. . . . The son, Mr. John Bell, waited about hoping it would be all
right. At last his father's well-known voice came through: 'Ye're wrong to rags,
Mr. Ruskin!' Then he knew it *was* all right, and went about his work. And
after that Ruskin and 'ald Will Bell' were firm friends" (W. G. Collingwood,
Ruskin Relics, pp. 23–24.]

[2] [See No. 44 in vol. iv. of Gould's *Birds of Great Britain*, where accounts of
the shooting of this rare visitor are given. "Meyer" refers to the *Taschenbuch
der Deutschen Vögelkunde*, by B. Meyer and J. Wolf, 1809, vol. ii. p. 328.]

water with his wings;—the great northern diver, at a pace which a well-manned boat can't keep up with. The stroke for progress, observe, is the same as in the air; only, in flying under water, the bird has to keep himself down, instead of keeping himself up, and strikes up with the wing instead of down. Well, the great divers hawk at fish this way, and become themselves fish, or saurians, the wings acting for the time as true fins, or paddles. And at the same time, observe, the head takes the shape, and receives the weapons, of the fish-eating lizard.

Magnified in the diagram to the same scale, this head of the saw-bill duck (No. 5[1]) is no less terrible than that of the gavial, or fish-eating crocodile of the Ganges. The gavial passes, by the mere widening of the bones of his beak, into the true crocodile,—the crocodile into the serpentine lizard. I drop my duck's wings off through the penguin; and its beak being now a saurian's, I have only to ask Professor Huxley to get rid of its feet for me, and my line of descent is unbroken, from the dove to the cobra, except at the one point of the gift of poison.

27. An important point, you say? Yes; but one which the anatomists take small note of. Legs, or no legs, are by no means the chief criterion of lizard from snake. Poison, or no poison, is a far more serious one. Why should the mere fact of being quadruped, make the creature chemically innocent? Yet no lizard has ever been recognized as venomous.

28. A less trenchant, yet equally singular, law of distinction is found in the next line of relationship we have to learn, that of serpents with fish.

The first quite sweeping division of the whole serpent race is into water serpents and land serpents.* A large

---

* Dr. Günther's division of serpents (*Reptiles of British India*, p. 166), the most rational I ever saw in a scientific book, is into five main kinds:

---

[1] [No. 5 in the list of diagrams below, p. 331. See now the upper figure on Plate XVII. The head of the gavial (central on the plate) is from Fig. 2 on Plate 9 in the volume of "Reptiles" in Cuvier's *Le Règne Animal*; the head of the common crocodile is from Fig. 1 on Plate 10 in the same.]

XVII

HEADS OF THE SAW-BILL DUCK
AND TWO CROCODILES

number, indeed, like damp places; and I suppose all serpents who ever saw water can swim; but still fix in your minds the intense and broad distinction between the sand asp, which is so fond of heat that if you light a real fire near him he will instantly wriggle up to it and burn himself to death in the ashes, and the water hydra, who lives in the open, often in the deep sea, and though just as venomous as the little fiery wretch, has the body flattened vertically at the tail so as to swim exactly as eels do.

29. Not that I am quite sure that even those who go oftenest to Eel Pie Island[1] quite know how eels *do* swim, and still less how they walk; nor, though I have myself seen them doing it, can I tell you how they manage it. Nothing in animal instinct or movement is more curious than the way young eels get up beside the waterfalls of the Highland streams.[2] They get first into the jets of foam at the edge, to be thrown ashore by them, and then wriggle up the smooth rocks—heaven knows how. If you like, any of you, to put on greased sacks, with your arms tied down inside, and your feet tied together, and then try to wriggle up after them on rocks as smooth as glass, I think even the skilfullest members of the Alpine Club will agree with me as to the difficulty of the feat; and though I have watched them at it for hours, I do not know how much of serpent, and how much of fish, is mingled in the

burrowing snakes, ground snakes, and tree snakes, on the land; and freshwater snakes and sea snakes, in the water.

All the water snakes are viviparous; and I believe all the salt-water ones venomous. Of the fresh-water snakes, Dr. Günther strongly says, "none are venomous," to my much surprise; for I have an ugly recollection of the black river viper in the Zoological Gardens, and am nearly certain that Humboldt speaks of some of the water serpents of Brazil as dangerous.[3]

---

[1] [The eyot on the Thames at Twickenham, once much frequented by boating parties.]

[2] [For a note on the climbing power of young eels, see Vol. XXV. p. 179.]

[3] ["Enormous water-snakes, in shape resembling the boa, are unfortunately very common, and are dangerous to Indians who bathe" (Humboldt's *Personal Narrative of Travels to the Equinoctial Regions of America*, English translation in Bohn's Series, vol. ii. p. 342).]

motion. But observe, at all events, there is no walking here on the plates of the belly; whatever motion is got at all, is by undulation of body and lash of tail: so far as by undulation of body, serpentine; so far as by lash of tail, fishy.

30. But the serpent is in a more intimate sense still, a fish that has dropped its fins off. All fish poison is in the fins or tail, not in the mouth. There are no venomous sharks, no fanged pikes; but one of the loveliest fishes of the south coast, and daintiest too when boiled, is so venomous in the fin, that when I was going eagerly to take the first up that came on the fishing-boat's deck with the mackerel line, in my first day of mackerel fishing, the French pilot who was with me caught hold of my arm as eagerly as if I had been going to lay hold of a viper.[1]

Of the common medusa, and of the sting ray, you know probably more than I do: but have any of us enough considered this curious fact (have any of you seen it stated clearly in any book of natural history?), that throughout the whole fish race,—which, broadly speaking, pass the whole of their existence in one continual gobble,—you never find any poison put into the teeth; and throughout the whole serpent race, never any poison put into the horns, tail, scales, or skin?

31. Besides this, I believe the aquatic poisons are for the most part black; serpent poison invariably white; and, finally, that fish poison is only like that of bees or nettles, numbing and irritating, but not deadly; but that the moment the fish passes into the hydra, and the poison comes through the teeth, the bite is mortal. In these senses, and in many others (which I could only trace by showing you the undulatory motion of fins in the seahorse, and of body in the sole), the serpent is a fish without fins.

32. Now, thirdly, I said[2] that a serpent was a honeysuckle

---

[1] [For Ruskin's experiences of mackerel fishing during his stay at Boulogne in 1861, see Vol. XVII. p. xxxviii.]

[2] [See above, p. 306.]

with a head *put on.* You perhaps thought I was jesting;
but nothing is more mysterious in the compass of creation
than the relation of flowers to the serpent tribe,—not only
in those to which, in *Proserpina,* I have given the name
Draconidæ,[1] and in which there is recognized resemblance
in their popular name, Snapdragon (as also in the speckling
of the Snake's-head Fritillary), but much more in those
carnivorous, insect-eating, and monstrous, insect-begotten,
structures,[2] to which your attention may perhaps have been
recently directed by the clever caricature of the possible
effects of electric light, which appeared lately in the *Daily
Telegraph.*[3] But, seven hundred years ago, to the Florentine, and three thousand years ago, to the Egyptian and
the Greek, the mystery of that bond was told in the
dedication of the ivy to Dionysus, and of the dragon
to Triptolemus.[4] Giotto, in the lovely design[5] which is
to-night the only relief to your eyes, thought the story of
temptation enough symbolized by the spray of ivy round
the hazel trunk ; and I have substituted, in my definition,
the honeysuckle for the ivy, because, in the most accurate
sense, the honeysuckle is an " anguis "—a strangling thing.
The ivy stem increases with age, without compressing the
tree trunk, any more than the rock, that it adorns ; but
the woodbine retains, to a degree not yet measured, but
almost, I believe, after a certain time, unchanged, the first
scope of its narrow contortion ; and the growing wood of
the stem it has seized is contorted with it, and at last
paralyzed and killed.

That there is any essential difference in the spirit of life
which gives power to the tormenting tendrils, from that
which animates the strangling coils, your recent philosophy
denies, and I do not take upon me to assert. The serpent

[1] [See Vol. XXV. p. 358.]
[2] [Compare *Proserpina,* Vol. XXV. p. 219.]
[3] [A leading article on March 15, 1880, discussing possible effects of the experiments made by Dr. Siemens, F.R.S., showing that electric light produces
" chlorophyl."]
[4] [Compare Vol. XXI. p. 115, and Vol. XX. p. 243 (with Plate V. there).]
[5] [See, again, Plate 43 in Vol. XXIII.]

is a honeysuckle* with a head put on; and perhaps some day, in the zenith of development, you may see a honeysuckle getting so much done for it.

33. It is, however, more than time for me now to approach the main parts of our subject, the characteristics of perfect serpent nature in pattern, motion, and poison. First, the pattern—*i.e.*, of their colours, and the arranged masses of them. That, the scientific people always seem to think a matter of no consequence; but to practical persons like me,[1] it is often of very primal consequence to know a viper when they see it, which they can't conveniently, except by the pattern. The scientific people count the number of scales between its eyes and its nose, and inform you duly of the amount; but then a real viper won't stand still for you to count the scales between his eyes and his nose; whereas you can see at a glance, what to us Londoners, at least, should surely be an interesting fact—that it has a pretty letter H on the top of its head (Diag. No. 10[2]). I am a true Cockney myself,—born within ring of Bow; and it is impressive to me thus to see such a development of our dropped Hs. Then, the wavy zigzag down the back, with the lateral spots—one to each bend, are again unmistakable; and a pretty general type of the kind of pattern which makes the poets and the story-tellers, when they need one epithet only, speak always of the "spotted snake."[3] Not but that a thrush or a woodpecker are much more spotty than any snakes, only they're a great deal more than that, while the snake can often only be known from the gravel he lies on by the comparative symmetry of his spots.

---

* Farther note was here taken of the action of the blossoms of the cranberry, myrtilla regina, etc., for more detailed account of which (useless in this place without the diagram) the reader is referred to the sixth number of *Proserpina.*[4]

---

[1] [Compare *Proserpina*, Vol. XXV. p. 407.]
[2] [See below, p. 331; the diagram is here reproduced (Fig. 40).]
[3] [See, for instance, the song in *A Midsummer Night's Dream*, Act ii. sc. 2 ("you spotted snakes with double tongue").]
[4] [See now Vol. XXV. p. 363, and Plate XXIV. there ("*Myrtilla Regina*").]

34. But, whether spotted, zigzagged, or blotched with reticulated stains, this, please observe, is constant in their colours: they are always, in the deadly serpents, lurid, or dull.

The fatal serpents are all of the French school of art,—French grey; the throat of the asp, French blue, the brightest thing I know in the deadly snakes. The rest are all gravel colour, mud colour, blue-pill colour, or in general, as I say, French high-art colour.[1] You will find this pointed out long ago in one of the most important chapters of *Modern Painters*,[2] and I need not dwell upon it now, except just to ask you to observe, not only that puff-adders and rattlesnakes have no resemblance to tulips and roses, but that they never have even the variegated greens and blues of mackerel, or the pinks and crimsons of the char or trout. Fancy

Fig. 40

the difference it would make in our general conception of creation, if peacocks had grey tails, and serpents golden and blue ones; or if cocks had only black spectacles on their shoulders, and cobras red combs on their heads,— if humming-birds flew in suits of black, and water-vipers swam in amethyst!*

* Had I possessed the beautiful volume of the *Thanatophidia*, above referred to [p. 297], before giving my lecture, I should have quoted from it the instance of one water-viper, Hydrophis nigrocincta[3] (*q.* purpurcocincta?),

---

[1] [See Vol. XXII. p. 202.]
[2] [In vol. iv. ch. iii. ("Of Turnerian Light"), § 23 (Vol. VI. p. 68).]
[3] [Plate xxv.; described at p. 27. The drawing of Hypnale Nepa is on Plate xvii.]

35. I come now to the fifth, midmost, and chiefly important section of my subject, namely, the manner of motion in serpents. They are distinguished from all other creatures by that motion, which I tried to describe the terror of, in *The Queen of the Air*—calling the Serpent "a wave without wind,—a current,—but with no fall."[1] A snail and a worm go on their bellies as much as a serpent, but the essential motion of a serpent is undulation,—not up and down, but from side to side; and the first thing you have got to ask about it, is, *why* it goes from side to side. Those who attended carefully to Professor Huxley's lecture, do not need to be again told that the bones of its spine *allow it* to do so; but you were not then told, nor does any scientific book that I know, tell you, why it *needs* to do so. Why should not it go straight the shortest way? Why, even when most frightened and most in a hurry, does it wriggle across the road, or through the grass, with that special action from which you have named your twisting lake in Hyde Park, and all other serpentine things? That is the first thing you have to ask about it, and it never has been asked yet, distinctly.

36. Supposing that the ordinary impression were true, that it thrusts itself forward by the alternate advance and thrust-backward of the plates of its belly, there is no reason why it should not go straight as a centipede does, or the

who *does* swim in amethyst, if the colouring of the plate may be trusted, rather than the epithet of its name. I should also have recommended to especial admiration the finishing of the angular spots in Dr. Shortt's exquisite drawing of Hypnale Nepa.

Mr. Alfred Tylor, on the evening when I last lectured, himself laid before the Zoological Society, for the first time, the theory of relation between the vertebræ and the succession of dorsal bars or spots, which I shall be rejoiced if he is able to establish; but I am quite ready to accept it on his authority, without going myself into any work on the bones.[2]

---

[1] [See Vol. XIX. § 63.]

[2] [The paper referred to does not appear in the Zoological Society's *Proceedings*, but see chapter xiv. of his posthumous work (edited by Sydney B. J. Skertchly), *Colouration in Animals and Plants*, 1886.]

more terrific scarlet centipede or millepede,—a regiment of soldiers. I was myself long under the impression, gathered from scientific books, that it moved in this manner, or as this wise Natural History of Cuvier[1] puts its, "by true reptation";* but, however many legs a regiment or a centipede may possess, neither body of them can move faster than an individual pair of legs can,—their hundred or thousand feet being each capable of only one step at a time; and, with that allowance, only a certain proportion of pace is possible, and the utmost rapidity of the most active spider, or centipede, does not for an instant equal the dash of a snake in full power. But you—nearly all of you, I fancy—have learned, during the sharp frosts of the last winters, the real secret of it, and will recognize in a moment what the motion is, and only can be, when I show you the real rate of it. It is not often that you can see a snake in a hurry, for he generally withdraws subtly and quietly, even when distinctly seen; but if you put him to his pace either by fear or anger, you will find it is the sweep of the outside edge in skating, carried along the whole body,—that is to say, three or four times over. Outside or inside edge does not, however, I suppose, matter to the snake, the fulcrum being according to the lie of the ground, on the concave or convex side of the curve, and the whole strength of the body is alive in the alternate curves of it.

37. This splendid action, however, you must observe,

---

* It cannot be too often pointed out how much would be gained by merely insisting on scientific books being written in plain English.[2] If only this writer had been forbidden to use the word "repo" for "crawl," and to write, therefore, that serpents were crawling creatures, who moved by true crawlation,—his readers would have seen exactly how far he and they had got.

---

[1] [*The Animal Kingdom*, of which vol. ix. in the English edition is *The Class Reptilia, arranged by the Baron Cuvier, with Specific Descriptions by Edward Griffith and Edward Pidgeon*, 1831, p. 287.]

[2] [See above, p. 261; and compare below, p. 540 n.]

can hardly ever be seen when the snake is in confinement.
Half a second would take him twice the length of his
cage; and the sluggish movement which you see there, is
scarcely ever more than the muscular extension of himself
out of his "collected"[1] coil into a more or less straight line;
which is an action imitable at once with a coil of rope.
You see that one-half of it can move anywhere without
stirring the other; and accordingly you may see a foot or
two of a large snake's body moving one way, and another
foot or two moving the other way, and a bit between not
moving at all; which I, altogether, think we may specifically call "Parliamentary" motion; but this has nothing in
common with the gliding and truly serpentine power of
the animal when it exerts itself.

38. (Thus far, I stated the matter in my lecture, apolo-
gizing at the same time for the incompleteness of demonstration which, to be convincing, would have taken me
the full hour of granted attention, and perhaps with small
entertainment to most of my hearers. But, for once, I
care somewhat to establish my own claim to have first
described serpent motion,[2] just as I have cared much to
establish Forbes's claim to have first discerned the laws of
glacier flow; and I allow myself, therefore, here, a few
added words of clearer definition.

39. When languidly moving in its cage—or stealthily
when at liberty,—a serpent may continually be seen to
hitch or catch one part of its body by the edge of the
scales against the ground, and from the fulcrum of that
fixed piece extend other parts or coils in various directions.
But this is not the movement of progress. When a serpent is once in full pace, every part of its body moves
with equal velocity; and the whole in a series of waves,
varied only in sweep in proportion to the thickness of the
trunk. No part is straightened—no part extended—no part

---

[1] [See above, § 12, p. 303.]
[2] [Compare Ruskin's letter to Acland, given in Vol. XXV. p. xxviii.]

stationary. Fast as the head advances, the tail follows, and between both—at the same rate—every point of the body. And the impulse of that body bears it against, and is progressively resilient from, the ground at the edge of each wave, exactly as the blade of the oar in sculling a boat is progressively resilient from the water. In swimming, the action is seen in water itself, and is partially imitated also by fish in the lash of the tail. I do not attempt to analyze the direction of power and thrust in the organic structure, because I believe, without very high mathematics, it cannot be done even for the inorganic momentum of a stream, how much less for the distributed volition of muscle, which applies the thrust at the exact point of the living wave where it will give most forwarding power.

I am not sure how far the water serpents may sometimes use vertical instead of lateral undulation; but their tails are I believe always vertically flattened, implying only lateral oar-stroke. My friend Mr. Henry Severn,[1] however, on one occasion saw a large fresh-water serpent swimming in vertically sinuous folds, with its head raised high above the surface, and making the water foam at its breast, just as a swan would.)

40. Adding thus much to what I said of snake action, I find myself enabled to withdraw, as unnecessary, the question urged, in the next division of the lecture, as to the actual pain inflicted by snake-bite, by the following letter,* since received on the subject, from Mr. Arthur Nicols:—

"With respect to your remark that there are no descriptions of the sensation produced by snake-poison, in the nature of things, direct evidence of this kind is not easy to get; for, in the first place, the sufferer is very soon past the power of describing the sensations; and, in the second, but a

* A series of most interesting papers, by Mr. Nicols, already published in *The Country*, and reprinted in *Chapters from the Physical History of the*

---

[1] [The late Mr. Henry Severn, brother of Mr. Arthur Severn, lived much in Australia and was greatly interested in scientific subjects.]

minute fraction of those who are killed by snakes in India come under the hands of medical men. A person of the better class, too, is rarely bitten fatally. The sufferers are those who go about with naked feet, and handle wood, and whose work generally brings them into contact with snakes.

"A friend brought me from India last year several specimens of Echis carinata, a species about nine inches long, whose fangs (two on one maxilla in one instance) were as large as this—(a quarter of an inch long, curved), and hard as steel.

"This Echis kills more people in its district than all the other snakes together; it is found everywhere. We must also remember how very few persons bitten recover. Indirect evidence seems to point to a comatose state as soon as the poison takes effect; and those writhings of bitten animals which it gives us so much pain to witness are probably not the expression of suffering. In one of Fayrer's cases the patient (bitten by a cobra) complained, when taken to the hospital, of a burning pain in his foot; but as no more is said, I infer he then became incapable of giving any further description. The 'burning' is just what I feel when stung by a bee, and the poison soon makes me drowsy. In one instance I lay for an hour feebly conscious, but quite indifferent to the external world; and although that is fourteen years ago, I well remember speculating (albeit I was innocent of any knowledge of snakes then) as to whether their poison had a similar effect. It should not, I think, concern us much to learn what is the precise character of the suffering endured by any poor human being whose life is passing away under this mysterious influence, but to discover its physiological action."

41. Most wisely and truly said: and indeed, if any useful result is ever obtained for humanity by the time devoted recently, both in experiment and debate, to the question of the origin of life, it must be in the true determination of the meanings of the words Medicine and Poison, and the separation into recognized orders of the powers of the things which supply strength and stimulate function, from those

Earth (Kegan Paul & Co., 1880), may be consulted on all the points of chiefly terrible interest in serpent life. I have also a most valuable letter describing the utter faintness and prostration, without serious pain, caused by the bite of the English adder, from Mr. Spedding Curwen, adding the following very interesting notes. "The action was, and, so far as I have seen, always is, a distinct hammer-like stroke of the head and neck, with the jaw wide open. In the particular case in question, my brother had the adder hanging by the tail between his finger and thumb, and was lowering it gradually into our botany-box, the lid of which I was holding open. There were already three adders in the box; and in our care lest *they* should try to escape, we did not keep enough watch over the new capture. As his head reached the level of the lid of the box, he made

which dissolve flesh and paralyze nerve. The most interesting summed result which I yet find recorded by physicians, is the statement in the appendix to Dr. Fayrer's *Thanatophidia* [p. 173] of the relative mortal action of the Indian and Australian venomous snakes; the one paralyzing the limbs, and muscles of breathing and speech, but not affecting the heart; the other leaving the limbs free, but stopping the heart.

42. But the most terrific account which I find given with sufficient authority of the effect of snake-bite is in the general article closing the first volume of Russell's *History of Indian Serpents*.[1] Four instances are there recorded of the bite, not of the common Cobra, but of that called by the Portuguese Cobra di Morte. It is the smallest, and the deadliest, of all venomous serpents known,—only six inches long, or nine at the most, and not thicker than a tobacco-pipe,—and, according to the most definite account, does not move like ordinary serpents, but throws itself forward a foot or two on the ground, in successive springs, falling in the shape of a horse-shoe. In the five instances given of its bite, death follows, in a boy, ten minutes after the bite; and in the case of two soldiers, bitten by the same snake, but one a minute after the other, in their guard-room, about one in the morning,—the first died at seven in the morning, the second at noon; in both, the powers of sight gradually failing, and they became entirely blind before

a side-dart at my hand, and struck by the thumb nail. The hold was *quite* momentary, but as the adder was suspended by the tail, that may be no guide to the general rule. The receding of the blood was only to a small distance, say a quarter of an inch round the wound. The remedies I used were whisky (half a pint, as soon as I got to the nearest inn, and more at intervals all day, also ammonia), both to drink and to bathe the wound with. The whisky seemed to have no effect: my whole body was cold and deathly, and I felt none of the glow which usually follows a stimulant."

---

[1] [*An Account of Indian Serpents, collected on the Coast of Coromandel*, by Patrick Russell, M.D., F.R.S., 1796. In 1801 appeared posthumously *A Continuation of An Account of Indian Serpents*. Ruskin here quotes from the former volume, pp. 79, 80.]

death. The snake is described as of a dark straw colour, with two black lines behind the head; small, flat head, with *eyes that shone like diamonds.*

43. Next in fatal power to this serpent,—fortunately so rare that I can find no published drawing of it,—come the Cobra, Rattlesnake, and Trigonocephalus, or triangle-headed serpent of the West Indies. Of the last of these snakes, you will find a most terrific account (which I do not myself above one-third believe) in the ninth volume of the English translation of Cuvier's *Animal Kingdom.* It is a grand book of fifteen volumes, copiously illustrated, and quite unequalled for collection of the things you do not want to know in the body of the text, and for ceasing to be trustworthy the moment it is entertaining. I will read from it a single paragraph concerning the Trigonocephalus, of which you may believe as much or as little as you like:—

"These reptiles possess an activity and vivacity of motion truly alarming. A ferocious instinct induces them to dart impetuously upon passengers, either by suddenly letting go the sort of spring which their body forms, rolled in concentric and superposed circles, and thus shooting like an arrow from the bow of a vigorous archer, or pursuing them by a series of rapid and multiplied leaps, or climbing up trees after them, or even threatening them in a vertical position."[1]

44. The two other serpents, one used to be able to study at our own Zoological Gardens ; but the cobra has now for some years had the glass in front of him whitened, to prevent vulgar visitors from poking sticks at him, and wearing out his constitution in bad temper. I do not know anything more disgraceful to the upper classes of England as a body, than that, while on the one hand their chief recreations, without which existence would not be endurable to them, are gambling in horses, and shooting at birds,[2] they are so totally without interest in the natures and habits of animals in general, that they have never thought

[1] [*The Animal Kingdom,* vol. ix. (as cited above, p. 317 *n.*), p. 352.]
[2] [For a collection of Ruskin's references to sport, see Vol. VII. p. 341 *n.*]

XVIII

STUDIES OF CERASTES CORNUTUS

of enclosing for themselves a park[1] and space of various kinds of ground, in free and healthy air, in which there should be a perfect gallery, Louvre, or Uffizii, not of pictures, as at Paris, nor of statues, as at Florence, but of living creatures of all kinds, beautifully kept, and of which the contemplation should be granted only to well-educated and gentle people who would take the trouble to travel so far, and might be trusted to behave decently and kindly to any living creatures, wild or tame.

45. Under existing circumstances, however, the Zoological Gardens are still a place of extreme interest; and I have been able at different times to make memoranda of the ways of snakes there,[2] which have been here enlarged for you by my friends, or by myself; and having been made always with reference to gesture or expression, show you, I believe, more of the living action than you will usually find in scientific drawings: the point which you have chiefly to recollect about the cobra being this curious one—that while the puff-adder, and most other snakes, or snakelike creatures, swell when they are angry, the cobra flattens himself; and becomes, for four or five inches of his length, rather a hollow shell than a snake. The beautiful drawing made by Mr. Macdonald in enlarging my sketch from life shows you the gesture accurately, and especially the levelling of the head which gives it the chief terror. It is always represented with absolute truth in Egyptian painting and sculpture ; one of the notablest facts to my mind in the entire history of the human race being the adoption by the Egyptians of this serpent for the type of their tyrannous monarchy,[3] just

---

[1] [Lord Rothschild and his son, the Hon. Walter Rothschild, have, however, done this in some measure in Tring Park. The recent institution of "Reserves" both in British South Africa and in British East Africa may also be mentioned in this connexion.]

[2] [Several of such drawings of snakes are now at Oxford ; Educational Series, Nos. 169–175 (Vol. XXI. pp. 89, 90). The enlarged drawings by Mr. Macdonald and Mr. Severn, which were exhibited at the lecture (see below, Nos. 6 and 7, p. 331), are not now available. The two plates here given (XVIII. and XIX.) are from Ruskin's sketches made from life at the Zoological Gardens.]

[3] [On this subject, see Fors Clavigera, Letter 26.]

as the cross or the lily was adopted for the general symbol of kinghood by the monarchs of Christendom.

46. I would fain enlarge upon this point, but time forbids me: only please recollect this one vital fact, that the nature of Egyptian monarchy, however great its justice, is always that of government by cruel force; and that the nature of Christian monarchy is embodied in the cross or lily, which signify either an authority received by divine appointment, and maintained by personal suffering and sacrifice; or else a dominion consisting in recognized gentleness and beauty of character, loved long before it is obeyed.

47. And again, whatever may be the doubtful meanings of the legends invented among all those nations of the earth who have ever seen a serpent alive, one thing is certain, that they all have felt it to represent to them, in a way quite inevitably instructive, the state of an entirely degraded and malignant human life. I have no time to enter on any analysis of the causes of expression in animals, but this is a constant law for them, that they are delightful or dreadful to us exactly in the degree in which they resemble the contours of the human countenance given to it by virtue and vice;[1] and this head of the cerastes, and that of the rattlesnake,[2] are in reality more terrific to you than the others, not because they are more snaky, but because they are more human,—because the one has in it the ghastliest expression of malignant avarice, and the other of malignant pride. In the deepest and most literal sense, to those who allow the temptations of our natural passions their full sway, the curse, fabulously (if you will) spoken on the serpent, is fatally and to the full accomplished upon ourselves; and as for noble and righteous persons and nations, the words are for ever true, " Thou art fairer than the children *of men:* full of grace are thy lips," so for the ignoble

---

[1] [On this subject, see *Modern Painters*, vol. ii. (Vol. IV. pp. 157–158).]

[2] [The head of the cerastes (showing "malignant avarice") may be the upper one on Plate XVIII.; "malignant pride" may be seen in the full face on Plate XIX., which, however, is a study of Vipera elegans, not of the rattlesnake.]

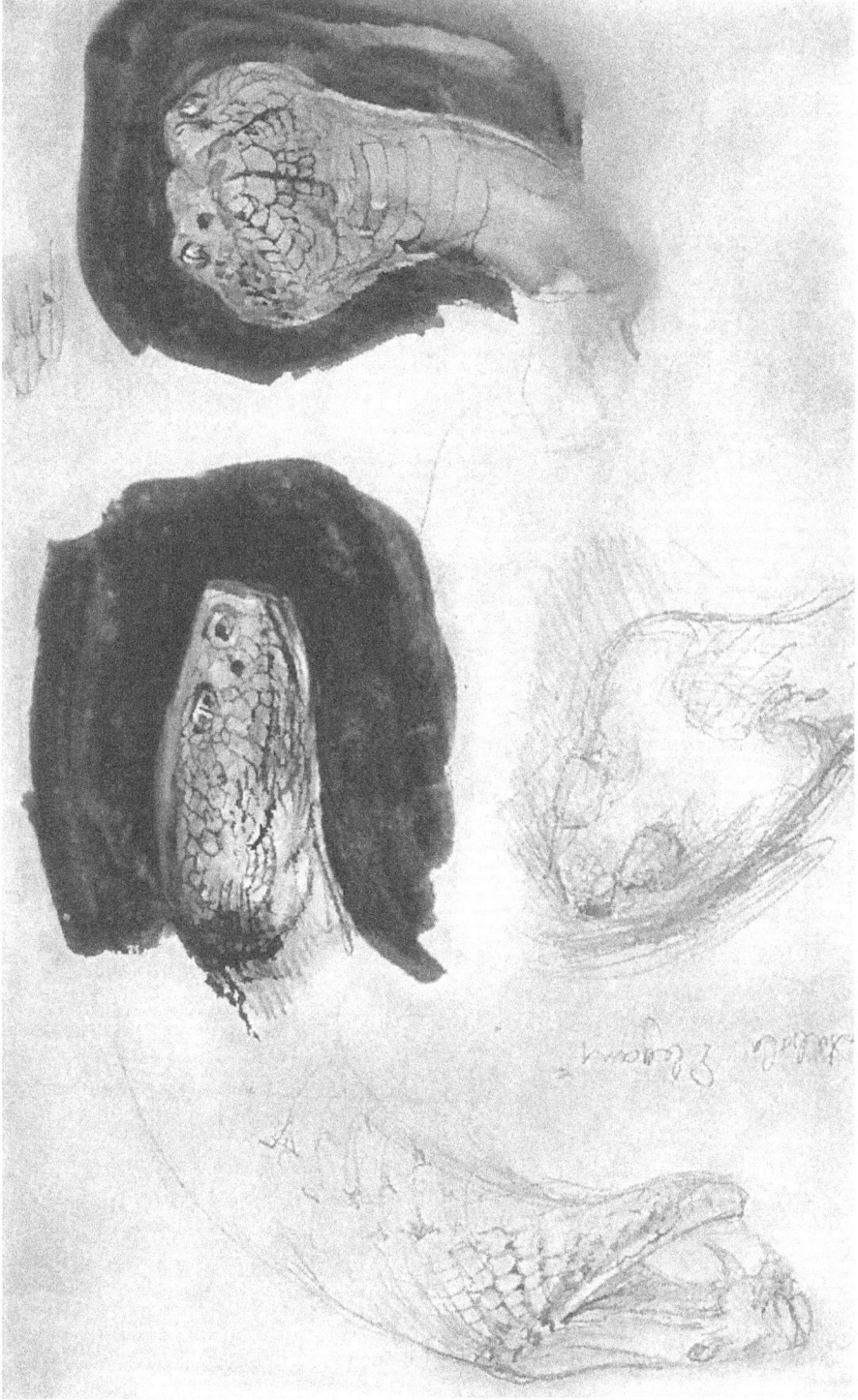

and iniquitous, the saying is for ever true, "Thou art fouler than the children of the Dust, and the poison of asps is under thy lips."[1]

48. Let me show you, in one constant manner of our national iniquity, how literally that is true. Literally, observe. In any good book, but especially in the Bible, you must always look for the literal meaning of everything first,—and act out that, then the spiritual meaning easily and securely follows. Now in the great Song of Moses, in which he foretells, before his death, the corruption of Israel, he says of the wicked race into which the Holy People are to change, "Their wine is the poison of dragons, and the cruel venom of asps."[2] Their wine,—that is to say, of course, not the wine they drink, but the wine they give to drink. So that, as our best duty to our neighbour is figured by the Samaritan who heals wounds by pouring in oil and wine,[3] our worst sin against our neighbour is in envenoming his wounds by pouring in gall and poison. The cruel venom of *Asps*—of that brown gentleman you see there!

49. Now I am sure you would all be very much shocked, and think it extremely wrong, if you saw anybody deliberately poisoning so much as one person in that manner. Suppose even in the interests of science, to which you are all so devoted, I were myself to bring into this lecture-room a country lout of the stupidest,—the sort whom you produce by Church of England education, and then do all you can to get emigrated out of your way; fellows whose life is of no use to them, nor anybody else; and that —always in the interests of science—I were to lance just the least drop out of that beast's tooth into his throat, and let you see him swell, and choke, and get blue and blind, and gasp himself away—you wouldn't all sit quiet there, and have it so done—would you?—in the interests of science.

[1] [Genesis iii. 14; Psalms xlv. 2; Romans iii. 13.]
[2] [Deuteronomy xxxii. 33.]
[3] [Luke x. 34.]

50. Well; but how then if in your own interests? Suppose the poor lout had his week's wages in his pocket—thirty shillings or so; and, after his inoculation, I were to pick his pocket of them; and then order in a few more louts, and lance their throats likewise, and pick their pockets likewise, and divide the proceeds of, say, a dozen of poisoned louts, among you all, after lecture: for the seven or eight hundred of you, I could perhaps get sixpence each out of a dozen of poisoned louts; yet you would still feel the proceedings painful to your feelings, and wouldn't take the sixpenn'orth—would you?

51. But how, if you constituted yourselves into a co-operative Egyptian Asp and Mississippi Rattlesnake Company, with an eloquent member of Parliament for the rattle at its tail? and if, brown asps getting scarce, you brewed your own venom of beautiful aspic brown, with a white head, and persuaded your louts to turn their own pockets inside-out to get it, giving you each sixpence a night,—seven pounds ten a year of lovely dividend!—How does the operation begin to look now? Commercial and amiable —does it not?

52. But how—to come to actual fact and climax—if, instead of a Company, you were constituted into a College of reverend and scholarly persons, each appointed—like the King of Salem [1]—to bring forth the bread and wine of healing knowledge; but that, instead of bread gratis, you gave stones for pay; and, instead of wine gratis, you gave asp-poison for pay,—how then? Suppose, for closer instance, that you became a College called of the Body of Christ, and with a symbolic pelican for its crest,[2] but that this charitable pelican had begun to peck—not itself, but other people,—and become a vampire pelican, sucking blood instead of shedding,—how then? They say it's

---

[1] [Genesis xiv. 18.]

[2] [For the Christian symbolism of the pelican, which arose from the belief that the bird, while really pruning its feathers, was feeding its young with its own blood, see F. E. Hulme's *Symbolism in Christian Art*, pp. 188 *seq.* Hence the title *The Pelican* for the magazine of *Corpus Christi* College, Oxford, from which recollections of Ruskin are reprinted in Vol. XX. p. xxx.]

an ill bird that fouls its own nest. My own feeling is that a well-behaved bird will neither foul its own nest nor another's, but that, finding it in any wise foul, it will openly say so, and clean it.

53. Well, I know a village, some few miles from Oxford, numbering of inhabitants some four hundred louts, in which my own College of the Body of Christ keeps the public-house, and therein sells—by its deputy—such poisoned beer that the Rector's wife[1] told me, only the day before yesterday, that she sent for some to take out a stain in a dress with, and couldn't touch the dress with it, it was so filthy with salt and acid, to provoke thirst; and that while the public-house was there she had no hope of doing any good to the men, who always prepared for Sunday by a fight on Saturday night. And that my own very good friend the Bursar, and we the Fellows, of Corpus, being appealed to again and again to shut up that tavern, the answer is always, "The College can't afford it: we can't give up that fifty pounds a year out of those peasant sots' pockets, and yet 'as a College' live."

Drive that nail home with your own hammers, for I've no more time; and consider the significance of the fact, that the gentlemen of England can't afford to keep up a college for their own sons but by selling death of body and soul to their own peasantry.

54. I come now to my last head of lecture—my caution concerning the wisdom which we buy at such a price. I had not intended any part of my talk to-night to be so grave; and was forced into saying what I have now said by the appointment of Fors that the said village Rector's wife should come up to town to nurse her brother, Mr. Severn, who drew your diagrams for you. I had meant to be as cheerful as I could; and chose the original title of my lecture, "A Caution to Snakes," partly in play, and partly in affectionate remembrance of the scene in

[1] [Mrs. Furneaux (sister of Mr. Arthur Severn), whose husband was at this time Rector of Heyford.]

New Men and Old Acres, in which the phrase became at once so startling and so charming, on the lips of my much-regarded friend, Mrs. Kendal.[1]

But this one little bit of caution more I always intended to give, and to give earnestly.

55. What the best wisdom of the Serpent may be, I assume that you all possess;—and my caution is to be addressed to you in that brightly serpentine perfection. In all other respects as wise, in one respect let me beg you to be wiser than the Serpent,[2] and not to eat your meat without tasting it,—meat of any sort, but above all the serpent-recommended meat of knowledge. Think what a delicate and delightful meat that used to be in old days, when it was not quite so common as it is now, and when young people—the best sort of them—really hungered and thirsted[3] for it. *Then* a youth went up to Cambridge, or Padua, or Bonn, as to a feast of fat things, of wines on the lees, well refined.[4] But now, he goes only to swallow,—and, more's the pity, not even to swallow as a glutton does, with enjoyment; not even—forgive me the old Aristotelian Greek, ἡδόμενος τῇ ἀφῇ[5]—pleased with the going down, but in the saddest and exactest way, as a constrictor does, tasting nothing all the time. You remember what Professor Huxley told you—most interesting it was, and new to me—of the way the great boa does not in any true sense swallow, but only hitches himself on to his meat like a coal-sack;—well, that's the exact way you expect your poor modern student to hitch himself on to *his* meat, catching and notching his teeth into it, and dragging the skin of him tight over it,—

---

[1] [Mrs. Kendal (Miss Madge Robertson) played Lilian Vavasour in the first representation of *New Men and Old Acres* (by Tom Taylor and A. W. Dubourg) at the Haymarket Theatre, 25th October 1869. Lilian remarks : "And his wife —well, she's a caution for snakes !" (Act i., p. 21 in Lacy's Acting Edition). Earlier in the act (p. 10) Lilian says of another girl that "in spite of her Ruskinism-run-mad she isn't half a bad sort."]

[2] [Matthew x. 16. On the wisdom of the serpent, compare Vol. VII. p. 353.]

[3] [Matthew v. 6.]

[4] [Isaiah xxv. 6.]

[5] [Aristotle : *Ethics*, iii. 9, 10 ("pleased with the touch").]

till at last—you know I told you a little ago[1] our artists didn't know a snake from a sausage,—but, Heaven help us, your University doctors are going on at such a rate that it will be all we can do, soon, to know a *man* from a sausage.

56. Then think again, in old times what a delicious thing a book used to be in a chimney corner, or in the garden, or in the fields, where one used really to read a book, and nibble a nice bit here and there if it was a bride-cakey sort of book, and cut oneself a lovely slice—fat and lean—if it was a round-of-beef sort of book. But what do you do with a book now, be it ever so good? You give it to a reviewer, first to skin it, and then to bone it, and then to chew it, and then to lick it, and then to give it you down your throat like a handful of pilau. And when you've got it, you've no relish for it, after all. And, alas! this continually increasing deadness to the pleasures of literature leaves your minds, even in their most conscientious action, sensitive with agony to the sting of vanity, and at the mercy of the meanest temptations held out by the competition of the schools. How often do I receive letters from young men of sense and genius, lamenting the loss of their strength, and waste of their time, but ending always with the same saying, " I *must* take as high a class as I can, in order to please my father."[2] And the fathers love the lads all the time, but yet, in every word they speak to them, prick the poison of the asp into their young blood, and sicken their eyes with blindness to all the true joys, the true aims, and the true praises of science and literature; neither do they themselves any more conceive what was once the faith of Englishmen; that the only path of honour is that of rectitude, and the only place of honour, the one that you are fit for. Make your children happy in their youth; let distinction come to them, if it will, after well-spent and

---

[1] [See above, p. 299.]
[2] [On the subject of competitive examinations, see Vol. I. p. 384 *n.*, and Vol. XVI. p. 469 *n.*]

well-remembered years; but let them now break and eat
the bread of Heaven with gladness and singleness of heart,
and send portions to them for whom nothing is prepared;[1]
—and so Heaven send you its grace—before meat, and
after it.

NOTE TO CHAPTER I[2]

THE following list of the subjects of the exhibited diagrams may perhaps
prevent the need of hurried explanations in the course of the lecture, but
I have not put numbers on the drawings; for, if large enough to be well
seen, they always spoil the look of a sketch (and also I am not sure that
all I have prepared can be shown); but I hope there will be no difficulty
found in identifying those arranged in the lecture-room.

On the second leaf the list of the names given to the serpent tribe
in the great languages saves the space of another diagram, and perhaps to
some careful students may also spare the trouble of copying it.

DIAGRAMS

1. PART OF GIOTTO'S SCULPTURE OF THE CREATION OF EVE. Her danger
foretold by the twining ivy above. (*Drawn by Mr. A. Severn.*[3])

2. THE CONSTELLATION DRACO; enlarged from a common celestial globe:
the colour altered, but the form retained, to show the smallness of the
assistance hitherto accepted by modern astronomy from modern zoology.
(*Drawn by Mr. L. Hilliard.*)

3. HEAD OF THE SAME, as it was really drawn on the globe—a general
type of modern dragon idealism. (*Drawn by Mr. A. Severn.*[4])

4. THE HYDRA OF LERNA. Greek ideal, from a coin of Phæstus. (*Drawn
by myself.*[5])

> "Non te rationis egentem,
> Lernæus turbâ capitum circumstetit anguis."[6]

> "Thee, God, no face of danger could affright,
> Nor huge Typhœus, nor the unnumbered snake
> Increased with hissing heads in Lerna's lake."
> —*Dryden.*

---

[1] [Nehemiah viii. 10. Compare Vol. XX. p. 377.]
[2] [Added in this edition from the fly-sheet circulated on the occasion of the
delivery of the lecture.]
[3] [For references to this, see above, pp. 298 and *n.*, 313.]
[4] [See above, Fig. 38, p. 299.]
[5] [Fig. 39, p. 300.]
[6] [*Æneid*, viii. 300; quoted also in *Queen of the Air*, § 4 (Vol. XIX. p. 299).]

5. HEADS OF THE GAVIAL OF THE GANGES AND THE COMMON CROCODILE. From the last edition of Cuvier. (*Drawn by Mr. L. Hilliard.*[1])

6. HEAD OF THE COBRA, THREATENING. Profile and front, *a* and *b*. (*Sketch from life by myself. Enlarged and drawn by Mr. A. Macdonald.*[2])

7. HEAD OF THE RATTLESNAKE, THREATENING. Profile and front, *a* and *b*. (*Sketch from life by myself. Enlarged and drawn by Mr. A. Severn.*[2])

8. SKULL OF THE RATTLESNAKE. (*Drawn by Mr. A. Burgess.*[3])

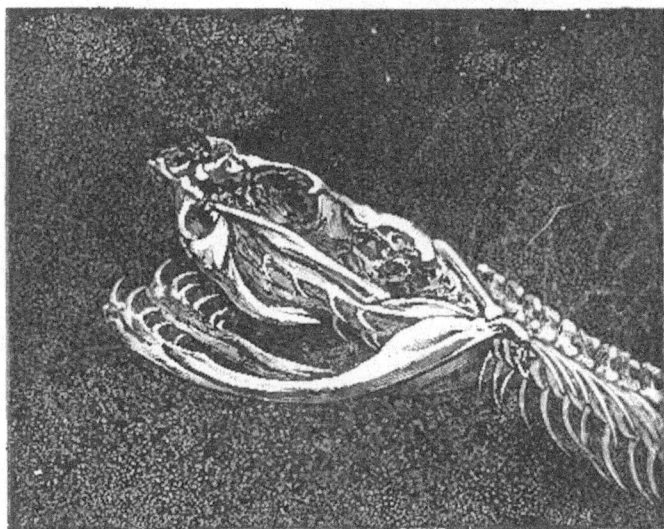

Fig. 41

9. BLACK VIPER. (*Sketch by myself, from Carpaccio. Enlarged and drawn by Mr. A. Burgess.*[4])

10. COLOUR PATTERN OF THE ENGLISH VIPER, with Greek ornament for comparison. (*Drawn by Mr. L. Hilliard.*[5])

11. EGYPTIAN ASP. (*Sketched from life, and drawn, by myself.*[6])

12. CRANBERRY BLOSSOM. (*Sketched from life, and drawn, by myself.*[6])

[1] [With the head of the saw-bill duck also; Plate XVII., p. 310.]
[2] [These drawings are not known to the editors.]
[3] [Fig. 41 here.]
[4] [This drawing, at one time in the Oxford Collection, is not now available: see Vol. XXI. p. 90 *n*, (No. 171).]
[5] [Fig. 40, p. 315.]
[6] [These drawings are not known to the editors.]

NAMES OF THE SNAKE TRIBE IN THE
GREAT LANGUAGES[1]

1. Ophis (Greek). "The seeing" (creature understood). Meaning especially one that sees all round it.

2. Dracon (Greek). Drachen (German). "The beholding." Meaning one that looks well into a thing, or person.

3. Anguis (Latin). "The strangling."

4. Serpens (Latin). "The winding."

5. Coluber (Latin). Couleuvre (French). "The coiling."

6. Adder (Saxon). "The grovelling."

7. Snake (Saxon). Schlange (German). "The crawling" (with sense of dragging, and of smoothness).

[1] [On these names, see above, pp. 301-304.]

CHAPTER II[1]

REVISION

1. IF the reader will look back to the opening chapter of *Deucalion*,[2] he will see that the book was intended to be a collection of the notices of phenomena relating to geology which were scattered through my former works, systematized so far as might be possible, by such additional studies as time permitted me.

Hitherto, however, the scattered chapters have contained nothing else than these additional studies, which, so far from systematizing what preceded them, stand now greatly in need of arrangement themselves; and still more of some explanation of the incidental passages referring to matters of higher science than geology, in which I have too often assumed that the reader is acquainted with—and in some degree even prepared to admit—the modes of thought and reasoning which have been followed throughout the general body of my writings.

I have never given myself out for a philosopher; nor spoken of the teaching attempted in connection with any subject of inquiry, as other than that of a village showman's "Look—and you shall see."[3] But, during the last twenty years, so many baseless semblances of philosophy have announced themselves; and the laws of decent thought and rational question have been so far transgressed (even in our universities, where the moral philosophy they once taught is now only remembered as an obscure tradition,

---

[1] [Issued in May 1883.]
[2] [See above, pp. 96-98.]
[3] [Compare *Love's Meinie*, § 122 (Vol. XXV. p. 112), where Ruskin speaks of his function as that of the Interpreter.]

and the natural science in which they are proud, presented only as an impious conjecture), that it is forced upon me, as the only means of making what I have said on these subjects permanently useful, to put into clear terms the natural philosophy and natural theology to which my books refer, as accepted by the intellectual leaders of all past time.

2. To this end, I am republishing the second volume of *Modern Painters*,[1] which, though in affected language, yet with sincere and very deep feeling, expresses the first and foundational law respecting human contemplation of the natural phenomena under whose influence we exist,—that they can only be seen with their properly belonging joy, and interpreted up to the measure of proper human intelligence, when they are accepted as the work, and the gift, of a Living Spirit greater than our own.

3. Similarly, the moral philosophy which underlies all the appeals, and all the accusations, made in the course of my writings on political science, assumes throughout that the principles of Justice and Mercy which are fastened in the hearts of men, are also expressed in entirely consistent terms throughout the higher—(and even the inferior, when undefiled)—forms of all lovely literature and art ; and enforced by the Providence of a Ruling and Judging Spiritual Power, manifest to those who desire its manifestation, and concealed from those who desire its concealment.

4. These two Faiths, in the creating Spirit, as the source of Beauty,—in the governing Spirit, as the founder and maintainer of Moral law, are, I have said, *assumed* as the basis of all exposition and of all counsel, which have ever been attempted or offered in my books. I have never held it my duty, never ventured to think of it even as a permitted right, to proclaim or explain these faiths, except only by referring to the writings, properly called inspired,[2] in which the good men of all nations and languages had

---

[1] [Compare the Preface to the rearranged edition of 1883 (Vol. IV. p. 3).]
[2] [Compare *Lectures on Art*, § 44 (Vol. XX. p. 54).]

concurrently—though at far distant and different times—declared them. But it has become now for many reasons, besides those above specified, necessary for me to define clearly the meaning of the words I have used—the scope of the laws I have appealed to, and, most of all, the nature of some of the feelings possible under the reception of these creeds, and impossible to those who refuse them.

5. This may, I think, be done with the best brevity and least repetition, by adding to those of my books still unfinished, *Deucalion*, *Proserpina*, *Love's Meinie*, and *Fors Clavigera*, explanatory references to the pieces of theology or natural philosophy which have already occurred in each, indicating their modes of connection, and the chiefly parallel passages in the books which are already concluded; among which I may name the *Eagle's Nest* as already, if read carefully, containing nearly all necessary elements of interpretation for the others.

6. I am glad to begin with *Deucalion*, for its title already implies (and is directly explained in its fifth page[1] as implying) the quite first principle, with me, of historic reading in divinity, that all nations have been taught of God[2] according to their capacity, and may best learn what farther they would know of Him by reverence for the impressions which He has set on the hearts of each, and all.

I said farther in the same place that I thought it well for the student first to learn the "myths of the Betrayal and Redemption" as they were taught to the heathen world; but I did not say what I meant by the "Betrayal" and "Redemption" in their universal sense, as represented alike by Christian and heathen legends.

7. The idea of contest between good and evil spirits for the soul and body of man, which forms the principal subject of all the imaginative literature of the world, has hitherto been the only explanation of its moral phenomena

---

[1] [Of the original edition; in this volume, p. 98.]
[2] [Isaiah liv. 13; John vi. 45.]

tenable by intellects of the highest power. It is no more a
certain or sufficient explanation than the theory of gravita-
tion is of the construction of the starry heavens; but it
reaches farther towards analysis of the facts known to us
than any other. By "*the* Betrayal" in the passage just
referred to I meant the supposed victory, in the present
age of the world, of the deceiving spiritual power, which
makes the vices of man his leading motives of action, and
his follies, its leading methods. By "*the* Redemption" I
meant the promised final victory of the creating and true
Spirit, in opening the blind eyes, in making the crooked
places straight and the rough plain, and restoring the power
of His ministering angels,[1] over a world in which there shall
be no more tears.

8. The "myths"—allegorical fables or stories—in which
this belief is represented, were, I went on to say in the
same place, "incomparably *truer*" than the Darwinian—or,
I will add, any other conceivable materialistic theory—
because they are the instinctive products of the natural
human mind, conscious of certain facts relating to its fate
and peace; and as unerring in that instinct as all other
living creatures are in the discovery of what is necessary
for their life: while the materialistic theories have been
from their beginning products, in the words used in the
passage I am explaining (page 6, line 8[2]), of the "*half* wits
of impertinent multitudes." They are half-witted, because
never entertained by any person possessing imaginative
power,—and impertinent, because they are always announced
as if the very defect of imagination constituted a superiority
of discernment.

9. In one of the cleverest—(and, in description of the
faults and errors of religious persons, usefullest)—books
of this modern half-witted school, *Une Cure du Docteur
Pontalais*,[3] of which the plot consists in the revelation by

---

[1] [Isaiah xlii. 7, xl. 4; Hebrews i. 14; Isaiah xxv. 8.]
[2] [Again of the original edition; in this volume, p. 99, line 5.]
[3] [By Robert Halt (pseudonym of Charles Vieu): first edition, 1865.]

an ingenious doctor to an ingenuous priest that the creation
of the world may be sufficiently explained by dropping oil
with dexterity out of a pipe into a wineglass,—the assumption that " la logique " and " la methode " were never applied
to theological subjects except in the Quartier Latin of Paris
in the present blessed state of Parisian intelligence and
morals, may be I hope received as expressing nearly the
ultimate possibilities of shallow arrogance in these regions
of thought; and I name the book as one extremely well
worth reading, first as such ; and secondly because it puts
into the clearest form I have yet met with, the peculiar
darkness of materialism, in its denial of the hope of immortality. The hero of it, who is a perfectly virtuous
person, and inventor of the most ingenious and benevolent
machines, is killed by the cruelties of an usurer and a
priest ; and in dying, the only consolation he offers his wife
and children is that the loss of one life is of no consequence in the progress of humanity.

This unselfish resignation to total death is the most
heroic element in the Religion now in materialist circles
called the Religion " of Humanity," and announced as if it
were a new discovery of nineteenth-century sagacity, and
able to replace in the system of its society, alike all former
ideas of the power of God, and destinies of man.

10. But, in the first place, it is by no means a new
discovery. The fact that the loss of a single life is of
no consequence when the lives of many are to be saved,
is, and always has been, the root of every form of beautiful courage; and I have again and again pointed out,
in passages scattered through writings carefully limited in
assertion, between 1860 and 1870,[1] that the heroic actions
on which the material destinies of this world depend, are
almost invariably done under the conception of death as a
calamity, which is to be endured by one for the deliverance of many, and after which there is no personal reward

---

[1] [See, for instance, *Crown of Wild Olive* (Vol. XVIII. p. 179), *Ethics of the Dust*
(*ibid.*, p. 204), and *Lectures on Art*, § 83 (Vol. XX. p. 86).]

to be looked for, but the gratitude or fame of which the victim anticipates no consciousness.

11. In the second place, this idea of self-sacrifice is no more sufficient for man than it is new to him. It has, indeed, strength enough to maintain his courage under circumstances of sharp and instant trial; but it has no power whatever to satisfy the heart in the ordinary conditions of social affection, or to console the spirit and invigorate the character through years of separation or distress. Still less can it produce the states of intellectual imagination which have hitherto been necessary for the triumphs of constructive art; and it is a distinctive essential point in the modes of examining the arts as part of necessary moral education, which have been constant in my references to them, that those of poetry, music, and painting, which the religious schools who have employed them usually regard only as stimulants or embodiments of faith, have been by me always considered as its *evidences*.[1] Men do not sing themselves into love or faith; but they are incapable of true song, till they love, and believe.

12. The lower conditions of intellect which are concerned in the pursuit of natural science, or the invention of mechanical structure, are similarly, and no less intimately, dependent for their perfection on the lower feelings of admiration and affection which can be attached to material things: these also—the curiosity and ingenuity of man—live by admiration and by love;[2] but they differ from the imaginative powers in that they are concerned with things seen—not with the evidences of things unseen[3]—and it would be well for them if the understanding of this restriction prevented them in the present day as severely from speculation as it does from devotion.

13. Nevertheless, in the earlier and happier days of

---

[1] [See, more especially, the Rede Lecture "On the Relation of National Ethics to National Arts": Vol. XIX. pp. 163 *seq.*]

[2] [See the quotation from Wordsworth in *Modern Painters*, vol. ii. (Vol. IV. p. 29 *n.*); and compare *Fors Clavigera*, Letter 5, § 14.]

[3] [Hebrews xi. 1.]

Linnæus,[1] de Saussure, von Humboldt, and the multitude
of quiet workers on whose secure foundation the fantastic
expatiations of modern science depend for whatever good or
stability there is in them, natural religion was always a
part of natural science; it becomes with Linnæus a part of
his definitions; it underlies, in serene modesty, the courage
and enthusiasm of the great travellers and discoverers, from
Columbus and Hudson to Livingstone; and it has saved
the lives, or solaced the deaths, of myriads of men whose
nobleness asked for no memorial but in the gradual enlargement of the realm of manhood, in habitation, and in social
virtue.

14. And it is perhaps, of all the tests of difference between the majestic science of those days, and the wild
theories or foul curiosities of our own,[2] the most strange and
the most distinct, that the practical suggestions which are
scattered through the writings of the older naturalists tend
always directly to the benefit of the general body of mankind;[3] while the discoverers of modern science have, almost
without exception, provoked new furies of avarice, and new
tyrannies of individual interest; or else have directly contributed to the means of violent and sudden destruction,
already incalculably too potent in the hands of the idle and
the wicked.

15. It is right and just that the reader should remember, in reviewing the chapters of my own earlier writings
on the origin and sculpture of mountain form, that all the
investigations undertaken by me at that time were connected
in my own mind with the practical hope of arousing the
attention of the Swiss and Italian mountain peasantry to
an intelligent administration of the natural treasures of their
woods and streams.[4] I had fixed my thoughts on these

---

[1] [For Linnæus in this connexion, see the words quoted in the Preface of
1883 to *Modern Painters*, vol. ii. (Vol. IV. pp. 4, 5).]

[2] [Compare Vol. XXV. pp. 56, 163.]

[3] [Compare Vol. IV. p. 34 (note of 1883), where this passage is referred to.]

[4] [On this subject, see *Modern Painters*, vol. iv. chaps. xix., xx. (Vol. VI. pp. 385
seq.); and compare Vol. XVII. pp. 547–552.]

problems where they are put in the most exigent distinctness by the various distress and disease of the inhabitants
of the valley of the Rhone, above the Lake of Geneva : a
district in which the adverse influences of unequal temperatures, unwholesome air, and alternate or correlative drought
and inundation, are all gathered in hostility against a race
of peasantry, the Valaisan,[1] by nature virtuous, industrious,
and intelligent in no ordinary degree, and by the hereditary and natural adversities of their position, regarded by
themselves as inevitable, reduced indeed, many of them,
to extreme poverty and woful disease ; but never sunk into
a vicious or reckless despair.

16. The practical conclusions at which I arrived, in
studying the channels and currents of the Rhone, Ticino,
and Adige, were stated first in the letters addressed to the
English press on the subject of the great inundations at
Rome in 1871 (*Arrows of the Chace*, vol. ii. pp. 160–174[2]),
and they are again stated incidentally in *Fors* (Letter 19),
with direct reference to the dangerous power of the Adige
above Verona. Had those suggestions been acted upon,
even in the most languid and feeble manner, the twentieth
part of the sums since spent by the Italian Government in
carrying French boulevards round Tuscan cities, and throwing down their ancient streets to find lines for steam tram-
ways, would not only have prevented the recent inundations
in North Italy, but rendered their recurrence for ever impossible.

17. As it is thus the seal of rightly directed scientific
investigation, to be sanctified by loving anxiety for instant
practical use, so also the best sign of its completeness and
symmetry is in the frankness of its communication to the
general mind of well-educated persons.

The fixed relations of the crystalline planes of minerals, first stated, and in the simplest mathematical terms

---

[1] [Compare Vol. VI. p. 410, and Vol. VII. p. 111.]
[2] [Now printed in Vol. XVII. pp. 547–552.]

expressed, by Professor Miller of Cambridge,[1] have been
examined by succeeding mineralogists with an ambitious
intensity which has at last placed the diagrams of zone
circles for quartz and calcite, given in Cloizeaux's mineralogy,[2] both as monuments of research, and masterpieces of
engraving, a place among the most remarkable productions
of the feverish energies of the nineteenth century. But in
the meantime, all the characters of minerals, except the
optical and crystalline ones, which it required the best instruments to detect, and the severest industry to register,
have been neglected;* the arrangement of collections in
museums has been made unintelligibly scientific, without the
slightest consideration whether the formally sequent specimens were in lights, or places, where they could be ever
visible; the elements of mineralogy prepared for schools have
been diversified by eight or ten different modes, nomen-
clatures, and systems of notation; and while thus the study
of mineralogy at all has become impossible to young people,
except as a very arduous branch of mathematics, that of its
connection with the structure of the earth has been post-
poned by the leading members of the Geological Society,
to inquire into the habits of animalculæ, fortunately for the
world invisible, and monsters, fortunately for the world un-
regenerate. The race of old Swiss guides, who knew the
flowers and crystals of their crags, has meanwhile been
replaced by chapmen, who destroy the rarest living flowers

---

* Even the chemistry has been allowed to remain imperfect or doubtful, while the planes of crystals were being counted: thus for an extreme
instance, the most important practical fact that the colour of ultramarine
is destroyed by acids, will not be found stated in the descriptions of that
mineral by either Miller, Cloizeaux, or Dana;[3] and no microscopic studies
of refraction have hitherto informed the public why a ruby is red, a
sapphire blue, or a flint black. On a large scale, the darkening of the
metamorphic limestones, near the central ranges of the Alps, remains
unexplained.

---

[1] [See above, p. 272 n.]
[2] [See above, p. 271 n.]
[3] [For Dana's book, see p. 376 n.]

of the Alps to raise the price of their herbaria, and pedestrian athletes in the pay of foolish youths; the result being that while fifty years ago there was a good and valuable mineral cabinet in every important mountain village, it is impossible now to find even at Geneva anything offered for sale but dyed agates from Oberstein;[1] and the confused refuse of the cheap lapidary's wheel, working for the supply of Mr. Cook's tourists with " Trifles from Chamouni."

18. I have too long hoped to obtain some remedy for these evils by putting the questions about simple things which ought to be answered in elementary schoolbooks of science, clearly before the student. My own books have thus sometimes become little more than notes of interrogation,[2] in their trust that some day or other the compassion of men of science might lead them to pause in their career of discovery, and take up the more generous task of instruction. But so far from this, the compilers of popular treatises have sought always to make them more saleable by bringing them up to the level of last month's scientific news; seizing also invariably, of such new matter, that which was either in itself most singular, or in its tendencies most contradictory of former suppositions and credences: and I purpose now to redeem, so far as I can, the enigmatical tone of my own books, by collecting the sum of the facts they contain, partly by indices, partly in abstracts, and so leaving what I myself have seen or known, distinctly told, for what use it may plainly serve.

19. For a first step in the fulfilment of this intention, some explanation of the circumstances under which the preceding lecture (on the serpent) was prepared, and of the reasons for its insertion in *Deucalion*, are due to the reader, who may have thought it either careless in its apparent jesting, or irrelevant in its position.

---

[1] [See above, p. 64 n.]

[2] [See, for instance, *Modern Painters*, vol. v. (Vol. VII. pp. 133 *seq.*), and in this volume, pp. 28, 41, 70: compare *Proserpina*, Vol. XXV. p. 335 ; and below, p. 386.]

I happened to be present at the lecture given on the same subject, a few weeks before, by Professor Huxley, in which the now accepted doctrine of development was partly used in support of the assertion that serpents were lizards which had lost their legs; and partly itself supported reciprocally, by the probability which the lecturer clearly showed to exist, of their being so.

Without denying this probability, or entering at all into the question of the links between the present generation of animal life and that preceding it, my own lecture was intended to exhibit another series, not of merely probable, but of observable, facts, in the relation of living animals to each other.

And in doing so, to define, more intelligibly than is usual among naturalists, the disputed idea of Species itself.

As I wrote down the several points to be insisted on, I found they would not admit of being gravely treated, unless at extreme cost of pains and time—not to say of weariness to my audience. Do what I would with them, the facts themselves were still superficially comic, or at least grotesque: and in the end I had to let them have their own way; so that the lecture accordingly became, apparently, rather a piece of badinage suggested by Professor Huxley's, than a serious complementary statement.

20. Nothing, however, could have been more seriously intended; and the entire lecture must be understood as a part, and a very important part, of the variously reiterated illustration, through all my writings, of the harmonies and intervals in the being of the existent animal creation— whether it be developed or undeveloped.

The nobly religious passion in which Linnæus writes the prefaces and summaries of the *Systema Naturæ*,[1] with the universal and serene philanthropy and sagacity of Humboldt, agree in leading them to the optimist conclusion, best, and unsurpassably, expressed for ever in Pope's *Essay*

---

[1] [See above, p. 339 n.]

on Man;[1] and with respect to lower creatures, epigrammatized in the four lines of George Herbert,—

> "God's creatures leap not, but express a feast
>   Where all the guests sit close, and nothing wants.
>   Frogs marry fish and flesh;—bats, bird and beast,
>   Sponges, non-sense and sense, mines,* th' earth and plants."[2]

And the thoughts and feelings of these, and all other good, wise, and happy men, about the world they live in, are summed in the 104th Psalm.

21. On the other hand, the thoughts of cruel, proud, envious, and unhappy men, of the Creation, always issue out of, and gather themselves into, the shambles or the charnel house: the word "shambles," as I use it, meaning primarily the battle-field, and secondly, every spot where any one rejoices in taking life;† and the "charnel house" meaning collectively, the Morgue, brothel, and vivisection-room.[3]

22. But, lastly, between these two classes, of the happy and the heartless, there is a mediate order of men both unhappy and compassionate, who have become aware of another form of existence in the world, and a domain of

---

* "Mines" mean crystallized minerals.

† Compare the Modern with the Ancient Mariner—gun versus cross-bow. —"A magnificent albatross was soaring about at a short distance astern, for some time in the afternoon, and was knocked over, but unfortunately not picked up" (*Natural History of the Strait of Magellan:* Edmonston and Douglas, 1871, page 225[4]).

---

[1] ["All nature is but art, unknown to thee;
     All chance, direction which thou canst not see;
     All discord, harmony not understood;
     All partial evil, universal good;
     And spite of pride, in erring reason's spite,
     One truth is clear, Whatever is, is right."

                                          Epistle I. 289.]

[2] [*The Temple* ("Providence," 133-135). The first two lines are quoted also in Vol. IV. p. 176, and Vol. IX. p. 307.]

[3] [See above, p. 179 *n.*]

[4] [*Notes on the Natural History of the Strait of Magellan and West Coast of Patagonia, made during the voyage of H.M.S. 'Nassau' in the years 1866-69,* by Robert O. Cunningham, M.D., naturalist to the expedition.]

zoology extremely difficult of vivisection,—the diabolic. These men, of whom Byron, Burns, Goethe, and Carlyle are in modern days the chief, do not at all feel that the Nature they have to deal with expresses a Feast only; or that her mysteries of good and evil are reducible to a quite visible Kosmos, as they stand; but that there is another Kosmos, mostly invisible, yet perhaps tangible, and to be felt if not seen.*

Without entering, with Dr. Reville of Rotterdam, upon the question how men of this inferior quality of intellect become possessed either of the idea—or substance—of what they are in the habit of calling "the Devil"; nor even into the more definite historical question, "how men lived who did seriously believe in the Devil"—(that is to say, every saint and sinner who received a decent education between the first and the seventeenth centuries of the Christian æra)—I will merely advise my own readers of one fact respecting the above-named writers, of whom, and whose minds, I know somewhat more than Dr. Reville of Rotterdam,—that *they*, at least, do not use the word "Devil" in any metaphorical, typical, or abstract sense, but—whether they believe or disbelieve in what they say—in a distinctly personal one:[1] and farther, that the conceptions or imaginations of these persons, or any other such persons, greater or less, yet of their species—whether they are a mere condition of diseased brains, or a perception of really existent external forces,—are nevertheless real *Visions*, described by them "from the life," as literally and straightforwardly as ever any artist of Rotterdam painted a sot—or his pot of beer: and farther—even were we at once to grant that all these visions—as for instance Zechariah's, "I saw the Lord sitting on His Throne, and Satan standing at His right

---

* *The Devil his Origin Greatness and Decadence* (*Sic*, without commas), Williams and Norgate, 1871.

---

[1] [On this subject, see Vol. XVII. p. 365, and Vol. XXII. p. 171; and on the element of true vision in imaginative work, Vol. IV. p. 222.]

hand to resist Him,"[1] *are* nothing more than emanations of the unphosphated nervous matter—still, these states of delirium are an essential part of human natural history: and the species of human Animal subject to them, with the peculiar characters of the phantoms which result from its diseases of the brain, are a much more curious and important subject of science than that which principally occupies the scientific mind of modern days—the species of vermin which are the product of peculiar diseases of the skin.

23. I state this, however, merely as a necessary Kosmic principle, without any intention of attempting henceforward to engage my readers in any department of Natural History which is outside of the ordinary range of Optics and Mechanics: but if they should turn back to passages of my earlier books which did so, it must always be understood that I am just as literal and simple in language as any of the writers above referred to: a d that, for instance, when in the first volume of *Deucalion*, p. 264, I say of the Mylodon—"This creature the Fiends delight to exhibit to you," I don't mean by "the Fiends" my good and kind geological friends at the British Museum, nor even the architect who made the drain-pipes from the posteriors of its gargoyles the principal shafts in his design for the front of the new building,[2]—be it far from me,—but I do mean, distinctly, Powers of supernatural Mischief, such as St. Dunstan, or St. Anthony, meant by the same expressions.

With which advice I must for the present end this bit of explanatory chapter, and proceed with some of the glacial investigations relating only to the Lakes—and not to the Inhabitants—whether of Coniston or Caina.[3]

---

[1] [Zechariah iii. 1.]

[2] [The Natural History Museum (a branch of the British Museum) at South Kensington, built (1873–1880) from the designs of A. Waterhouse, R.A.]

[3] [For the ice of Caina, see Dante's *Inferno*, xxxii.; and compare Vol. XVIII. p. 99.]

Bruno Artifex.

of crystalline understated depend on, what appears
we can call of the coprices, I of aggregation

55 . Take one of the simplest. Look at the back of a fallen
oakleaf, covered by the first snows — since I winter
you never saw the structure so displayed in the
living leaf. In exact is proportion to the frosseteen
of the ribs, the therefore shorts higher from them.
and shows them were in a finished trembling,
on a beautifully exaggerated scale — finest with
finest, but strong, and strong, then the strongest
scarcely magnify the finest — but use pounds
magnify the strongest.
But why should the needles shoot higher from the
ribs, than the interstices?
Answer — in the beginning I conceive the whole

A PAGE OF THE MS. OF "DEUCALION" (vol. ii, ch. iii.)

CHAPTER III

BRUMA ARTIFEX[1]

1. The frost of 9th March, 1879, suddenly recurrent and severe, after an almost Arctic winter, found the soil and rock of my little shaded hill garden, at Brantwood, chilled underneath far down; but at the surface, saturated through every cranny and pore with moisture, by masses of recently thawed snow.

The effect of the acutely recurrent frost on the surface of the gravel walks, under these conditions, was the tearing up of their surface as if by minutely and delicately explosive gases; leaving the heavier stones imbedded at the bottom of little pits fluted to their outline, and raising the earth round them in a thin shell or crust, sustained by miniature

---

[1] [In the MS. material (at Brantwood) for the second volume of *Deucalion* there is a note headed "Bruma Artifex" which deals with another phenomenon of frost:—

"Look at the back of a fallen oak-leaf, covered by the first sugary rime of winter. You never saw the structure as displayed in the living leaf. For exactly in proportion to the projection of the ribs, the hoar-frost shoots higher from them and draws their network now in a finished bas-relief, on a beautifully exaggerated scale — scarcely magnifying the finest—but very greatly magnifying the strongest.

"But why should the needles shoot higher from the ribs than the interstices?

"Again, in the beginning of a very pretty modern French novelette ('La Bûche,' the first part of the story of Sylvestre Bonnard), an old philosopher (who is amazingly like one I really know of) comforts himself at his fireside with the sight of the 'feuilles de fougère' of hoar-frost on his window-pane. The usual form of such hoar-frost is better described, by much, in its likeness to fern-leaves than by our common word, 'arborescent.' But a lovelier though rarer form is that of bending moss— every fibre of which will be exquisitely curved, while in the ordinary type it was rigidly straight. But what has been the change in the conditions?"

For another reference to Anatole France's *Le Crime de Sylvestre Bonnard* (published in 1881, "La Bûche" being the title of the first part of the novel), and for Ruskin's sympathy with the "old philosopher," see Vol. XVIII. p. lxxi. *n.*

The MS. of a first draft of the present chapter (which was ultimately rewritten) is in possession of Mr. F. W. Hilliard. The chapter is there called "*Pruina Arachne.*" Passages from that MS. are given at p. 202 *n.*]

ranges of basaltic pillars of ice, one range set above another, with level plates or films of earth between; each tier of pillars some half-inch to an inch in height, and the storied architecture of them two or three inches altogether; the little prismatic crystals of which each several tier was composed being sometimes knit into close masses with radiant silky lustre, and sometimes separated into tiny, but innumerable shafts, or needles, none more than the twentieth of an inch thick, and many terminating in needle-*points*, of extreme fineness.

2. The soft mould of the garden beds, and the crumbling earth in the banks of streams, were still more singularly divided. The separate clods,—often the separate *particles*, —were pushed up, or thrust asunder, by thread-like crystals, *contorted* in the most fantastic lines, and presenting every form usual in twisted and netted chalcedonies, except the definitely fluent or meltingly diffused conditions, here of course impossible in crystallizations owing their origin to acute and steady frost. The coils of these minute fibres were also more parallel in their swathes and sheaves than chalcedony; and more lustrous in their crystalline surfaces: those which did not sustain any of the lifted clods, usually terminating in fringes of needle-points, melting beneath the breath before they could be examined under the lens.

3. The extreme singularity of the whole structure lay, to my mind, in the fact that there was nowhere the least vestige of *stellar* crystallization. No resemblance could be traced,—no connection imagined,—between these coiled sheaves, or pillared aisles, and the ordinary shootings of radiant films along the surface of calmly freezing water, or the symmetrical arborescence of hoar-frost and snow. Here was an ice-structure wholly of the earth, earthy; requiring for its development, the weight, and for its stimulus, the interference, of clods or particles of earth. In some places, a small quantity of dust, with a large supply of subterranean moisture, had been enough to provoke the concretion of masses of serpentine filaments three or four inches

long; but where there was no dust, there were no filaments, and the ground, whether dry or moist, froze hard under the foot.

4. Greatly blaming myself for never having noticed this structure before, I have since observed it, with other modes of freezing shown in the streamlets of the best watered district of the British Islands,—with continually increasing interest: until nearly all the questions I have so long vainly asked myself and other people, respecting the *variable* formations of crystalline minerals, seem to me visibly answerable by the glittering, and softly by the voice, of even the least-thought-of mountain stream, as it relapses into its wintry quietness.

5. Thus, in the first place, the action of common opaque white quartz in filling veins, caused by settlement or desiccation, with transverse threads, imperfectly or tentatively crystalline (those traversing the soft slates of the Buet and Col d'Anterne are peculiarly characteristic, owing to the total absence of lustrous surface in the filaments, and the tortuous aggregation of their nearly solidified tiers or ranks), cannot but receive some new rays of light in aid of its future explanation, by comparison with the agency here put forth, before our eyes, in the early hours of a single frosty morning; agency almost measurable in force and progress, resulting in the steady elevation of pillars of ice, bearing up an earthy roof, with strength enough entirely to conquer its adherence to heavier stones imbedded in it.

6. Again. While in its first formation, lake or pool ice throws itself always, on calm water, into stellar or plumose films, shot in a few instants over large surfaces; or, in small pools, filling them with spongy reticulation as the water is exhausted, the final structure of its compact mass is an aggregation of vertical prisms, easily separable, when thick ice is slowly thawing: prisms neither formally divided, like those of basalt, nor in any part of their structure founded on the primitive hexagonal crystals of the ice; but starch-like, and irregularly acute-angled.

7. Icicles, and all other such accretions of ice formed by additions at the surface, by flowing or dropping water, are always, when unaffected by irregular changes of temperature or other disturbing accidents, composed of exquisitely transparent vitreous ice (the water of course being supposed transparent to begin with)—compact, flawless, absolutely smooth at the surface, and presenting on the fracture, to the naked eye, no evidence whatever of crystalline structure. They will enclose living leaves of holly, fern, or ivy, without disturbing one fold or fringe of them, in clear jelly (if one may use the word of anything frozen so hard), like the daintiest candyings by Parisian confectioner's art, over glacé fruit, or like the fixed juice of the white currant in the perfect confiture of Bar-le-Duc;[1]—and the frozen gelatine melts, as it forms, stealthily, serenely, showing no vestige of its crystalline power; pushing nowhere, pulling nowhere; revealing in dissolution, no secrets of its structure; affecting flexile branches and foliage only by its weight, and letting them rise when it has passed away, as they rise after being bowed under rain.

8. But ice, on the contrary, formed by an unfailing supply of running water over a rock surface, increases, not from above, but *from beneath*. The stream is never displaced by the ice, and forced to run over it, but the ice is always lifted by the stream; and the tiniest runlet of water keeps its own rippling way on the rock as long as the frost leaves it life to run with. In most cases, the tricklings which moisten large rock surfaces are supplied by deep under-drainage which no frost can reach; and then, the constant welling forth and wimpling down of the perennial rivulet, seen here and there under its ice, glittering in timed pulses, steadily, and with a strength according to the need, and practically infinite, heaves up the accumulated bulk of chalcedony it has formed, in masses a foot or a foot and a half thick, if the frost hold; but always more

---

[1] [Currant-jam and other confitures are still largely exported from Bar-le-Duc. The town was a favourite stopping-place with Ruskin : see Vol. VII. pp. xxvii.–xxviii.]

or less opaque in consequence of the action of the sun and
wind, and the superficial additions by adhering snow or
sleet; until the slowly nascent, silently uplifted, but otherwise motionless glaciers,—here taking casts of the crags, and
fitted into their finest crannies with more than sculptor's
care, and anon extended in rugged undulation over moss or
shale, cover the oozy slopes of our moorlands with *statues*
of cascades, where, even in the wildest floods of autumn,
cascade is not.

9. Actual waterfalls, when their body of water is great,
and much of it reduced to finely divided mist, build or
block themselves up, during a hard winter, with disappoint-
ingly ponderous and inelegant incrustations,—I regret to
say more like messes of dropped tallow than any work of
water-nymphs. But a small cascade, falling lightly, and
shattering itself only into *drops*, will always do beautiful
things, and often incomprehensible ones. After some fortnight or so of clear frost in one of our recent hard winters
at Coniston, a fall of about twenty-five feet in the stream of
Leathes-water, beginning with general glass basket-making
out of all the light grasses at its sides, built for itself at
last a complete veil or vault of finely interwoven ice, under
which it might be seen, when the embroidery was finished,
falling tranquilly: its strength being then too far subdued to
spoil by over-loading or over-labouring the poised traceries
of its incandescent canopy.

10. I suppose the component substance of this vault to
have been that of ordinary icicle, varied only in direction
by infinite accidents of impact in the flying spray. But
without including any such equivocal structures, we have
already counted five stages of ice familiar to us all, yet not
one of which has been accurately described, far less explained. Namely,

(1) Common deep-water surface ice, increased from beneath, and floating, but, except in the degrees of its own
expansion, not uplifted.

(2) Surface ice on pools of streams, *exhausting* the water

as it forms, and adherent to the stones at its edge. Variously increased in crusts and films of spongy network.

(3) Ice deposited by external flow or fall of water in superadded layers—exogen[1] ice,—on a small scale, vitreous, and perfectly compact, on a large one, coarsely stalagmitic, like impure carbonate of lime, but I *think* never visibly fibrous-radiant, as stalactitic lime is.

(4) Endogen ice, formed from beneath by tricklings over ground surface.

(5) Capillary ice, extant from pores in the ground itself, and carrying portions of it up with its crystals.

11. If to these five modes of slowly progressive formation, we add the swift and conclusive arrest of vapour or dew on a chilled surface, we shall have, in all, six different kinds of—terrestrial, it may be called as opposed to aerial —congelation of water : exclusive of all the atmospheric phenomena of snow, hail, and the aggregation of frozen or freezing particles of vapour in clouds. Inscrutable these, on our present terms of inquiry; but the six persistent conditions, formed before our eyes, may be examined with some chance of arriving at useful conclusions touching crystallization in general.

12. Of which, this universal principle is to be first understood by young people;—that every crystalline substance has a brick of a particular form to build with, usually, in some angle or modification of angle, quite the mineral's own special property,—and if not absolutely peculiar to it, at least peculiarly used by it. Thus, though the brick of gold, and that of the ruby-coloured oxide of copper, are alike cubes, yet gold grows trees with its bricks, and ruby copper weaves samite with them. Gold cannot plait samite, nor ruby copper branch into trees; and ruby itself, with a far more convenient and adaptable form of brick, does neither the one nor the other. But ice, which has the same form of bricks to build with as ruby, can, at its pleasure, bind

---

[1] [On the terms "exogen" and "endogen," compare *Proserpina*, Vol. XXV. p. 321.]

them into branches, or weave them into wool; buttress a polar cliff with adamant, or flush a dome of Alp with light lovelier than the ruby's.

13. You see, I have written above, "ruby," as I write "gold" or ice, not calling their separate crystals, rubies, or golds, or ices. For indeed the laws of structure hitherto ascertained by mineralogists have not shown us any essential difference between substances which crystallize habitually in symmetrical detached figures, seeming to be some favourite arrangement of the figures of their primary molecules; and those which, like ice, only under rare circumstances give clue to the forms of their true crystals, but habitually show themselves in accumulated mass, or complex and capricious involution. Of course the difference may be a question only of time; and the sea, cooled slowly enough, might build bergs of hexagonal ice-prisms as tall as Cleopatra's needle, and as broad as the tower of Windsor; but the time and temperature required, by any given mineral, for its successful constructions of form, are of course to be noted among the conditions of its history, and stated in the account of its qualities.

14. Neither, hitherto, has any sufficient distinction been made between properly crystalline and properly cleavage planes.* The first great laws of crystalline form are given by Miller[1] as equally affecting both; but the conditions of substance which have only so much crystalline quality as to break in directions fixed at given angles, are manifestly to be distinguished decisively from those which imply an effort in the substance to collect itself into a form terminated at symmetrical distances from a given centre. The distinction is practically asserted by the mineral itself, since it is seldom that any substance has a cleavage parallel to more than one or two of its planes: and it is forced farther on our notice by the ragged lustres of true cleavage

* See vol. i., chap. xiv., §§ 20–22.

---

[1] [See pp. 4 seq. of the *Elementary Introduction to Mineralogy* cited above, p. 272 n.]

planes like those of mica, opposed to the serene bloom of the crystalline surfaces formed by the edges of the folia.

15. Yet farther. The nature of cleavage planes in definitely crystalline minerals connects itself by imperceptible gradations with that of the surfaces produced by mechanical separation in their masses consolidating from fusion or solution. It is now thirty years, and more, since the question whether the forms of the gneissitic buttresses of Mont Blanc were owing to cleavage or stratification, became matter of debate between leading members of the Geological Society; and it remains to this day an undetermined one! In succeeding numbers of *Deucalion*,[1] I shall reproduce, according to my promise in the introduction, the chapters of *Modern Painters* which first put this question into clear form;[2] the drawings which had been previously given by de Saussure and other geologists having never been accurate enough to explain the niceties of rock structure to their readers, although, to their own eyes on this spot, the conditions of form had been perfectly clear. I see nothing to alter either in the text of these chapters, written during the years 1845 to 1850, or in the plates and diagrams by which they were illustrated; and hitherto, the course of geological discovery has given me, I regret to say, nothing to add to them: but the methods of microscopic research originated by Mr. Sorby,[3] cannot but issue, in the hands of the next de Saussure, in some trustworthy interpretation of the great phenomena of Alpine form.

16. I have just enough space left in this chapter to give some illustrations of the modes of crystalline increment which are not properly subjects of mathematical definition; but are variable, as in the case of the formations of ice above described, by accidents of situation, and by the modes and quantities of material supply.

---

[1] [No more numbers of *Deucalion* were, however, issued; but Ruskin began to fulfil his promise (compare p. 98) by the publication in 1884 of the reprints entitled *In Montibus Sanctis*.]

[2] [See Vol. VI. pp. 219 *seq.*]

[3] [See above, p. 207.]

17. More than a third of all known minerals crystallize in forms developed from original molecules which can be arranged in cubes and octahedrons; and it is the peculiarity of these minerals that whatever the size of their crystals, so far as they are perfect, they are of equal diameter in every direction; they may be square blocks or round balls, but do not become pillars or cylinders. A diamond, from which the crystalline figure familiar on our playing cards has taken its popular name, be it large or small, is still a diamond, in figure as well as in substance, and neither divides into a star, nor lengthens into a needle.

18. But the remaining two-thirds of mineral bodies resolve themselves into groups, which, under many distinctive conditions, have this in common,—that they consist essentially of *pillars* terminating in pyramids at both ends. A diamond of ordinary octahedric type may be roughly conceived as composed of two pyramids set base to base; and nearly all minerals belonging to other systems than the cubic, as composed of two pyramids with a tower between them. The pyramids may be four-sided, six-sided, eight-sided; the tower may be tall, or short, or, though rarely, altogether absent, leaving the crystal a diamond of its own sort; nevertheless, the primal separation of the double pyramid from the true tower with pyramid at both ends, will hold good for all practice, and to all sound intelligence.

19. Now, so long as it is the law for a mineral, that however large it may be, its form shall be the same, we have only crystallographic questions respecting the modes of its increase. But when it has the choice whether it will be tall or short, stout or slender, and also whether it will grow at one end or the other, a number of very curious conditions present themselves, unconnected with crystallography proper, but bearing much on the formation and aspect of rocks.

20. Let *a*, Fig. 1, Plate XX., be the section of a crystal formed by a square tower one-third higher than it

is broad, and having a pyramid at each end half as high as
it is broad. Such a form is the simplest general type of
average crystalline dimension, not cubic, that we can take
to start with.

Now if, as at *b*, we suppose the crystal to be enlarged
by the addition of equal thickness or depth of material on
all its surfaces,—in the figure its own thickness is added
to each side,—as the process goes on, the crystal will gradually lose its elongated shape, and approximate more and
more to that of a regular hexagon. If it is to retain its
primary shape, the additions to its substance must be made
on the diagonal lines dotted across the angles, as at *c*, and
be always more at the ends than at the flanks. But it
may chance to determine the additions wholly otherwise,
and to enlarge, as at *d*, on the flanks instead of the points;
or, as at *e*, losing all relation to the original form, prolong
itself at the extremities, giving little, or perhaps nothing,
to its sides. Or, lastly, it may alter the axis of growth
altogether, and build obliquely, as at *f*, on one or more
planes in opposite directions.

21. All the effective structure and aspect of crystalline
substances depend on these caprices of their aggregation.
The crystal of amethyst of which a longitudinal section is
given in Plate XX., Fig. 2, is more visibly (by help of
its amethyst staining), but not more frequently or curiously,
modified by accident than any common prism of rough
quartz will be usually found on close examination; but in
this example, the various humours, advances, and pauses of
the stone are all traced for us by its varying blush; and
it is seen to have raised itself in successive layers above
the original pyramid—always thin at the sides, and oblique
at the summit, and apparently endeavouring to educate
the rectilinear impulses of its being into compliance with a
beautiful imaginary curve.

22. Of prisms more successful in this effort, and constructed finally with smoothly curved sides, as symmetrical
in their entasis as a Greek pillar, it is easy to find

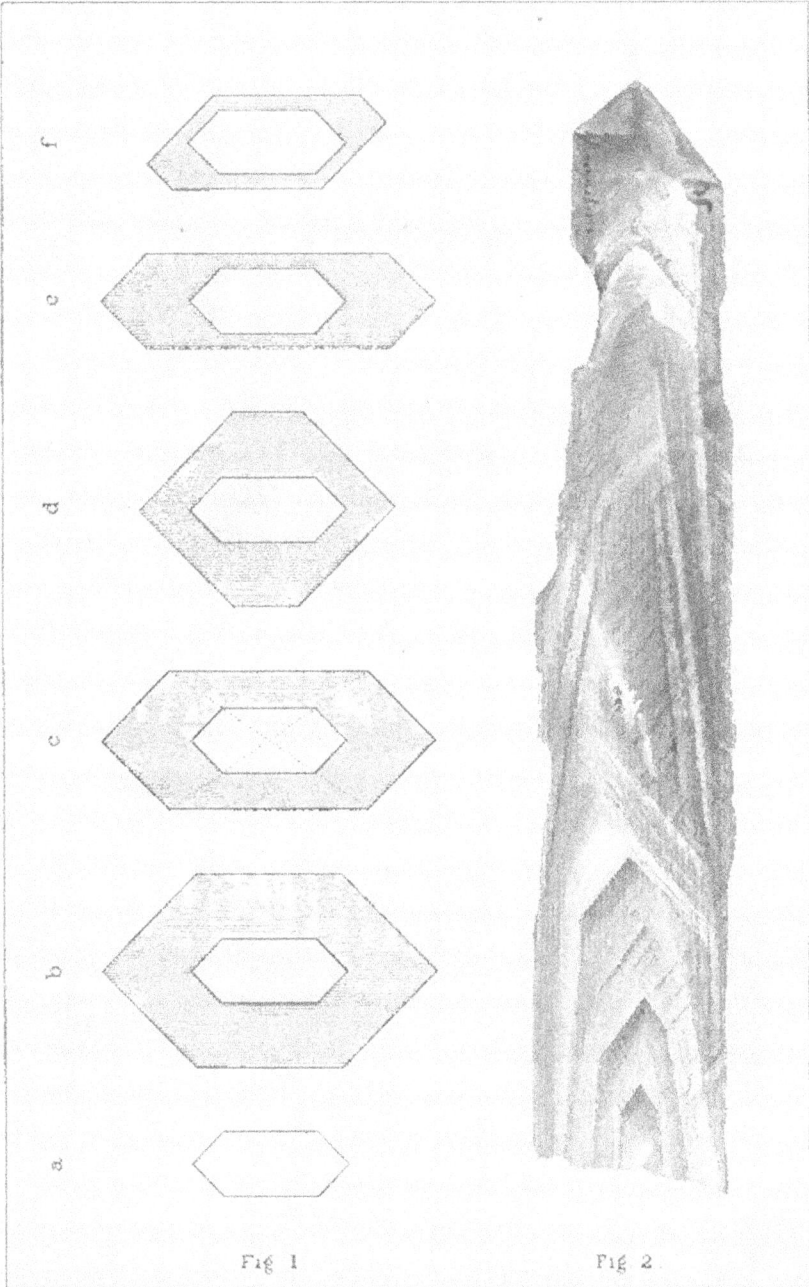

XX

Fig 1

Fig 2

MODES OF CRYSTALLINE INCREMENT

examples in opaque quartz—(not in transparent*)—but no quartz crystal ever *bends* the vertical axis as it grows, if the prismatic structure is complete; while yet in the imperfect and fibrous state above spoken of, § 5, and mixed with clay in the flammeate forms of jasper, undulation becomes a law of its being!

23. These habits, faculties, and disabilities of common quartz are of peculiar interest when compared with the totally different nature and disposition of ice, though belonging to the same crystalline system. The rigidly and limitedly mathematical mind of Cloizeaux passes without notice the mystery, and the marvel, implied in his own brief statement of its elementary form " Prisme hexagonal *regulier*."[1] Why "regular"? All crystals belonging to the hexagonal system are necessarily regular, in the equality of their angles. But ice is regular also in *dimensions*. A prism of quartz or calcite may be of the form

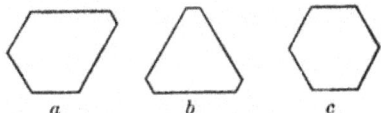

Fig. 42

a on the section, Fig. 42,† or of the form *b*; but ice is always true—like *c*, as a bee's cell—" prisme regulier."

So again, Cloizeaux tells us that ice habitually is formed in "tables hexagonales *minces*." But why thin?—and *how* thin? What proportion of surface to edge was in his mind as he wrote, undefined? The square plates of uranite, the hexagonal folia of mica, are "minces" in a quite different sense. They can be seen separately, or in masses which are distinctly separable. But the "prisme hexagonale mince,

---

* Smoky quartz, or even Cairngorm, will sometimes curve the sides parallel to the axis, but (I think) pure white quartz never.

† I think it best to number my woodcuts consecutively through the whole work, as the plates also; but Fig. 37 is a long way back,[2] p. 231, vol. i. Some further notes on it will be found in the next chapter.[3]

---

[1] [*Manuel de Minéralogie*, by A. Des Cloizeaux, vol. i. p. 7.]

[2] [In the original edition "Fig. 42" was "Fig. 6," and "Fig. 37" "Fig. 5," Figs. 38–41 being added in this edition.]

[3] [No other chapter was published, however, nor does the MS. material contain the notes here promised.]

regulier" of ice cannot be split into thinner plates—cannot
be built into longer prisms; but, as we have seen, when it
builds, is fantastic in direction, sudden in force, endlessly
complex in form.

24. Here, for instance, Fig. 43, is the outline of one of
the spicula of incipient surface ice, formed by sharp frost
on calm water already cooled to the freezing-point. I have
seen literally clouds of surface ice woven of these barbed
arrows, shot,—or breathed, across half a mile of lake in
ten minutes. And every barb of them *itself* a miracle of
structure, complex as an Alpine peak.

These spicula float with their barbs downwards, like
keels, and form guiding ribs above like those of leaves,

Fig. 43

between which the entire surface of the water becomes
laminated; but, as it does so, the spicula get pushed up
into little mountain ridges, always steeper on one side than
the other—barbed on the steep side, laminated on the other
—and radiating more or less trigonally from little central
cones, which are raised above the water-surface with hollow
spaces underneath.

And it is all done with "prismes hexagonales reguliers"!

25. Done,—and sufficiently explained, in Professor Tyndall's imagination, by the poetical conception of "six poles"
for every hexagon of ice.* Perhaps!—if one knew first

---

* *Forms of Water,* in the chapter on snow. The discovery is announced,
with much self-applause, as an important step in science.[1]

---

[1] [See Chapters IX. and X. in *Forms of Water.* The discussion is thus concluded
(§ 95): "Our first notions and conceptions of poles are obtained from the sight
of our eyes in looking at the effects of magnetism, and we then transfer these
notions and conceptions to particles which no eye has ever seen. The power by
which we thus picture to ourselves effects beyond the range of the senses is what
philosophers call the Imagination, and in the effort of the mind to seize upon the
unseen architecture of crystals, we have an example of the 'scientific use' of this
faculty. Without imagination we might have *critical* power, but not *creative* power
in science."]

XXI

THE OLYMPIAN LIGHTNING

what a pole was, itself—and how many, attractive, or repulsive, to the east and to the west, as well as to the north and the south—one might institute in imaginative science—at one's pleasure;—thus also allowing a rose five poles for its five petals, and a wallflower four for its four, and a lily three, and a hawkweed thirteen. In the meantime, we will return to the safer guidance of primal mythology.

26. The opposite plate (XXI.) has been both drawn and engraved, with very happy success, from a small Greek coin, a drachma of Elis,[1] by my good publisher's son, Hugh Allen. It is the best example I know of the Greek type of lightning, grasped or gathered in the hand of Zeus. In ordinary coins or gems, it is composed merely of three flames or forked rays, alike at both extremities. But in this Eleian thunderbolt, when the letters F.A. (the old form of beginning the name of the Eleian nation with the digamma) are placed upright, the higher extremity of the thunderbolt is seen to be twisted, in sign of the whirlwind of electric storm, while its lower extremity divides into three symmetrical lobes, like those of a flower, with spiral tendrils from the lateral points: as constantly the honeysuckle ornament on vases, and the other double groups of volute completed in the Ionic capital, and passing through minor forms into the earliest recognizable types of the fleur-de-lys.

27. The intention of the twisted rays to express the action of storm is not questionable—"tres imbris torti radios, et alitis austri."[2] But there can also be little doubt that the tranquillities of line in the lower divisions of the symbol are intended to express the vital and formative power of electricity in its terrestrial currents. If my readers will refer to the chapter in *Proserpina* on the roots of plants,[3]

---

[1] [An electrotype of the coin is II. B. 33 in the exhibition at the British Museum.]

[2] [Virgil: *Æneid*, viii. 430 ; quoted also in the lecture on "The Eagle of Elis" (Vol. XX. p. 400).]

[3] [See Vol. XXV. p. 223.]

they will find reasons suggested for concluding that the root is not merely a channel of material nourishment to the plant, but has a vital influence by mere contact with the earth, which the Greek probably thought of as depending on the conveyance of terrestrial electricity. We know, to this day, little more of the great functions of this distributed fire than he: nor how much, while we subdue or pervert it to our vulgar uses, we are in every beat of the heart and glance of the eye, dependent, with the herb of the field and the crystal of the hills, on the aid of its everlasting force. If less than this was implied by the Olympian art of olden time, we have at least, since, learned enough to read for ourselves his symbol, into the higher faith, that, in the hand of the Father of heaven, the lightning is not for destruction only; but glows, with a deeper strength than the sun's heat or the stars' light, through all the forms of matter, to purify them, to direct, and to save.

NOTES FOR
THE INTENDED CONTINUATION OF
"DEUCALION"

1. BRUMA INERS

2. THE ENIGMA OF DENUDATION

3. THE FORMATION OF MALHAM COVE

4. MOUNTAIN FORM IN THE ALPS

NOTES FOR THE INTENDED CONTINUATION
OF "DEUCALION"

1. BRUMA INERS[1]

1. SEVEN days before the frost described in last chapter,[2] on Sunday, the 2nd March 1879, a wild south-west wind was blowing, while the ice on Coniston Water was still unbroken at the Waterhead, except round the edges at the shore, and formed a united field about a quarter of a mile in length, and nearly as much in main width, varying in thickness—though it had been wasted by nearly a fortnight's thaw—from two inches, on the average, to an inch and a half, or less, near the sheltered shore, and to three or four inches in places where it had first frozen stoutly.*

Against this field of ice, practically fixed (though not attached to the shore, yet jammed against it), came the full force of the waves from the south. Four miles and a half of open water due south enabled the wind to raise rollers at the northern extremity of the lake, which, where they came against the ice, yet in deep water still unbroken, were about five paces apart from crest to crest, and two to three feet vertical in rise from trough to ridge; that is to say, very sufficiently formidable waves even for sea-surface, and driven before a gale which no small boat could have been long rowed against.

2. They came full against the ice at its southern edge; and I ask the reader, first, to guess what would under such circumstances happen. I had vainly, myself, tried to think it out, before the frost broke it out, feeling more and more the helplessness of thought *versus* sight; and when the frost did, day by day of absolute calm, with grey cloud and drizzling rain, effected ignoble dissolution of all but the obstinate ice-field at the Waterhead. Only at the very final moment came the long-wished-for gale—but in time to show me much of what I wanted. I had expected the stormy water to break at, and over the edge of the fixed ice, as against that of a large raft, with infinite shattering and upthrowing of the immediate border opposed to it; and though I have been a good deal surprised in my life by things turning out other than I expected, I never was more so than when, first getting down to the Waterhead beach, I saw the whole field of ice, two inches thick, following the forms of the waves as quietly and softly as if it had been a tapestry of white silk, and the great rollers declining under it in gently concentric modulation of curve to the farthest shore.

---

* During the previous frost it will be remembered that the lake had been frozen entirely across, forming a perfectly safe field for traverse from shore to shore (the ice being at least eighteen inches thick) for a distance of three miles from the head of the lake. There was some unsafe ice, and, I think, even to the last, a patch or two of open water near the foot, where the stream enters from Goat's Water.

[1] [From a printed proof, corrected by Ruskin. This is the beginning of a chapter intended to follow the last.]

[2] [See above, p. 347.]

3. At first I could not believe the ice was indeed unbroken, but thought it must all have been divided into fine jointings. That question was soon settled. I found it was not only in the broad field unbroken, but still so sound that I could not with all my strength break a piece of it by direct force; but the transition to this continuity from the really shattered edge was effected through three stages. Where the waves first struck it, for about two fathoms breadth, the ice was entirely broken up, and fell into a close-packed crowd of fragments only from three or four inches to a foot square, clashing against and shattering each other into smaller pieces, or forced up upon the surface of the yet undivided ice, with a steady hissing sound, as of breakers on a distant beach. Once or twice during the preceding thaw, under a lighter breeze, the ice edge—in this same comminuted state, looking at a distance like a sheet of foam—was driven against the quite fixed field beyond it * with a musical ring, as of broken glass, tinkling sometimes far into the distance like the bells of an infinite flock of sheep, or melting between the two shores into a vague and indescribable murmur, mixed with the sound of the wind.

4. Under the stronger storm, beyond this two or three fathoms width of floating fragments, was a belt of ice, broken like a dissecting map by complex fissures, into pieces three or four feet wide, but not disconnected. This part of the frozen surface of course took the swells of the rollers with ease, its fissures allowing a certain jointed play of the pieces on each other, grinding slightly at their edges; but the strange thing was that no jets or gushes of water whatever took place at the fissures, nor could their jointed action be traced by the eye. Absolutely and intimately flexible, they took the forms of the waves without the slightest appearance of strain, merely repressing their minor ripples; but fitting themselves easily to the run of the cross surge, so that the ice in this part of it rose and fell in variously divided hills and valleys, just as the rollers themselves were divided in the open water.

5. But, within and beyond this belt—some twenty or thirty yards broad of jointed or dissected ice—the visible fissures entirely disappeared, and practically ceased to take place. The ice was as perfectly united as on calm water, and was able to reduce the already checked strength of the rollers into long and consistent swells, declining gradually in their sequent pulses till, at the extremity of the Waterhead bay, the gentle undulation of the surface could scarcely be traced in profile from the shore.

6. Through the whole of this ice-field, therefore, to its jointed edge there was shown a perfectly easy and harmonious flexibility in all directions, within certain limits, the fractures taking place when these limits were exceeded. The possible degree of this curvature depends on the precise nature of the substance of the thawing ice—question which I will not enter

* I had like to have passed the night on the lake in my first investigation of this comminuted ice-band, for having pulled the *Jumping Jenny*[1] through it, right up to the solid ice, I could not for some time get her out again, the wind taking the little blunt-headed boat hard against it, and the oars refusing to break through the wedged fragments, for their stroke—while yet once a foot or two back from the solid ice there remained no firm edge to thrust again.

[1] [A boat built for Ruskin, from his own design, by Mr. Bell (see above, p. 309), and named after Nanty Ewart's brig in *Redgauntlet* (chaps. xii. *seq.*).]

upon, till I have had an opportunity of seeing strong ice broken up by storm in frost, or at least before thaw has much reduced it.   The actual mass of the ice-field in the present instance was composed of vertical prisms like starch, and the play of these on each other might be conceived as enough to admit of all the flexibility of the united tissue required under the above conditions.[1]

2. THE ENIGMA OF DENUDATION[2]

2. All geologists are using the term in wild confusion of thought, never attempting to define it, even in their own minds.   They will find it needful to define both word and thought, and that clearly in these following respects.

(A.) There is a certain modification of the surface of dry land, produced by changes of temperature and by decomposition, which goes on during periods when no rain falls.   The removal of the produced sand or mould can under such conditions be effected by gravity only, or by the wind, which, if it cause the denudation of one spot, will cause the investiture of another, and build a drift for every rock it bares.   This operation must be described distinctly, and the forms of terrestrial surface which it produces made clearly known, in their localities.

(B.) The same operation must be described, with the additional action of rain and streamlets, in countries without watershed.

(C.) The same operation must be described, with the additional action of rain and streamlets, in countries that *have* watersheds.

3. After these three definitions and descriptions have been rightly given, will come the question, Why some countries have watershed and others not ? and why it is that rivers so rarely branch so that one could connect itself with another.

And on this question, the analyses of the form of river-beds given for the first time by Mr. Alfred Tylor,[3] and the analyses of the curves of hill surface from the summit to the base produced by pure and simple denudation on a mass of homogeneous ground, given also first by Mr. Tylor with mathematical accuracy, must be clearly learned by the student before he proceeds a step further in geological inquiry.

But the next step is a vast one.

4. Denudation by presently existing causes, of homogeneous ground, will produce forms capable of analysis under Mr. Tylor's laws.   But denudation of ground with hard and soft places in it will develop the knots and remove the soft rock, and yield, the longer it is continued, more clear sight of the earth's interior structure.

Of *this* action, which is the entirely vital and notable one, for all purposes of geologic investigation, I find scarcely any notice — certainly no accurately descriptive notice—taken by geologists at all! and therefore I

---

[1] [Here the proof-sheets end.]

[2] [Printed from the author's MS. at Brantwood ; headed by him as being intended for "*Deucalion*, vol. ii.," "place of chapter not yet fixed."   The MS. begins with § 2. Section 1 would apparently have consisted of passages from *Modern Painters*, a note on the top of the first sheet being, "Of Torrents correct p. 269, *M. P.*—cleavings, quote 405."   The references are to the original edition of vol. iv.   See now Vol. VI. p. 320 (where Ruskin writes of "torrent sculpture"), and p. 479 (on Rock Cleavage).]

[3] [See above, p. 290.]

here once for all put the questions before the young and earnest student in clear form for him.

5. He will find in any soft sea cliff, or at the curves of any rapid hill stream, entirely characteristic instances of cliffs produced by erosion. And he cannot walk half-a-dozen miles in any district of moderately hard rock without coming to some piece of cliff which has not been produced by erosion, but by disclosure.

Note the opposition in the two words. An eroded or "gnawed out" cliff is, typically, a hollow place dug out of a bank; a disclosed cliff is a hard knob projecting from a bank. And where such knob or knot occurs there must be cause for it in the structure or hardness of the beds—cause of which the discovery will be real and useful geology.

6. And thus, broadly, the cliffs and crags of any raised district divide themselves into those which in the process of their formation are moved back, and those which in process of formation are never moved at all, but stand in stronger and stronger development from the ground that is removed around them.

And of these stable cliffs one vast class is formed at places where some former agency has torn, and perhaps dislocated, the beds vertically, with conditions of cleavage wholly peculiar to such places.

7. All these conditions of denudation must be first described, as they may be traced in actual operation under existing influences. But the passages, to which the reader will find reference in the index[1] under the head "Sculpture of Mountains," have shown, and without any possibility of dispute,* not only that the existing forms of rocks were developed under different powers from any now in operation, but that those which are now active undo instead of continuing the work of the past, filling the former valley and abasing the former crest.

8. To trace the course, and prove the action, of such past, and now wholly ceased, agency, is still within the range of the student's wise effort, so long as he can reason concerning the masses of rocks now rising around him; can measure the bulk of débris fallen from the visible cliff, though under processes of decay long since ended; and sound the depth of the waters he can still launch upon, though their abysses were cloven by forces of which he has no present conception or example. But with the *visibility* of his subject terminates the scope of the student's useful inquiry. What Savoy was like before Mont Blanc lighted its clouds, and England like

---

* Let it also in justice be remembered that this demonstration was begun while the advocates of continuous and similar natural force, forming the school of Sir Charles Lyell, were yet in the first enthusiasm of their theory. The subject was first touched in the 96th page of the fourth volume of *Modern Painters*. I quote here the opening sentence, in itself conclusive of more than I then knew, or even guessed :—

"Only give a river some little sudden power in a valley, and see how it will use it. Cut itself a bed? Not so, by any means, but fill up its bed, and look for another, in a wild, dissatisfied, inconsistent manner. Any way, rather than the old one, will better please it; and even if it is banked up and forced to keep to the old one, it will not deepen, but do all it can to raise it, and leap out of it."[2]

---

[1] [See below, p. 587; under "Mountains."]

[2] [Vol. VI. p. 122. See there the note of 1885, where Ruskin rejoices in having thus early perceived a fact to which, in his later geological studies, he attached very great importance.]

under the seas which laid the sand of Ingleborough, advanced geologists may advise in their own council chambers ; but they only violate the modesties and sully the honour of science in proclaiming their imperfect thought to the careless curiosity of the vulgar and the easily persuaded imaginations of the young.

9. To take for their subject of study a single square inch of mountain ground, to describe it, draw it completely and with precision, and separate by unmistakable barrier the known concerning it from the unknown, will do more to advance, not merely the science of geology, but the entire capacity for any sort of science in their own and their pupils' minds, than to announce the most splendid imaginations, even supposing them (and how much is granted in such supposition!) afterwards to prove in any extended measure true concerning any number of immeasurable worlds through any cycle of innumerable ages.

10. Here is my good and kind friend, Mr. Clifton Ward, for instance, a man making the most brilliant and constant discoveries respecting the structure of rocks, yet he cannot yet be content with the mass of his sound work, but must needs * " wish to make an experiment," namely, " whether it be possible to draw up a Chronology of the Lake District to place beside the ascertained facts of History !" though at the same time he is "sensible how uncertain are many of the data." And so we launch into the extremely profitable and practical inquiry, whether "we may assume perhaps that the 10,000 feet of Skiddaw Slates were laid down at the rate of $\frac{1}{30}$th of an inch per annum, and took six million years to lay down," or whether we should "take the rate at $\frac{1}{10}$th instead of $\frac{1}{30}$th, and conclude that they only took $1\frac{1}{2}$ million."

And all this time, much as he has learned, my friend cannot tell me how long it took to crystallize one of the little garnets in the Borrowdale Greenstone, nor even the mode of that process,† nor can he tell me so much as how one of the little agates of Wallacrag was inserted in the crag at all; and after I have spent I don't know how much good time over the section in the Geological Survey of Maiden Moor (page 257, above), and got at no result from it but puzzlement, he serenely writes to me that such sections of contorted strata are merely conventional signs of the general facts, and are necessarily done to order by the assistants in any large scheme of survey.

Now I must very solemnly assure my friend, and all energetic and able geologists now at work, that they will have to stop the practice of letting assistants do things to any kind of order, except that of their own eyes and wits ; and that if they only get over a fathom of ground a day, they must give irrefragable account of the same, in what they say of it, and say nothing of any inch of it that they have not seen.

So, in time, they will fathom out some few of the ways of the world and wills of its Maker, but not otherwise.

---

* I have opened his last pamphlet on the geology of the Lake District at random—at page 19 !

† Compare Chap. IX. § 11 [p. 210].[1]

---

[1] [On the Physical History of the English Lake District, with Notes on the Possible Subdivision of the Skiddaw Slates, by the Rev. J. Clifton Ward, 1879 : a pamphlet of 28 pp., extracted from the Geological Magazine, February and March 1879.]

3. THE FORMATION OF MALHAM COVE[1]

I find I must let the glaciers go on at their own quiet pace, for a while, having been set upon some other work which I may not be able to return to.

My challenge, to all geologists, for explanation of the cliff at Fluelen,[2] includes two questions. That cliff is made of a coiled limestone, especially violent and varied in contortion. And it is cut splendidly steep, by powers which, as far as I am aware, no human being has ever attempted to explain.

I put the debate as to the contortion of beds aside for the time, and take up only the question of cutting, and with that I am minded to proceed, because I have been drawing ragged-robin leaves in Malham Cove, under a limestone cliff, which "*if* not," as the old Malham guide modestly admits may be indeed found on investigation, "the highest in the known world," is yet a very stalwart precipice of some, as I guess, three hundred feet high,

---

[1] [This MS. fragment is headed "*Deucalion*, No. 3. Bolton Abbey, 25th Sept., '75." The following letter (Settle, September 11), written a fortnight earlier to Mr. Tylor, F.G.S., shows that at first Ruskin was inclined to admit to the river the action which, in this latter passage, he seems to have intended to refute :—

"I have read with extreme interest the greater part of your paper in the *Geological Magazine*, and found some things in it quite new to me and of extreme value. The suggestion of the formation of cliffs by subterranean streams, is, I think, quite conclusively true in some rocks and localities ; and it happened by most grotesque and opportune Fors that I read it yesterday in the carriage returning from the foot of the cliff at Malham Cove, over the production of which I had been puzzling myself the whole afternoon, and which I felt convinced, the moment I read that passage in your paper, had been produced exactly as you say by the River Aire, which runs out of its foot, thus [rough sketch] (the stamp on my paper represents charmingly the weathering by the rain on the upper parts of the cliffs), out of a low chink rather than cave into a gorge, softly declining to the plains.

"The cliff is about 350 to 400 feet high, and this scene with that of the Peak Cavern are the best, I believe, in England to prove the truth of this theory of yours, in a great extent, and perhaps a much greater one than I am yet prepared to admit. Your experiments on ice, and your measures of rivers, are also of extreme value. But, my dear Mr. Tylor, why *will* you mix up talk of things which you have not investigated, with talk of what you have? Keep to your own knowledge and your own conclusions from it, and your papers would soon be the only authorities on pluvial action. But you flounder into a mere ash-hole of half-burnt cinder of theory when you meddle with Charpentier and Agassiz. Agassiz was not a great man, but a merely industrious counter of fish-bones and fish-scales. Charpentier's theory has long—so far as I have heard from chemists of authority—been known to be false, and for my own part, having read the whole book with extreme care (which I don't believe *you* have), I entirely reject it as ridiculous.

"It gives me extreme pleasure to see that Arthur Severn can assist you in your work ; and above all things I'm glad you have taken up Saddleback. It is a great escarpment, and I believe entirely owing to denudation. The difference of its form from that of Skiddaw would furnish you with quite a mine of study in action of pluvial and torrent power. You have not yet at all enough studied dry decomposition and the action of frost on surfaces unaffected by flowing water, and from which rain does but wash down their natural ruin."

For Mr. Tylor's paper, see above, p. 290 *n*.]

[2] [See above, p. 155.]

and cut here and there, even into overhanging brows and cornices, under which the face of it falls with a directness very unusual to the grass at its foot, except only that a terrace or gallery three or four feet wide projects towards its base. This precipice bends round into a hollow "cove" or sweep of interior Coliseum wall, terminating a valley which runs up to it from the lowlands, just as a railway cutting enters the piece of high ground which leads to a tunnel; and at the foot of the cliff, where in the railway cutting the tunnel would be, wells out of a low horizontal cleft, not a foot above the water in any place, the innocent little River Aire. This merely explanatory figure will give you a rough idea of the arrangement [rough sketch]. What can be more evident, say the geologists, than that the River Aire is responsible for the whole arrangement; that it has excavated the valley, dissolved the tunnel, cut the cliff, and will in time cut away all York-shire; and to begin with, at least, a large part of my friend Mr. Morrison's estate.

[Here the fragment breaks off.]

4. MOUNTAIN FORM IN THE ALPS [1]

Ice, if it be deep, may have tremendous force by its, not hydrostatic, but—if I allow myself a fine word for once—crystallostatic pressure.

And therefore ice ten thousand feet deep, if once you can set it a-running, may for aught I know do great things. But how are you to set it a-running? That it may move at all, you must have a hill for it to come down, and a hole for it to flow into. What I want to know is, first of all, how you get your hill and your hole. That's the gist of the matter.

But, secondly, I know nothing about ice ten thousand feet deep; I've never seen ice, to my knowledge, so much as ten hundred feet deep, and not often more than two hundred. Nor do I know anything about the way that the hills and the holes were made for the ice to run about, two hundred thousand years ago. But I know the hills and holes it runs about now, and I know they are made *for* the ice, and not by it.

And that, for instance, here at Chamouni, the ravines in the Montagne de la Côte are soft in outline, not because glaciers have rounded them, but because their rock is soft to the weathering. And here, at Carrara,[2] the ravines are sharp edged, not because glaciers have not rounded them, but because the rock is hard to the weathering.

And, broadly, in all cases now, hills receive their shape by the action of the weather on their sides, of which the débris is *washed* down by rain, or CARRIED down by ice, and is more or less removed and spread below by the stream or glacier, which therefore appears indeed to be gnawing them away, but is only sweeping down their ruins, itself checked and encumbered in its course annually more and more, and rising in its bed on the choked bottom of the ravine, which it is filling with a plain of gravel, and effacing the ancient furrow below, as the rain and the frost efface the ancient eminence above.

---

[1] [This fragment is among the MSS. which are headed by Ruskin "Deucalion Material." The passages form part of the lectures which he delivered at Oxford and London : see above, p. 89.]

[2] [Here Ruskin showed his drawings of " Crest of La Côte," engraved in *Modern Painters*, vol. iv. (Vol. VI. pp. 259, 260) ; and the drawing, " At Carrara," which is given in Vol. II. p. 208, and which he described as an illustration of " Ravines cut by weather and wavy structure in hard rock."]

You may still in places find a chasm here, still in places find a splinter there, petty processes of excavation, tiny emancipations of precipice, but on the whole denudation means ultimate loss of form.

> [Here the MS. takes the form of notes, and the passage seems to have been worked up into p. 149 above, the notes being to the effect that students are too much given to "looking at things with which you have no business—abroad instead of home—Niagara (compare p. 254), instead of the good old High Force of Tees."]

Now, the force which cut that ravine, whatever it was, was also the force which let out the high waters all over the Alps, and had no power only to touch the pools at their bases.

At first sight you imagine it is a force going on at the present day. It is nothing of the sort. No existing power can account for the cutting of a ravine like the Gorge of Trient, or of Gondo, or of this, at Faido on the St. Gothard.

The force which cut these ravines gave their utmost sublimity to the hill forms, shaped these peaks, hewed out this overwhelming precipice.

All forces now at work are subduing the forms so produced—rounding off these peaks ; breaking down this precipice ; filling up the hollow between with gravel ; filling up the lakes with silt ; subduing, melting, effacing ; never sculpturing.

You see what a vital, what a mighty point of separation this is between the two powers. The mountain block is given, as the marble to Michael Angelo, to the first gigantic forces which develop form. They sculpture its features, they ordain the strength and shape the contour of its limbs, they leave it in utmost majesty of mountain birth. With the existence of man begins, so far as we can discern, on the hills of his nativity, the weakness of his disease and the terror of his death.

Thenceforward they receive no more any lordliness of height, no more any perfectness of form ; they also with us groan and travail, with us are abased. Cliff totters, crag moulders, knoll and meadow crumble to dust ; the frost gnaws them, the floods rave at them, their own ruins encumber, and, even beneath the domes of their once called eternal chain, their flowers and forests are failing, and drought and heat consume the snow waters as the grave those which have sinned.

And yet by subtle changes the school of geologists, to which I have already referred, will tell you that snows, or rains, or ice have carved upon it no subtle or slight change, but all the main features of its contours. They imagine that this valley, now filled by a glacier, has been cut by the flux of the glacier that fills it, and that the ridgy buttress has been wasted to its edge by the fury of the clouds which it gathers. I think I can show you the fallacy of their conceptions — show you that both valley and ridge were created under conditions of which nothing is known. But since the world was fit for human habitation they have stood in their appointed—that is, in their existing—form, the lapse of their ice and the drift of their tempests pass over them as though they knew it not, and the utmost changes wrought, to our knowledge, upon their being are only as to some heroic human frame, the wasting of the limb and the furrowing of the brow, which scarcely diminish the majesty of the body, though they may be the signs of its doom to dissolution.

V

OF THE DISTINCTIONS OF FORM
IN SILICA

(1884)

[*Bibliographical Note.*—This paper, contributed by Ruskin, was read at Edinburgh (Professor Geikie being in the chair) by the Rev. W. W. Peyton, Local Secretary for Scotland, on July 24, 1884. The following abstract of it appeared in the *Mineralogical Magazine and Journal of the Mineralogical Society,* vol. vi., 1886, pp. i.–ii. :—

"In this paper the attention of mineralogists was directed to some points in the crystallization of silica, and to the relations of colour to the state of substance of minerals, which the author considered had been hitherto entirely neglected. He asked why 'large and well-developed quartz crystals . . . in proportion to its purity' [§ 4].

"Referring to certain structures in the specimens of agates exhibited by him, the author affirmed that all the types of undulated structure were due to crystallization only, and expressed his opinion that the contortions of the gneiss of the Alps are not due to wrinkling by pressure, but to crystallization. He recognizes six distinct states of siliceous substance, viz. flint, jasper, chalcedony, hyalite, opal, and quartz."

A long report of the paper appeared in the *Edinburgh Courant,* June 25, 1884.

The paper was published by Ruskin in October 1884, as Chapter I. in Part I. of *In Montibus Sanctis* (see Bibliographical Note in Vol. III. p. lxii.). The Preface to *In Montibus Sanctis,* which occupied pp. iii.–vii., has already been printed in Vol. III. pp. 678, 679. The present paper ("Chapter I.") occupied pp. 1–25; the Postscript, which follows (here pp. 386–391), occupied pp. 26–40. In this volume the paragraphs are numbered. In § 28, line 9, "phenomena" is here a correction for "phenomenon."

The catalogue of minerals which Ruskin sent in illustration of the paper is printed below, pp. 520–526.]

OF THE DISTINCTIONS OF FORM
IN SILICA

(*Read before the Mineralogical Society, July 24th, 1884*)

1. As this paper, by the courtesy of the secretaries, stands first on the list of those to be read at the meeting, I avail myself of the privilege thus granted me of congratulating the Society on this occasion of its meeting in the capital of a country which is itself one magnificent mineralogical specimen, reaching from Cheviot to Cape Wrath; thus gathering into the most convenient compass, and presenting in the most instructive forms, examples of nearly every mineralogical process and phenomenon which have taken place in the construction of the world.

2. May I be permitted, also, to felicitate myself, on the permission thus given me, to bring before the Mineralogical Society a question which, in Edinburgh of all cities of the world, it should be easiest to solve, namely, the methods of the construction and painting of a Scotch pebble?[1]

3. I am the more happy in this unexpected privilege, because, though an old member of the Geological Society, my geological observations have always been as completely ignored by that Society, as my remarks on political economy by the Directors of the Bank of England; and although I have repeatedly solicited from them the charity of their assistance in so small a matter as the explanation of an agate stone on the forefinger of an alderman,[2] they still, as I stated the case in closing my first volume of *Deucalion*,

---

[1] [Compare *Two Paths*, §§ 151, 152 (Vol. XVI. pp. 382-383).]
[2] [*Romeo and Juliet*, Act i. sc. 4, 56.]

discourse on the catastrophes of chaos, and the processes of creation, without being able to tell why a slate splits, or how a pebble is coloured.[1]

4. Pebble,—or crystal; here in Scotland the main questions respecting these two main forms of silica are put to us, with a close solicitude, by the beautiful conditions of agate, and the glowing colours of the Cairngorm, which have always variegated and illuminated the favourite jewellery of Scottish laird and lassie.

May I hope, with especial reference to the

> "favourite gem
> Of Scotland's mountain diadem,"[2]

to prevail on some Scottish mineralogist to take up the hitherto totally neglected subject of the relation of colour in minerals to their state of substance: why, for instance, large and well-developed quartz crystals are frequently topaz colour or smoke colour,—never rose-colour; while massive quartz may be rose-colour, and pure white or grey, but never smoke colour;—again, why amethyst quartz may continually, as at Schemnitz and other places, be infinitely complex and multiplex in crystallization, but never warped; while smoky quartz may be continually found warped, but never, in the amethystine way, multiplex;—why, again, smoky quartz and Cairngorm are continually found in short crystals, but never in long slender ones,—as, to take instance in another mineral, white beryl is usually short or even tabular, and green beryl long, almost in proportion to its purity?

5. And, for the better solution, or at least proposition, of the many questions, such as these, hitherto undealt with by science, might I also hope that the efforts of the Mineralogical Society may be directed, among other quite feasible objects not yet attained, to the formation of a museum of what might be called mineral-geology, showing examples

---

[1] [See above, p. 292.]

[2] [*Rokeby*, canto iii. 8; of the Greta, "Matching in hue the favourite gem Of Albin's mountain-diadem."]

of all familiar minerals in association with their native rocks, on a sufficiently large and intelligible scale? There may be, perhaps, by this time, in the museum of Edinburgh,[1]— but there is not in the British Museum, nor have I ever myself seen,—either a specimen of pure Cairngorm in the gangue, or a block of trap containing agates of really high quality, whether from Scotland, Germany, or India.

6. Knowing the value of time to the meeting, I leave this, to my thinking, deeply important subject of the encouragement of geognostic mineralogy, to their own farther consideration ; and pass to a point of terminology which is of extreme significance in the study of siliceous minerals, namely, the desirableness, and I should myself even say the necessity, of substituting the term "spheroidal" for "reniform"[2] in mineralogical description. Every so-called "kidney-shaped" mineral is an aggregate of spheroidal crystallizations, and it would be just as rational and elegant to call sea-foam kidney-shaped, as to call chalcedony so. The word "Botryoidal"[3] is yet more objectionable, because it is wholly untrue. There are many minerals that resemble kidneys; but there is no substance in the whole mineral kingdom that resembles a bunch of grapes. The pisolitic aggregations which a careless observer might think grape-like, are only like grape-*shot*, and lie in heaps, not clusters.

7. But the change I would propose is not a matter of mere accuracy or elegance in description. For want of observing that the segmental surfaces of so-called reniform and botryoidal minerals are spheroidal, the really crystalline structure producing that external form has been overlooked, and in consequence, minerals have been continually described either as amorphous, or as mixtures of different substances, which are neither formless nor mingled, but are

---

[1] [It may be mentioned that there is now in the Edinburgh Museum of Science and Art a collection of Scottish Agates, formed by Dr. Heddle of St. Andrews, and presented in 1898. An *Explanatory Guide* to the collection was written in 1899 by Mr. J. C. Goodchild.]

[2] [Compare *Banded and Brecciated Concretions*, § 14 (above, p. 49).]

[3] ["Minerals presenting an aggregation of large sections of small globes are called botryoidal" (Trimmer's *Practical Geology*, 1841, p. 74).]

absolutely defined in structure, and absolutely homogeneous
in substance.

8. There are at least six states of siliceous substance
which are thus entirely distinct,—flint, jasper, chalcedony,
hyalite, opal, and quartz. They are only liable to be confused with each other in bad specimens; each has its own
special and separate character, and needs peculiar circumstances for its production and development. The careful
history of the forms of these six minerals, and the careful
collection of the facts respecting the mode of their occurrence, would require a volume as large as any that are
usually issued by way of complete systems of mineralogy.
Whereas, sufficient account is usually supposed to be rendered of them in a few sentences, and moreover every
sentence of these concise abstracts usually contains, or implies, an unchallenged fallacy.

9. I take, for example, from the account of "chalcedonic
varieties of quartz" given in Dana's octavo of 456 close-
printed pages (Trübner, 1879),—the entire account occupies
no more than a page and three lines,—the following sentences :[1]—

"Chalcedony often occurs lining or filling cavities in amygdaloidal rocks,
and sometimes in other kinds. These cavities are nothing but little caverns,
into which siliceous waters have filtrated at some period. The stalactites
are 'icicles' of chalcedony, hung from the roof of the cavity.

"Agate, a variegated chalcedony. The colours are distributed in clouds,
spots, or concentric lines. These lines take straight, circular, or zigzag
forms, and when the last, it is called fortification agate, so named from
the resemblance to the angular outlines of a fortification. These lines are
the edges of layers of chalcedony, and these layers are successive deposits
during the process of its formation.

"Mocha stone, or moss agate, is a brownish agate, consisting of chalcedony with dendritic or moss-like delineations, of an opaque yellowish-brown
colour."

10. Now, with respect to the first of these statements,
it is true that cavities in amygdaloidal rocks are nothing
but little caverns, just as caverns in any rocks are nothing
but large cavities. But the rocks are called "amygdaloidal,"

[1] [*Manual of Mineralogy and Lithology*, by James D. Dana, 3rd edition, pp. 235–
236.]

because their cavities are in the shape of almonds, and there must be a reason for this almond shape, which will bear on the structure of their contents. It is also true that in the rocks of Iceland there are cavities lined with stalactites of chalcedony. But I believe no member of this Society has ever seen a cavity in Scotch trap lined with stalactites of chalcedony; nor a Scotch pebble which gave the slightest evidence of the direction of its infiltration.

11. The second sentence is still more misleading, for in no sense is it true that agate is a " variegated chalcedony." It is chalcedony separated into bands of various consistence, and associated with parallel bands of jasper and quartz. And whether these bands are successive deposits during the process of formation or not, must be questionable until we produce the resemblance of an agate by a similar operation, which I would very earnestly request some of the members of the Mineralogical Society to do, before allowing statements of this positive kind to be made on the subject in popular text-books.

12. The third sentence confounds Mocha stone with moss agate, they being entirely different minerals. The delineations in Mocha stone are dendritic, and produced by mechanical dissemination of metallic oxides, easily imitable by dropping earthy colours into paste. But moss agates are of two kinds, brown and green, the one really like moss, the other filiform and like seaweed; and neither of them is at present explicable or imitable.

13. The inaccuracy of the statements thus made in so elaborate a work on mineralogy as Dana's, may, I think, justify me in asking the attention of the Mineralogical Society to the distinctions in the forms of silica which they will find illustrated by the chosen examples from my own collection, placed on the table for their inspection.[1] I place, first, side by side, No. 1, the rudest, and No. 7, the most delicate, conditions of pure chalcedony; the first, coarsely spheroidal, and associated with common flint; the

[1] [The catalogue which Ruskin sent to explain the examples is given below: see p. 520.]

second, filiform, its threads and rods combining into plates, —each rod, on close examination, being seen to consist of associated spheroidal concretions.

14. Next to these I place No. 2, a common small-globed chalcedony formed on the common quartzite of South England, with opaque concentric zones developing themselves subsequently over its translucent masses. I have not the slightest idea how any of these three specimens can have been formed, and simply lay them before the Society in hope of receiving some elucidatory suggestions about them.

15. My ignorance need not have remained so abject, had my other work left me leisure to follow out the deeply interesting experiments instituted by Mr. E. A. Pankhurst and Mr. J. I'Anson, of which the first results, being indeed the beginning of the true history of silica, were published by those gentlemen in the *Mineralogical Magazine* for 1882.[1] I have laid their paper, kindly then communicated to me, on the table, for immediate comparison of its plates with the specimens, and I have arranged the first two groups of these, adopting from that paper the terms exogenous and endogenous, for the two great families of agates, so as to illustrate the principal statements made in its pages.

16. It would materially facilitate the pursuit of their

---

[1] ["On some Artificial Forms of Silica, illustrative of the Structure of Agates, Chalcedonies," etc., a paper read September 2, 1881, and printed (with three plates) in the *Mineralogical Magazine and Journal of the Mineralogical Society of Great Britain and Ireland*, February 1882, vol. v. pp. 34–40. "The infinite variety in the great group of banded agates falls readily into two types. The first in which the solidification appears to have taken place in successive layers from within *outwards*; and the second, in which the impression is conveyed that the order of growth has been from without *inwards*. In both cases, beyond the successive bandings, there is very frequently observed a crystalline mass of silica, which does not present the banded structure. In the first class this crystalline mass is to the *outside* of the banded part; in the second, it is towards the *inside*, as though there had been a central cavity within the bands which had become filled with crystalline unbanded silica. Assuming for the convenience of nomenclature, that these two types are the results of the two methods of growth indicated—a view which we hope to justify in the course of this paper—we may name these the *exogenous* and *endogenous* types respectively." The writers go on to explain the experiments by which the two methods of growth were obtained, as shown in the plates accompanying the paper. The distinction between solidifications "from within outwards" and "from without inwards" is Ruskin's: see above, pp. 57–58.]

discoveries if some of the members of the Society would
register and describe the successive phenomena of crystallization in any easily soluble or fusible minerals. The
history of a mineral is not given by ascertainment of the
number or the angles of the planes of its crystals, but by
ascertaining the manner in which those crystals originate,
increase, and associate. The ordinary mineralogist is content to tell us that gold, silver, and diamond are all cubic ;
—it is for the mineralogist of the future to say why gold
associates its countless cubes into arborescent laminæ, and
silver into capillary wreaths ; while diamond condemns its
every octahedron to monastic life, and never, except by
accident, permits one of them to crystallize beside another.

17. At pages 5 and 6 of Mr. J. I'Anson's paper [1] will
be found explanations, more or less complete, of the forms
which I have called " folded " agates and " lake " agates,[2]
reaching to No. 40. The specimens from 40 to 60 then
illustrate the conditions of siliceous action which I am still
alone among modern mineralogists in my mode of inter-
preting.

The minor points of debate concerning them are stated
in the descriptions of each in the catalogue ; [3] but there are
some examples among them from which branch lines of
observation leading far beyond the history of siliceous
pebbles. To these I venture here to direct your special
attention.

18. No. 3 is a fragment of black flint on which blue
chalcedony is deposited as a film extending itself in circles,
exactly like the growth of some lichens. I have never seen
this form of chalcedony solidify from circles into globes, and
it is evident that for this condition we must use the term
" cycloidal," instead of " spheroidal." I need not point out
that " reniform " would be here entirely absurd.

---

[1] [That is, at pp. 5, 6 of a reprint of the paper from the *Mineralogical Magazine*;
at pages 37, 38 of the Magazine itself.]

[2] [See above, pp. 65 ("folded" agates), 178, § 24 ("lake" agates).]

[3] [See below, p. 522.]

This apparently common specimen (and, as far as regards frequency of occurrence, indeed common enough) is nevertheless one of the most profoundly instructive of the whole series. It is, to begin with, a perfect type of the finest possible *flint*, properly so called. Its surface, eminently characteristic of the forms of flint-concretion, is literally a white dust of organic fragments, while the narrow fissure which has opened in it, apparently owing to the contraction of its mass, is besprinkled and studded, as closely, with what might not unfitly be called pearl-chalcedony, or seed-chalcedony, or hail-chalcedony; for seen through the lens, it exactly resembles the grains of minute hail, sticking together as they melt; in places, forming very solid crests,—in others, and especially in the rifted fissure, stalactites, possibly more or less vertical to the plane in which the flint lay.

19. In No. 5 the separation into concentric films is a condition peculiar to flint-chalcedony, and *never found in true agates*.

In No. 6 (chalcedony in stalactitic coats, on amethyst) the variation of the stalactites in direction, and their modes of agglutination, are alike unintelligible.

No. 8 is only an ordinary specimen of chalcedony on hæmatite, in short, closely combined vertical stalactites, each with a central stalactite of black iron-oxide; but it is to be observed, in comparing it with No. 6, that when chalcedony is thus formed on rods of hæmatite, the stalactites are almost unexceptionally vertical, and quite straight. The radiate ridge at one side of this example is, however, entirely anomalous.

20. No. 9. The succeeding specimen, though small, is a notable one, consisting of extremely minute and delicate shells or crusts of spheroidal hæmatite, establishing themselves in the heart of quartz. I have no idea of the method, or successions in time, of this process. These I leave to the consideration of the Society, but I point to the specimen as exquisitely exhibiting the laws of true spheroidal crystallization, in a mineral which both in its massive and

crystalline state is continually associated with quartz. And it cannot but be felt that this spheroidal structure of hæmatite could as little be explained by calling or supposing it a mixture of micaceous hæmatite with amorphous hæmatite, as that of chalcedony by calling it a mixture of hexagonal with amorphous quartz.

21. No. 10. Next follows a beautiful and perfectly characteristic example of massively spheroidal agate, in which first grey and then white chalcedony, peculiarly waved and faulted by a tendency to become quartz, surrounds earthy centres, and is externally coated with pure quartz. And here I must ask the Society to ratify for me the general law, that in all solid globular or stalactitic conditions of chalcedony, if any foreign substance occurs mixed with them, it is thrown to their centres, while the pure quartz is always found on the outside.* On the other hand, the usual condition of geodes of chalcedony found in the cavities of rocks, is to purify themselves towards the interior, and either coat themselves with quartz on the interior surface, or entirely fill the central cavity with quartz.

22. No. 46 is a most literally amygdaloidal,—almond-shaped—mass of silica; only, not poured into an almond-shaped cavity in basalt, but gathered into a knot out of Jurassic limestone, as flint is out of chalk.

It is, however, banded quite otherwise than flint, the bands giving occasion to its form, and composed of different substances. Whereas those of flint are of the flint itself in different states, and always independent of external form.

Secondly. It seems to me a question of considerable interest, why the coarse substance of flint and of this dull hornstone can be stained with black, but not chalcedony, nor quartz. The blackest so-called quartz is only a clear umber, and opaque quartz is never so stained at all. Natural black onyx is of extreme rarity, the onyx of commerce being

---

* It is to be noticed also that often in stalactitic or tubular concretions the purest chalcedony immediately surrounds the centre.

artificially stained;[1] the black band in the lake agate, No. 32, is probably bituminous. And in connection with this part of the inquiry, it seems to be the peculiar duty of the mineralogist to explain the gradual darkening of the limestones towards the central metamorphic chains.

Thirdly, and principally. This stone [No. 46] gives us an example of waved or contorted strata which are unquestionably produced by concretion and partial crystallization, not compression, or any kind of violence. I shall take occasion, in concluding, to insist farther on the extreme importance of this character.

The specimen was found by my good publisher, Mr. Allen, on the southern slope of the Salève; and it is extremely desirable that geologists in Savoy should obtain and describe more pebbles of the same sort, this one being, as far as my knowledge goes, hitherto unique.

23. (71–77.) These seven examples of opal have been chosen merely to illustrate farther the modes of siliceous solution and segregation, not with that of illustrating opal itself,—every one of the seven examples presenting phenomena more or less unusual. The two larger blocks, 71, 72 (Australian), give examples in one or two places of obscurely nodular and hollow concretion, before unknown in opal, but of which a wonderful specimen, partly with a vitreous superficial glaze, has been sent me by Mr. Henry Willett, of Arnold House, Brighton, a most accurate investigator of the history of silica.[2] It is to be carefully noted, however, that the moment the opal shows a tendency to nodular concretion, its colours vanish.

24. No. 73 is sent only as an example of the normal state of Australian opal, disseminated in a rock of which it seems partly to have opened for itself the shapeless spaces it fills. In No. 71, it may be observed, there is a tendency in them to become tabular. No. 74, an apparently once fluent state of opal in veins, shows in perfection the

---

[1] [At Oberstein : see above, p. 64.]
[2] [For his contributions on the subject to *Deucalion*, see above, pp. 206, 211–218.]

arrangement in straight zones *transverse* to the vein, which I pointed out in my earliest papers on silica as a constant distinctive character in opal-crystallization.[1] No. 75 is the only example I ever saw of stellar crystallization in opal. No. 76, from the same locality, is like a lake agate associated with a brecciate condition of the gangue; while No. 77, though small, will be found an extremely interesting example of hydrophane. The blue bloom seen in some lights on it, when dry, as opposed to the somewhat vulgar vivacity of its colours when wet, is a perfect example of the opal's faculty of *selecting* for its lustre the most lovely combinations of the separated rays. A diamond, or a piece of fissured quartz, reflects indiscriminately all the colours of the prism; an opal, only those which are most delightful to human sight and mental association.

25. (78–80.) These three geological specimens are placed at the term of the series, that the importance of the structure already illustrated by No. 46 may be finally represented to the Society; No. 46 being an undulated chalcedony; No. 78, an undulated jasper; No. 79, a hornstone; and No. 80 a fully developed gneiss.

I have no hesitation in affirming,—though it is not usual with me to affirm anything I have not seen, and seen close, —that every one of these types of undulated structure has been produced by crystallization only, and absolutely without compression or violence. But the transition from the contorted gneiss which has been formed by crystallization only, to that which has been subjected to the forces of upheaval, or of lateral compression, is so gradual and so mysterious, that all the chemistry and geology of modern science is hitherto at fault in its explanation; and this meeting would confer a memorable benefit on future observers by merely determining for them the *conditions* of the problem.

26. Up to a certain point, however, these were determined by Saussure, from whose frequent and always acutely

----

[1] [See above, p. 48.]

distinct descriptions of contorted rocks I select the following, because it refers to a scene of which the rock structure
was a subject of constant interest to the painter Turner ;[1]
the ravine, namely, by which, on the Italian side of the
St. Gothard, the Ticino escapes from the Valley of Airolo.

"At a league from Faido the traveller ascends by a road carried on a
cornice above the Ticino, which precipitates itself between the rocks with
the greatest violence. I made the ascent on foot, in order to examine
with care the beautiful rocks, worthy of all the attention of a rock-lover.
The veins of that granite form in many places redoubled zigzags, precisely
like the ancient tapestries known as point of Hungary, and there it is
impossible to say whether the veins of the stone are, or are not, parallel
to the beds ; while finally I observed several beds which in the middle of
their thickness appeared filled with veins in zigzag, while near their
borders they were arranged all in straight lines. This observation proves
that the zigzags are the effect of crystallization, and not that of a compression of the beds when they were in a state of softness. In effect,
the middle of a bed could not be pushed together ('refoulé') unless the
upper and lower parts of it were pushed at the same time."[2]

27. This conclusive remark of Saussure renders debate
impossible respecting the cause of the contortions of gneiss
on a small scale ; and a very few experiments with clay,
dough, or any other ductile substance, such as those of
which I have figured the results in the sixth plate of
Deucalion,[3] will prove, what otherwise is evident on sufficient reflection, that minutely rhythmic undulations of beds
cannot be obtained by compression on a large scale. But
I am myself prepared to go much farther than this.
During half a century of various march among the Alps,
I never saw the gneiss yet, which I could believe to have
been wrinkled by pressure, and so far am I disposed to
carry this denial of external force, that I live in hopes
of hearing the Matterhorn itself, whose contorted beds I
engraved thirty years ago in the fourth volume of *Modern
Painters* (the book is laid on the table, open at the

---

[1] [See Vol. VI. pp. xxv., xxvi., 33 *seq.*]
[2] [Summarised from § 1802 of Saussure's *Voyages dans les Alpes*, vol. iv. pp. 7, 8
(1796 edition).]
[3] [Plate XV. in this volume (p. 257).]

plate¹), pronounced by the Mineralogical Society to be nothing else than a large gneissitic crystal, curiously cut!

Whether this hope be vain or not, I believe it will soon be felt by the members of this Society, that an immense field of observation is opened to them by recent chemistry, peculiarly their own: and that mineralogy, instead of being merely the servant of geology, must be ultimately her guide. No movement of rocks on a large scale can ever be explained until we understand rightly the formation of a quartz vein, and the growth, to take the most familiar of fusible minerals, of an ice-crystal.*

28. And I would especially plead with the younger members of the Society, that they should quit themselves of the idea that they need large laboratories, fine microscopes, or rare minerals, for the effective pursuit of their science. A quick eye, a candid mind, and an earnest heart, are all the microscopes and laboratories which any of us need; and with a little clay, sand, salt, and sugar, a man may find out more of the methods of geological phenomena than ever were known to Sir Charles Lyell. Of the interest and entertainment of such unpretending science I hope the children of this generation may know more than their fathers, and that the study of the Earth, which hitherto has shown them little more than the monsters of a chaotic past, may at last interpret for them the beautiful work of the creative present, and lead them day by day to find a loveliness, till then unthought of, in the rock, and a value, till then uncounted, in the gem.

* A translation into English of Dr. [G. F.] Schumacher's admirable essay, *Die Krystallisation des Eises*, Leipzig, 1844, is extremely desirable.

---

¹ [Plate 39; at p. 290 of Vol. VI. in this edition.]

POSTSCRIPT

29. I BELIEVE that one of the causes which has prevented my writings on subjects of science from obtaining the influence with the public which they have accorded to those on art, though precisely the same faculties of eye and mind are concerned in the analysis of natural and of pictorial forms, may have been my constant practice of teaching by question rather than assertion.[1] So far as I am able, I will henceforward mend this fault as I best may; beginning here with the assertion of the four facts for which, being after long observation convinced of them, I claim now, as I said in the Preface,[2] the dignity of Discoveries.

I. That a large number of agates, and other siliceous substances, hitherto supposed to be rolled pebbles in a conglomerate paste, are in truth crystalline secretions out of that paste in situ, as garnets out of mica-slate.

II. That a large number of agates, hitherto supposed to be formed by broken fragments of older agate, cemented by a gelatinous chalcedony, are indeed secretions out of a siliceous fluid containing miscellaneous elements, and their apparent fractures are indeed produced by the same kind of tranquil division which terminates the bands in banded flints.

III. That the contortions in gneiss and other metamorphic rocks, constantly ascribed by geologists to pressure, are only modes of crystallization.

And IV. That many of the faults and contortions produced on a large scale in metamorphic rocks, are owing to the quiet operation of similar causes.

---

[1] [Compare above, p. 342.]
[2] [The Preface to *In Montibus Sanctis*, now printed as an Appendix to the first volume of *Modern Painters* (Vol. III. p. 678).]

30. These four principles, as aforesaid, I have indeed worked out and discovered for myself, not in hasty rivalry with other mineralogists, but continually laying before them what evidence I had noted, and praying them to carry forward the inquiry themselves. Finding they would not, I have given much time this year to the collection of the data in my journals, and to the arrangement of various collections of siliceous and metallic minerals, illustrating such phenomena, of which the primary one is that just completed and catalogued in the British Museum (Nat. Hist.), instituting there, by the permission of the Trustees, the description of specimens by separate numbers: the next in importance is that at St. George's Museum in Sheffield; the third is one which I presented this spring to the Museum of Kirkcudbright; the fourth that placed at St. David's School, Reigate; and a fifth is in course of arrangement for the Mechanics' Institute here at Coniston; the sixth, described in the preceding chapter, may probably, with some modification, be placed at Edinburgh, but remains for the present at Brantwood, with unchanged numbers.[1]

31. The six catalogues describing these collections will enable any student to follow out the history of siliceous minerals with reference to the best possible cabinet examples; but for a guide to their localities and the modes of their occurrence, he will find the following extracts from Pinkerton's *Petralogy** more useful than anything in modern books; and I am entirely happy to find that my

---

* Two vols. 8vo, Cochrane & Co., Fleet Street, 1811.[2] A quite invaluable book for clearness of description, usefulness of suggestion, and extent of geognostic reference. It has twenty beautiful little vignettes also, which are models of steel engraving.

---

[1] [For (1) the British Museum Catalogue, see below, pp. 397–414; (2) the Sheffield Catalogue (unfinished), pp. 418–456; (3) the Kirkcudbright Catalogue, pp. 458–486; (4) the St. David's School Catalogue, pp. 491–513; (5) the Coniston Catalogue, pp. 516–518; and (6) the Edinburgh Catalogue, pp. 520–526.]

[2] [*Petralogy: a Treatise on Rocks*, by J. Pinkerton. Ruskin's quotations are from vol. i. pp. 135, 136–138, 223–225, 225–227, and 291–293. Dots have here been inserted to mark the breaks between the several passages. Ruskin's copy of the book is now in the Ruskin Museum, Coniston Institute.]

above-claimed discoveries were all anticipated by him, and are by his close descriptions, in all points confirmed. His general term "Glutenites," for stones apparently formed of cemented fragments, entirely deserves restoration and future acceptance.

"The division of glutenites into bricias and pudding-stones, the former consisting of angular fragments, the latter of round or oval pebbles, would not be unadvisable, were it in strict conformity with nature. But there are many rocks of this kind; as, for example, the celebrated Egyptian bricia, in which the fragments are partly round and partly angular; while the term glutenite is liable to no such objections, and the several structures identify the various substances.

"The celebrated English pudding-stone, found nowhere in the world but in Hertfordshire, appears to me to be rather an original rock, formed in the manner of amygdalites, because the pebbles do not seem to have been rolled by water, which would have worn off the substances in various directions; while, on the contrary, the white, black, brown, or red circlets, are always entire, and parallel with the surface, like those of agates. Pebbles therefore, instead of being united to form such rocks, may, in many circumstances, proceed from their decomposition; the circumjacent sand also arising from the decomposition of the cement.

"Mountains or regions of real glutenite often, however, accompany the skirts of extensive chains of mountains, as on the north-west and south-east sides of the Grampian mountains in Scotland, in which instance the cement is affirmed by many travellers to be ferruginous, or sometimes argillaceous. The largeness or minuteness of the pebbles or particles cannot be said to alter the nature of the substance; so that a fine sandstone is also a glutenite, if viewed by the microscope. They may be divided into two structures: the large-grained, comprising bricias and pudding-stones; and the small-grained, or sandstones.

"At Dunstaffnage, in Scotland,* romantic rocks of a singularly abrupt appearance, in some parts resembling walls, are formed by glutenite, in which the kernels consist of white quartz, with green or black trap porphyries, and basalts.

"In the Glutenite from the south of the Grampians, from Ayrshire, from Inglestone bridge, on the road between Edinburgh and Lanark, the cement is often siliceous, as in those at the foot of the Alps, observed by Saussure.

"Another Glutenite consists of fragments of granite, cemented by trap.

"Siderous glutenite, or pudding-stone of the most modern formation, is formed around cannons, pistols, and other instruments of iron, by the sand of the sea.

"Glutenite of small quartz pebbles, in a red ferruginous cement, is found in the coal-mines near Bristol, etc.

* For convenience in quotation, I occasionally alter Pinkerton's phrases, —but, it will be found by reference to the original, without the slightest change in, or loss of, their meaning.

"Porphyritic bricia (*Linn. a Gmelin*, 247), from Dalecarlia in Sweden, and Saxony. Calton Hill, Edinburgh ? . . .

"The entirely siliceous glutenites will comprehend many important substances of various structures, from the celebrated Egyptian bricia, containing large pebbles of jasper, granite, and porphyry, to the siliceous sandstone of Stonehenge. These glutenites are of various formations ; and the pudding-stone of England would rather seem, as already mentioned, to be an original rock, the pebbles or rather kernels having no appearance of having been rolled in water. Patrin* has expressed the same idea concerning those pudding-stones which so much embarrassed Saussure, as he found their beds in a vertical position, while he argues that they could only have been formed on a horizontal level. This curious question might, as would seem, be easily decided by examining if the kernels have been rolled, or if, on the contrary, they retain their uniform concentric tints, observable in the pudding-stone of England, and well represented in the specimen which Patrin has engraved. But the same idea had arisen to me before I had seen Patrin's ingenious system of mineralogy. In like manner rocks now universally admitted to consist of granular quartz, or that substance crystallized in the form of sand, were formerly supposed to consist of sand agglutinated. Several primitive rocks contain glands of the same substance, and that great observer, Saussure, has called them Glandulites, an useful denomination, when the glands are of the same substance with the rock ; while Amygdalites are those rocks which contain kernels of quite a different nature. He observes, that in such a rock a central point of crystallization may attract the circumjacent matter into a round or oval form, perfectly defined and distinct ; while other parts of the substance, having no point of attraction, may coalesce into a mass. The agency of iron may also be suspected, that metal, as appears from its ores, often occurring in detached round and oval forms of many sizes, and even a small proportion having a great power.

"On the other hand, many kinds of pudding-stone consist merely of rounded pebbles. Saussure describes the Rigiberg, near the lake of Lucerne, a mountain not less than 5800 feet in height above the sea, and said to be eight leagues in circumference, which consists entirely of rolled pebbles, and among them some of pudding-stone, probably original, disposed in regular layers, and imbedded in a calcareous cement. The pudding rocks around the great lake Baikal, in the centre of Asia, present the same phenomenon ; but it has not been observed whether the fragments be of an original or derivative rock. . . .

"The siliceous sandstones form another important division of this class. They may sometimes, as already mentioned, be confounded with granular quartz, which must be regarded as a primary crystallization. The sand, which has also been found in micaceous, schistus, and at a vast depth in many mines, may be well regarded as belonging to this formation ; for it is well known, that if the crystallization be much disturbed, the substance will descend in small irregular particles.

"Siliceous sandstones are far more uncommon than the calcareous or

* i. 154 [*Histoire Naturelle des Mineraux*, par Eugène Melchior Louis Patrin, Paris (an. ix.), 1801].

argillaceous. The limits of the chalk country in England are singularly marked by large masses of siliceous sandstone, irregularly dispersed. Those of Stonehenge afford remarkable examples of the size and nature of those fragments, but the original rock has not been discovered. Trap or basaltin often reposes on siliceous sandstone.

"But the most eminent and singular pudding-stones are those occurring in Egypt, in the celebrated bricia of the Valley of Cosseir, and in the siliceous bricia of the same chain, in which are imbedded those curious pebbles known by the name of Egyptian jasper; and which also sometimes contain agates. Bricias, with red jasper, also occur in France, Switzerland, and other countries; but the cement is friable, and they seldom take a good polish. All these rocks present both round and angular fragments, which shows that the division into bricias and pudding-stones cannot be accepted: a better division, when properly ascertained, would be into original and derivative glutenites. In a geological point of view, the most remarkable pudding-stones, which might more classically be called Kollanites, from the Greek,* are those which border the chains of primitive mountains, as already mentioned. The English Hertfordshire pudding-stone is unique; and beautiful specimens are highly valued in France, and other countries. It is certainly an original rock, arising from a peculiar crystallization, being composed of round and oval kernels of a red, yellow, brown, or grey tint, in a base consisting of particles of the same, united by a siliceous cement. . . .

"Of small-grained argillaceous glutenite, the most celebrated rock is the Grison, or Bergmanite, just mentioned, being composed of grains of sand, various in size, sometimes even kernels of quartz; which, with occasional bits of hard clay slate, are imbedded in an argillaceous cement, of the nature of common grey clay slate. When the particles are very fine, it assumes the slaty structure, and forms the grauwack slate of the Germans. It is the chief of Werner's transitive rocks, nearly approaching to the primitive; while at the same time it sometimes contains shells, and other petrifactions.

"This important rock was formerly considered as being almost peculiar to the Hartz, where it contains the richest mines; but has since been observed in many other countries. The slaty grison, or Bergmanite, has been confounded with a clay slate; and we are obliged to Mr. Jameson for the following distinctions: 1. It is commonly of a bluish, ash, or smoke grey, and rarely presents the greenish or light yellowish grey colour of primitive clay slate. 2. Its lustre is sometimes glimmering from specks of mica, but it never shows the silky lustre of clay slate. 3. It never presents siderite nor garnets. 4. It alternates with massive grauwack. But is not the chief distinction its aspect of a sandstone, which has led to the trivial French name of *grès-gris*, and the English *rubble-stone*, which may imply that it was formed of rubbed fragments, or of the rubbish of other rocks? The fracture is also different; and three specimens of various fineness, which I received from Daubuisson at Paris, could never be confounded with clay slate.

* Κόλλα, cement; the more proper, as it also implies iron, often the chief agent.

"This rock is uncommonly productive of metals, not only in beds but also in veins, which latter are frequently of great magnitude. Thus almost the whole of the mines in the Hartz are situated in greywack. These mines afford principally argentiferous lead-glance, which is usually accompanied with blend, fahl ore, black silver ore, and copper pyrites. A more particular examination discloses several distinct venigenous formations that traverse the mountains of the Hartz. The greywack of the Saxon Erzgebirge, of the Rhine at Rheinbreidenbach, Andernach, etc., of Leogang in Salzburg, is rich in ores, particularly those of lead and copper. At Vorospatak and Facebay, in Transylvania, the greywack is traversed by numerous small veins of gold."

32. These passages from Pinkerton, with those translated at p. 384 from Saussure, are enough to do justice to the clear insight of old geologists, respecting matters still at issue among younger ones; and I must therefore ask the reader's patience with the hesitating assertions in the following chapters[1] of many points on which a wider acquaintance with the writings of the true Fathers of the science might have enabled me to speak with grateful confidence.

[1] [That is, of *In Montibus Sanctis*—the chapters in question being "The Dry Land" (*Modern Painters*, vol. iv. ch. vii.), and "Of the Materials of Mountains" (*ibid.*, ch. viii.): see in this edition Vol. VI. pp. 115–135.]

VI

CATALOGUES OF MINERALS

1. CATALOGUE OF SELECTED EXAMPLES OF NATIVE SILICA IN THE BRITISH MUSEUM (1884)

2. CATALOGUE OF MINERALS IN ST. GEORGE'S MUSEUM, SHEFFIELD (1877–1886)

3. CATALOGUE OF FAMILIAR MINERALS IN THE MUSEUM OF KIRKCUDBRIGHT (1884)

4. CATALOGUE OF SILICEOUS MINERALS FOR ST. DAVID'S SCHOOL, REIGATE (1883)

5. CATALOGUE OF MINERALS PRESENTED TO THE CONISTON INSTITUTE (1884)

6. CATALOGUE OF MINERALS SHOWN AT EDINBURGH (1884)

7. NOTES ON MINOR COLLECTIONS:—

    (1) WHITELANDS COLLEGE (1883)

    (2) CORK HIGH SCHOOL FOR GIRLS (1889)

CATALOGUE

OF

A SERIES OF SPECIMENS

IN THE

BRITISH MUSEUM (NATURAL HISTORY)

ILLUSTRATIVE OF THE MORE COMMON FORMS OF

NATIVE SILICA.

ARRANGED AND DESCRIBED BY

JOHN RUSKIN, F.G.S.,

HONORARY STUDENT OF CHRIST CHURCH, HONORARY FELLOW OF CORPUS CHRISTI
COLLEGE, AND SLADE PROFESSOR OF FINE ART, OXFORD.

GEORGE ALLEN,
SUNNYSIDE, ORPINGTON, KENT.
1884.

[*Bibliographical Note.*—This Catalogue was published in pamphlet form, with the title-page as shown here on the preceding page.

Octavo, pp. viii. + 29; Half-title ("Catalogue of Selected Examples of Native Silica"), pp. i.–ii.; Title-page (with imprint at the foot of the reverse, "Printed by Hazell, Watson & Viney, Limited, London and Aylesbury"), pp. iii.–iv.; Preface (here pp. 397–398), pp. v.–viii.; Text, pp. 1–29.

Issued in September 1884, in buff-coloured paper wrappers, with the title-page (enclosed in a double-ruled frame) repeated upon the front cover. Price 1s. (1000 copies.)

A notice of the Catalogue appeared in the *Pall Mall Gazette* of September 30, 1884, which on October 13 contained the following letter from Ruskin :—

"MR. RUSKIN'S CATALOGUE OF SILICAS

"*To the Editor of the 'Pall Mall Gazette'*

"SIR,—I did not till to-day notice the reference to my Catalogue of British Museum Silicas in your issue of September 30, with the added question why none are on sale at the Museum. I believe the authorities, and with great reason, hesitate in admitting the precedent of the sale there of any unofficial catalogue or statement respecting the collections; but mine may be obtained, as all my other books, from Mr. Allen, of Orpington.—I am, Sir, your obedient servant,   J. RUSKIN.

"BURGATE HOUSE, CANTERBURY, *Oct.* 11."

The delay in selling the Catalogue in the Museum was, however, only caused by the necessity of obtaining official sanction at a meeting of the Trustees : see above, p. liv.]

PREFACE

This series of specimens has been selected to illustrate the more frequent varieties of Native Silica. One of these, quartz, is the most common of minerals; it is almost the only component of most gravels, sands, and sandstones, while it enters largely into the composition of many of the metamorphic schists and crystalline rocks : others, as flint and jasper, though not so plentiful, are still important constituents of the earth's crust; while chalcedony, the principal substance of agates, from early periods has been an important material in the arts. These varieties are in most works on mineralogy treated as accidental conditions of one and the same substance. But they are in this carefully chosen series exhibited in their essential distinctions, and their gradated phases of connecting state ; and they may be studied in these generally occurring forms with the greater facility, because all those siliceous minerals have been excluded which appear to have been produced by narrowly local circumstances. Thus chalcedony involved in bitumen found in Auvergne, and nearly all the forms of opal, including hyalite and cacholong, must be looked for in their proper places in the great gallery ; few minerals being shown in this selected series but those which, though here seen in their finest conditions, are in their less striking forms of frequent occurrence, and of extreme importance in the structure and economy of the world.

The authorities of the Museum are not responsible for any speculative statement or suggestion made in the following catalogue, but the description of each specimen has been submitted for modification or correction, and may, therefore, be received with perfect confidence; while, on my own part, the attention which I have given to this department of mineralogy for upwards of fifty years may, I think, justify me in claiming the reader's attention to statements which may at first seem to him, on the mere evidence presented in this single series, daring, or even indefensible. He may, at least, rest assured that they are in no case prompted by the desire of gaining credit for originality; my conviction being that there is nothing in my views on the subject of siliceous construction which may not be found already formalized by mineralogists of the last century.

A considerable number of the specimens here described have been presented to the Museum out of my own chosen examples at Brantwood (or, in some instances, directly purchased by me for this series), in order to fill gaps in its order which could not be supplied from the National collection without loss to the beauty and completeness of the series in the great gallery. The pieces numbered 7, 20, 21, 24, 28, 38, 52, 80, 90, 91, 95, 97, 98, 101, 103, 104, 116, 117, 118, 126, may be particularized, but it may perhaps be permitted me to suggest that the names of donors should be merely registered in the historical account of the British Museum and its collections, and should cease to encumber either the cases, or the scientific guides to them.

JOHN RUSKIN.

BRANTWOOD,
August 1st, 1884.

CATALOGUE

OF

A SERIES OF SPECIMENS

IN THE

BRITISH MUSEUM (NATURAL HISTORY)

ILLUSTRATIVE OF THE MORE COMMON FORMS OF

NATIVE SILICA

1. COMMON nodular flint, showing the distinctly concentric structure of many so-called flint "pebbles." The term "pebble" should always be restricted to those produced by friction on beaches or in streams; whereas, in this example (as also in all cases of amygdaloidal agate concretion), the form of the stone is owing either to its own manner of coagulating or crystallizing, or to the shape of the cavity it was formed in. This example is curious only in the demonstration of its structure by a loose smaller nodule in the middle.

   Part of the surface is artificially polished; the rest reticulated, like that of nearly all flint pebbles, rolled or not (compare my F. 1 at Sheffield[1]), the reticulation being structural and not due to impact.

2. Common branchiate flint.

   In its secretion from chalk, flint often assumes very strange branching or even bone-like forms, quite distinct from those of all other minerals. This is a small but interesting type. (Isle of Wight.)

   Allan-Greg Collection, 1860.

3. Black flint; banded; extremely fine specimen. This banding is the first and rudest condition of agatescent structure. See the paper on the subject in the *Geological Magazine*, vol. i., 1864, p. 145, by Mr. S. P. Woodward, who was the first to explain the structure.[2] (Banks of the River Samara, Russia.)

   Presented by Count Apollos de Moussin Poushkin.

---

[1] [See below, p. 419.]

[2] [For another reference to this paper, and particulars of it, see above, p. 47 n.]

4. Common flint, coarsely amygdaloidal, determining itself (primarily?) into zones parallel to its surface, and (secondarily?) into porous white or grey cloudings, tending to apparent brecciation. Very characteristic. (Basel.)

5. Common flint, coated with a thin film of blue chalcedony; determining itself (by alteration?) into white zones, transverse to the coating; but changing its colour only, not its structure, the alteration seeming in places to be arrested by the minute fissures. Beautiful.

This example is put side by side with Nos. 11 and 19, to show the general types of nascent flint-chalcedony. (Croydon.)

Purchased, 1861.

6. Sausage-shaped nodule of flint, replacing the stem of a sponge. Coated with chalcedonic film. (English.)

7. Almond of pure chalcedony enclosed in flint. Unique, in my experience. For comparison with Nos. 1, 5, and 6. (English.)

8. Flint altered by contact with basalt: red and in flaky disintegration, passing into an amorphous white mass, like the exterior of a common flint. There may, perhaps, be some clue in this rude example to the processes at work on fine material in No. 15. (Antrim.)

Allan-Greg Collection, 1860.

The first seven specimens are all white or black, or greyish blue. This one introduces the question of the red colour of jasper, and of the level bedding of lake-agate.

9. White jasper, passing into beautifully banded brick-coloured jasper; exquisitely spotted, as the latter also, with dendritic oxide of manganese, of microscopic delicacy; the mass, here and there, retreating to form cells filled with bluish chalcedony, transitional to quartz, while at the outside it is in some parts brecciate to extreme minuteness: on one side is a little of the melaphyre, in a cavity of which it was formed.

Very lovely, but not to be seen in its full beauty without a lens.* (Oberstein.)

10. Rounded pebble of white jasper, in flammeate and writhed bands, exactly intermediate between the bands of flint and those of folded agate. Stained in centre by oxide of iron like Nos. 8 and 9. Superb.

11. Small stalactitic chalcedony in flint. Very pretty. (Sussex.)

Mantell Collection.

---

* These tantalizing statements are of course only made to direct the student in the examination of similar specimens elsewhere.

12. Almond-shaped flint pebble, probably dropped out of such a rock as No. 31, and showing the outer yellow band which resulted either from its contact with the matrix or the action of water, or weather, when the pebble was loose. (Subsequently?) banded with bands extending to the surface.

13. Small nodule of finely-zoned agate, showing very remarkable fractures. (Scotland.)

14. Agate. Salmon colour; amygdaloidal, small, compact, and of extreme fineness, showing orbicular concretion at the exterior and a nucleus of exquisitely levelled beds of two orders. (Scotland.)
    Many of these small nodules out of the Scottish trap are inestimable in exhibition of fine siliceous structure.
    Presented by Benjamin Bright, Esq., 1873.

15. Egyptian jasper, faulted, for comparison with other examples of definite fault. Whether actually shifted, or independently banded on opposite sides of the vein, is for the present, to me, questionable. (Near Cairo.)

16. Red, or dark subdued crimson, jasper, arranging itself in eddied bands, which look faulted in their sudden undulation, traversed by others less distinct and transverse, which will be seen under the lens to be distinctly brecciate at one part of the stone, giving one of the most subtle examples of incipient brecciation. (Urals.)

17. Portion of a vein of irregularly banded pink jasper, with traversing ferruginous stains. Fine; but at the back, showing straight divisions across the beds. (Urals.)

18. Flint formed round sponge and passing into recumbent chalcedony, a kind of pebble extremely common on the beaches of the south coast of England (out of the greensand formation?) (Sussex.)

19. Common chalk flint, with spongiform chalcedony replacing sponge partially filling the interior hollow. A fragment of an Echinus with a small attached serpula at one extremity is on the outside. (Near Croydon.)
    Purchased, 1861.

20 Pudding-stone, so called, but I believe, concretionary. The upper
and surface of 21 shows at one extremity, new "pebbles," forming in
21. the old ones. (Hertfordshire.)

22. Fragmentary flint, in siliceous paste.

23. Common fragmentary natural mosaic, seemingly formed by contraction of yellow jasper, leaving fissures like those in drying clay, afterwards filled by siliceous paste. Compare note on No. 30.

24. Flint passing into jasper, seemingly brecciate. This material forms huge masses of the coast-rocks at Sidmouth, and the low stone walls of the fields are mostly built of it. Conf. 37.

25. Block of pure yellow mossy jasper, passing into reddish-brown chalcedony, in some parts tinged with purple. The form associated with sponge flints, the veins of chalcedony isolating portions of paler jasper. (Ekaterinburg, Russia.)
Presented by Count Apollos de Moussin Poushkin.

26. Common flint, apparently crushed and recemented ; but the structure has never been properly studied, and is in some of its conditions at present inexplicable. Cut and polished under my own direction.

27. Pink opal, exhibiting resemblances of brecciation. (Quincy, near Bourges.)

28. Chalcedonic flint, confused in aspect between a breccia and a conglomerate : and stained (by iron oxide ?) of the most brilliant scarlet I ever saw in the material.[1]

29. Yellow opaque ferruginous silica, enclosing fragments of crystallized quartz, and traversed at one side by irregular veins of grey chalcedony. (Zweibrücken.)
Beroldingen Collection, 1816.

30. Chalcedony in horizontal layers of slightly varying substance, passing by irregular alteration into opaque conditions, first yellow, then white, which must be carefully distinguished from true white jasper. Seen on the polished surface, they seem to be partly related to the fissures caused by contraction during (desiccation ?). (Faroe Islands.)

31. Boulder rock of the southern drift (slice of), presenting the most interesting phenomena of siliceous pebble-beds. (Hertfordshire.)

32. Jasper, an enormous nodule in three bands, grey, purple, and paler purple, round a sandy nucleus : the grey band becoming brown at its exterior; and the entire mass determining itself into incipiently porphyritic conditions. At one point the grey band gathers into small spiral or shell-like forms.
Wonderfully interesting. Presented by Sir Richard Owen, who brought it from Cairo.

33. Purple chalcedony, coating quartz; only noticeable for its fine colour. (Near St. Austell ?)
Purchased, 1856.

[1] [Compare No. 41 in the Reigate Catalogue, below, p. 502.

34. Purple chalcedony in lifted crusts, associated with chlorite and cassiterite.
    Very singular, though scarcely seeming so at first glance. (Wheal Maudlin, Lanlivery, Cornwall.)
    Purchased, 1851.

35. Common flint - chalcedony, the external iron - stain more delicately applied, and the pores of the chalcedonic crust very peculiar. (Flonheim, Hesse.)
    Beroldingen Collection, 1816.
    The three examples 33 to 35 show the most beautiful purple colours reached by common flint-chalcedony. They are always a little more rusty or red than the more delicate bloom of the purer varieties of opaque-surfaced chalcedony.

36. Grey flint-chalcedony of the south coast, with spongy or mossy ochreous secretions.

37. Part of No. 24.

38. Common red flint-chalcedony, richly developed in the hollow of a flint.

39. Chalcedony in crusts, with emergent or inflowing stalactites. (Aden?)

40. Another variety of the same state. (Aden?)
    Both 39 and 40 presented by the Hon. Robert Marsham, 1877.

41 Examples of chalcedonic "nuts" formed in trap rocks.
and The former in diabase from Montrose, the latter in basalt from Co.
42. Derry: both presented by Benjamin Bright, Esq., 1873.

43. Lake-chalcedony traversed by chloritic filaments.

44. Another example,—both singularly fine.

45. Slice of a large block of lake-chalcedony, with dispersed chlorite.

46. Chalcedony with inclined stalactites, like 39 and 40. (Iceland.)

47. Brown compactly-knitted chalcedony; very rare.

48. Black recumbent chalcedony. (Redruth.)
    Purchased, 1859.

49. Chalcedony associated with chrysoprase. (Baumgarten.)
    Aylesford Collection.

50. White flint-chalcedony in crusts; wonderful. The separation by crevasses, apparently opening gradually, of chalcedonic films and crusts in this grand specimen, is a structure peculiar to flint chalcedony. It never occurs in true agates.

51. Portion of a nodule of lake-chalcedony in which opaque white masses are separated by clear currents which, one by one, join an increasing current descending at the side. Unique, so far as I know, in this resemblance to a river and its tributaries. (Faroe Islands.)
      Allan-Greg Collection, 1860.

52. A larger slice from the centre of the same nodule, formerly one of the most valued pieces in my own collection.

53. Common lake-chalcedony of Iceland, in level beds, traversed by stalactitic tubular layers. The museum is curiously poor in specimens of this character: but the surface of the single tube, seen in the polished section, is of extreme beauty.
      Beroldingen Collection, 1816.

54. Chalcedony in beds evidently shattered and faulted, afterwards re-cemented, with a kind of ripple mark instead of their natural reniform structure, on their external surfaces. The most wonderful and inexplicable piece I ever saw.
      Allan-Greg Collection, 1860.

55. Chalcedonic geode, traversed by straight beds or laminæ of fine chalcedony, with separating cavities which have the aspect of moulds of tabular crystals now fallen out or dissolved. "The only one I have seen with these impressions" (W. G. L.).
      Looked at from the interior cavity of the geode, the separate mass round the great laminar impression has the common look of a crust on a tabular crystal. (Dept. of Salto, Uruguay.)
      Presented by W. G. Lettsom, Esq., 1863.

56. Chalcedony. Pseudomorph, after calcite? Note in the interior of its cavity the enclosed laminæ with oblique terminations. Superb. (Uruguay.)
      Presented by W. G. Lettsom, Esq., 1877.

57. Common grey flint, or semi-flint, passing into opaque blue semi-chalcedony, forming a cell, lined with pure common chalcedony half an inch thick, across which cell are formed one single and two conjunct cylinders of solid chalcedony, the conjunct one terminated spherically as usual, but the single one simply traversing the cell. (English.)
      Not easily to be matched in its strangeness and simplicity.

58. Confluent recumbent chalcedony. (Iceland.)
      Purchased, 1837.

59. Geode of chalcedony, very large, with vertical stalactites of the same material as its walls. Superb. (Iceland.)

60. Flamboyant* black chalcedony on crystallized quartz, magnificent. (Pednandrea mines, near Redruth.)
    Purchased, 1868.

61. Chalcedony in a level field, with rods irregularly recumbent on it, each apparently composed of two segments soldered together, and forming sometimes an extremely sharp ridge at the junction. Very curious. With groups of yellow dolomite crystals.

62. Blue chalcedony in vertical walls and rods, each of the latter having a minute central rod of iron-oxide; in a cavity of iron-oxide. Beautiful. (Ruskowa, Hungary.)

63. Chalcedony in vertical stalactites dependent from a thin crust of the same material; each enclosing mossy filaments of chlorite, and coated with small crystals of quartz. Very beautiful.

64. Mural chalcedony, that is to say, chalcedony in which the rods or other reniform processes collect laterally into walls or tablets of a fairly uniform thickness. On the grandest scale.
    Purchased, 1851.

65. Fine white chalcedony, in crusts of extreme delicacy, developing themselves into groups of straight rods, which in places distinctly affect a trigonal arrangement; of extreme beauty and rarity, yet in its encrusted structure having something in common with the ordinary spongiform states like those of No. 19. (Guanaxuato, Mexico.)
    Heuland Collection.

66. Chalcedony, common grey, in prostrate rods formed of globules adherent round a fine thread of some central substance. (Faroe Islands.)

67. Chalcedony in spiral whorls, encrusted with crystals of quartz, partially filling a cavity in a slate veinstone containing dispersed copper pyrites and dolomite. Superb. (Cornwall.)
    Purchased, 1851.

68. Recumbent rod-chalcedony, fine, but much injured by fracture. (Trevascus mine, Cornwall.)
    Greville Collection, 1810.

---

* I take leave to use this word as best descriptive of these forms, peculiar to chalcedony, though sometimes partially imitated in Aragonite and a *few other* minerals when obscurely crystalline.

69. Portion of a geode of amethystine quartz coated by brown chalcedony. (Oberstein.)
    Greville Collection, 1810.

70. Dove-coloured flamboyant chalcedony, on quartz. Loveliest form of this mineral. Trevascus mine, Cornwall.

71. White flamboyant chalcedony, of unique beauty on quartz. Trevascus mine, Cornwall.
    Greville Collection, 1810.

72. Another portion of the geode, No. 69, partly filled by flammeate chalcedony. Unequalled, I believe, in Europe.

73. Heliotrope, pisolitic; though not easily seen to be so: with quartz, semi-crystalline, forming an agatescent series of irregular bands in the centre of an amorphous mass. (Banda, India.)
    Purchased, 1867.

74. Jasper, dull red and green, obscurely banded, with pale brown orbicular concretions, ugly, but very instructive in their method of formation. (Isle of Rum.)

75. Heliotrope, in riband beds, with two elongated white spaces formed by minute quartz. The mass of it shown on the rough edge to be minutely pisolitic. Extremely fine. (Banda, India.)
    Purchased, 1865.

76. Heliotrope, the red forming a compact and united mass, in the middle of which are finely agatescent bands of blue chalcedony round a small cavity. (Banda, India.)
    Heuland Collection.

77. Heliotrope, indistinctly pisolitic in the manner of No. 73, but having the quite opaque portions subdivided by a spongy structure of microscopic fineness. This structure, however, exists, though less apparently, both in 73 and 75. (Banda, India.)
    Purchased, 1882.

78. Heliotrope, pisolitic, extremely clear and fine, and of unusual size. The slab is 5 inches long by $3\frac{1}{4}$ inches wide. (Banda, India.)

79. Heliotrope, distinctly pisolitic in the green mass, leaving the white spots in the form of a paste, filling the cavities between the spheres. (Banda, India.)
    Purchased, 1882.

80. Heliotrope, massive, partly degenerating into chert or flint; divided by broad veins of chalcedony and milky quartz, in which it is to be observed that the layers are arranged differently on opposite sides of the vein.

Splendid, and peculiarly illustrative of veined structure. (Banda, India.)

81. Agate in grand mass, of the fine beds usually found at Kunnersdorf in Saxony brecciate or inlaid,* here in the order of their lines; locally faulted, but not consistently—*i.e.*, the faults not going through all the beds. Of consummate interest. (Kunnersdorf.)

82. Brown, yellow, and purple agate, the purple space developing across the concentric beds. Wonderful.

83. Oval slab of amethystine agate, exactly in the transitional state between common amethystine quartz-rock and inlaid agate. A perfect and marvellous type of incipient inlaying.

Purchased, 1882.

84. Exquisitely delicate amethystine inlaid agate, containing hollows with peculiar surfaces. (Kunnersdorf.)

85. Inlaid agate, amethystine, finest kind. The spot of quartz developed in the midst of the white banded bed is very rare. (Kunnersdorf.)

86. Jasperine agate, the form of the first layers being that of the crystals of quartz, partly amethystine, upon which they are based. Magnificent. (Kunnersdorf.)

Purchased, 1883.

87. Agate, a portion of an amygdaloidal nodule with jasperine bands of exquisite beauty, illustrating nearly every phenomenon of folding, and crystalline interference. The minute cones of quartz locally traceable with a lens along the white, and the finely-veined innermost bed, exactly like tents of a camp in the desert, are extremely rare; but the most peculiar feature in the stone is the jagged red crystalline formation filled up with spotted white, on its lower side † (left hand of spectator), totally absent on the other side. (Oberstein.)

Presented by Benjamin Bright, Esq., 1873.

---

* I shall in general use the term "inlaid" of stones consisting of apparent fragments embedded in a crystalline matrix, respecting which I am in doubt if the fragments be really broken or not. The term "inlaid" is descriptive, and involves no theory.

† In all cases when agates are convex on one side and flat on the other, it may be assumed with probability that the flat side was the bottom.

88. Inlaid agate on the grandest scale: superb. (Kunnersdorf.)
Purchased, 1883.

89. Inlaid agate, a thin slab, polished on both sides. The most interesting piece of faulted bedding I ever saw. (Kunnersdorf.)
Bequeathed by the Rev. C. M. Cracherode, 1799.

90. Jasperine agate developing transverse bands. Wonderful.

91. Inlaid agate, consisting of opaque shell-like bands, embedded in pure chalcedony; part of a rolled pebble from the east coast. Unique in my experience.

92. Jasper, in concentric bands, apparently determined by hæmatite. Unique in my experience.

93. Reddish-brown "semiopal," in singularly-faulted beds.

94. Jasper, deep red, in beds, more or less faulted and distorted, the interstices filled by milk-white quartz and chalcedony, and the whole seeming to form a vein in a chloritic rock. (India.)
Purchased, 1874.

95. Jasper, in beds arranged at more or less sharp angles.

96. Red mural agate.

97. Inlaid agate, with the zones in some places continuous round the apparent fragments. The most interesting piece I ever saw.

98. Mural agate, in crossing plates and walls. Unique.

99. Inlaid agate, divided by straight fissures, the beds concurrent on opposite sides. Superb. (Kunnersdorf.)

100 Inlaid agate, with portions of involved calcite. Wonderful.

101. Agate in perfect development by two steps only out of compact silica. Unique in my experience. It will be seen that there are two states of chalcedonic secretion, one traversed by irregular traces of fissure —the other zoned.

102. Agate feebly zoned, but of beautiful substance, developed in a mass with precise edges, almost rectilinear in the section, out of a mixture of dolomite and chalcedony. Unique also in my experience, though in nearer relation to known structures than No. 101.

103. Conchoidal agate; so called by me on first describing it, from its resemblance to fragments of shells, by which certainly some varieties have been produced. (Oberstein.)

104 and 105. Two pieces of an agate developing itself by writhed contraction out of white semiopal.

106. Grey agate, stalactitic, in part, and partly crystalline. Very wonderful. (Wheal Friendship, Tavistock.)
Sloane Collection, 1753.

107. Common lake-agate (artificially stained), with hollow in centre. (Uruguay.)
By exchange, 1863.

108. Common lake-agate, with its centre filled. (Uruguay.)
Purchased, 1874.

109. Oval-domed agate, with lifted lake-bed. Superb. (Uruguay.)

110. Folded agate, involving a small tabular agate in its outer layer.

111. Half of a nodule of extremely interesting lake-agate; its level beds twice interrupted by elevations towards the left hand, as it now lies. (Uruguay.)

112. Lake-agate, not nodular, but of irregular external form. (Uruguay.)

113. Agate, amygdaloidal with (pendant?) stalactites of chalcedony, filling the upper part of its cavity, the rest being occupied by quartz, while the base is composed of a ragged jasperine concretion, presumably related to the condition especially indicated in No. 87. (Scotland.)
Presented by Benjamin Bright, Esq., 1873.

114. Chalcedony, common, massive, in extremely flat reniform concretion, and drawn into quite marvellous complexity of irregularly bent and involved zones, formed apparently by a new development of structure, more or less following the original larger zones. (Uruguay.)
Purchased, 1872.

115. Larger portion of a divided nodule of folded agate; the best example of the structure I ever saw.
Purchased, 1872.

116   Parts of a large nodule of rock-crystal, the summits of the individual
and   crystals being directed inwards; the central cavity afterwards filled
117.   with a bluish-white agate, of which the bands follow the contours
       of the crystal.[1]

118.   Small white sparkling quartz crystals grouped so as to form recum-
       bent and intermingled rods, an extremely beautiful example of a
       very unusual structure.[2]

119.   Hemisphere of quartz formed by radiating crystals, of which the pro-
       jecting summits are remarkable for a peculiar play of colour.[3]
       (India.)

120.   A clear tapering rock-crystal,[4] with the usual striations on its faces.
       By exchange, 1868.

121.   A strange sheaf-like group of amethystine-tinged crystals, with three-
       sided summits, resting on a base of chalcedony: the lesser indi-
       viduals of the upper part of the sheaf are all nearly parallel to
       the central large one. (Elba?)[5]
       By purchase, 1870.

122.   A group of several large white crystals, each of them compound, and
       analogous in structure to 121, but having a more simple summit
       with six sides: with adherent chalybite (carbonate of iron). (Vir-
       tuous Lady Mine, near Tavistock.)
       Purchased, 1870.

---

[1] [The proof-sheets add :—
    "In the common form, agate is *always external*, and the *quartz formed
    in the centre*. This condition is rare, but typical; not a local accident;
    mineralogists have, however, as yet given no account of the two exactly
    contrary actions of secretion, which the two modes of formation infer."]
[2] [The proof-sheets add :—
    "Generally, when quartz takes this form, the stalactitic rods have
    centres of hæmatite, or Millerite, or some other metallic oxide, but under
    no condition is the form frequent."]
[3] [The proof-sheets add: "Unique and inestimable." Compare the "Postscript," p. 413.]
[4] [Compared, in the proof-sheets, "with the *diverging* cluster, 121."]
[5] [The proof-sheets have a different, and longer, note :—
    "121. Sheaf-quartz. I give this name to groups of imperfectly formed
    crystals, arranged in a sheaf, narrower at the base than the apex; in the
    best examples two such groups are set base to base, and the united cluster
    has the aspect of a sheaf, bound in the middle.
    "This form, which occurs sometimes in other minerals capable of
    radiation, as arragonite, stilbite, epidote, Millerite, and hæmatite, when
    filiform in the interior of quartz, is to be carefully distinguished, on the
    one hand, from true doubly terminated crystals, like many of the complex
    forms of calcite; and, on the other, from merely interrupted or broken
    portions of spherically radiating masses. The true *sheaf* group is to be

123. Rock-crystal enclosing long slender crystals of rutile, some of them showing the characteristic red colour, and also some mica.
Sloane Collection, 1753.

124. A curious specimen of rock-crystal, with remarkable striations: at first sight the shape of the specimen appears to be due to fracture, but closer examination reveals the crystalline faces on the edges.[1] (La Gardette, Dauphiny.)
Purchased, 1837.

125. A clear transparent rock-crystal with peculiar impressions. (Savoy.)

126. A group of crystals illustrating a previous stage in their growth by the enclosed foreign matter which has been deposited on the faces of the earlier individuals.[2]

127. A remarkable growth of crystals disposed parallel to each other in such a way as to indicate an approach to a single compound crystal: the jagged saw-like individuals, lengthened parallel to an edge formed by the meeting of a pyramid-face with a prism-face, are very noteworthy.
Sloane Collection, 1753.

128. Very similar to 127 in structure, but the resulting individual more complete: enclosed layers of foreign matter, arranged parallel to the faces of the crystal, again illustrate a previous stage of growth.

129. Rock-crystal enclosing thick and thin crystals of rutile.
Sloane Collection, 1753.

130. Beautifully clear rock-crystal, enclosing moulds due to four-sided prisms, and tabular crystals or fragments, of some mineral since removed. (Brazil.)
Heuland Collection.

recognized by its repetition in different places of the specimen, either in always similar proportions, as in sheaf, quartz, or in filiform hæmatite, radiating at a *never-exceeded fixed angle*.
"In the present example the quartz group [is] slightly amethystine, and terminated in a trigonal pyramid, and based on a coarse chalcedony in a manner wholly unexampled in my experience."]
[1] [The proof-sheets add :—
"The fracture shows exquisite prismatic colours in some lights."]
[2] [The note in the proof-sheets is :—
"126. Quartz, showing progressive crystallization at two periods, the interior dark crystals of the first period being sprinkled with iron glance on their surfaces, and detached from the upper layer by minute cavities, which tell as white. Superb."]

131. Rock-crystal enclosing chlorite, with pink and brown altered conditions of the same mineral.[1] (Minas Geraes, Brazil.)
   Purchased, 1838.

132. A large fragment of rock-crystal with vermicular chlorite dispersed throughout its mass, and some small plates of hæmatite: beautifully iridescent. (Brazil.)

133. Probably a portion of 132 (see farther observations on this, and the following specimen, in the Postscript, p. 413).

134. A large polished slab of green avanturine-quartz.[2] (India.)
   Presented by Colonel C. S. Guthrie, 1865.

[1] [The proof-sheets add : "Lovely."]
[2] [The proof-sheets add :—
   "A much valued precious stone, this slab being of extraordinary size."]

POSTSCRIPT

THE manner in which No. 132 is placed and levelled permits the spectator, standing between it and the window, to see by vivid reflection its splendid iridescence. In quartz this iridescence is always owing to irregular fissures with close surfaces,—*flaws*, that is to say, in parts of the stone, and not conditions of its proper structure. In the varieties of Felspar known as Labradorite and Moonstone, various colours are structurally reflected from different parts of the stone, or a pale blue light from the whole of it, but there is no opalescent interchange of hues anywhere. On the contrary, in opal, the cause of the colours pervades the whole structure of the stone, and opal is not perfect opal unless it is iridescent *throughout*: there is a difference, too, in the spectrum of the colours reflected from those given by fissured crystal, which is as yet a matter of unexplained mystery, and will always be one of extreme interest.

Specimen No. 119, in which the exterior surfaces of the radiating quartz crystals are opalescent, is (hitherto) unique. They are truly *opalescent*, not merely splendent in the manner of No. 132, and they enable the observer at once to recognize the essential difference between the colour-tones of opal and of fissured quartz. The colours of opal are always of a subdued tone, and of perfect purity,—no mixture of hue ever takes place which dulls or corrupts; but in fissured quartz the colours are unsubdued, being only those obtainable in the common spectrum of the prism; and the colours are often blended so as to detract from each other's purity, and give coppery or bronzed combinations of red and green, which would never be allowed by a good painter; while the blue chiefly reflected by quartz is only that which is produced by the pigments formed of prussiate of iron, the blues reflected by opal are, on the contrary, always those produced by smalt and ultramarine.

I need not insist on the singularity of this distinction in hues of reflected light which are absolutely unaffected by coloured chemical elements in the substances exhibiting them, and are produced only by different structures in clear, or translucent, silica.

It is true that a certain quantity of water is always engaged in, or combined with, opal, while there is none in compact (it is possible there may be sometimes in fissured) quartz. But, singularly, in hydrophane opal, of which the colours are greatly increased in power by the absorption of water on immersion, what the colours gain in power they lose in purity, and the hues of dipped hydrophane are vulgarized down to almost the level of those of quartz.

I would also direct the observer's attention, in the beautiful specimen 132, to the form of the contained chlorite, described as "vermicular."

413

Chlorite, which ought to be more simply termed " Greenite " or " Greeny," [1] is a combination of silica, alumina, magnesia, protoxide of iron, and water, in approximately the proportions of 25, 20, 20, 25, 10, in the hundred parts, or in this altered order easily memorable.

Silica,	Alumina,	Water,	Magnesia,	Iron.
25	20	10	20	25

And worth memory, for chlorite is the colouring matter of almost innumerable varieties of green stone.   It is extremely desirable that mineralogists should distinguish in all catalogues the silicas coloured by, or involved with, this mineral, from the numerous conditions of heliotrope and agate in which the green may be owing to other constituents.

I may permit myself, in conclusion, to observe that the stones in this case having been all placed so as both to exhibit their peculiarities with distinctness, and to admit of convenient comparison with each other, where comparison was desirable, I have hope that their present order may be a permanent one ; and perhaps lead to similar arrangements of other groups in which perfect exhibition of character is more desirable than multiplication of examples or consistency of theoretical system.   In a museum intended primarily for the instruction of the general public, it is not of the least consequence whether silicates come after carbonates or oxides after sulphides : but it is of vital and supreme importance that specimens whose beauty is in their colour should be put in good light, and specimens whose structure is minute, where they can be seen with distinctness.

J. R.

[1] [On this passage compare, above, pp. liii.–liv., 51.]

2

CATALOGUE OF MINERALS IN ST. GEORGE'S MUSEUM, SHEFFIELD

(1877–1886)

[*Bibliographical Note.*—The Sheffield Catalogue has never been published, in the sense of being placed on sale with Ruskin's works. But copies of it in both the forms here described have been in private circulation. They are among the rarest of Ruskiniana :—

First Edition (1877).—An octavo pamphlet of 64 pages, in a wrapper, without imprint. The wrapper is plain, and there is no title-page. Page 1 has the following "drop-title": "St. George's Museum, Sheffield. | Mineralogical Department. | Substance I. Silica."

Then follows the same tabular arrangement of eight classes of siliceous minerals as is given in *Deucalion* (above, p. 200). But in the body of the Catalogue "Class F. Flint" takes the place of "Class J. Jasper," and "Class H. Hyalite" is not given.

"First Class. A. Agate" occupies pp. 1–29; "Class C. Carnelian," pp. 30–31; "Class F. Flint," pp. 32–34; "Class L. Chalcedony," pp. 35–44; "Class M. Amethyst," pp. 45 (the page is partly blank and begins with "M. 6") –47; "Class O. Opal," pp. 48, 49; "Class Q. Quartz," pp. 50–55.

The Catalogue then proceeds to the second of the twenty substances enumerated in *Deucalion* (above, p. 199), namely, "Oxide of Titanium," under which head "Class A. Acicular Rutile" occupies pp. 56, 57. "Substance III. Oxide of Iron. Class A. General Group of the Black, Red, and Brown Oxides" follows, pp. 58–64. Headlines on the left-hand pages indicate the *substance*, while those on the right-hand pages indicate the *class*.

The only copy of this Catalogue which the editors have seen is in the Ruskin Museum, Sheffield. It contains several dates in the text (recording the purchase of specimens), the latest of them being July 1877.

Later in the year Ruskin wrote a Preface (here p. 418), dated "18th October, 1877," in which he apologises for "the broken form in which I permit its publication." The Catalogue, however, as printed and placed in the Museum, did not contain this Preface. Early in 1878 Ruskin fell ill, and the Museum copy represents the state of the Catalogue as he left it before his illness.

It was not until eight years later that Ruskin took its revision in hand for publication. He had then been busy with several other Catalogues of Siliceous Minerals; and the next edition was confined to that class.

Second Edition (1886).—An octavo pamphlet, pp. vi.+54. The title-page is as follows :—

Catalogue | of | Siliceous Minerals. | Permanently Arranged in | St. George's Museum, Sheffield. | By | John Ruskin, LL.D., | Honorary Student of Christ Church, | and Honorary Fellow of Corpus Christi College, Oxford. | George Allen, Sunnyside, Orpington, Kent. | 1886.

Imprint at the foot of the reverse: "Printed by Hazell, Watson, & Viney, Ld., London and Aylesbury." Preface, pp. iii.–vi. This is the Preface above referred to, which was written in 1877, and intended for the First Edition. Catalogue, pp. 1–54. Headlines as before.

The arrangement of the classes (see p. 419 here) differs from that given in *Deucalion*, and in the First Edition of the Catalogue.

"Class I. Flint. F." (pp. 1–10) comprises (as F. 1–8) the specimens which in ed. 1 were "A. 1–8"; and then (as F. 9–18) those which in ed. 1 were "F. 1–10."

416

"Class II. Jasper" (pp. 11–13) comprises (with an added introductory sentence) what in ed. 1 was "Class C. Carnelian."

"Class III. Chalcedony. C." (pp. 13–21) coincides with "Class L. Chalcedony" in ed. 1.

"Class IV. Opal. O." (pp. 21–23) corresponds with the same class in ed. 1, but enumerates some additional specimens.

"Class V. Hyalite. H." (pp. 23–24) is not contained in ed. 1.

"Class VI. Quartz. Q." (pp. 24–30) coincides with the same class in ed. 1.

"Class VII. Amethyst. M." (pp. 30–32) corresponds with the same class in ed. 1, but "M. 1–5," which were left blank in that edition, are noted.

"Class VIII. Agate. A." (pp. 33–54) contains, first, A. 1–4, which are A. 9–12 in ed. 1; then A. 5–12, which are not included in ed. 1; and afterwards A. 13–60, which correspond with the same numbers in ed. 1.

This Second Edition was, as already stated and as appears from its title-page, limited to Siliceous Minerals. In Mr. Wedderburn's possession there are several proof-sheets containing (1) some of the matter, dealing with other substances, held over from the First Edition; (2) additional matter of a like kind; and (3) notes on additional specimens under various heads of the Siliceous Minerals.

The *present edition* as given in this volume gives (i.) the text of the Second Edition, described above, adding in footnotes several additional passages and different readings which occur in ed. 1 (see, *e.g.*, pp. 419, 421, 422), and additional notes (from Mr. Wedderburn's proofs) on O. 12, 13 (p. 429 *n.*); and (ii.) adding to the text (again from Mr. Wedderburn's proofs) notes on A. 63, 70–75; next (iii.) "Substance II. Oxide of Titanium," which appears in ed. 1 only; (iv.) "Substance III. Oxide of Iron. Class A," which appears both in ed. 1 and on Mr. Wedderburn's proofs; and (v.) notes for other classes (pp. 451–456), which do not appear in ed. 1, being here added from Mr. Wedderburn's notes.

The principal *Variæ Lectiones* between eds. 1 and 2 have been already described, or are noted below the text. The Museum copy of ed. 1 has the lines numbered in print at the side of each page ("5," "10," "15," "20," "25"). It also contains in print several notes of prices, etc., for the author's convenience. Occasionally these are noted in the present text (*e.g.*, p. 444); the others are: A. 51, "Wright, mixed specimens bought 10th July, 1877"; A. 52, "Wright, mixed specimens"; A. 53, "Wright, mixed specimens, 5s. First specimen in account, July, 1877"; A. 59, "Wright, 6s. Second specimen, July, 1877"; A. 60, "Wright, mixed specimens"; L. 1 (C. 1 here), "Bought of Mr. Richard Talling, 17s."; Q. 27, "Wright, 35s. 1877"; Q. 28, "Wright, 5s. Bought 4th July, 1877."

At A. 34 (p. 437) the Catalogue adds: "the other half of this piece, presented to the British Museum, is No.   of their Select Silicas." The words are now omitted, because in fact the other half was not included in the British Museum specimens.]

PREFACE

THE student using the following catalogue is generally referred to the eighth chapter of *Deucalion* for explanation of its method;[1] but a few words are still needed to justify the broken form in which I permit its publication.

The object of all books used in St. George's schools will be simply educational, not scientific. That is to say, they will never be abstract statements of science generally known, but practically explanatory statements of the small portion of science which it is thought desirable that the pupil should know.

And these explanations will always be given in the way which I think likeliest to make the matter clear to a young reader; and not at all in the systematic way which would appear fittest to a person, acquainted with things of which the ordinary student is, and must for ever remain, ignorant.

But more especially, and pointedly, the books used in the schools of St. George will avoid any attempt at scientific *classification*, because, as I have already explained again and again, in my Oxford lectures,[2] no existing scientific classification can possibly be permanent. The only systems yet of any real value, are those which, founded on easily visible phenomena, enable the young student with least pains to gather for himself the materials of future labour.

Thus I arrange Amethyst under a separate head from Quartz, in order to direct the student's attention to the fact, of extreme importance to all his future reasoning on the subject, that amethyst, though indistinguishable unless by some minute admixture of iron or manganese from common rock-crystal, is yet always, by the practical powers of nature, distinguished in this notable manner, that when the two minerals occur together the amethyst is characteristically superposed on the quartz, and only in the rarest instances quartz on amethyst. And again, Carnelian[3] is separated from Agate, in which, nevertheless, it constantly occurs as a component material, in order to direct the student's observation to the special manner in which ferruginous stains are diffused in gelatinous silica, forming a substance so pleasant to the eye and so permanent in its quality as to have affected in no small degree the entire current of the arts of humanity, by the occupation of its most industrious and skilful sculptors in the engraving of this stone.

Of course, a collection admitting continual increase cannot be catalogued unless with irregular dispersion of the heads of its nomenclature. But it is necessary for the right use of this catalogue that its pages should not be altered in successive editions; the form in which it is now issued will therefore be permanent; all additions subsequently made to the collection will be indicated in supplementary sheets.

BRANTWOOD, 18*th October*, 1877.

[1] [See above, pp. 197 *seq.*]
[2] [See, for instance, *Eagle's Nest*, § 186 (Vol. XXII. p. 248); and compare *Proserpina*, Vol. XXV. p. 359.]
[3] [Ultimately, however, carnelian was to be arranged with jasper: see below, p. 423.]

CATALOGUE

OF

SILICEOUS MINERALS

PERMANENTLY ARRANGED IN

ST. GEORGE'S MUSEUM, SHEFFIELD

Siliceous minerals are in this catalogue arranged, and lettered for convenience of study, under the following eight divisions :—

1. Flint.	Lettered F.	5. Hyalite.	Lettered H.
2. Jasper.	,, J.	6. Quartz.	,, Q.
3. Chalcedony.	,, C.	7. Amethyst.	,, M.
4. Opal.	,, O.	8. Agate.	,, A.

CLASS I. FLINT. F.

F. 1. Flint pebble,[1] the first I could find of the size I wanted (about that of a walnut) in the loose mould of the flower-beds of Hyde Park (presumably out of the Kensington gravel).[2]

Waterworn (I suppose), but not so much as to round it completely; so that it retains one of its original cavities. These cavities in flint will be found subjects of much interest, hereafter.[3]

On observing its surface carefully, part will be found, practically, quite smooth, while the rest is covered with a kind of reticulation, ineffaceable by ordinary friction, and indicating some interior structure in the stone. Compare British Museum Select Silicas, 1 [p. 399].

F. 2. Piece of a flint pebble from the gravel of Kent, precisely similar to F. 1 in external character; cut and polished. It is seen to consist of three different flinty substances: the principal mass yellow, and rudely disposed in straight bands; the second, dull red, disposed in bands which separate so as to form irregular cells, enclosing the third component substance, common grey *flint*, properly so called. The roughly reticulated surface may be seen to belong to this common flint; the smoother, to the yellow and red portions. Both the yellow and red are properly to be

---

[1] [This specimen is mentioned in *Deucalion*: see above, p. 167.]

[2] [The Museum copy reads, ". . . Hyde Park, by way of the simplest beginning of geological investigation in the capital of England."]

[3] [See, for instance, F. 6 (p. 421) and C. 5 (p. 425).]

419

called "Jasper." The yellow often occurs in large masses, and is very often banded in this more or less straight way, right across. The red rarely and partially occurs banded; but when in bulk, is mossy, or compact.

F. 3. The central piece of the pebble from which F. 2 was cut, showing on one side the natural colour of the cut surface, unpolished; on the other, a natural fracture of the stone weathered; the bands of the yellow jasper being more clearly shown by the weathering; but I do not know what has become of the red.

F. 4. Banded flint, passing into yellow jasper, with a concretion of red earth, in a form which may be plainly called "almondine" instead of in Latin "amygdaloidal"; meaning that the concretion is more or less in the shape of an almond.[1] The frequency of this form of siliceous concretion, and the reasons of it, will be a subsequent matter of study.

This flint contains many interesting phenomena, of which the principal are [2]—

α. The formation of its outer coat with small circular cavities, and a larger irregular, yet smooth, depression.

β. The secretion of a porous white mass, partially intervening between the outer coat and the interior, and containing two quite regularly and smoothly formed almondine cavities.

γ. The secretion of jasperine substance,—it is scarcely distinct enough from white flint to be called jasper,—in a double series of bands, one series intersecting the other.

δ. The almondine secretion of red earth, mostly siliceous, but not fine enough to take polish.

F. 5. Common pale brown flint, presenting casts of marine organisms.

This is a rolled pebble, whose surface, though a little wider in reticulation, closely resembles that of F. 1; while the mass is here composed of the ordinary substance of which F. 4 is a more finely collected or secreted state. The fineness of this siliceous paste is sufficiently shown by the extremely delicate impression of a plant, and a small shell; the former, under a good pocket lens, being a more beautiful example of a cast of vegetable form than can usually be found in larger specimens.

F. 6. Common grey flint, with white coat exhibiting cave-structure. I shall use this term to express the formation of hollow spaces, without traceable cause. The inner mass containing organisms, and showing incipient conditions of brecciation.

The fossil sponge contained in this stone is merely another and clearer example of the complete transformation into silica seen in the last specimen; but the inorganic actions are of more complex kind, and may be described as follows—

α. The white coat is, in many respects, like that of F. 4; but red jasperine secretions variegate it like marble, where its outer surface is broken; and, instead of being sharply separated from the inner mass, as

[1] [Compare above, p. 210.]
[2] [For Ruskin's questions, numbered with Greek letters, see above, p. 203.]

in F. 4, it penetrates the stone in irregular spaces, of which a certain number follow a fine tortuous fissure, traced, like a red thread, by oxide of iron, and connected with the formation of a cavernous hollow, large in proportion to the size of the stone, and partly surrounded by what we shall afterwards recognize to be *agatescent* bands : a smaller hollow, with fine granulated interior surface, occurs deeper in the stone.

β. On the polished side of the stone, holding the oval fossil upright, and the white coating above, the grey part of the flint will be seen to consist of three different states of substance : a dark grey ground, with white spots (half of a very large one is conspicuous low down on the right); a paler grey mass surrounding the fossil, not spotted ; and, lastly, two paler grey masses, not spotted, on the left. Each of these portions of the stone are noteworthy. Under our present letter, β, we will class the dark grey mass illuminated by the white spots, of which the large one on the edge (that portion of the stone showing us two sections of it) is the most characteristic. It has what we shall afterwards recognize as a partially reniform or chalcedonic outline, and is, in common with the other spots under consideration, perfectly united with the substance of the stone, showing no tendency to form cavities.[1]

γ. The paler grey mass detaching itself from the dark ground round the fossil appears produced by organic influence ; and I cannot show it as a purely structural condition of silica.

δ. But the two spaces on the left are the first instances we have seen of a most important structural state. They are separated, by lines so sharp as to look like fractures, from the formation γ ;—by lines less sharp, but still distinct, from the formation β ;—and, finally, the one highest on the left enters the formation a, with two rounded points, like a grey cloud rising against a white one. The significance of this minute piece of inlaying[2] will be only understood as we farther learn the system of flint architecture.

F. 7. Portion of brown agate, showing the first structural conditions of agate in simplicity.

In the last specimen, the central sponge shows, in combination, the two principal colours of flint-chalcedony, brown and grey. Nos. F. 6 and F. 7 show these colours in separate mass. The brown is the prevailing colour given to imperfect chalcedony by oxide of iron ; and is especially characteristic of the chalcedonies of chalk, greensand, and other geological formations which have not been subjected to extreme heat. In volcanic chalcedony it is seldom, so far as I know, completely diffused ; but associates itself with purer colours.

I do not know the locality of this example ; but it is of great interest, —first, in its exhibition of the pure almondine outside form (which is so liable to be confused with that of a rolled pebble) in its clear relation to the internal structure of the stone ; secondly, in the manner of the interruption of that structure at one point in the oval ; thirdly, in the four distinct courses of formation,—A, brown agate ; B, coarse white quartz ;

---

[1] [The Museum copy adds, "like the white spots connected with its coating."]

[2] [The Museum copy adds, ". . . this minute piece of structure, like the inlaying of the carved keystone in a Lombardic arch, will be only . . ."]

C, white jasper coating this quartz; D, fine translucent quartz filling the interior. The example is placed, however, in this part of the series merely for its colour, and oval form, together with a flinty surface.

F. 8. Portion of grey agate, showing agate structure of ordinary character.

The somewhat monotonous silver-grey may be considered the natural colour of true chalcedony, passing, in more beautiful examples, into various tints of milk-blue, violet-blue, and purple. The fine layers of which its masses are usually composed cannot be clearly seen in its translucent state; but they may be seen defining themselves clearly at the edge of the fracture on the unpolished side, otherwise interesting because illustrative of the opacity which seems to be produced in stones of this kind by weathering.

All stones of fine siliceous substance, disposed in well-defined layers, are properly called Agate. The *fine substance is essential;* a stratified chert or banded flint would not be called an agate.

From this specimen, up to which we have traced the gradual refinement of common flint, the study of agate must be pursued in the eighth section of silica [p. 435]. We here return to states of pure flint.

F. 9. Massive flint, entirely filling and partly enclosing a large fossil shell, with fragments of others. Accurately representative of the substance properly called flint in its sincerity.[1] See the references to this specimen in the *Grammar of Silica*,[2] of which I repeat here one or two for the student's convenience.

The broken *surface* of perfect flint is feebly lustrous, passing into entirely dull smoothness on one side, and into a moderately bright polish on the other; but it is never perfectly dim, as jasper is; nor perfectly lustrous, as glass is. For its capacity of artificial polish, see F. 12.

The *colour* of perfect flint is the dark grey of this specimen, passing into jet black on one side, and dull white on the other.

The external coat of common flint is white, more or less thick. I have chosen this one for the extreme simplicity of its substance, and thinness of its coat.

Flint has also a way of lapping and slopping itself about the bodies it encloses, like dropped plaster, of which aspect we have here a fine example.

---

[1] [The Museum copy does not contain the reference to the *Grammar of Silica*, and reads :—

"... sincerity. It is at once tough and brittle, so that, though thus violently broken, it shows no flaws in its mass, and the fractures show curved and zoned surfaces, which from their resemblance to shells are called 'conchoidal.' Thus the broadest fracture here, ending in a short point, looks almost like the impression of an oyster shell. The surface . . ."

The Museum copy also adds these "Questions respecting this sample :—α. Reason of the black colour. β. Reason of the white coat. γ. Reason of conchoidal fracture. δ. Reason of dead surface."]

[2] [See below, p. 534. The specimen is also referred to in *Deucalion :* see above, p. 210.]

F. 10. Common flint, showing a tendency to become stratified, and an increased thickness of its white coat.

In all rocks secreting flint, organic bodies are so frequently present, that it is next to impossible to find a piece of flint clear of them: but I believe the form of this example to be essentially inorganic, and not constructed upon any flat fossil. But the projecting nodule is, I doubt not, an attached sponge, or some such thing.

F. 11. Common flint in a double nodule formed on a sponge, enclosing the cast of the shell of an echinus in carbonate of lime, showing various phenomena in the white coats, though still thin, and the yielding nature of the sponge when the shell was imbedded in it.

F. 12. Splinter of flint, polished on one side.

F. 13. Flint containing crystalline iron pyrites.

The iron is here, I doubt not, secreted from, or through, the flint, as any crystal of one substance is found in another. The flint is not gathered round, or poured over it, afterwards.

F. 14. Flint passing into chalcedony, with branching sponges. The lapping wave-like lines of chalcedony are, however, an inorganic structure. The forms of granular chalcedony and minute quartz in this stone are of exquisite interest.

F. 15. Second portion of F. 14, showing the lapping waves to have been formed over some imbedded shell, as ripples over a sandbank.

F. 16. Flint passing into chalcedony, with imbedded sponges, seen clearly in its outer surface.

F. 17. Flint with imbedded echini, and their spines, of extreme beauty.[1]

F. 18. Flint with imbedded fragments of Inoceramus.

CLASS II. JASPER. J.

In this class I shall arrange, with the ordinary forms of Jasper, also the semi-translucent rose or carnation-coloured varieties of siliceous substance called by jewellers "carnelian."

J. 1. Carnelian in semi-chalcedonic concretion, forming a mass with level beds at the base, a rudely obtuse chalcedonic surface round a hollow in the centre, and concave-chalcedonic external surface, with darker knots, to me at present quite inexplicable.

I believe quite natural in colour, and very fine. Polished on the flat under surface. (India.)

---

[1] [The Museum copy adds "Examine with lens," and to F. 18 adds "Splendid."]

J. 2. Portion completing the mass of J. 1, polished on the vertical fracture, which has been singularly even. The whole of extreme interest in relation to the structure of level-bedded or "lake" agates.[1]

J. 3. Red Jasper, or Sard, rudely chalcedonic in central cavity, placed here for comparison with the structure of J. 1 in the central cavity.

J. 4. False (dyed) Carnelian, showing the way in which the artificial colour pervades and pollutes the entire substance of the stone.

J. 5. Exterior portion of J. 4, with natural fracture, showing the manner in which the concave-chalcedonic surface of J. 1 has been produced.

J. 6. True Carnelian, forming bands in moss agate.

J. 7. Portion of J. 5, cut thin, to show the mossy structure of the exterior.

J. 8. Portion of a large ball of amethyst quartz, formed round a nucleus of agate, veined with vermilion jasper, and dark carnelian, traversed by cross veins of brown jasper and more or less corresponding spotted clouds of vermilion jasper.
Exquisite; and of extreme importance for study of structure.

J. 9. Stalactitic agate with mossy central nuclei; passing in its outer coats from iron-brown to a purple carnelian colour of extreme rarity (unless it has been artificially modified).

J. 10. Scarlet Flint (semi-carnelian), secreting itself in bands within white external bands, round much-fissured orbicular centres, out of a quartzose paste. It is possible this specimen may be a conglomerate, but I believe very firmly not.

CLASS III. CHALCEDONY. C.

C. 1. Chalcedony, first coating quartz, then dripping over the coat, the drip being in various directions, appearing partly as if directed by weight, partly as radiating from a centre, or irregularly impulsive. The quartz on which it is poured seems interiorly massive, but it is not: it radiates rudely from a rocky centre, purifying itself in zones, to the exterior, where it forms small perfect apices of crystals to receive the chalcedony.
(Redruth, Cornwall.[2])

---

[1] [See above, p. 379.]
[2] [There is a reference to this specimen in the *Grammar of Silica*: see below, p. 536. The Museum copy of the Catalogue adds :—
"Questions. α. Original state of radiating quartz. β. The nature of the pause between its formation and the pouring of the chalcedony on its surface. γ. The condition of the chalcedony when it formed the coat only. δ. The condition of the chalcedony when it began to drop."]

C. 2. Stalactitic chalcedony, becoming agate by development of interior zones (see the polished portion), and coated by finer stalactites dripping at a different angle, the whole having been formed on a rock which seems to have had influence on the formation of the stalactites by causing honeycomb-like cavities in their centres.

The apparently loose white earth which clogs the centre of this specimen, will be found, on examination, arranged in concentric coats of radiating fibres round the stalactites; and suggests the idea that the whole mass was once solid, and the stalactites formed out of it by contraction. The clogging earth is siliceous like the rest.[1]

C. 3. Coarse grey chalcedony, with white concretions in the centres of its masses, and associated with porous rock, confusedly suggestive of action like that in C. 2.[2]

C. 4. Chalcedony forming itself in detached concretions within flint.

Look, with a lens, at the little grotto just above the number. It is formed of the white substance above spoken of, and contains a shell-like piece of it in the centre. With this compare the formation of the large grey detached, or semi-detached, coats on the left.[3]

C. 5. (Two pieces, marked a and b, placed with C. 6 in this part of the series for immediate comparison with higher forms of chalcedony; compare also C. 16, C. 17.) Roundish nodule of flint, with cavity (left by a perished sponge?), partially filled by chalcedony, strewn with small quartz crystals.[4]

Secretion of flint is here unusually simple and massive; notable as not following the outline of the sponge, but taking a rudely rounded contour of its own.

---

[1] [The Museum copy adds :—
"Questions. α. Nature of rock causing cavities in the mass. β. The nature of the change which causes the stalactites to become concentrically zoned. γ. The nature of the two forces which direct the stalactites at different angles."]

[2] [The Museum copy adds :—
"Questions. α. The relation of the white centres of concretion to the external white coating of flint. β. The mode of association with the rock which prevents the chalcedony from forming itself purely and finely."]

[3] [The Museum copy adds :—
"Questions. α. That common to nearly all flints : the nature of their white coating, and its cause. β. The manner of contraction in the chalcedony if the mass was once solid. γ. The manner of deposition in the chalcedony if the mass was once hollow. δ. Has organic matter been partly the origin of either structure?"]

[4] [The Museum copy adds :—
"Questions. α. Mode of original secretion of the flint from the chalk, round the sponge. In this specimen, the secretion . . . its own. β. Mode of disappearance of the organic matter of sponge. γ. Mode of secretion of chalcedony, which appears in the thinner fragment of the stone to be stalactitic. δ. Mode and period of crystallization which equalizes and limits the size of the quartz crystals."]

C. 6. Nodule of solid flint, with fully developed external coats, apparently detaching themselves.[1]

Is the thick outer white coat broken away accidentally in some parts, or originally unformed round the oval concretions? In either case, what determines the distinct separation of this outer coat from the inner mass?

(From the chalk of Abbeville, France; taken by myself from its site in the cliff to the south-west of the town.)

C. 7. Pure chalcedony, forming nodules in volcanic rock.

This is an entirely characteristic specimen of the mineral properly called chalcedony, being massive; translucent, but not transparent; not traversed by bands, and externally presenting the appearance of bubbles on a boiling liquid. But they are not produced by ebullition, but by a fine spherical crystallization, the fibres of the spherical masses being, in chalcedony, too delicate to be visible.

Questions. α. Are the cavities in the volcanic rock first formed by gas, and the chalcedony afterwards introduced, or is the chalcedony secreted at once as the cavities form?

β. In either case, what must have been the condition of the chalcedony at the time of its introduction?

γ. Supposing it introduced at a high temperature, would it contract on cooling?

δ. If it was so introduced, and so contracted, why are there no cavities in the masses occupying smaller hollows, while the larger cavity is only coated to the depth of half an inch, thus leaving a large space in the interior unfilled?

C. 8. Agatescent chalcedony, closely crowded in its spherical crystallization, and beautiful in its blue colour: the bands, by which it is distinguished from common chalcedony, appearing in process of formation.[2]

C. 9. Perfectly agatescent chalcedony, showing the position of a stalactite with an external coating, the interstices of the whole having been filled with massive quartz.

Questions. α. The distinction between the perfectly formed bands in this specimen and the imperfect ones in C. 8.

β. The distinction between chalcedony and quartz, here exquisitely defined.

γ. The mode of separation between these two minerals, producing the stalactitic form of chalcedony.

[1] [The Museum copy reads :—
    "Questions. α. and β, as α and β in C. 5; the secretion here being not apparently determined by organic bodies. γ. Is the . . . concretions? δ. In either . . . mass? ε. The cause of the formation of the elongated nodules and tongues in this inner mass."]
[2] [The Museum copy adds :—
    "Questions. α. The cause of spherical crystallization in rapid interstices as opposed to the bolder forms in C. 8 (?). The masses in that would, however, look bolder in a larger piece, being subdivided on their surfaces

C. 10. Perfectly agatescent chalcedony, apparently breaking up the enclosing rock into tabular splinters, and enclosing straight beds in its interior, probably *in situ*, parallel with the horizon.

Questions. α. Nature of the tabular fragments of rough rock, and mode of their separation.

β. Nature of the straight and probably horizontal beds, better traceable by comparison with following examples.

C. 11. Perfectly agatescent chalcedony, filling a cavity in a brown rock, which it seems partially to have torn up, as in C. 10, into slaty splinters, but itself enclosing crystalline quartz, instead of the straight beds of C. 10.

C. 12. Perfectly agatescent chalcedony, a portion cut from a mass filling a cavity in brown rock, which it is apparently dividing into fine dendritic or moss-like branches and films, beautifully seen when the transparent stone is held up against candlelight.

Questions. α. Mode of division or secretion of the mossy portions (to me, at present, wholly inscrutable).

β. Manner of formation of the bands of the chalcedony, after these mossy fragments were in their present or an analogous position.

γ. Relation of these mossy fragments to the splinters in C. 11, on one side, and to the moss, so called, of dendritic or common moss agates, on the other.

C. 13. Common chalcedony, entirely filling small cavities in volcanic rock (trap), each cavity being primarily coated by fine dark bands forming spherical concretions, apparently filled with, and formed out of, the substance of the enclosing rock. Taking a lens of ordinary power, and beginning with the little jag in the outside edge of the stone to the left of the number, the bands may be seen in that jag, and the three chalcedonic concretions above, in perfection, and will be recognized as often appearing in the stone, where there is no room for the interior chalcedony.

Questions. α. The consistence of the stone when the cavities of this irregular form are developed, as compared with the state producing cavities like those of C. 12.

β. Nature and formation of the spherical bands, which may be studied to advantage on the weathered side of the stone.

(Broken by myself from the trap-veins in Wallacrag, above Derwentwater, Cumberland.)

C. 14. Amethyst quartz, with a dull band of chalcedony at the roots of it, apparently modifying the structure of the enclosing rock, but arranging itself also so as to produce cellular spherical cavities.

> like large waves into small ones. β. The manner of development in the bands. γ. The nature of the inner rough surface frequent in this kind of chalcedony.
> "I believe from Iceland, but perhaps as probably from Felsobanja, Transylvania."]

C. 15. Amorphous chalcedony, dimly banded when seen against light, traversed by stalactitic concretions, rudely stellar in transverse section, and which appear to modify or direct the agatescent structure of the mass, so that the bands form round the extremities of the stalactites.[1]

C. 16. Chalcedony with milky surface, in partial stalactitic development in white flint, full of inexplicable phenomena.

C. 17. Minute chalcedonic formation in apparently calcined flint, rent by fissures.
No account has, I believe, been given of this deceptive aspect in flint.[2]

C. 18. White chalcedony, perfectly opaque, and disposed in reniform concretions, beautifully smooth in dome, having all the interstices filled with quartz, which radiates in harmony with the chalcedonic structure, while a fresh concretionary action, in spheres penetrated by minute vesicular cavities filled with clear chalcedony, takes place in the opaque chalcedony.[3]

C. 19. Flint containing fossil (sponges?) filled by blue chalcedony. Placed in this series, not for its fossils, but its chalcedonic development.

C. 20. Disturbed concretion, of the basest kind of chalcedony, in pale flint.
The formative action here is to me wholly inexplicable. (From Watford Tunnel.)

C. 21. Translucent, chalcedony coating amethyst quartz, which itself is formed on a nucleus of green fluor spar: all the three minerals delicate and refined in their kind. The dropped concretions, partly in rods, on the surface of the first chalcedony, are of extreme interest.

C. 22. Pure chalcedony, with milky surface, containing bands of moss jasper (not organic, I believe), and with vesicular under surface of extreme interest.

C. 23. Brown agate, connecting the chalcedonic with simply banded silica. Against the light it will be seen to be reniform on the flat side (but not on the arched one), over marvellous shell-like concretions of extremely rare fineness and beauty forming its outer coat.

[1] [The Museum copy adds:—
    "Questions. a. Direction of stalactitic action. β. Mode of stellar, as opposed to smooth, concretion in stalactitic masses.
    "Specimen part of a large one in my own collection,—I believe Indian."]
[2] [The Museum copy reads, "of this form of apparently fired flint."]
[3] [The Museum copy adds:—
    "I very much grudge this specimen, having no other like it, and will in future substitute another for it, if I can find one exhibiting the same phenomena as distinctly on a larger scale; the extreme delicacy of this making it less likely to be studied."]

C. 24. Brown agate, examined by Mr. Clifton Ward. Section A. See *Deucalion*, vol. i. p. 268.

C. 25. Brown agate, examined by Mr. Clifton Ward. Sections B. 1 and B. 2 were taken from the opposite extremities. *Deucalion*, vol. i. p. 270.

C. 26. Wax chalcedony, examined by Mr. Clifton Ward. Section C. [*Deucalion*, vol. i. p. 270.]

C. 27. Stellar Indian Chalcedony, like that of Auvergne.

C. 28. Pure stalactitic chalcedony. (Cornwall.)

C. 29. Pure branched (or bird's nest) chalcedony. (Cornwall.)

C. 30. Trickling chalcedony in primary and secondary formation over amethyst quartz, the whole on an agatescent base. Of extreme interest.

CLASS IV. OPAL. O.

O. 1. Noble opal in the matrix. Polished on the edge; the red and fiery parts especially are of highest possible jewellers' quality; and the rough parts, unpolished, at the other end, show the intrinsic beauty of the stone.[1] (Hungary.)

O. 2. Variety of noble opal, called hydrophane, not showing its full colour until wetted.

O. 3. Noble opal, of fine jewellers' quality in veins, very characteristic. (Hungary.) Wright, £3.[2]

O. 4. Noble opal, in dispersed gelatinous mass. (Hungary.)

O. 5. Noble opal, in veins passing into common opal. (Brazil.)

O. 6 to O. 11. Noble opal, level barred, in black matrix. (Brazil.)

O. 12. Noble opal, level barred, in black matrix. Magnificent. (Brazil.)

O. 13. Noble opal, dispersed in brown matrix. (Hungary.) The eight pieces, O. 6 to O. 13, bought of Mr. Bryce Wright, £6, 10th July, 1877.[3]

---

[1] ["Unexcelled in England," says Ruskin of this specimen (*Master's Report on St. George's Guild, 1881*, § 3).]

[2] [Mr. Wright, of Great Russell Street: see *Fors Clavigera*, Letter 70, § 13, for this purchase.]

[3] [Mr. Wedderburn's proofs contain the following fuller notes on O. 12 and 13 :—

"O. 12. Precious opal in a film on dark rock, the film itself, I believe, a thin stratified plate or vein, transversely barred into four bars :—

"A. A dark and most precious bar (with irregular edge caused by rounded cuttings), itself barred transversely again by green and blue colours. I

O. 14. Noble opal, straight barred, in porphyritic matrix. Superb.

O. 15. Piece cut from O. 14, to show porphyritic base.

O. 16. Opal passing into chalcedony, an extremely rare condition. Wetted, and under lens, the crimson fire in portions of this stone is quite of unsurpassable brilliancy.

O. 17. Opal, transparent, throwing beautiful green light, forming the centre of a brown siliceous nodule of quite anomalous formation.

O. 18. Opal, throwing feebly green and red light; but partially agatescent, forming a nodule of undulated and irregularly extended zones, in a sandstone matrix. Of extreme rarity.

O. 19. Milky opal, throwing blue, green, and fiery lights, disseminated in a brown siliceous matrix. A newly discovered condition.

O. 20. Dark blue opal, passing into green, forming veins in ferruginous jasper. (Australia.) Presented by Mrs. R. C. Nockold.

CLASS V. HYALITE. H.

The first seven specimens in this series are described at length in *Deucalion*, vol. i. pp. 235-238.

H. 1. Hyalite, on cellular lava. (Probably from Auvergne.)

H. 2. Compact globular Hyalite. (India.)

H. 3. Wax chalcedony, probably formed by dropping water at high temperature. (India.)

> have not yet seen this condition of double or woven bar in any piece I know.
> "B. A dark grey colourless bar.
> "C. A broad white bar, becoming gradually filled with blue, red, and green pie.
> "D. Another grey bar, of which only a small portion is seen.
> "Unique in beauty and regularity, as far as I have seen.
> "Wright. Finest piece of the eight, bought altogether for £6.
> "O. 13. Precious opal, finely dispersed in porphyry. A square-cut slice injured by further trace of the slitting-mill, and cement on edges, but of rare beauty in its substance, and of more interest than most specimens of this formation in showing the termination of the opalescent mass by the compact brown rock.
> "Question. a. Is this the edge of a vein?
> "Of opal generally, see the stupidest book I know of in the world, but containing a quantity of facts in the lumber of it which are useful: Bischof's *Chemical Geology*, vol. ii. pp. 459-464. Note especially the statement at p. 459 that noble opal, when heated, evolves ammonia, and that banded opal is full of infusoria!"]

H. 4. Sugar chalcedony, *i.e.*, wax chalcedony, with minute external crystallization of quartz. (India.)

H. 5. Star chalcedony, passing into Hyalite. (Auvergne.)

H. 6. Wax chalcedony with spherical cavities, on lava. (Auvergne.)

H. 7. Fluent chalcedony of unvarying depth, on brown iron-ore : a rare condition.

H. 8. Hyalite, partly white, partly clear, on lava. (Hanau.)

H. 9. Hyalite, a newly discovered form, transitional to chalcedony, forming geodes containing air and water, in basalt. (Uruguay.)

H. 10 (A and B). The two halves of such a geode, cut open and polished on the sectional surfaces : exposing the crystallization in the interior hollow, and the modes of chalcedonic aggregation.

CLASS VI. QUARTZ. Q.

Q. 1. From the beach at Barmouth. Common white quartz pebble, such as may be picked up in any district, formed of débris of quartzose mountains. Feebly translucent, and with structural, not accidental, flaws, indicated by red oxide of iron. Immense masses of the mountainous body of the earth are formed of quartz in this state. It differs from flint in being semi-crystalline, and free of the blackening substance, whatever that is ; and, I believe, by having no water in it.[1]

Q. 2. Quartz of exactly the same consistence as Q. 1, but rising into agatescent crystals, coated with red oxide of iron, with finely superimposed concurrent quartz and fluate of lime.
Marvellous throughout, and illustrative of more phenomena than can be counted.

Q. 3. Cap quartz, formed on common yellow quartz, as Q. 2 on common white. Three pieces detach themselves from the matrix in this example, showing the banded structure of the crystals.

Q. 4. Banded quartz, forming pyramidal cumuli, limited by concurrent arcs. I have never seen quartz in this form detach itself in caps. The coarse superimposed fluor, sprinkled with iron pyrites, has unusual characters at the angles.

---

[1] [The Museum copy adds, "Compare Question *a*, F. 1"; that is, F. 9 in the present Catalogue. See above, p. 422 *n*.]

Q. 5. Portion corresponding to Q. 4.

Q. 6. Quartz terminated by straight planes (compare the convex ones of I. Q. 5), morbidly * interrupted in formation towards the same side in all the crystals.

Q. 7. A single crystal of straight-planed quartz, morbidly thrown into small crystals on one side. I have had this stone cut at the base, to show the successive or concretionary lines of crystallization, which, it will be seen, vary their angles constantly.

Q. 8. Quartz with sulphuret of silver, itself formed in a spherical concretion of short crystals on a brown rock, which, through the lens, will be seen to be singularly divided, containing small portions of sulphuret of iron.

By comparing the smaller crystals, all in confusion, in Q. 7, with the comparatively symmetrical setting of points outwards in Q. 8, resembling the jagged surface of a knight's mace, the structures henceforward to be called "Solute" quartz, and "Mace" quartz, will be understood sufficiently.

Q. 9. Quartz shooting into long crystals out of a solid mass containing darker substance, with sulphuret of iron, and finally superimposed carbonate of iron.

The crystals in this example will be seen to taper towards their apices. Quartz of this form will be hereafter called "Spire" quartz.

Q. 10. Quartz in an aggregate of crystals, forming themselves into a mass partly covered with solute quartz. The crystals themselves are notable for their approximately equal size, and for being formed each of two clusters, joined at a kind of waist in the middle, and expanding to the extremities. This form will be called "Sheaf" quartz.

Q. 11. Common pure quartz, with attached crystals on its flank.[1]

Q. 12. Flute-beak quartz;[2] a single crystal showing the mode of terminal truncation. (Dauphiné.)

Q. 13. Quartz ill made (American) to oppose to fine Dauphiné structure; cut across to show mode of aggregation. Compare Q. 25.

---

* I use this word when crystallization is interrupted in a persistent manner, and not by a single accident.

---

[1] [For a reference to this specimen, see *Deucalion*, i. ch. vii. § 20 (above, p. 175).]
[2] [See *Deucalion*, p. 204.]

Q. 14. Companion piece of Q. 13.

Q. 15. Fine quartz, with interior crystal having equal planes. The outer planes polished to let it be seen.

Q. 16. Quartz of same fine quality; outer planes natural, to be compared with the polished ones of last example. The external hollows show pretty surfaces inside.

Q. 17. Crystal with outer planes polished, to show surfaces of two successively rising crystals; and cut and polished through the centre to show interior structure.

Q. 18. Companion piece to Q. 17.

Q. 19. White quartz, with planes of minor concurrent crystal, cut transversely, and through apex, to show agatescent structure.

Q. 20. Companion piece to Q. 19.

Q. 21. White quartz in a perfectly typical crystal, of average form, moderately long, symmetrical on all its sides, and terminated at both extremities. All other forms of quartz are to be considered as derived from this, by elongating, or obliquely extending their planes, or unequally truncating their angles; and then attracting themselves variously to the matrix.

Q. 22. Black quartz, with its lateral planes expanded, and curved into a screw-surface as they extend. Superb.

Q. 23. Screw-quartz on matrix. Very fine.

Q. 24. Quartz in crystals, complete on one side, and confusedly aggregate on the other, internally developed in agatescent coats, and, apparently, sprinkled with crystals of black oxide of iron before the superimposition of their last and purest coat. The polished section cuts neatly through four of these crystals of oxide of iron, showing their exact position on the inner plane. But I have no doubt that this aspect of having been first deposited, and then glazed by the final coat, is entirely deceptive. Had that coat been laid over the iron crystals, it would have gone more or less up and down over them, like cement poured over loose stones. But it lies absolutely even, letting the iron crystals emerge through it when large enough; and they are without question formed in precisely the same manner as the sheaves of brown oxide in M. 14, or of actynolite in Q. 29, whatever that manner may be.

Polished on the base, and vertical section, of one of the crystals, and of supreme interest.

Q. 25. Portion corresponding to the vertical section, in Q. 23.

Q. 26. Coarse crystal, cut transversely, and polished on the section. America.

Q. 27. Portion corresponding to Q. 25, but sloped on the section to show a curious modification of the clear central mass at one of the angles.

Q. 28. Orbicular agate (compare A. 42), surrounded by two belts of crystalline quartz, the inner amethystine, the outer green, and gradually arranging themselves into the outline of a large hexagonal crystal.

Q. 29, 30. Broken pieces of a very thin slice of rock crystal, containing green (actynolite?) in radiating sheaves and detached crystals, of singular clearness and beauty. The two pieces are to be kept together for the more convenient study of this form of capillary crystal.

CLASS VII. AMETHYST. M.

M. 1. Group of dark amethyst, reddened by points of oxide of iron, and therefore unfit for jewellery; but as a specimen of the mineral itself, not to be surpassed for size and crystalline beauty.

M. 2, 3, 4, 5. Examples of degrees in amethystine colour.

M. 6. Pale amethyst developing itself over grey folded agate in irregular cells in trap rock.

M. 7. Dark amethyst developing itself above apparently broken fragments of a matrix entirely inexplicable to me.

M. 8. Amethyst developing itself in coats above common white quartz, with uniform crystalline structure, transverse to the bands. Polished on the vertical fracture.

M. 9. Amethyst developing itself in coats and nests, in white quartz, showing warped faults of concretion. Engraved in *Deucalion*, Plate IV.[1]

M. 10. Amethyst developing itself in bands with white quartz; showing warped faults in concretion. The lower portion engraved in *Deucalion*, Plate IV.

M. 11. Amethyst developing itself in concentrated colour on some planes of its crystallization. The smaller group is exquisitely defined when seen with lens on its exact profile.

[1] [Where the plate was repeated from the *Geological Magazine*; in this volume, Plate VII. p. 55.]

M. 12. Amethyst developing itself gradually over white quartz in a single crystal: exquisite.

Polished on the basic transverse section.

M. 13. Base of the crystal M. 12, polished on the correspondent section.

M. 14. Amethyst developing itself in stains of lovely colour through common quartz, such stains being apparently in some degree connected with the presence of fine sheaves of acicular crystals of oxide of iron, of a beautiful golden colour within the crystals, and black where their extremities emerge on the surfaces.

Very beautiful, but only to be seen with lens, and in good light.

M. 15. Amethyst in crystals coated on one side with fine sugary white quartz, developing themselves over white quartz out of a grey matrix showing warped faults of concretion. Both matrix and first secretion of white quartz containing small finely dispersed crystals of sulphuret of iron.

Polished on the basic, and transverse sections, which are both of extreme interest.

M. 16. Amethyst, pure purple, semi-opaque, in radiating crystallization; the most beautiful condition of amethyst quartz used for jewellery.

CLASS VIII. AGATE. A.

A. 1. Opaque agate in defined bands: standard example of structure.

A. 2. Translucent agate in defined bands: standard example of structure.

These two examples are given as types of the most exquisite conditions possible in opaque and translucent agate. Larger and more showy stones are common; but more perfect and accomplished agates than these two do not exist.

A. 3. Portion of agate, itself half jasperine half chalcedonic, exhibiting level lake of pure blue chalcedony; and banded structure proceeding from two foci. This example presents in a single group the principal questions relating to the structure of the finer orders of agates, not folded. (From the Hill of Kinnoull, Perth.)

A. 4. Companion piece of A. 3.

A. 5–12. Seven chosen and typical examples of Scottish pebbles, from the trap-rocks of Perthshire.[1]

---

[1] [For a reference to A. 8, see *Deucalion*, i. ch. vii. § 21 (above, p. 176).]

A. 13. Extremity of a banded agate; exhibiting the level and concentric structures in complex repetition.

A. 14. The portion of same agate sequent on A. 13

A. 15. The portion sequent on A. 14.

A. 16. Opposite extremity of the same agate.

A. 17. Jasperine Fort * agate in union with Conchoidal agate.
The red substance seen in this agate forms, when in less perfect concretion, a large quantity of common gravel-flints. I shall call it flinty jasper, as opposed to the finer kinds, which, I believe, have more alumina in them, and are tougher; this kind being liable to the perpetually occurring flaws which here disfigure an otherwise most precious specimen.

A. 18. Piece of a true Conchoidal agate.

A. 19. Second piece of A. 18.

A. 20. Transverse portion of Conchoidal agate, joining the preceding two pieces. This is a very marvellous example; the little white wheel-like concretion on its convex face, just below the number, contains every phenomenon of significance in the structure of Brecciate agates.

A. 21. Crystalline and agatescent white quartz, with green fluor and amethyst quartz, forming bands in a compact greenstone, itself semi-crystalline, with dispersed copper pyrites. (Cornwall.)

A. 22. Quartz, semi-crystalline, semi-agatescent, in bands, with superimposed fluor, consisting of purple cubes, with a pale green transparent coating.

A. 23. Quartz (with superimposed carbonate and sulphuret of iron) forming itself in light and dark bands round an irregular cavity in a grey rock full of dispersed sulphuret of iron. Typical, for banded structure.

A. 24. Quartz with hornblende (?) in agatescent concretions, sometimes called "Orbicular granite." The canon of agatescent and concretionary structure. (Elba.[1])

A. 25. Orbicular and stalactitic agate, developing itself in quartz. Slice of large block. (Scotland?)

* For explanation of term "Fort" Agate, see A. 40, page 438; for "Conchoidal," see A. 71, page 445; "Brecciate," see A. 35, and the notes on A. 21, 22, 23, in *Deucalion*, vol. i. p. 211.]

[1] [The Museum copy adds, "Compare III. A. 2": see p. 449.]

A. 26. Stalactitic agate, with compact milky agate in beautiful development. Look at it with lens. (Scotland.)

A. 27, 28, 29, 30. Four pieces of one grand ball of "Orbicular agate." See A. 41 and 42, p. 439.

A. 31. Chalcedonic silica in connection with fluor.

A. 32. Companion piece to A. 31.[1]

A. 33. Quartz and amethyst in connection with fluor: a similar (not companion) example to A. 21. Polished to show the concretionary structure of the dark rock.

A. 34. Quartz and chalcedony in concretion with calcareous rock.

A. 35. Crucial example of (so called) "brecciate" or "broken up" siliceous formation. See A. 45, and the sequent examples, with final comment on A. 60.

A. 36. Typical and lovely example of "folded" agate, brown and grey shown on two sections, and having a concave chalcedonic surface exposed on the exterior (compare C. 1 and C. 5), which, it is especially to be observed, does not coincide with the interior bands, but interferes with and truncates them.

These interior bands consist of four white ones, which can be traced in continuity all round, here approximately touching each other, and there separating, with intermediate dark spaces, which, sometimes of great width, become so narrowed or effaced where the white bands are in apparent contact, that one cannot conceive the total number of bands to be all represented where they seemingly disappear; yet they gradate and diminish truly towards those points.* This is the characteristic of true folded agate: that some of its bands shall be visibly continuous, and the rest *narrowed harmoniously* to the points where they vanish—the whole being cast into sweeping curves or folds as of drapery, or of a flowing stream. With this perfect example, the next two are to be compared with extreme care; for these, though they must be included in the general class of "folded" agates, present transitional phenomena which connect them with other groups.

A. 37. Folded agate in transition to Fort agate. Shown on two sections. This example presents the already described essential character of folded agate—that the bands shall be continuous, and cast into folds like

---

* Microscopic diagrams of sections made at such points are extremely desirable.

[1] [The Museum copy adds, "Goes face to face."]

those of drapery; but in this case the bands do not flow *easily*, but are seemingly crumpled and crushed, or uncomfortably stretched by the interference with them of masses of grey quartz, which push violently and irregularly for their places. And, on the larger section, near the, much to be regretted, fracture, the bands, elsewhere curved, are bent into a sharp angle, or rather series of successive angles; which indicate that at this point the interfering quartz is trying to throw itself into a right-planed crystal; which if it had succeeded in doing, it would have formed the bands into true "fort" agate; compare A. 40.

A. 38, A. 39. Correspondent pieces of folded agate in transition to Island agate. Artificially stained black [1] on the outer surfaces; but seen in its natural colour on the section.

It will be seen that the bands in this stone are indeed for the most part concurrent in their course, like those of true folded agate; but that, instead of flowing like drapery or water, they are cramped into more or less angular and broken outlines, resembling those of a rocky coast. And on the outer stained surface of A. 39 the appearance of folding nearly ceases, and the bands take the form common to an immense number of agatescent masses, to which, I think, the obviously descriptive name of "Island" agates may be conveniently given.

It is to be noticed also in this example that the principal white band (see the stained and polished surface of A. 38) appears to be almost violently interrupted by the interior black mass. It is never, in spite of this, abruptly terminated, but contracts itself with an exquisiteness of adaptation like that of organic tissue. Hold the specimen with its rough and opaque edge towards you, and examine with a rather strong lens the diminution of this white band among its companions, over the back of the dark emerging mass.

You must also note, in this piece, the interference of small orbicular or semi-orbicular concretions along its dark lines, and on the outer edge of the white one.

Lastly, on the unpolished surface of it, you will see that a triangular portion of its brown rock matrix has been *caught* in the agate, as if broken away. You are to note this for comparison with the group of brecciate agates.

A. 40. Fort agate, in perfect development. Spoiled by artificial staining, but of extreme value as an example of construction.

A large group of stones, more or less resembling the lines of fortified towns, were once called popularly "fortification" agates. I shall contract this term into "Fort.," and limit it to those stones which are arranged, as in this example, in straight lines, forming clearly pronounced angles.

Such forms are always conditions of compromise between agate and crystalline quartz: this example being of peculiar interest because it shows the separation of the planes on two natural fractures with subdued chalcedonic surfaces.

[1] [At Oberstein: see above, p. 64.]

A. 41. Orbicular agate, taking the pseudomorphic aspect of Fort agates

At the first glance, the central structure of this stone might be confused with that of A. 40. But it has nothing in common with it. The agatescent substance here has been merely superimposed on a crystal of which it took the cast, being itself entirely otherwise minded, recusant of rectilinear limit, and resolved to throw itself, as soon as possible, into *spherical* concretions. Of these it will be seen on the rough side of the stone to have nearly accomplished the formation of one, resembling an imbedded pebble; and that so closely, that no one would be likely to imagine, from that side of the stone, the whole to be concurrent and simultaneous concretion, as in the next example.

A. 42. Orbicular agate in pure structure. (Compare the series from A. 27 to A. 30, p. 437, and Q. 28.) You may at once compare and distinguish it from Folded agate, by its essential character of *parallelism* in the coats. You see that here the coats are all carried through with mathematical precision of parallelism, as if drawn with compasses. Where they appear broader or narrower, it is only that the polished section is not vertical to them. They do not yield to each other, nor interfere with each other; but when there is no more room to pass, stop, with the precision of a Gothic moulding. Compare the manner in which the quartz interstices stop, in this, and in A. 37.

Orbicular agate, therefore, you will observe, is not so called as being always arranged in orbs or spheres; but only as being drawn with the precision of compasses in parallel coats, which *tend* to form perfect spheres as far as the nucleus at their centre permits them. On the other side of the stone, its bands follow the extended nucleus in wide segments of circles; but they never lose their parallelism.

A. 43. Almondine agate in trap-rock.

We must now take up a series of forms which lead to a collateral branch of inquiry. As there is no question whatever that in A. 41 the apparently imbedded pebble is *not* a pebble, but only a compact centre of concretionary energy;—so, in this example, there is no doubt that the apparently imbedded purple agate is not imbedded in, but secreted out of, the circumfluous paste. Nor is there any question in this case whether the cavity has been formed first and filled afterwards. No cavities have existed at any time, but the pure siliceous substance has withdrawn itself from the felspathic paste in larger or smaller masses of various shape, in which one condition is unfailing—that they shall be all surrounded by a green coat, lined with a white one; these two films of green and white appearing together in the formation of innumerable small spots throughout the rest of the stone, without the interior silica. On the opposite side of the stone, the green substance forms a narrow continuous vein, nearly an inch long.

I hope to receive from my chemical friends, some day, complete analysis and account of this green coating of volcanic agates.

A. 44. Almondine (?) agate in quartzose paste.

This example of what is ordinarily called by geologists "pudding-stone," is the first of the present series respecting which there is open question as to the manner of its origin. I *know* that the apparent pebbles in A. 42 and A. 43 are not pebbles. But I am only under strong conviction that the apparent pebbles in A. 44 are not pebbles.

I therefore put a note of interrogation to my word "almondine." Observe, I propose no theory. Here is simply a thing to be ascertained left open for ascertainment.* But, respecting the terms of question, note—

First. Seen through a good lens, you will find the quartzose paste cellular, the cells for the most part angular in outline, filled with transparent silica, and in places enlarging so as to admit red or grey secretions of silica. Hold the piece in your left hand, with the largest of the pebble-forms to your left, in the hollow between your thumb and forefinger. Then examine the paste at the right-hand bottom corner, and on the right of that largest pebble-form (where there is a little jagged fracture) examine the three small grey secretions.

This cellular paste is a common constituent of so-called pudding-stones, and I want carefullest sections and descriptions of it. Compare A. 60.

Secondly. You see that nearly all the pebble-forms in this stone are surrounded by a *black* coat, as the pebble-form in A. 43 with a *green* one.

Supposing these are indeed rolled pebbles and not secretions, this black coat is the effect of weathering on the former beach.

Supposing them secretions, it is a primary agatescent band; which I believe it to be. Examine, for its typical state, the largest pebble-form on the right side, upper corner, where it is dendritically ramified into interior jasper; and then examine the wedge-shaped pebble-form below, where a portion of the black substance is taken up by the jasper, and included as a distinct pebble.

A. 45. Brecciate agate. Primal form; jasperine paste, secreting common flint. (Coast of Devonshire.)

Immense masses of this, ordinarily called, "conglomerate," exist in the neighbourhood of Sidmouth. I do not believe it to be a conglomerate at all, but a marvellous jasperine paste, secreting angular spaces of flint. Hold the piece with its scarlet edge downwards, and examine the horizontal secretion of grey opaque silica running in from the mass on the right, and then that of transparent flint above. There is no end to the wonder of this piece, well examined.

A. 46. Jasperine agates in porphyritic paste.

The four specimens, A. 43–46, are to be arranged in the cases together, A. 43 as a stem, and A. 44–46 as three branches: 44 central, 45 on the left, 46 on the right.

---

* For more positive statements on all these points I now refer the reader to my paper read before the Mineralogical Society, this year, at Edinburgh, and published as the introduction to my reprint of the geology in *Modern Painters*, "In Montibus Sanctis."[1]

(BRANTWOOD, 26th *Sept.*, 1884.)

---

[1] [See now, above, p. 386.]

Then we have in the complete group, A. 43, indubitably secreted agate in trap rock; branching into indubitable brecciate agate, A. 45, on the left, and indubitable porphyritic agate on the right; while in the centre, the questionable piece, A. 44, connects the entire system with true breccias, or agglutinated diluvia.

There is no question, therefore observe, whatever, respecting the origin of the agates in the example A. 46. They are just as essential an element of the whole rock as the porphyritic crystals in the paste enclosing them, and they contain examples of nearly every method of stellar and orbicular concretion, but no levelled lines. On the rough surface, the resemblance to a breccia is complete.

Holding the specimen with its longest edge downwards, and its chief red agates at the top, and drawing a line through the two on the right of the red cluster downwards, it will nearly pass through three small white ones, of which the uppermost is a beautiful example of semi-orbicular concretion, begun at the side of the stone.

We will now follow out each of these branches with such other illustrative examples as we can.

A. 47. Brecciate agate: fine jasperine cellular paste; secreting angular and almondine flint. Rare, and beautiful.

Compare the flat angular pieces in this example near its centre with the flat angular pieces in A. 45, and the almondine and pebble-like secretions with A. 44.

I state this piece, without any note of interrogation, to be segregate; but I admit a shade of doubt relating to it. It stands, in fact, as far as my present knowledge extends, in clearness of evidence, about midway between the indubitably segregate structure of A. 45, and the admittedly dubitable one, A. 44. The jasperine paste in its cellular structure, and the dark-coated almondine secretions, are exactly correspondent to this last-named example.

A. 48. Brecciate agate in secondary development; as a white jasperine paste, secreting chalcedonic and quartzose veins.

I hesitate whether[1] to say, of such examples as this, that the jasper is secreted by the quartz, or the quartz by the jasper. But, broadly, a rock must always be considered as secreting the material which fills its veins. The opaque mass of all rocks, generally speaking, secretes the clearest and purest element of its substance in separate veins. And as argillaceous limestone secretes pure carbonate of lime,—or black slate, quartz,—or altered gneiss, quartz and gold,—so here, the white jasper, retreating into agatescent coils, secretes the transverse veins of crystalline or compact quartz.

Hold the specimen with the pointed end of it downwards and the veins upright. Then, with a lens of considerable power, begin examining it on the left-hand side.

α. You find first an upright vein of grey quartz, semi-chalcedonic at the edge;—interrupted and shifted by a transverse wedge of the same substance,

---

[1] [The Museum copy reads, "It has been matter of some difficulty to me to determine whether to say . . ."]

which pushes the white jasper away to the left, like a stalactite;—while, above and below, the same formation runs easily into little inlets, and as it were land-locked harbours, in the white agate, appearing again in the middle of it in pools, and being throughout concurrent with it, as its basic formation, under the white band which throughout follows what we thus conceive as the coast line.

β. Secondly. Two upright veins of brown crystalline quartz, concurrent at the bottom, interrupt the jasperine formation by sharp truncation; whereas the grey vein is in harmony with it. I should once have described these veins merely as filled fissures. We shall see better what they are as we go on examining the stone; but be quite clear first about the difference between the concurrence of the grey quartz with the coast line of the jasper, and the abrupt cutting through it of the brown quartz.

γ. Thirdly. Going on towards the right, you find the white-zoned agate interrupted by two exceedingly delicate veins, again totally different in character from the formations a and β.

You see there is a horizontal flaw, more or less structural, but not to be called a vein, running across the stone at one-third of its height, as you hold it point downwards; and just above that, a yellow space of well-defined agate in the middle of the white jasperine bands.

The right-hand corner of this space is cut off by the principal of the two fine veins I am now speaking of; tracing this upwards, you will find you are like to lose it in a lake of grey quartz, which, nevertheless, it traverses distinguishably, and then, forming itself into exquisitely fine agatescent lines, cuts right through, shifting, as it cuts, the three agatescent bands that border the lake; there becoming white in apparent sympathy with the white jasper, it traverses that with equal independence, and stops suddenly at the edge of a new formation (which we shall find presently to be β repeated). Tracing our fine vein downwards you will find it cuts the white jasper again to the edge of this β formation, by a vein of which it is in return cut across itself, and presently lost on the other side. Looking then to the left, you will find another smaller vein of the same formation; which, traced downwards, cuts the agate bands clear across, and then loses itself like the other in formation β.

The shifting of the three bands at the lake border by the first of these veins, though on so minute a scale, is to be carefully noted for a perfect example of crystalline "fault."

δ. Fourthly. We have just seen that the white veins of formation γ are cut across by a brown one, which now examining, we shall find, wherever its course is distinct, to be precisely similar to the vein β, and to a third vein, of which a portion is seen on the right, near the edge of the stone—all these three veins being composed of transparent brown walls with a crystalline centre.

The entire course of the one on the left, and the short portion of that on the right, are perfectly distinct: but the course of the central one, which we are now examining, is quite other than distinct. It is best seen under the horizontal flaw, dividing itself into two branches, of which one cuts the formation γ across, as above described; but above this horizontal flaw, it involves itself in a confused crystalline mass, in which the edge of the white jasperine agate is, as it were, carried away, as a piece of crumbling coast by inundation; and while retaining some consistency in its bands near the top of the stone, is yet on the whole broken up into a quite indescribable confusion of fragments, like ice partly broken, partly melting, but more or less governed in their disorder by a system of fine vertical veins connected with formation γ. The greater part of this broken system is finally terminated and enclosed by the main branch of the formation we are now examining, δ, which separates the agate in this place from the pure quartz of the exterior; but below (about the horizontal flaw), the white agatescent formation rallies and re-asserts itself with a final effort; cut fiercely through by the third change of formation δ, and again by a most subtle and fatal one of formation γ, it expires, or diffuses itself, at last in the terminal quartz.

ε. It remains only to be noticed that the horizontal flaw so often spoken of, though in places seeming a mere crack made visible by the lodgment in it of the black polishing powder, is nevertheless connected with a minute series of disturbances in the white jasper of the central mass.

There are few phenomena of agatescent structure, or of mineral veins in general, on which the stone we have thus far examined does not give important evidence. But the especial condition which it is placed in this part of the series to introduce, and primarily illustrate, is that of so-called "brecciation"; the apparently violent disruption and displacement of jasperine formations by interferent quartz.

The simplicity of this disruption, in the first and third occurrence of formation β (seen in its clear and natural quartzose fracture at the back of the stone), and the complexity of the disruption in the central course of formation, render the specimen more than ordinarily instructive. It was numbered 1707*b* of my own collection; in which the correspondent piece, 1707*a*, yet remains.

A. 49. Brecciate agate in perfect development. I shall not attempt to describe this example; but content myself with making, respecting it, the quite positive statement, that no fracture *by violence* has taken place at any time in any portion of it whatever; that it is an entirely quiet crystalline formation of quartz and jasper under conditions hitherto unexplained; and every appearance it presents of sudden disruption and dislocation is deceptive.[1]

A. 50. Brecciate agate in perfect development, with amethyst secreted in the veins. The largest mass I have myself seen of such stone, and one of the most valuable gifts I have made to Sheffield.

A. 51. A most notable piece of levelled "Lake agate," with lateral bands developed gradually from the beds as they are superimposed,— these lateral bands coating small crystals of quartz in a manner wholly new to me.

Below these crystals, a small portion of the stone, looking exactly like

---

[1] [The Museum copy adds, ". . . deceptive—a piece of natural imposture which is really like the work of a lying spirit." Compare above, p. 70.]

the eddies of a little channel at the sea-shore which a wave has just flowed into, must be noted as the first example we have seen of its kind.[1]

At the upper fracture the surface of the stone shows the separation of one of its level zones from another. (Brazil.)

A. 52. Green moss agate, with white and red jasperine agate, in bands of unrivalled fineness. Note especially with a good lens those in the largest white spot round the minute green ones. (India.)

A. 53. Transitional state between folded and fort agate, with some extremely minute, but singular conditions of levelled beds in an external hollow. Chalcedonic in its larger cavities, and generally worth attentive study.

A. 54, 55. Pieces of a concretionary brecciate agate, secreting flint in angular forms. (Spain?)

A. 56, 57. Pieces of, I believe, concretionary almondine grey jasper, but of a kind new to me.[2]

A. 58. Concretionary almondine jasper; the yellow paste secreting flint. The broken outline of the butterfly-like mass is most notable, and the apparently straight cut edge across the largest oval is a natural fracture, slightly concave, at the real edge of the concretion. Very precious. (Berkshire.[3])

A. 59. Concretionary almondine jasper, with secondary grey and yellow jasper within the almondine masses.

A. 60. Cellular jasper secreting angular flint—an entirely characteristic specimen of the kind of paste that forms these almondine pudding-stones of finest kind.

The paste in this stone is remarkably crystalline, and a section of it should be cut for microscopic examination, while the fissured state of the secreted masses in some cases, and their compactness in others, make this example one of the crucial specimens in distinguishing true from concretionary breccia.[4]

[1] [The Museum copy reads, ". . . we have seen of a formation which I shall call 'eddied' agate."]

[2] [The Museum copy adds, as a note on the source of this specimen, " Wright, out of the cabinet bought for Hull." The reference is to a box of agates, and to other stones, given by Ruskin to Miss Annie Somerscales, for study by her pupils at Hull.]

[3] [The Museum copy adds that this specimen, bought from Wright, was "called by him ' Plum-pudding' stone."]

[4] [The Museum copy adds:—

"The group from A. 54 to A. 60 is now to be compared with A. 35, which has, it will be observed, the blackest spot in the obtusest corner. Next to that black spot, towards the interior of the stone, examine the white space with a grey centre, in its mode of interference with the two grey ones.

"Finally, examine with extreme care C. [now J.] 10, placed among the carnelians; because if, as I believe, it is not a pudding-stone, but an orbicular concretion, the mode of limitation in its scarlet masses, and the nature of the scarlet infusion become of extreme interest in relation to the group of carnelians."]

A. 63. Agate belonging to the same class as A. 52, but showing with peculiar simplicity and clearness the formation of white transverse bands across some of the beds, while they are themselves elsewhere interrupted by others. At one place it is easy to imagine a likeness to a segment of an eclipse of the sun with emergent halo. (India.)

A. 70. Stalactitic agate, in finest possible development. Apparently a filled cave, from the roof of which once hung stalactites of agate, falling to a lake or pool of it on the floor. Unique, as far as I have seen, and explanatory of the terms "stalactitic" and "lake" agate in the clearest manner. But wholly inexplicable itself.[1]

A. 71. Red and yellow jasper in conchoidal concretion. This stone, of a rare species (1700 in my own collection), introduces us to the most wonderful group of all the agates—that which, from its resemblance to a paste with embedded shells, I have called "conchoidal"; but the student must be careful to distinguish the term in this use from the same term as applied to certain conditions of fracture. Conchoidal fracture means that the stone breaks into the curvilinear forms of shells, but I mean by conchoidal agate one which is variegated by zones like actually embedded shells.

In this example the resemblance is vague enough, the structure being exactly intermediate between conchoidal and almondine, so that on one side the piece may be conceived as passing into pudding-stone, and on the other, into true conchoidal agate. I don't suppose it would ever have become either, but it is admirably illustrative of both. If you compare it with the last example, you will feel at once that the characteristic condition in this one is the tendency of its yellow concretions to break themselves up, and let the red paste, as it were, flow into them.

In the corner diagonally opposite to that where this ticket is, you see one apparently dividing like a polype into two; and if then you turn to the other side of the stone, you will see that this tendency to part in the middle is constant, giving a look to many of the concretions like that of the two sides of a double seed or fruit, and that the small one near the centre of the stone shows the incipient division by extremely delicate transverse fissures. Also the greater part of the red paste being finely speckled, distinctly red portions are here and there separated, giving another phase of brecciate aspect to the whole.

In real conchoidal agate, the separated portions of the bands form themselves into convolute spirals with mechanical precision; but before examining them we must further study the separate formation of fragmentary zones less singular in contour.

A. 72. White quartz rock, forming fragmentary zones in the substance of it.

Unique in my own collection, and I am very sorry to part with it; but it is needful to complete our evidence at present. The specialty of the piece is in the formation of the zones with the dark belt in them all on the same side, and their quite arbitrary and irregular position.

[1] [For another reference to this specimen, see No. 146 in the Kirkcudbright collection (below, p. 478).]

A. 73. Conchoidal agate, second stage of development.

1678 of my own collection,[1] and another heavy loss to me, for this stone is not only singularly clear in the internal structure, but exquisitely almondine in its entire mass, and of quite supreme general interest.

It is a mass of white jasper, which contains, as nuclei of its secretion, greenish bands perfectly definite in structure, of constant thickness, interrupted by sharp truncations, and bent into more or less conchoidal courses. These nuclei are surrounded by a white, fine agate paste, and the interstices filled with a yellower one slightly more crystalline. With a good lens exquisite chains of small spherical concretions are seen formed along the outer surface of the white paste, projecting into the yellow one.

The white coat, when uninterrupted—

Here I interrupt myself!—having suddenly made the discovery (Brantwood, July 17th), in examining the rough fracture, that they are shells indeed!—which gives me ever so much fresh work to do on other stones.

A. 74. Almondine agate, containing apparently coralline flints. This title is a somewhat too bold generalization, for there is only one (apparently coralline flint) visible on one side of the stone; nor am I sure that the little stellar white spot thus referred to, in the centre of the largest almond on that side, is a coral at all. It may turn out to be a merely deceptive stellar concretion. But on the opposite side the piece is of extreme interest. Hold it with the shorter side, that has the large truncated crescent in the corner, downwards. Then on the right side, half-way up, is a beautiful red jasper concretion, with black coating, and next, on the left of it, a bright little orange jasper, also with black coating. Going on to the left of this the paste will be seen gradually changing into the grey masses or effacing their edges, and I am prepared to state almost positively that the whole is concretionary, but yet hesitate, because, if it be so, we ought to find in different places a perfect series of transitional states, from the mere massive siliceous paste to the accurately defined pudding-stone, and I can't yet show anything approaching to such a series. There are plenty of varieties of state under other conditions, more or less resembling the pudding-stone, and yet certainly concretionary, but I can't find pieces that some day would have become pudding-stone if they hadn't been stopped half-way and put in the Sheffield Museum; and that's the sort of piece I want to find. Thus the Sidmouth piece[2] is certainly concretionary, and shows two or three well-defined angular flints. But these are not round flints, nor ever would have become round; nor, as far as I know, would any further change have taken place in that piece, if it had been let alone, any more than in this. See the note on A. 71.

A. 75. Shells turned into jasper, common condition.

[1] [In the MS. catalogue of the Brantwood collection, the description is: "Brecciate green chains: the most wonderful that I have. The chains all consistent."]

[2] [For Sidmouth specimens, see pp. 402 (24) and 440 (45).]

SUBSTANCE II.[1]—OXIDE OF TITANIUM

CLASS A. ACICULAR RUTILE

II. A. 1. Perfect small crystal of quartz with exquisitely fine acicular rutile.

I have never seen in any museum a more exquisite example of the association of the oxides of silica and titanium in their finest states. The symmetry of the rock crystal and the almost invisible fineness of traversing threads, are consummate art-work in their kind.

The specimen is also of great value in exhibiting the contemporaneous formation of the crystals. No one will suppose threads of rutile so fine as these could have been fixed in air or water till the quartz enveloped them, and still less that they could have penetrated it afterwards.

II. A. 2. Acicular rutile, forming continuously inside of quartz, and outside of it.

A most notable specimen, examined with the lens. It must, however, never be held for examination but at the sides marked by the purple line, or there will be a chance of breaking some of the finer crystals.

I need not say that threads such as those in II. A. 1 are never found thus external to the quartz. But this concurrent formation of one group of crystals traversing another is not of unfrequent occurrence, and is of extreme marvellousness. The rutile cannot have been formed before the quartz, for many of its crystals are held by the rock-crystals as separately as Punch holds his stick.

II. A. 3. Rock crystal, containing rutile, in large crystals.
Wright, July 10th, 1877.

II. A. 4. Rock crystal, containing tourmaline.
Placed in this series for comparison of the mode of development of the two minerals in pure quartz. Bought with II. A. 3, July 10th, 1877.

---

[1] [See the list of twenty substances, which Ruskin proposed to illustrate in the Museum, and each of which was to have had "a separately bound portion of catalogue": *Deucalion*, above, p. 199.]

SUBSTANCE III.—OXIDE OF IRON

CLASS A. GENERAL GROUPS OF THE BLACK, RED, AND BROWN OXIDES

Before entering on the examination of these groups, the student should observe that the brown oxide of iron is the protoxide, containing one equivalent of iron and one of oxygen; and the red oxide, or hæmatite, is the peroxide, containing two equivalents of iron to three of oxygen; while the black, or magnetic oxide, is, I believe,[*] a mixture of the two in equal proportions, and would therefore consist of three equivalents of iron to four of oxygen, having thus less oxygen than the peroxide, and more than the protoxide.

Fractionally expressed, a given bulk of protoxide, or brown oxide, contains half its bulk of pure iron; deutoxide, or black oxide, three-sevenths; peroxide, or red oxide, two-fifths; or, reducing the fractions, in 70 cubic feet of brown oxide there are 35 cubic feet of iron and 35 of oxygen; in 70 cubic feet of black oxide there are 30 cubic feet of iron and 40 of oxygen; in 70 cubic feet of red oxide there are 28 cubic feet of iron and 42 of oxygen.

Now a cubic foot of iron weighs, roughly, $3\frac{1}{2}$ cubic feet of oxygen. If, therefore, we cut out a cube of the red oxide weighing 140 pounds, it will contain 98 pounds of iron and 42 of oxygen; an equal-sized cube of black oxide will weigh 145 pounds, and contain 105 pounds of iron and 40 of oxygen; and an equal cube of brown oxide will weigh $157\frac{1}{2}$ pounds, and contain $122\frac{1}{2}$ pounds of iron and 35 pounds of oxygen.

Arranging these numbers in tabular form :—

Equal-sized Cubes of	Weighing in Pounds.	Contain	
		Of Iron in Pounds.	Of Oxygen in Pounds.
Brown oxide . .	$157\frac{1}{2}$	$122\frac{1}{2}$	35
Black oxide . .	145	105	40
Red oxide . .	140	98	42

From these figures it appears that in all brown ores of iron at least half their bulk, and approximately a third or more of their weight, is

[*] As I learned my chemistry thirty years ago, whatever I say on the subject must be tested by modern books.

composed of *pure air*. This power of the air to become a solid in combination with metals is the cause of the greater number of important phenomena on the earth's surface. It will be observed that the specimen III. A. 1 contains the three oxides in agatescent concretion—that which contains most air being inside, and that which contains least air being outside—while the intermediate, or black oxide, forms an intermediate band. Nevertheless, the quartz which fills the cavities contains more air than any of the ores of iron, three-fourths of its bulk being pure air, as we shall see in its future analysis.

III. A. 1. Red, black, and brown oxides of iron, forming agatescent bands, with quartz filling their interstices, holding exactly the same relation to the bands which it does to those of the chalcedony in the examples banded with quartz.

Questions. α. State of the quartz and iron oxides before crystallization. β. Reason of a portion of the interior quartz being stained with iron while the rest is white.

(Bought of Mr. R. Talling, 4s. 6d.   From the Royal Iron Mine, Cornwall.)

III. A. 2. Orbicular granite, showing quartz and a black or blackish-green mineral called hornblende, arranging themselves in banded concretions, such as those of the iron oxides in III. A. 1. This specimen is placed at this point of the series because I believe there can be no question that the mass of it was once homogeneous, and that the orbicular bands are owing to a gradual rearrangement of the elements of it during crystallization—no cavities, however, being anywhere formed. The comparison of this structure with the cavity-forming concretions of the iron-oxide is of great interest.

The section of this specimen produced by the fracture above the ticket is traversed by a white vein, which passes through the concretions without interruption, and shows that after these were solidified, farther contraction of the rock took place. Compare I. A. 24, p. 436.

(Isle of Elba.)

III. A. 3. Black and brown oxides of iron in confused concretion, the brown oxide crystallizing in bands, here and there so fine in fibre as to break into mere dusty ochre: the black oxide mixed with films of red, and entirely broken up and disturbed in crystallization.

Questions. α. Relation of this form of the protoxide to common rust, and structure of common rust in relation to the iron it consumes, as compared with the roots of lichen, consuming stone. β. The cause of difference between the confused arrangements of the radiating masses of black oxides and the orderly arrangement of radiating masses in No. 16.

(From the Royal Iron Mine, Cornwall. Bought of Mr. R. Talling. No. 248 in his list.)

III. A. 4. Red oxide of iron in chalcedonic coats associating itself with quartz; and black oxide crystallized on the quartz surface. The quartz crystals in the cavity above the ticket are partially stalactitic. The greater part seen of the surface of the red oxide is a mere film, the basis of the quartz. It has been formed on a reniform mass, of which a portion is seen towards the centre.

Questions. α. The nature of solution from which the quartz and iron oxides were, it seems, at the same time crystallized. β. The reason of the distinction between the red and black oxides, that the first is generally (always?) *under* quartz, if quartz be present ; and the black oxide superimposed on quartz, if quartz be present.

(Probably from the iron mines of Cumberland. Bought at Keswick.)

III. A. 5. Chalcedonic quartz in belts and laminæ, coated on the surface by minute crystals of black oxide of iron.

By chalcedonic quartz, I mean quartz which, without losing its own character of glittering instead of gelatinous fracture, takes either a stalactitic form, or arranges itself, as here, in winding or otherwise curved masses. In the centre of each of these winding walls may be traced a fine brown film of less crystalline character, while the thick walls split into extremely delicate ones on the edge of the specimen. Finally, where the crystals of black oxide are most minute, their fibres seem to unite with the quartz, and to compel it to join in their own reniform structure, unnatural to the quartz itself.

Questions. α, as α in III. A. 4. β. The nature of this "mural" or walled structure, which takes place, usually with enclosed cells, in many most interesting and various ways when quartz is associated with metals.

(From the Royal Iron Mine, Cornwall. Bought of Mr. R. Talling. No. 466 in his list.)

III. A. 6. Quartzose oxide of iron, in walls, forming stalactites, coated with quartz in their cavities. By quartzose oxide of iron, I mean quartz and oxide so intimately mixed that it is impossible to say in what proportion they are joined, but not chemically affecting each other : otherwise the mineral would be a silicate of iron, not a quartzose oxide. The walled structure is to be compared with III. A. 5. The stalactites have all a central rod or thread of the iron oxide, though concealed by a coating of quartz crystals, and appear to be approximately vertical between horizontal walls. But this vertical direction may be only a mode of crystallization, as we shall see in other specimens.

Questions. α. The mode of adoption, as it were, by which the quartz follows the form and involves the substance of the iron. β. The method of vertical crystallization.

(From the Royal Iron Mine, Cornwall. Bought of Mr. Richard Talling.)

III. A. 7. Oxide of iron, with, I believe, also oxide of manganese, in stalactitic formations closely resembling that of III. A. 6, but developing themselves from two foundational coatings laid over walls of quartz. These coatings at one place, just by the torn remnant of a former ticket, seem to be themselves torn and displaced, a separate concretion of the quartz walls being formed below.

Questions. α. The stalactitic development of apparently more liquid from less liquid oxide. β. Manner of displacement of coats ; one of the most difficult questions presented by the reniform formations of chalcedony, oxide of iron, sulphuret of iron, and carbonate of copper.

I omitted in describing the specimen to notice the fused and slag-like appearance of the under surface with lustrous copper-coloured cavities, the nature and cause of which will be our Question γ.

(Royal Iron Mine. Bought of Mr. Talling.)

CLASS M. MICACEOUS IRON

III. M. 1. Micaceous iron suspended in flakes, in the interior of rock crystal with acicular oxide of iron.

III. M. 2. Micaceous iron set in quartz, itself containing embedded rutile.

III. M. 3. Micaceous iron in rounded flakes of approximately equal size, set by their edges on the surface of quartz crystals. The effort of the iron flake is to set itself perpendicular to the surface of the crystal, but I cannot make out if they penetrate it.

The crystals themselves are opaque white, agatescent at apex, where they are constructed of three bands—translucent, white, and translucent.

III. M. 4. Micaceous iron beginning to arrange itself in rosette clusters on a thin coat of bright quartz, overlaid on silicate (?) of iron, beginning to be stalactitic, but mainly arranged in curved layers. Of supreme interest.

III. Y. 1. Meteoric iron. Ziquipilco, Mexico.

III. Y. 2. Meteoric iron. Zacatecas, Mexico.
(No. 1711 of the Stowe Collection.)

XIX. Y. 1. Native platinum in mass.

V. F. 1. Labrador felspar, given me by my old friend and master in mineralogy, Mr. James Tennant,[1] whose work at King's College has been the best practical teaching I know of, in these days of idle nomenclature.

CLASS P. PLUMOSE OXIDE OF IRON

III. P. 1. Plumose oxide of iron in exquisite development, showing the planes of crystallization in a series of clear quartz crystals, forming a ball of stellar quartz, by an agatescent series of three or four extremely fine bands of quartz coloured by the red oxide of iron; the whole enveloped by a coating of rude brown quartz, in which the smooth planes and interrupting

---

[1] [James Tennant (1808-1881), dealer in minerals, also teacher of mineralogy and of geology at King's College : see Vol. XII. p. 438.]

surfaces, rough with smaller crystals, are interchanged in a quite inexplicable manner.

Unique, as far as I know; other half of the piece in my own Brantwood series. Locality unknown to me.

III. P. 2. Plumose oxide of iron alone, in ordinary agatescent development, cut and polished, a few small crystals of quartz occurring in centres.

III. P. 3. Plumose oxide in exquisitest possible formation in crystals of amethyst quartz (examine the cut side with lens), shooting in sheaves from perfectly fine central points to the surface of the crystals, and coming out externally in more confused fibres, coated with earthy yellow oxide (yellow ochre), this entire quartz and iron formation being interrupted by carbonate of lime in a way which would be called brecciate in a common agate.

I may possibly have passed by, or thrown away without notice, many specimens of this kind, having only found out the beauty of this one by cutting it to show the stratification.

Questions. α. What angle may sheaves of plumose oxide reach when involved in quartz? I have never seen them reach a quadrant, still less become stellate. β. Where the plumes emerge, I think they will be found running into plates, and connecting themselves with the forms of micaceous iron. I want larger examples of this intermediate structure.

III. P. 4. Plumose oxide passing into common fibrous oxide, involved with quartz in a soft earth which puzzles me.

A formation especially illustrative of so-called brecciate structure, the group of plumose oxide looking exactly like a broken fragment. It will be seen to be composed of minor plumose groups in confused aggregation, extending themselves into a mass with rounded surface.

III. P. 5. Part of a crystal of amethyst containing plumose oxide of iron.

III. P. 6. White quartz, with plumose oxide of paler colour.

SUBSTANCE IV.—ALUMINA

IV. A. 1. Pure alumina crystallized (sapphire), showing on the plane surfaces most subtle lines of crystallization, with bands of colour partly consistent, partly transverse, and on the rough surfaces all round various conditions of form, some lustrous and approximating to ordinary conchoidal fracture, but some dull, and, like the surfaces of slag, running into pores, often very small. The whole structure totally inexplicable in every way. Practical question about it only, what rock it comes out of, this, I fancy, under a rolled piece. Stowe Collection, 175.

IV. A. 2. Ruby, hexagonal crystal in imperfect development beginning to take pure form and clearness at the edge opposite the brown ochreous one, with beautiful tints of violet in some lights, and showing diagonal cleavage, with felspathic lustre all through the mass. Stowe, 757.

IV. A. 3. Piece of hexagonal crystal of brown sapphire, exquisitely defined in the bands parallel to sides of hexagon, with beautiful crystalline fractures on the transverse planes of the crystal, and conchoidal fractures on their edges. Stowe, 756.

IV. A. 4. Pale brown ruby, with rough fractures showing the banded structure perfectly. Stowe, 761.

IV. A. 5. Transverse slice of large hexagonal crystal of brown sapphire, embedded in a red paste (what?), and containing apparently a metal in pseudo-brecciate portions. Neither the red paste nor this metallic substance yield to the knife. Stowe, 754.

SUBSTANCE V.—POTASSA

V. C. 1. Silicate of potash (glass), slowly cooled, in bottom of glass furnace, with stellar reniform crystallization developing itself in the mass. Cut and polished.

V. C. 2. Glass, slowly cooled, with stellar crystallization developing itself in the mass.

Question. a. Is the white substance in V. C. 1 and V. C. 2 a mineral of fixed equivalents, formed out of the disproportionately combined mass, or a mere incipient crystallization of the mass itself?

453

SUBSTANCE XII.—SULPHUR

Classes
{
A. Sulphur native, pure.
B. Sulphur with iron.
C. Sulphur with copper.
D. Sulphur with lead.
M. Sulphur with molybdena.
N. Sulphur with antimony.
R. Sulphur with quicksilver.
S. Sulphur with silver.
T. Sulphur with tellurium.
}

XII. B. 1. Sulphuret of iron in complex, more or less globular, crystals with triangular facets, strewed on one of the sub-planes of the pyramid of a crystal of quartz.

Unique in my own collection, and as far as I have seen, though the black oxide often forms itself in this manner.

XII. B. 2. Sulphuret of iron (?) in globular concretions, composed of minute cubes on quartz, with cellular interior filled with ?   .[1]   I want a mineralogist's account of this piece, which seems partly combined with grey copper.

XII. B. 3. Sulphuret of iron (?), minute ball in carbonate of lime (?), exquisitely neat and finished in the insertion.

XII. B. 4. Sulphuret of iron, dispersed in a crystalline mass of quartz, with reniform brown oxide of iron in chalcedonic bands and stellar quartz outside.

Unique, as far as I have seen, with no end of wonder in it.

[1] [Thus left by Ruskin in the proof.]

GEOLOGIC SERIES[1]

CLASS Q. QUARTZITE
LOCAL GROUP I. IDRIS QUARTZITE

I. Q. 1. Quartzite from St. George's ground, Barmouth.

I. Q. 2. Veined quartzite passing into apparent conglomerate. Splendid piece from hillside above Barmouth.

I. Q. 3. Finer grained quartzite, in distinct masses between quartz veins, from same spot—the side of the high summit (900 feet) above the town to the north, on its south-east slope.

I. Q. 4. Transitional between quartzite and slate; same spot; a bent piece with narrow slab, straight-planed, attached.

I. Q. 5. Block showing weathered surfaces and ferruginous decomposition, pebbles of quartz falling out.

I. Q. 6. Weathered surface, detaching itself in curved layers, which separate by a cleavage running through the apparent pebbles. I believe an entirely impossible structure in true conglomerate.

I. Q. 7. Finer grained, taking porphyritic aspect.

I. Q. 8. Finer grained yet, with quartz veins meeting at an acute angle, and crystallized at an oblique angle to planes of cleavage, giving a false appearance of shift to the fracture.

I. Q. 9. Finest grained, passing into slate, with iron pyrites.

I. Q. 10. Finest of all grained, now a true slate n.e.

I. Q. 11. Detached pebbles of quartzite.

---

[1] [For this intended series, see *Deucalion*, above, p. 203.]

CLASS H. HORNBLENDE ROCK

LOCAL FAMILY R. ROSA HORNBLENDE

R. H. 1. Pebble from the eastern shore of the Lake of Orta.

Quite as instructive as a large piece which I had no room to carry, and showing the peculiar luck of hornblende, and speckled grain of its development on the white sand matrix (not yet examined).

Immense masses of the Alps are formed of aggregates of this kind variously confused with gneiss. I hope, as *Deucalion* goes on, to put some order into the confusion.

CATALOGUE OF FAMILIAR MINERALS IN THE MUSEUM OF KIRKCUDBRIGHT

(1884)

[*Bibliographical Note.*—This Catalogue of minerals, presented by Ruskin in May 1884, was first printed in the *Kirkcudbright Advertiser*, 1884, in the issues of May 30; June 6, 13, 20, 27; July 4, 11, 18, 25; and August 1.

It was next printed as a pamphlet uniform with the Catalogues of Silica in the British Museum and at St. David's School, Reigate. No title-page was given; the "drop-title" on page 1 being as follows:—

Catalogue | of | Two Hundred Specimens of Familiar Minerals | arranged by | Professor Ruskin | for | The Museum of Kirkcudbright.

Octavo, pp. 59. Introductory remarks, pp. 1–2; Catalogue, pp. 2–59. There is no imprint. The headline on the left-hand pages is "Catalogue"; those on the right-hand pages correspond with the divisions of the Catalogue ("Silver," "Tellurium," etc.).

This pamphlet is among the rarest of Ruskiniana.

Mr. Wedderburn possesses a copy of the first part of the Catalogue (Nos. 1–28), with some corrections in Ruskin's hand. These, which have here been made, are: No. 40, line 6, comma deleted after "minerals"; No. 55, "octahedrons" for "octhedrons"; "Carbonates of Metals," page 469, line 4, "(the metal's)" inserted.

In the Museum of Kirkcudbright there is a MS. copy of the Catalogue (not in Ruskin's hand), which shows many variations from the text in the *Advertiser* and in the printed Catalogue. The variations are of little importance, but they show, as usual, Ruskin's careful revision before printing. Two of the variations are noted below (pp. 453, 464).]

CATALOGUE

OF

TWO HUNDRED SPECIMENS OF FAMILIAR MINERALS

ARRANGED BY

PROFESSOR RUSKIN

FOR

THE MUSEUM OF KIRKCUDBRIGHT

The arrangement of minerals adopted in the following Catalogue, though unsystematic according to the views of modern mineralogists, is an old-fashioned one, which will be found far more useful, in familiarizing the student quickly and easily with the general aspects of the mineral kingdom. These he will find himself at liberty, as his knowledge advances, to systematize either at his own pleasure or under the direction of his tutors;—but I would request that the numbers, attached to my specimens, be preserved: because I am at present endeavouring to organize a system of mineralogical instruction for schools, in which the accessible specimens to which it will refer, in provincial towns, may be permanently connected by their numbers, both with each other, and with the great central examples of mineralogical structure, which have just been so admirably arranged [1] under the windows of the north side of the mineral gallery in the British "Natural History Museum" at Kensington.

PART I.—METALLIC MINERALS

GROUP I

Metals found in the Native State

Gold, Silver, Copper, Tellurium

I. GOLD

1. **Native gold,** dispersed in quartz, and associated with the sulphurets of lead and iron. This specimen is quite characteristic of the mode in which gold generally occurs in Australia, California, and Nova Scotia, but this particular specimen is, I believe, an unusually rich one, from Dolgelly in North Wales.

[1] [The MS. copy adds, "by Mr. Fletcher" (Keeper of the Mineralogical Department). The collection referred to by Ruskin is contained in the first four window-cases. It affords, in conjunction with Mr. Fletcher's *Introduction to the Study of*

2. Native gold, delicately crystallized, sprinkled on the wall of a vein, previously lined by quartz. I have no doubt that this example is from Boitza in Transylvania. See notes on Transylvanian gold in the St. David's Catalogue [below, p. 496, No. 35].

3. Native gold, richly dispersed in white quartz, which has been cut and polished to show the distribution of the metal. (Nova Scotia.)

4. Native gold, three small specimens, in white quartz; originally crystalline, but rounded by being rolled in streams, in two of the specimens; the third uninjured, in smaller crystallization. (All three specimens from South Australia.)

5. Small nugget of solid rolled gold, a little quartz remaining in it, here and there. (Gold fields of South Australia.)

6. Ordinary form of gold procured in stream-washings. (Bendigo, South Australia.)

7. Ordinary form of Scottish gold procured in stream washings in Ross-shire, etc., usually recognizable by the flatness of its flakes.

II. SILVER

8. Native silver, occupying the cavity of a vein, chiefly formed by sulphuret of silver in a metamorphic rock—presumably from Chili.

9. Native silver, crystallized in plates and branches; associated with carbonate of lime; very massive and of extreme interest. (Peru·or Chili.)

10. Branched native silver, with carbonate of lime; extremely beautiful. I am not sure of the locality. The crystalline forms, in the little untarnished terminal cluster, are extremely well defined.

11. Native silver, mossy—in extremely fine twisted fibres, mixed with grains of chalcedonic quartz; an entirely marvellous and beautiful formation under the lens, though, to the naked eye, spoiled by the inevitable tarnishing: the whole in a vein of siliceous rock; and to my mind of extraordinary interest.

12. Native silver, massive, associated with native copper. (Lake Superior.) The lustrous surfaces are, I suppose, marks of the cutting tool, by which the specimen was detached, but there is something in their surfaces which I do not understand.

Minerals, a most ingenious and beautiful introduction to the science. Ruskin was often at the Museum while these cases were being arranged, and took the liveliest and most appreciative interest in the work. Since his time, six other cases (V.–X.) have been arranged, as an introduction, in a similar fashion, to the study of rocks.]

III. COPPER

13. Native copper, mossy, and crystallized in rods. For comparison with mossy silver, No. 11; of extreme beauty and interest. Stowe Collection, No. 2248. (Moldawa, Hungary.)

14. Native copper. Beautifully crystalline in spires and bands. Locality uncertain.

15. Native copper. Branched, in extremely bold crystallization. Compare its cleavages with the surfaces of the silver in No. 12.

16. Native copper. Crystallized in a fissure of calcareous rock. (Siberia.) 2249 of the Stowe Collection.

17. Native copper, associated with carbonate of lime, and forming itself in purity within the crystals of the latter.

18. Native copper, similarly associated with crystals of quartz, and occasionally formed in their interior. Extremely rare. (Cornwall.)

IV. TELLURIUM

19. I believe this to be native tellurium; but request its examination by any passing mineralogist.

20. Nagy-agite, so called from the locality where it is almost exclusively found, Nagy-ag in Transylvania. It is placed with the native metals because it consists of three pure metals in combination, about 30 per cent. tellurium to 10 of gold, and 60 of lead; but, this proportion being fixed, and the combination chemical, it is more properly to be called an ore than an alloy.

Next to the native metals I place the sulphurets of metals, which may, in general, be recognized with the eye, as metallic minerals, and which I am in the habit of dividing, for convenient memory, into three groups. (1) The transparent red sulphurets of silver, antimony and arsenic, quicksilver and arsenic. (2) The opaque grey sulphurets of silver, lead, molybdenum, and antimony. And (3) the opaque yellow sulphurets of copper and iron together, and of iron alone. I begin with the red sulphurets of silver.

GROUP II

Sulphurets of Metals

I. RED SULPHURETS OF SILVER

21. Proustite. Sulphuret of silver and arsenic, in perfect colour and crystallization.

   I have just said that the sulphurets may generally be recognized by the eye as metallic minerals. This first example seems a direct contradiction, for, when first raised from the mine, it looks more like a group of rubies than a metallic ore, but in most cases, I am sorry to say, when exposed to the light, or even by the mere lapse of time, the transparency and colour depart, and the mineral looks like most other sulphurets, a lustrous grey metal. This group of crystals, for which I paid Mr. Bryce Wright I forget whether £10 or £15, is in beautiful state at present, and I hope may long preserve its colour, but even if that fades, will remain an exquisite example of perfectly terminated crystallization. The proportion of elements in proustite is roughly—silver, 65; arsenic, 15; and sulphur, 20.

22. Proustite, dispersed in calcite, in veins in metamorphic rock. (Mexico.)
   I have great hopes that this beautiful specimen may retain its colour, as it has been already full twenty years in my own collection, without perceptible diminution in beauty. It has, however, been protected from the light, but had better now take its chance, and be seen, while it *can* be. There is a beautiful small crystal perfectly terminated, in the cavity of calcite at one end. Stowe Collection, 2200.

23. Pyrargyrite. Sulphuret of silver and antimony, usually called "dark red silver," being deep crimson when first raised from the mine, but I have never known a specimen keep its colour. The crystallization of this example is very good. Stowe Collection, 2206. (Locality—Andreasberg, Hartz.)

24. Pyrargyrite, in fine minute crystallization with calcite,—on a base of galena?

25. Pyrargyrite, with calcite, galena, and a slight coating in one place of native arsenic. Stowe Collection, 2005. (Andreasberg, Hartz.)

II. GREY SULPHURET OF SILVER

26. Argentite. Pure sulphuret of silver, in exquisite fringes of pointed crystals, like hoar frost. I believe this sulphuret, containing neither arsenic nor antimony, shows no red colour, but is always grey and opaque, though when found it has a brilliant lustre. (This lovely example is from Chañarcillo, Chili, whence also the fine Proustite, No. 21.)

III. GREY SULPHURET OF LEAD

27. Cubic galena, formed in a block of calcite. I am extremely sorry that this wonderful specimen was spoiled by dust and friction before I became possessed of it, for it is impossible to see anything more characteristic of the formation of two of the most interesting minerals developed in the veins of the earth. Their varieties and caprices of crystallization are quite infinite, and any man's life might be happily spent in merely describing and illustrating the various forms of calcite and galena. The composition of sulphuret of lead—lead 87, sulphur 13—is within a decimal or two proportioned like that of sulphuret of silver, Argentite being in like manner—silver 87, and sulphur 13. Some small crystals of sulphuret of zinc occur in this specimen at the base of the galena.

28. Galena. Cubic, showing the construction of the cube out of minor crystals. On pearl-spar. Beautiful.

29. Galena in splendidly various crystallization, tending gradually to the formation of cubes, with here and there a little sulphuret of iron. Quite wonderful.

30. Galena on quartz, having its cavities filled by sulphuret of iron. I leave the description of the forms of the galena to better crystallographers.

31. Galena in cubes modified on the angles, set on chalcedonic quartz. Very interesting.

32. Galena, in complex crystallization, with quartz superimposed.

IV. GREY SULPHURET OF ANTIMONY

33. Pure sulphuret of antimony, "antimonite," in flattish rods, more or less thrown into radiating sheaves. Stowe Collection, 2526. (From Felsöbanya, Hungary.)

34. Antimonite in radiating sheaths, on the wall of a quartz vein.
    In nearly all radiating formations of mineral substance, the radiating groups are rooted like trees, and expand upwards ; but the sulphuret of antimony, together with one or two ores of lead, has the curious fancy to reverse this usual and rational arrangement, and to set all its radiating groups to stand on their heads. This specimen is a beautiful example of the gymnastic performance, of which, I believe, no mineralogist, except myself, has taken any notice.

35. Antimonite, irregularly radiating on sugary chalcedonic quartz—extremely beautiful, and showing every possible arrangement of transverse and radiating crystals.

36. Antimonite, in finer crystallization. This mineral can go on, increasing the subtlety of its formation, until it resembles the finest wool or down, but such specimens are of small use in the collection of a museum, as they can only be seen with a lens.

37. Antimonite (with small interspersed sulphuret of iron), in exquisitely fine, though confused, crystallization, between films of the most delicate quartz—which seems inspired by the antimony to show what *it* also can do in crystalline needlework. Entirely marvellous and lovely. (Cornwall.)

V. GREY SULPHURET OF LEAD AND ANTIMONY

38. Bournonite on quartz.

With the grey sulphurets of single metals we must associate this mixed sulphuret of lead, antimony, and copper—about 20 per cent. of sulphur to 40 of lead, 25 of antimony, and 15 of copper. The specimens found in Cornwall, of which this is a characteristic one, are remarkable for their beautiful association with sugary quartz. See the next example.

39. Bournonite with the finest tabular quartz—this formation of the latter being, so far as I know, peculiar to its association with Bournonite, and looking exactly as if it had made up its mind to imitate the tabular crystals of the latter mineral, just as, in No. 37, it imitated the needles of the antimony. Of course such an idea is wholly imaginary; but the fact of the influence of minerals upon each other, so as to induce different modes of crystallization, is a most important one; and, as yet, wholly neglected by mineralogists.

40. Bournonite, with common quartz and sulphuret of iron. This specimen, giving us a grey sulphuret and a yellow sulphuret in association, may be taken as an introduction to the yellow sulphurets, and exhibits the sulphuret of iron (commonly called iron pyrites) in one of its most curious habits, namely, that of covering other minerals with a coat of itself *on one side only*, the determination of that side being, by the direction of the crystal to be covered, and not at all by any direction of current or vapour, as may be completely ascertained by the careful study of this pretty specimen.

VI. YELLOW SULPHURET OF IRON

41. Octahedric crystals of iron pyrites. Of all metallic minerals, next to the oxide of iron, this sulphuret, continually mistaken for *gold*, is the most generally distributed, and of all minerals known to me,

is the most singular in crystallization. I do not know how dissolved before its formation, but it seems to crystallize quicker than most other minerals, and to be extremely disturbing in its formation with them. It is described by mineralogists simply as *cubic*, with the varieties of form belonging to the cubic system, but no notice seems ever to be taken by them of the special powers and caprices which influence its formation. This crystal, No. 41, looks externally as if composed with a thousand separate efforts in every plane. Were it broken, however, the substance of it would appear perfectly compact.

42. Iron pyrites in curvilinear cubes. This specimen, which was one of the most interesting in my collection, exhibits the peculiar power of iron pyrites, in the greatest perfection. Under common conditions, its cubes are simple, like those of fluor-spar, but, when it chooses, it can warp their sides into a double curvature, which is convex on two opposite sides, and concave on the two other ones, and produced, apparently, by the adjustment of component crystals, just as a Gothic builder produces his vault. Two or three other minerals, especially quartz, mica, and uranite, share with sulphuret of iron this power of warping themselves, but as far as I know none of the conditions inducing the action have ever been explained.

43. Iron pyrites in octahedrons with *concave* sides. Very beautiful.

44. Iron pyrites in octahedrons with *convex* sides, forming the base of a group of beautiful white crystals of quartz, which, wherever broken, show that they are composed of a hexagonal crystal of clear quartz, with a white coating to finish them.

45. Iron pyrites in narrow octahedrons;[1] I believe this is the variety of the sulphuret of iron, in which there is a mixture of arsenic, which alters the mode of its crystallization from cubic to prismatic. The tabular character which it affects is better seen in the next specimen.

46. Marcasite, or white iron pyrites. This mineral is said to have the same composition as common pyrites, yet it is white in colour, or feebly brass yellow—and prismatic, instead of cubic in crystallization. Traces of manganese and arsenic occasionally found in it seem too slight to account for the change. A similarly inexplicable distinction exists between calcite and aragonite.

47. Common iron pyrites, in richly composed masses coating quartz. I cut this specimen, in hopes of ascertaining the reason of the projecting stalactite; but the section shows only the mode of attachment of the pyrites.

[1] [The MS. copy adds, "—a form not, so far as I am aware, described by mineralogists, but produced by a peculiar humour in this mineral for throwing itself into wedge-shaped crystals rather than square ones."]

48. Common iron pyrites, coating quartz, and itself encrusted with calcite —an example of the anomalous formations which this singular mineral continually induces in all it associates with.

49. Common iron pyrites, in complex groups, associated with fluor-spar, on the outside of a crust of quartz—which has thrown itself into a dome, and beneath that, into angular ridges—under the various provocations and persuasions of its eccentric friend. Whether the dome has ever been filled by sulphurous gas, or built over some now vanished mineral, which has left the lace-like impressions on its interior surface, I must leave better mineralogists than myself to decide.

50. Common iron pyrites, in extremely minute crystallization, involved with quartz upon a crust of orange-red quartz. Very beautiful.

51. Common iron pyrites, involved with quartz in tabular crusts, associated with unfinished crystals of fluor-spar on a basis of galena. The fluor is very remarkable in its record of process; the crystals seem to have been stopped in their growth, in order to be sprinkled over with fine quartz, after which they have been allowed, but only over portions of their surfaces, to begin building again.

52. Common iron pyrites in cubic crystallization, with a tendency to curvature, and involved with calcite in methods wholly indescribable.

53. Common iron pyrites, forming stalactites out of exquisitely sharp cubic crystals, penetrating and sometimes sustaining calcite. Quite wonderful.

54. Common iron pyrites in coats in the interior of a crystal of calcite— this is a frequent form of its occurrence in Derbyshire.

55. Common iron pyrites in pinnacles composed of octahedrons, occupying a cavity in quartz rock, through which the mineral is generally dispersed.

56. Common iron pyrites, formed in the interior of quartz crystals— a rare condition. No. 1454 of my own former collection.

VII. SULPHURETS OF COPPER

My collection has never been rich in this mineral; but the four examples I can spare are characteristic and pretty.

57. Sulphuret of copper and iron, commonly called copper pyrites, in vaguely tetrahedric crystals, black on the surface, with beautifully developed white quartz.

58. Copper pyrites—peacock variety—exquisitely crystallized among clear white quartz, on which, I think, it is for the most part super-imposed, and may be studied as an example of subsequent, instead of contemporary, crystallization. In one part of the specimen, a single cube of white fluor is suspended on the quartz, itself sustaining together yellow crystals of iron pyrites, and blue crystals of copper pyrites, of extreme beauty under the lens; the specimen altogether is one quite typical of crystallization done with the best spirit, and in the best style.

59. Sulphuret of copper and antimony, commonly called Grey copper, its real colour, seen on the fracture, being a yellowish grey, but in this form of crystallization it has beautifully golden red and blue iridescence on its surfaces; which, however, I fear somewhat lose their brilliancy by exposure. The mineral is one of the most complex of metallic sulphurets; principally, indeed, composed of copper and antimony, roughly in the proportion of 40 per cent. copper to 20 of antimony and 25 of sulphur; but the remaining 15 parts of the mineral contain, in the best and most characteristic specimens, silver, quicksilver, zinc, iron, and arsenic. I find that Miller[1] gives for its chemical symbol (I suppose founded on analysis of English specimens) 4 equivalents of lead, iron, zinc, and copper: but certainly zinc, copper, and antimony are essential in Hungarian specimens; this beautiful iridescent one is from Cornwall.

60. Grey copper in association with tabular quartz. I do not know the dark grey metallic mineral which is grouped with the iridescent crystals, nor have I the least idea under what conditions these tabular forms are developed.

GROUP III

Oxides of Metals

Next to the sulphurets, I place the oxides of the metals, but as the greater number of these have that alkaline property, which consists in forming salts with acids, few of them are of mineralogical importance in their uncombined form, except the oxides of iron and copper. The first of these diffused everywhere, in the form of iron ochre, gives all the most beautiful colours to the various jaspers, and is in general the source of all the dark and warm colours in rocks.[2] In its purity, it forms the group of minerals, typically represented by the following twelve specimens.

---

[1] [*Elementary Introduction to Mineralogy*, by the late William Phillips, new edition by Brooke and Miller, 1852, pp. 205 *seq.*]

[2] [See *The Two Paths*, Vol. XVI. pp. 376 *seq.*]

II. RED OXIDE OF IRON

61. Common Hematite. Oxide of iron in the proportion of 70 iron to 30 oxygen, in finely developed reniform masses. The mode of crystallization by radiating fibres, which induces this form, is exactly the same as that of chalcedony.

62. Hematite with still more distinctly fluent chalcedonic structure.

63. Fibrous hematite, showing the stellar arrangement to which the globular form is owing.

64. Portion of a more delicately fibrous mass, showing the interlacing of the fibres at their extremities.

65. Hematite, forming stalactitic rods, externally coated by quartz, a frequent form, but one of which the production is hitherto unexplained, since, in many cases, such formations do not admit the idea of the quartz having been deposited after the iron, but the entire stalactite seems to have been formed at once.

66. Hematite, passing from its common form into its state of perfect crystallization, in which its becomes black, instead of red, takes an extremely bright metallic lustre, and usually presents itself in the form of thin plates, whence it is often called micaceous iron, the plates being sometimes frequently grouped so as to resemble small roses. It generally displays itself to the greatest advantage in association with quartz, as in the present example, where some of the groups of crystals are extremely finished and beautiful.

67. Hematite in the same transitional state, interfering with and modifying the crystallization of the quartz above it, in which it produces entirely irregular and grotesque varieties of form. In the interior of the cave, the red portions of it show themselves through the quartz.

68. Hematite, interrupting the crystallization of quartz by agatescent zones. Extremely rare. I have one or two other specimens of the same kind, in my own collection, but never saw one in any other.

69. Hematite, in minute crystals, modifying quartz, so as to throw it into thin plates, to me, at present, inexplicable.

70. Hematite in its own simple and regular crystallization, in fine plates.

71. Hematite in complex crystallization, of extreme beauty.

72. Hematite in hexagonal roses, the prettiest of its conditions.

II. RED OXIDES OF COPPER

It has always been to myself one of the most troublesome points to remember in mineralogy that the *oxides* of copper are transparently red, like the *sulphurets* of silver. They are in like manner called "ruby copper," as the other "ruby silver," and have the same bad habit of losing their colour, in process of time. Curiously, however, the solid crystals lose their colour more easily than the slender ones, so that while cubes and octahedrons, a quarter of an inch thick, become gradually black, instead of crimson, acicular crystals, as fine as threads, remain, so far as my own experience goes, perfectly clear and bright!

73. Oxide of copper in cubes modified by large octahedric planes. This beautiful specimen has darkened since I got it (I think about ten years ago), but still shows much of its original colour.[1] Its crystallization cannot be surpassed.

74. Oxide of copper; octahedric in a cavity of ferruginous rock. Extremely fine.

75. Oxide of copper in confused crystallization, forming spheres. Beautiful.

76. Oxide of copper in confused crystallization, throwing itself into acicular crystals, as fine as hairs which are transitional from the common cubic forms to those of chalcotrichite. This exquisite specimen must be kept with the greatest care from the dust, as it never can be washed or cleaned, without destruction. The structure can only be seen now and then with a lens, by some favoured pupil or visitor.

77. Oxide of copper, presenting all the most complex forms of its crystallization with mingled lustrous and bloomed surfaces. Quite exquisite, but can only be seen under the lens.

78. Chalcotrichite, in confused mossy groups, associated with fibrous malachite.

79. Chalcotrichite, in richest form, beginning to weave itself into silken tissue.

80. Chalcotrichite in its quite delicatest form, the tissue woven smooth.

GROUP IV

Carbonates of Metals

Next to the oxides of metals, I place their carbonates, which are combinations of carbonic acids with their oxides. For the sake of brevity in common parlance, the combinations of acids with metallic

---

[1] [This, or No. 74, may be the piece mentioned by Ruskin, as "grieving him poignantly by losing its colour," in *Fors Clavigera*, Letter 4, § 5.]

oxides are called simply carbonates of iron, carbonates of copper, sulphates of iron, sulphates of lead, etc.; but it is always to be understood that these expressions mean carbonate of the oxide of iron, sulphate of oxide of lead, etc. No acid is capable of combining with a metal in its (the metal's) pure state.*

The carbonates of iron are ugly brown crystals of no general importance, and I do not think it worth while to send specimens of them. Carbonates of copper are, on the contrary, of extreme beauty and commercial value, and the following ten specimens of them will, I think, be found generally interesting.

I. CARBONATES OF COPPER

81. Malachite. Green carbonate of copper, associated with the earthy red oxide of copper, and with massive native copper. (Lake Superior.) I assume the red intermediate band to be the red oxide, but have not analyzed it. It is extremely hard, and seems to be in combination with siliceous earth, forming a kind of jasper.

82. Malachite, zoned and massive, of the fine quality preferred for ornamental work. The structure of the zones in malachite is exactly similar to that of the zones in spherical agate, and I discovered the first principles of agate formation chiefly by a comparison of the two minerals.

83. Malachite, stalactitic, correspondent in structure to the simple states of stalactitic chalcedony.

84. Malachite in wrinkled masses—which have the appearance of having been squeezed together—in a pasty state, but the specimen is a warning against trusting to first appearances in mineralogy; it will be found on examination that the forms are entirely concretionary, though under extremely singular and exceptional conditions.

85. Malachite, in partially isolated concretions, resulting in the appearance of being formed out of compressed fragments. Such a structure, when it presents itself in silica, is at once called a conglomerate, by mineralogists, without further examination; but I believe at least half the so-called conglomerates of siliceous pebbles have originally been formed, as I believe this malachite is;—only with a slightly stronger power of spherical crystallization. This specimen ought to be examined on both sides.

86. Malachite, fibrous, of great beauty, in association with a brown ochre, of which I do not know the nature. This specimen ought to be looked at through the lens.

---

* My chemistry is forty years old, but sound as far as I venture it, though the facts receive often, from modern chemistry, different explanations or expressions. Thus I believe it is now considered accurate to speak of salts of metals, one or more equivalent of water being taken up by the combining acid.

87. Malachite, partly fibrous, in chalcedonic concretion, on the ochreous substance seen in the last specimen.

88. Malachite in wrinkled concretions (to be compared with No. 84), which seem to have formed an almost hemispherical cake, of which we have here the broken half, looking as if it was meant to be toasted, and first salted outside with small crystals of the blue carbonate !

89. Malachite, on a siliceous rock, studded, externally, with pretty groups of foliated crystals of the blue carbonate.

90. Blue carbonate of copper, finely crystallized, from Chessy, near Lyons.

II. CARBONATES OF LEAD

Next to the carbonates of copper, I should like to see arranged at the museum an equal number of the carbonates of lead, which are extremely beautiful in silky and fibrous crystallization, but I have never entered myself upon their study, and can only continue the metallic series, at present, with a group of mixed ores of lead, which I will not venture to define, but which will afterwards find their place in the systematized collection; and with one or two specimens of the beautiful Cornish cupreo-uranite, which is, I believe, a phosphate of the oxides of uranium and copper.

91. White carbonate of lead, with a green mineral, which I presume to be an ore of copper.

92. Carbonate of lead, massive, with a green ore of lead, I believe a phosphate. (Caldbeck Fells, Cumberland.)

GROUP V

Salts of Various Metals

93. Muriate, I believe, of lead, in brownish yellow crystals, on a basis of quartz; forming one side of a wall, with a corresponding bed of quartz on the other, covered by large brown crystals, apparently deposited on one side of the wall, while the lead was being laid on the other. A most singular specimen.

94. Chromate, I believe, of lead; but have never studied this group of minerals. Whatever the crystals are, they appear to have been deposited by sublimation in the pores of a quartz, from which another bed of sustaining quartz had been first dissolved.

95. Arseniate (?) of lead, in beautiful acicular crystals, in cavities of white quartz—see both sides of the specimen.

96. Blue sulphate (cupreo-sulphate?) of lead, colouring quartz, and crystallizing in its cavities. (Caldbeck Fells, Cumberland.)

97. Uranite in obscure crystallization, on a micaceous rock. (Cornwall.)

98. Uranite, in beautifully formed square tables in the cavities of ochreous quartz. (Cornwall.)

99. Uranite—pale green—in clustering plates. (Cornwall.)

100. Tungstate of lime, well crystallized, in a quartz cavity. (Caldbeck Fells, Cumberland.) The semi-metal, tungsten, becoming here, under oxidation, an acid which unites with the alkaline earth, forms an exact transition between metallic and earthy minerals.

PART II.—EARTHY MINERALS

GROUP I

Quartz

101. Brown quartz of Switzerland, a fairly characteristic crystal; respecting which note, of quartz in general, that in its purity it is essentially a mineral which grows in cavities, and therefore characteristically does not terminate its crystals at both ends, but is attached by one to the rock; while on the contrary, the greater number of stones, to which we give the name of gems, form themselves equally all round, and often, as peculiarly the garnet, within the substance of solid rocks, while the diamond and ruby seem to collect themselves out of old gravels, or clays. It is therefore extremely difficult to find a quartz crystal well terminated at both extremities; but the modes of aggregation by which it is connected with the surfaces on which it is formed are an essential and deeply interesting part of its crystalline power. I have only by me a few examples, interesting rather by their peculiarities than beauty, which I can spare from my own collection, a large part of which has been already given to Sheffield and the British Museum; but the twenty specimens I have chosen will nevertheless form a suggestive nucleus for the arrangement of future pieces of more general character.

102. Pure rock crystal, probably from Schemnitz in Hungary, and interesting in three ways: first, that it is doubly terminated; secondly, that it indicates, by various interruptions, the process of its crystallization; and thirdly, that it indicates a tendency to oblique arrangement, which is peculiarly characteristic of quartz. A diamond, if it has not its points opposite each other, is an ill-made diamond; it seldom quite succeeds in getting them so, but it is always trying to do so: a crystal of quartz, on the contrary, is often as cross-tempered as a crab; and uses half its powers at one end, to express general dissent from the opinions of the other.

103. Tabular quartz, oblique in aggregation; it is impossible in this instance to say whether we are dealing with one crystal, or with two or three dozen.

104. Tabular quartz throwing itself into a uniform ridge; this may, I think, be considered a single crystal in which the ridge is attained by the prolongation of two opposite sides of the pyramid, of which the intermediate sides are seen at the lower extremity of the crystal.

105. Pure white quartz, in regular hexagons, their sides composed of or coated by minute pyramids; the whole specimen is part of a crust, formed on a lower bed of quartz.

106. Two unfinished crystals of quartz, probably from Schemnitz, and showing the tendency, frequent in fine crystals, to form their hexagon with three sides smaller than the other three.

107. White quartz, arranging itself in a tabular form, in which it uses for the most part the planes of its pyramid, as those of chief construction. It may be observed in passing that the pyramidal planes are easily distinguishable from the sides of the crystal by being set together in complex planes like the surface of ice, while the lateral planes are striated across with approximate straightness; noting this difference, it may be seen at a glance that the planes out of which the wall is constructed are nearly all pyramidal.

108. Fragment of the apex of a large quartz crystal, in which the pyramidal planes are so irregularly developed, as at first to give the impression that the crystal is a parallelopiped.

109. Tabular quartz, in an irregular group on mica slate; this curious form, in which two sides of each hexagonal crystal are widened at the expense of the four, is, so far as I am aware, peculiar to the Pass of the Tête Noire, near Chamouni.

110. Sheaf quartz. I attach this name to crystals which are composed towards their base of a radiating group of smaller ones. In accuracy I should have said towards their centre, instead of towards their

base, for crystals of this form always propose to be doubly terminated, and when they succeed, look as if their two extremities
emerged out of a sheaf of corn, tied tight in the middle; but it
is extremely difficult to find complete examples. In this specimen
one crystal only, that lying on its side, shows part of the opposite
extremity.

111. Taper quartz, the exactly opposite condition, in which a crystal tapers,
instead of enlarging, towards its end.   I believe I may generally
state that crystals which enlarge toward their extremities are composed of apparent clusters of minor radiating ones, as in number
110, whereas the tapering of this form of crystal is by retiring *steps*
in the transverse striæ.

112. Minute quartz crystals, in elongated aggregation, suggesting the structure of chalcedony. The form is rare, and I have never seen it on
a large scale.

113. Hacked quartz. The term is usually attached to groups of crystals,
closely aggregate in flat planes, a structure which, so far as I know,
is never found except in association with the oxide of iron: but no
mineralogist has explained what the iron has to do with it.

114. Foliated quartz. The planes here are not crystalline, but seem to be
the remnants of casts in fissures; the entire specimen has been
originally an incrustation upon fluor, but I have no idea how these
folia have been produced.

115. Foliated quartz, in zigzagged ridges upon common quartz, which is on
one side of the specimen smooth as usual; on the other, fretted
and fringed by minute processes of lace-like foliation, only to be
well seen with a lens.
These first fifteen specimens exemplify a few of the caprices of
this extraordinary mineral, perhaps in association with others, but
not mixed with, or penetrated by, them,—the last five show it in
closer modes of association.

116. Quartz containing mossy filaments of chlorite, with yellowish white
amianthus, extremely interesting; the largest plane is of course a
section polished.

117. Part of a quartz block, cut to the best advantage, to show its contents,
a beautiful congeries of crystals of rutile (oxide of titanium), one
very large, and another group of very unusual size; the base of the
crystal is composed chiefly of mica, the tables of which are deeply
embedded in the quartz. This specimen is an extremely fine one;
I believe from Madagascar.

118. Slice of a large block of quartz, extremely pure in substance, though flawed, not always to its disadvantage, as it exhibits charming iridescences in some lights. It contains beautiful long rods of tourmaline.

119. Quartz, in its natural crystallization with superficially embedded topaz. It seems to me extremely curious, as an exposition of mineral character, that while rutile, tourmaline, amianthus, chlorite, and oxide of iron are allowed to amuse themselves in the inside of rock crystals, just as if they were in the open air, topaz is never on any occasion allowed to get inside, but only plunged to a certain depth, as if it had fallen in by accident.

120. Amethyst quartz, with a curious flaw, or vein, across the base of its crystal. Amethyst is said to owe its colour to ferric acid, and it is notable respecting it that it seems a mineral of later parentage than common quartz, for if formed in association with crystals of the latter, it is invariably on the outside of them, never their nucleus.

The above twenty examples being enough to illustrate the ordinary characters of crystalline silica, that is to say, of the element in its purity, the next twenty represent its normal conditions when partially mixed with clay or other foreign matter, and when the mode of its solution, or circumstances of its deposit, have been the cause of its consolidation in more or less compact or irregularly fluent forms. The entirely commonest of these, the black or grey flint of our chalk formation, common though it be, is one of the most delicate of mineral substances, taking the casts of fossils, with a precision which the most exquisite cameo could not rival, and even in its most familiar aspect, presents intricacies of structure and varieties of form hitherto unexplained by any mineralogist.

121. I begin the series with a specimen of its banded structure, first observed by Mr. Henry Woodward,[1] although of no very rare occurrence, in Kentish chalk pits. It has nothing to do with any organic substance, and remains unexplained, but I have no doubt that it is an incipient condition of the banding which constitutes the most beautiful agates, and that the nature of agates will be best understood by watching their gradual development from this not unfrequent state of flint.

122. Flint, apparently broken and re-cemented, containing probably, also, vestiges of organism; the appearance of fracture is, I believe, wholly deceptive, but will say nothing positive respecting this example of complex problematic flint.

23. Zoned black flint, becoming white at the edges by weathering. The fragment has probably been part of a large bed, of which the smooth bluish-grey surface was the top; the uppermost bed will be seen to be altogether of finer material, and consists of what is properly called, not flint, but chalcedony.

[1] [A slip for S. P. Woodward: see above, p. 47.]

124. Black flint, containing sponges embedded in blue chalcedony ; a beautiful example, showing the difference between the substance of the two minerals with interesting precision.   For their usual relations, see my *Grammar of Silica* [below, p. 536].

125. Part of a nodule of massive chalcedony, formed in a volcanic rock, which it did not entirely fill, the central hollow being coated with small crystals of quartz.   Very characteristic of simple chalcedonic substance.

126. Chalcedonic agate, formed, I do not know where or how, but showing in the polished sections one of the modes of aggregation of this mineral, to which I beg the observer's close attention.   He will find, in the first place, that the rock on which the grey chalcedony is laid is not a rough substance, over which the finer one is poured, but, on the contrary, one affected by curious powers of minute crystallization, producing phenomena which, in a slice seen with a microscope, would be found of the most extraordinary variety and beauty.   In the ordinary accounts of such pieces of agate as this, the student is told, with vulgar confidence, that the chalcedony is merely a stalactitic deposit on the interior rock.[1]   It will be found, by more attentive observation, to be arranged in concentric masses, which meet each other at their edges with the precision of bubbles in foam, and are in all parts of the stone of exactly equal depth.   The difference between this structure and a real stalactitic deposit, such as that of a rivulet coating the stones at its edge with ice in hard frost, may be conclusively demonstrated on the first wintry day ; but it will be worth while also to take a bit of rough stone and dip it at successive intervals into any gelatinous substance, of which the layers would quickly congeal. Their gradually simplifying contours would soon form a striking contrast to those of this or any other chalcedony.

127. The reason of which difference may be seen at once by observing this next example of perfectly pure grey chalcedony, deposited in depth enough to allow of its reaching the full development of its proper form.   It would be well to smooth and polish the broader side of the large grey sphere of agate, of which the blundering lapidary has nearly destroyed the symmetry by uselessly polishing the edge.

128. White opaque chalcedony on, I believe, oxide of manganese, an example of extreme interest in showing the resolute way in which the chalcedony, compelled to form on the surface of a substance crystallized in sharp angles, yet within the never-exceeded depth of a quarter of an inch, rounds all these into its own normally spherical contours.

[1] [On this subject, see above, p. 376.]

129. Yellowish grey chalcedony, partly stalactitic, with internal tubular concretions, and with spherical concretions on its surface, singularly illustrative of its own crystalline action.

130. Fine chalcedony, of an indescribable colour, for which I can find no better term than the not very intelligible one, "violet-yellow"; partly chalcedonic, encrusted primarily on I know not what crystalline mineral, and beginning to throw itself into lambent layers, approximating to a form which I generally call "mural" chalcedony.[1]

131. Dark yellow chalcedony, partly stalactitic, but a beautiful example of the wilfulness of this extraordinary mineral, in that one-half of the stalactites drop crossways to the rest: the mineral is quite capable in some of its humours of dropping upside down.

132. Yellow chalcedony, stalactitic in radiating branches; very beautiful.

133. Pure chalcedony, in concentric layers, of which the interstices are partly filled by the thinnest possible films of iron-oxide; very beautiful.

134. Milky chalcedony, in a level bed on quartz, with incipient ramifications of the oxide of iron or manganese, wholly inorganic, but presenting a singular resemblance to moss or seaweed.

135. Mocha stone, so called from Mocha in Arabia, where the best specimens are found; an exquisite example of the structure above described.

136. Scarlet and grey jasper. We here begin the examination of siliceous earth definitely mixed with clay. The only difference between fine chalcedony and quartz seems to be structural: that of the first being gelatinous, of the second crystalline; but to form jasper, clay must be definitely mixed with the silica, and an appreciable quantity of oxide of iron added to give the crimson, scarlet, or orange-yellow hues, which constitute the value of the stone. The result of this admixture of clay is, first, that jasper is absolutely opaque, not even translucent on the edges, and can be used therefore for the ornamentation of agates, literally as a rich pigment perfectly distinct at the edge of the touch from the chalcedony in which it is laid or involved; though, when they are finely broken together, the effect of the opacity may be modified.

Secondly, jasper cannot crystallize, nor does it ever show the feeblest approach to crystalline structure. It is as homogeneous as a well-made brick.

Thirdly, and in consequence of its non-crystalline nature, it cannot be arranged in globular masses like chalcedony; a fact in itself

---

[1] [See above, p. 450.]

sufficiently demonstrative of the crystalline nature of those globes. In the present specimen there is an admixture of chalcedony through all the grey portions, but the red spaces are well representative of jasperine colour.

137. Dark red and grey jasper, variously marbled and combined, traversed by very slender chalcedonic veins. A generally characteristic example of ordinary jasper.

138. Red and white jasper, involved with agatescent chalcedony, full of most singular phenomena, and changing states of form ; illustrative of nearly every possible arrangement under which these two minerals can be seen.

139. Jasperine agate, that is to say, agate in which the passive jasper is arranged by the constructive chalcedony in regular zones. The transverse formations of deeper crimson lines, especially the one near the outer edge of the nodule, in which may be fancifully traced the outline of a bird's head and beak, is of extreme interest under the lens, in the small detached " faults" of stratification at the end of the beak, and the perfectly unfaulted curve of the line forming the bird's throat between the beak and breast, in which some more faults give a little the look of feathers.

140. Perfectly fine full-red jasper, involved with crystalline agate ; the jasper under the lens will be found composed of minute spots, and the substance of it is entirely passive, raised or retracted by the thrusting or receding quartz, which encloses the less definitely crystalline centre of the stone.

141. We now conclude our series of siliceous minerals by a series of ten examples of agate—properly so called—*nodules ;* that is to say, of quartz, amethyst, chalcedony, and jasper, variously grouped in the hollows of volcanic rocks. The first five examples exhibit developments of structure, and the last five the finished qualities of good stones. This first example, 141, is simply a bit of basaltic rock, in which a more or less cylindrical cavity has been coated with a bed first of grey chalcedony, then of brown chalcedony, and then of crystalline quartz, becoming slightly amethystine on the final surface. This is the commonest and poorest condition under which agatescent secretion occurs.

142. Part of an agate nodule, detached from the enclosing rock, resembling No. 141 in its general arrangement, but the quartz finer in arrangement, and the chalcedony dividing into very pretty agatescent lines. The interior hollow is partly filled by dirty and ill-shaped crystals of carbonate of lime, the presence of which substance has generally disturbed and degraded the action of the agate.

143, Part of a stalactite of quartz and dull-coloured amethyst, externally
A & B. coated with chalcedony. These conditions of crystallization outwards, instead of inwards, are comparatively of rare occurrence, and
have not yet been sufficiently studied by any mineralogist.

144. Part of a rather fine agate nodule, reversing the action of the previous
specimen, and completing that of 142 by adding a coat of chalcedony on the top of its interior crystals; a very perfect and interesting example.

145. Part of a nodule of perfectly developed agate, in which the cavity in
the centre has been very nearly filled. The external chalcedonic
bands show, towards the surface of the stone, the interruptions by
apparent gathering or folding which were first described by me in the
Geological Magazine,[1] though they are of perfectly common occurrence
in agates of certain localities. I have found it convenient to distinguish the stones presenting this appearance by the title of "folded"
agates.

146. An entirely superb example of the rarest condition of agate, that in
which stalactitic structure has been produced *through* a gelatinous
paste, which is itself, also, where it has room, arranging itself in
agatescent bands. On the flat side of the stone it will be seen that
this structure exists throughout its whole mass, and the sections of
the stalactites on both sides will, I think, ultimately show that they
are provoked in the substance of the chalcedony by the injection,
or partial crystallization, of the surrounding rock substance. This
was the best specimen of the kind in my own collection, next to
the uniquely beautiful one given to Sheffield,[2] to which, in absolute
mineralogical interest, this specimen is not inferior.

147. Chalcedonic agate, with interferent beds modifying the spherical ones.
The nature of the external coating of this stone, and of the irregular
action which it has induced within the grand spherical bands of
external chalcedony, are quite beyond my powers of either description or explanation.

148. Part of a nodule (or perhaps bed) of brown agate, quite abnormal in
the structure of both its surfaces. I mean, of course, by surfaces,
the unpolished portions of the narrower sides. Many agate nodules
have surfaces like the deeply channelled black and brown one, but
such a surface as that here formed above the fine quartz is entirely
abnormal.

149, Two slices of an exquisitely finished nodule of chalcedonic agate, with
A & B. partial traces between its films in the nature of Mocha stone.

[1] [See above, p. 65.]
[2] [A. 70; see above, p. 445.]

150, Entire nodule of perfect agate, giving the association of level with
A & B. concentric beds, which is continually found in agates of the highest
quality, and which is not one of the least mysteries of their construction. The flat fracture on the larger half of the nodule shows
the exact plane of these level beds; those which form the general
mass of the stone are delineated with exquisite precision, and present
for solution nearly every question relating to the banded structure
of these stones. However, I am sorry to be obliged to limit my
illustration of agates to the forms in which the zones are continuous.
Of those, in which they are interrupted so as to suggest the idea
of re-cemented fracture, all my good specimens have been already
given either to Sheffield or the British Museum.

GROUP II

Fluor Spar

151. Pale green fluor spar, two corners of tableted cubes. My own collection, next to silica, consists chiefly of fluor and calcite, minerals
which present themselves in peculiar beauty among our British
hills. Fluor spar is indeed the British mineral *par excellence*, the
cubic varieties of it being presented on a scale, and with a precision of crystallization, in Cornwall and Derbyshire, elsewhere
absolutely unrivalled. I think it better, therefore, to limit myself
in this small collection to the illustration of these simple and
indigenous minerals, adding only in the last ten of the series one
or two specimens necessary for reference and illustration. The
example with which I begin the fluors I have called tableted,
because on the broken sides of the cubes they seem to have been
finished by the imposition of external veneers or tablets. On
primary crystals the structure is not infrequent, and in the finest
examples of fluor the growth of the crystal is expressed also by
interior lines; but until I am myself entrusted with the nurture
and education of a crystal of fluor, I will not pledge myself to any
positive statements on the subject.

152. Fluor spar, transitional from purple to green, and from green to
purple, in confused cubic crystallization. This specimen is much
more interesting, structurally, than it at first appears. In the larger
cube the broad arrangement is of a lilac sub-crystal, with a green
outer one, turning purple at the corners; but on looking closely at
these corners with a lens, the purple layers will be found reduplicate
with green ones; and since the scientific people would probably
tell us, that seen by the aid of proper instruments, the purple
parts would be green and the green purple, and that if seen by
fluorescent, polarized, or epipolic light, they would be neither the
one nor the other, I can only leave the question of the colour of
this mineral-chameleon to the discernment and determination of
the qualified observer.

153. Fluor spar, which I think we may venture to call, for the present, purple, with extremely pretty calcite superimposed.

154. Fluor spar, an interesting cube, apparently built of smaller ones, in which I do not know the reason of the apparently porous whiteness inside, associated with, or perhaps I might more accurately say, as in some cases of human association, impaled upon, a lively group of crystals of white quartz—an extremely pretty and interesting illustration of the mode of companionship usually preferred by these minerals.

155. Fluor spar, pale green, lodged upon quartz, with superimposed calcite. The substructure of quartz seems itself to encrust inferior fluor, and the whole thing seems to be very like a child's game of " which hand's uppermost."

156. Pale green fluor, set into crusts of quartz crystals, which appear to have lost their heads in endeavouring to get out of the way of some dirty iron pyrites. Numbers of minute quartz crystals are seen crowding themselves under the ridges of the greater ones, thus disturbed, while the fluor, both below and above the agitated crust, takes apparently such place and room as suits it. How the business began, and how it would have ended, if this specimen had been let alone and not brought to Brantwood, mortal ingenuity may not say.

157. Fluor spar, pale green, forming itself in octahedrons built of small cubes on a crust of compact quartz. Extremely pretty.

158. Fluor spar, with interspersed galena, showing the manner of association of two friendly cubic minerals, both of them sprinkled with sulphuret of iron. The deceptive sprinkling of the surface of the cube of galena with small cubes of fluor, and the intimate subdivision of the lead between them, make this a very interesting specimen under the lens.

159. Yellow fluor, thoroughly sprinkled with sulphuret of iron, calcite superimposed to finish. I do not usually note localities of minerals unless there is some special reason, but it may be generally noted that these confused arrangements of cubic fluor with quartz and calcite are almost always Cornish, while the masses of blue and purple fluor out of which ornaments can be cut, are exclusively to be found in Derbyshire.

160. Green and purple fluor, developing into octahedric crystallization by myriads of small cubes, whose surfaces, and especially those of the green bed, which they allow partly to be seen through them, are marked with the most strange circular flaws and hollows. Deeply interesting.

161. Green fluor, of similar construction to the last specimen, but in purer substance and without the superficial cavities. Each of the small cubes is beautifully modified on the edges; floating crystals of dark iron pyrites add to the beauty of this wonderful specimen.

162. Green fluor, becoming purple at the surface, with which it constructs two pretty octahedral pyramids sprinkled with fine quartz; after which interruption, the purple fluor, greatly disturbed in its mind, sets to work again to build more octahedrons out of little cubes, but only makes a mess of it. Some iron pyrites, which has no business there at all, takes possession of the quartz on the other side, and spoils what would have otherwise been a most exquisite example of fluor structure.

163. Clear octahedric fluor, variously transfixed and beset by white quartz, with intrusive pyrites as usual, but I think, in this case, copper instead of iron. An extremely rare and valuable specimen.

164. Pure green fluor, partially covered by carbonate of lime; the angles of its own cubes exquisitely modified. It is to be noticed, as a constant habit in fluor, that if it means to modify the solid angles, as in this instance, it always polishes itself close up to them, exactly as if it had been touched with a mill. The most honest dealers are sometimes quite unable to persuade inexperienced purchasers of the genuineness of these specimens. The naturally polished planes are indeed always recognizable under the lens by their crystalline zones, but it is sometimes pleasant to show a tyro such a *pièce de conviction* as the crystal here above the ticket, over which the projecting carbonate of lime shows that no lapidary's wheel could have got at the polished surface.

165. Full-coloured green fluor in complete cubes, extremely fine and sharp in the edges, and entirely explanatory and illustrative of what fluor spar is in its simplicity. The student is to observe that though a cubic mineral is bound to try, and does conscientiously try, to be as much of a cube as it can, there is not the slightest chance of its ever succeeding. No cube of any cubic mineral ever existed in the world, that had all its sides equal.

166. Deep green fluor, set into a group of quartz crystals, founded on a bedding of slate. The deep plunging of the fluor into the quartz is, in this case, extremely remarkable, and the specimen is otherwise of value in showing the nature of contemporaneous formation. These crystals certainly were not stuck into the quartz after the quartz was complete, still less could they be suspended in the air till the quartz enveloped them. Both crystals are formed at the same moment and built in unison, but I cannot think why *our* chemists do not make solutions of things, which will show the

student how this unbelievable arrangement takes place, and let him see it going on. The angles of the fluor are slightly modified, as in the last specimen. The pretty white triangle seen in the angle of the main crystal in some lights, seems to mark the spot at which the modifying action began in the interior.

167. Octahedric fluor, like 162, but more involved with beds of quartz, showing the complex relations of the two minerals.

168. Cubic fluor, partly embedded in quartz, partly encrusted by it, the crust-laying in this case being of course a subsequent operation ; but the odd thing is that the quartz in this final action does not come down like snow on everything, but crusts all the fluor on every side to the same depth, and crusts none of itself anywhere.

169. The two closing specimens of our fluor series are chosen to show the difference between indecisive and decisive *style* in mineral formation. This piece, No. 169, looks at first like a fragment of a cube, but it is not a fragment at all, but a crystal which has never made up its mind what it would be ; never been able to persevere in smoothing a single plane, and broken itself off anyhow in a series of nondescript forms, for which there is neither rhyme nor reason. Very few minerals ever present themselves in such disorder, if crystalline at all.

170. Here, on the contrary, we have exquisitely decisive and accomplished crystals of three minerals,—quartz, fluor, and tetrahedric copper pyrites,—every one of them alike exemplary.

It has always seemed to me very singular that minerals, so far as I know them, are for the most part moral in company. One rarely sees correct quartz associating with misconducted fluor, or the reverse ; but one does not always see why the circumstances which are best for the education of one mineral should alike be favourable to that of another.[1]

GROUP III

Calcite

171. Carbonate of lime, in its elementary crystalline form, extremely pure, and showing the association of two directions of cleavage in a compact block. The perfection of the crystallization is so great that in some portions it breaks with a nearly conchoidal fracture across the proper cleavage. The substance cannot be seen in greater beauty or simplicity.

[1] [Compare *Ethics of the Dust*, §§ 62, 63 Vol. XVIII. pp. 279, 280).]

172. Dog-tooth spar, of Derbyshire, showing at its base, with great clearness, the mode of its construction out of the rhomboidal masses of the compact calcite. I could easily have sent groups of these crystals, weighing twenty or thirty pounds, but for purposes of study, this, which can be easily handled, is much more useful. To see the spar in perfection, it is well worth while stopping on the way to or from London, and to explore the cave in the High Tor of Matlock.[1]

173. Calcite, in flat or "nail-head" crystals, beautifully showing their relation to the rhomboidal mass out of which they are produced.

174. Calcite, in an obscurely globular crystal, beautifully showing its rhomboidal fracture.

175. Calcite, in a superbly constructed and modified crystal, on a mixed mass of calcite and native copper, in which the latter for the most part occupies the interior of the crystals. Extremely fine.

176. Calcite, in a beautifully pure crystal, highly modified.

177. Calcite, partly compact, passing into very interesting "nail-head" crystals, apparently built out of triangular plates. This is, however, merely an external appearance. They are not laminated on the fracture.

178. Calcite, with galena, the calcite peculiar in showing unfinished terminations of crystal.

179, 180. Other conditions, of more or less interrupted crystallization.

181. To be compared with 178 for the white plates at the terminations of the crystals, and with 182 for their position.

182. Calcite, in extremely short hexagonal crystals, more or less endeavouring to connect themselves into long ones. The blundering way in which, both in this example and in 181, the prisms seem to slip aside and off each other, is one of the most grotesque aspects of crystalline structure.

183. Calcite, in "nail-head" crystals of great beauty, but entirely confused in arrangement.

184. Calcite, in "nail-head" crystals, trying to constitute themselves into a radiating prism.

[1] [As Ruskin had often done: see *Præterita*, i. §§ 83, 106.]

185. Calcite, in extremely beautiful prismatic crystallization, lightly sprinkled with sulphuret of iron. With this example we begin a series of specimens, to which I request the student's close attention. The larger crystals in this example all present an appearance of enlarging towards the top, but this tendency is developed by their infinite cunning and multiplied adjustment of crystals, whose sides are really parallel, and whose real form, under the perfect law of its construction, is seen only in the smaller ones, of which I fear several may be broken on their journey; but I hope the perfect long one, in the recess under the two largest, may be spared. I have left, as a sort of index to it, a former number in my own collection, 860. Now the student will please observe that prisms of this form, though they continually associate themselves in groups with widening tops, only do so in one direction; they shoot and expand upwards from the base of the specimen, but they do not construct opponent prisms radiating the other way. In this respect the specimen is entirely different, and constitutionally different, from those which we have henceforward to examine.

186. If this specimen bears its carriage with any safety, it is an extremely pretty one, of what is called twin-crystallization, in which the halves of two crystals are symmetrically associated under a fixed law of rotation on their axes; but these are carefully to be distinguished as accidental forms from crystals in which the symmetry is constant, whether it be by apparent rotation or by balance in direction.

187 is an extremely pretty example of a crystal whose proper form is of two smooth-cut triangular towers, each terminating in triangular pyramids, which are always jagged, never in smooth planes; the two towers thus constructed being fitted together at their bases into one crystal, so that the sides of the one shall fit to the corners of the other. Calcite is, of all minerals that I know, the prettiest in its adjustments of form in this kind. This example is not a perfect one, but the more interesting, from its suggestion of the way the crystal is built.

188. Calcite, in one of its common doubly pyramidal crystals, in which there is no intermediate tower, as in the last example; but the edge of one pyramid immediately fits the side of the other on a base of thin bedded galena.

189. Calcite in complex double pyramidal crystals, which, when they came out of the mine, must have looked as if cut out of snow,—they are still beautiful, and perhaps a clever mineral dealer could partly redeem them from the stains of their twenty years' sojourn in this sinful world; but I am afraid the sullying brown of smoke on calcite, not to say on many other precious things, is virtually ineffaceable.

190. Calcite, in the prettiest of all its crystals, that in which groups of flat "nail-heads" place themselves in a double symmetry, in which they arrive at some resemblance to the form of a rose. The varieties of impulse, and applied law, in the construction of any one of these calcareous petals passes all calculation, and would surpass belief if we had not seen the thing done.

GROUP IV

Mixed Crystalline Earthy Minerals

191. I have chosen the last ten specimens in mixed minerals merely for the sake of illustration for modes of crystallization, and to be the beginnings of other groups in the future collection, if by good fortune it may be enlarged. This example is an extremely pretty one of the amicable crystallization of four minerals—beryl, quartz, felspar, and mica—formed in a cavity of granite in the Mourne Mountains in Ireland. The student is doubtless aware that granite, properly so called, is always composed of the three minerals—quartz, felspar, and mica—and the beryl is almost always found in association with some of the three; but its occurrence is extremely rare in the British Islands. It is here very neatly formed and terminated, and the white felspar is also seen in great perfection.

192. Beryl, in very large and very small crystals, all ill-made, and blunderingly stuck together, with black quartz from the Ural Mountains, in which very beautiful specimens of crystallization are often found; but in the plurality of cases they pitch and glue their work together in this unseemly way.

193. Garnets, in fine crystallization, though of dull brown colour, in the trap rock of Cumberland, beautifully illustrative of the formation of partly embedded, partly superficial crystals.

194. Massive garnet, of good rich brown colour, and unusual size, showing various cleavages.

195. Rubellite, a mineral belonging to the group of the schorls, but the prettiest of them, and extremely rare in crystals of this size. In felspar.

196. Topaz, embedded in black quartz, with well-formed planes and terminations.

197. Common mica, in extremely thin folia, with traces of I know not what mineral arranged in triangles; a newly-discovered mineral. It must be close on a white ground to be seen clearly.

198. Green mica, in obscurely hexagonal crystals, beautifully illustrative of micaceous aggregation.

199. Chlorite in segregation out of quartz, or into it; a most interesting piece of structure.

200. Graphic granite, composed chiefly of felspar, with dark quartz, and a peculiar variety of yellow mica; the whole collected and congealed under those great forces of crystallization which do not develop themselves merely in veins and in cavities, but pass through the entire mass of substance of the great mountains themselves.

CATALOGUE

OF THE

COLLECTION OF SILICEOUS MINERALS

GIVEN TO AND ARRANGED FOR

ST. DAVID'S SCHOOL, REIGATE.

BY

JOHN RUSKIN,

HONORARY STUDENT OF CHRIST CHURCH, OXON,
HONORARY FELLOW OF CORPUS-CHRISTI COLLEGE, OXON,
AND
SLADE PROFESSOR OF FINE ART, OXON.

MDCCCLXXXIII.

[*Bibliographical Note.*—This is a pamphlet uniform with the British Museum Catalogue. The title-page is as shown on preceding page.

Octavo, pp. ii. + 50. There is no imprint. The headline throughout is "Catalogue of | Siliceous Minerals." Issued in buff-coloured paper wrappers, with the title-page (enclosed in a double-ruled frame) reproduced upon the front, the rose being added above the date. The Catalogue was reserved for private circulation.

Copies of it were struck off in two stages; the *first edition* differs from the second (which is the one here reprinted) in the following respects: It does not contain the headings "Section I." to "Section V.," nor the lines in italic describing the contents of the several sections.

No. 16, the words "Compare . . . page 19" were added in ed. 2.

No. 17, the words "Cut and polished under my own direction" came, in ed. 1, after "finer condition."

No. 21, lines 2–4, ed. 1 reads ". . . all perfect, and it is extremely difficult . . . knocked off; but . . ."; lines 11–13, ed. 1 reads "Quite seriously, there are . . . drop and flow; but no mineralogist . . ."

No. 22, line 3, the words "but see notes on No. 48" were added in ed. 2; line 9, ed. 1, after "small pins," reads "This specimen is an extremely fine one, cut and polished under my own direction"; some of these words were in ed. 2 transferred to the end of the note.

No. 23, for "pulverulent," ed. 1 reads "calcareous (white)."

No. 24, the words "But see . . . page 27" were added in ed. 2.

No. 28, line 4, for "minor ripples," ed. 1 reads "divisions."

No. 30, line 5, "Frankenstein, in Silesia" added in ed. 2.

No. 31, lines 4 and 7, "native" and "crystals" were italicised in ed. 2; line 14, "(See one . . . 35)" added in ed. 2; line 17, "(See . . . 40)" added in ed. 2.

No. 31, the words "(See one in No. 34, and another in 35)" were similarly added.

No. 31, the passage at the end, "You will observe . . . below," was similarly added.

No. 32, line 2, "this" added in ed. 2.

No. 39, line 9, "once" added in ed. 2.

No. 40, line 11, for "sprinkling," ed. 1 reads "shake of it"; "the like of it" added in ed. 2; the italics in lines 31, 32, 46, 47 were introduced in ed. 2; line 42, for "*with* either," ed. 1 reads "either with"; line 51, for "on," ed. 1 reads "back"; line 63, for "pasty white pigment," ed. 1 reads "Chinese white."

No. C. 14, line 2, after "manganese," ed. 1 reads "(I think it is not heavy enough for iron)."

Section III., ed. 1 reads "These fifteen examples . . . finest jasper. It must be noted, in beginning to study these, that jasper is an opaque . . ."; lines 5 and 6, the quotation marks were introduced in ed. 2; line 2, "occasional" added in ed. 2; line 13, see p. 501 *n*.

No. 41, the words "(I have now . . . series)" were added in ed. 2.

No. 42, last words, for " puzzling interest," ed. 1 reads "puzzle and interest."

No. 43, the italics were introduced in ed. 2.

No. 45, last line, "of it" added in ed. 2.

No. 47, line 2, "hitherto" inserted in ed. 2; last line, ed. 1 reads "the more compact forms of it."

No. 48, A and B, the italics were introduced in ed. 2; line 7, "either of rod or film" added in ed. 2; line 14, "the first group of" added in ed. 2; line 15, "No. 24" added in ed. 2.

No. 51, line 1, "which I" before "submitted" in ed. 1; line 2, for "matter," ed. 1 reads "subjects."

No. 52, line 2, for "stolidity," ed. 1 reads "apathy."

No. 56, line 5, for "make anything of," ed. 1 reads "understand."

No. 58, line 1, for "Our three," ed. 1 reads "The three."

No. 59, last line, "p. 21" [now p. 500] added in ed. 2.

No. 61, "Quartz vein in Coniston Grit" added in ed. 2; line 17, "No. 61" was inserted after "built" in ed. 1.

No. 62, the last words, "Locality . . . consequence," were added in ed. 2.

No. 63, line 4, see p. 507 n.

No. 64, line 3, "combined" added in ed. 2.

No. 65, line 2, for "minor," ed. 1 reads "more divided"; line 2, "again," and line 3, "finally," added in ed. 2.

No. 67, line 2, ed. 1 reads ". . . nearly bothered out of its life—all, seemingly, because of some nasty earthy sulphuret of iron. The forms . . ."; line 6, for "unfretted," ed. 1 reads "unbothered."

No. 68, line 1, for "vexed," ed. 1 reads "bothered."

No. 69, line 6, for "these friends," ed. 1 reads "the parties."

No. 71, "Quartz with Tourmaline" added in ed. 2; line 7, see p. 509 n.

No. 75, last words, for "looking like a piece of intended jugglery," ed. 1 reads "seemingly all Maskelyne-and-Cookery."

No. 77, line 1, for "rutile (oxide of titanium)," ed. 1 reads "titanium"; line 3, for "whichever," ed. 1 reads "whatever."

No. 81, line 1, "white quartz coated with amethystine quartz" added in ed. 2; line 3, for "permit . . . work," ed. 1 reads "have taken an established position, as gems, or precious stones, for engraving"; last lines, ed. 1 reads "I have seen exceptional cases, but they are very rare. This No. 81 is a very notable crystal of white quartz coated with amethystine quartz, in which, however, the purple colour . . ."

No. 82, ed. 1 reads "Pale . . . interesting. I am afraid, however carefully packed, some of it may break on the way, but I think the remnant will always be pretty."

No. 83, line 3, for "all amethyst," ed. 1 reads "otherwise."

No. 84, last line, ed. 1 adds "like that of mother-of-pearl" after "dependent."

No. 88, last lines, the italics are introduced in accordance with a note in Ruskin's copy.

No. 90, line 3, for "It is more rarely," ed. 1 reads "It is sometimes, though more rarely"; for "seldom," it reads "not"; for "since the year 1878," "about five years."

Page 513, line 8, the words "(not . . . column)," and in line 15, the words "my old friend and Knight of the Hammer," were not in ed. 1; last words, ed. 1 reads ". . . will join on properly to the finer examples which I sent originally."

The result of the alterations and additions in ed. 2 was that whereas ed. 1 ended on p. 49, ed. 2 extended to p. 50.

In Mr. Wedderburn's possession there is a copy of the second edition with some further revisions in Ruskin's hand, which have here been made. These are :—

Nos. 17, 18, "Flint-chalcedony" for "Flint chalcedony."

No. 21, line 3, "more than" for "extremely"; lines 5 and 6, "down" and "up" italicised; line 11, "are" not italicised.

No. 31, line 3 from end, "tetherium" corrected to "tellurium"; last line, "additional observations at" for "farther notes."

No. 33, the last words italicised.

No. 37, last words, in eds. 1 and 2, ". . . beauty and illustrative pheno-mena respecting the occurrence . . ."

Section III., "and forms" added after "nature."

No. 44, last words, eds. 1 and 2, ". . . jasper, extremely fine."

No. 45, last line, "of it" added after "enough."

No. 47, last line, "its more compact forms" for "the more compact forms of it."

No. 70, last line, "properly" for "prettily."

No. 88, italics introduced.

No. 90, "I believe" inserted.]

CATALOGUE

OF

SILICEOUS MINERALS[1]

SECTION I

Nos. 1 *to* 30, *illustrating the nature and relations of*
FLINT *and* CHALCEDONY

1. Pure black flint with exquisitely characteristic "conchoidal" fracture, looking exactly like the cast of a shell.

2. Black flint less pure (with some admixture of chalk or, perhaps, clay), developing agatescent bands in purifying itself. The bands faulted and terminated in the unaccountable way which, with finer material, forms so-called "Brecciate" Agates. (Compare Nos. 10, 11, 12.) A small fossil (sponge?) is embedded in the angle of this specimen, which is a most finished and comprehensive example of flint-structure. Polished on two sides.

3. Two pieces, *a* and *b* (2070 of my old collection). A superb example of finely delineated and terminated agatescent structure, in dark grey flint.

4. Rolled pebble of coarse brown flint, with agatescent structure mimicking a fossil. The central division most curiously faulted.
   Ground down and polished on one side to show structure.

5. Black flint, full of somebody, I don't know who, gone to pieces. The (weathered or decomposing?) surface showing the forms projecting.

6. Common grey flint, with very unusual condition of surface. *I believe,* inorganic, and merely mimicking a fossil; but there may be much disguised organism provoking the forms.

[1] [This collection is now at Stone House, Broadstairs: see above, p. lix.]

7. Grey flint, with many enclosed organisms, one cruciform, going quite through;—the cross on one side diminishing to a quatrefoil round a small cavity on the other.
   Ground down and polished on the quatrefoil side.
   Extremely rare and fine.

8. Grey flint of like character, with many fossils, totally incomprehensible.
   Cut into two pieces, each polished.

9. Agatescent chalcedony (i.e., chalcedony throwing itself into bands), showing both the straight-levelled and concentric forms.
   An altogether exquisite example.
   Cut into four pieces, and polished, under my own direction.

10. Dark flinty agate, developing brecciation.
    A rectangular slice, polished.

11. Jasperine agate, in developed brecciation.

12. Jasperine agate, in perfect brecciation.
    A small slice off the best piece I have in my own collection.

13. Fawn-coloured flint, becoming chalcedonic on the inner surface. The outer surface, to me, inscrutable; the smaller and porous parts showing every state of incipient chalcedony.

14. Finer grey flint passing into finer chalcedony. The flint almost entirely made up of organisms.
    Two pieces, A and B, cut and polished under my own direction.

15. Still finer flint, with finer chalcedony. The latter, however, broken on the surface, showing delicate varieties of conchoidal fracture.
    Look with lens at those on the smallest piece above my old ticket, where the new number 15 should be put for indication. These fractures are, however, partly, where so finely rippled, indicative of the interior structure of the stone. See especially the sharp apex to the left hand of the ticket. The structure of the opaque part is mostly inorganic, founded on effaced shells. See near the red arrow.
    Two pieces, A and B, cut and polished under my own direction.

16. Common mossy flint of the south coast, passing into pure chalcedony, in one part showing black arborescences of oxide of iron (rare). No one has yet given any account of the nature of the mossy matrix, which is extremely common, but extremely curious. (Compare Nos. 22, 33, and C. 1 to 12, page 500, with notes at page 499.) Where the chalcedony is white, it begins to throw itself into

sausage-like forms, which might be mistaken for stalactites, but are nothing of the sort.

Their decomposition at the edge into coats (see the shortest polished side) is extremely notable.

This piece was originally twice as large, and broken; the other half is in my own collection.

17. Flint-chalcedony, making the best it can of itself. It cannot be seen in finer condition, showing the jasperine, spongy interior, edged with extremely minute sparkling quartz. Examine carefully with lens. I do not know how far these red parts are organic. It is very rare to find chalcedony of the two sides of the wall, as here; and still more rare to find it, as on the blue side, with superimposed pseudo-stalactites (to my great regret, broken short off).

Reference number in my old collection, 1591.

Cut and polished under my own direction.

18. Flint-chalcedony, perfectly pure, enclosing sponge changed into yellow jasper; frequent on the south coast, but this a very beautiful specimen. It is part of a rolled pebble; see next number.

19. Slice of black and grey flint, enclosing sponges saturated with blue chalcedony. The pebble No 18 is a rolled fragment out of such a space. This specimen, though, like No 18, of a kind frequently occurring, is extremely beautiful as an example; and especially interesting because the black and grey matrix is itself full of all kinds of organisms.

20. Highest quality of purple chalcedony, forming agatescent bands, but I do not think this has been formed on flint; it introduces us to the group of agates properly so called.

21. Chalcedony on amethyst-quartz. Extremely interesting, both because the points of its pseudo-stalactites are nearly all perfect (it is more than difficult to obtain specimens in which the points have not been knocked off), and also because they show their peculiar difference from real stalactites in dropping, half of them *down* and half of them *up*. Modern geology would, of course, not scruple to explain this phenomenon by the theory that the world had been turned upside down in the interval. I prefer, myself, to direct attention simply to the circumstance, and to the farther interesting fact, that one stalactite in the middle dropped *sideways*.

Quite seriously, though there are forms of chalcedony, like those deposited by the Iceland Geysers, which do really drop, and flow, no mineralogist has yet explained the intermediate incrustations, of which this is a notable example, whose forms seem to be totally independent of the action of gravity.

22. Chalcedony in agatescent deposit, partly stalactitic in central films and rods; a structure extremely frequent, but hitherto unexplained, and, I believe, for the present inexplicable. (But see notes on No. 48.) It is to be noticed especially that the basic films never exceed a certain thickness, nor the basic rods a certain diameter. All through the stone, whatever the thickness of imposed chalcedony, the walls are here always composed of one greenish-brown line, with two external white ones, and the rods are of the diameter of extremely small pins. It is often thought that the interior substance in chalcedonies of this kind has been originally moss; but I believe it is entirely inorganic, and, in this case, as in the next example, chiefly chloritic earth. Extremely fine, cut and polished under my own direction.

23. Moss agate, so called, being irregular chalcedony, traversed by a network of chloritic earth, and forming itself into level lines in the larger spaces. An extremely curious example, inexplicable as No. 22. Part of the cellular mass is more or less pulverulent, and yields easily to the knife.

24. The same kind of chalcedony in bolder development, probably Icelandic, showing connection with some kind of floor and roof, and inconceivably peppered over the surface,—with what, I don't believe any mortal can find out. (But see notes on No. 48, page 503.)

25. Icelandic chalcedony, showing the levelled or lake structure in an open space, and a more closely sprinkled external surface, of nameless ugliness, with two mysterious little pits in it, which are part of the chalcedonic structure. The whole, one petrified enigma.
    Cut and polished under my own direction. The other half, one of the choicest pieces in my own cabinet.

26. Level Icelandic chalcedony, traversed by pseudo-stalactites, each with its proper rod of nucleus. Two pieces, A and B, cut and polished under my own direction, and extremely instructive, if one could only understand a single word of what they say. Look at them against the light, and take care to keep the two pieces together.

27. Icelandic chalcedony, passing into quartz, which traverses a level lake, in the hollow of the interior, and crystallizes, not in the usual form of quartz, but in triangular pyramids. This, like No. 26, is an extremely rare specimen.

28. Grey chalcedony, forming an agatescent ball, showing a perfectly conchoidal fracture on the broken surfaces, distinguished from that of flint by its comparative dulness and absence of the ribbed minor ripples, which show the greater elasticity both of flint and glass.

29. Brownish-red chalcedony (carnelian), coating oxide of iron. The finest carnelian is rose-colour or crimson, but it is almost impossible now to find a piece that has not been dyed. The colour of this is very good by transmitted light.

30. Chrysoprase. A condition of silica, intermediate between flint and chalcedony, but varying in colour from white to green; dull in fracture, as the broken piece will show, better than the old surface; and never throwing itself into globular or stalactitic forms. It is found, I believe, only at one place in Europe, Frankenstein in Silesia, and is extremely respected by me because it is found nowhere in America.

It is the most valuable form of silica except opal; but it is properly connected with flint and chalcedony, opal forming an entirely distinct family of minerals. These thirty specimens, therefore, contain complete illustrations of the materials of which the transparent or translucent portions of agates are composed. We next enter on a series illustrating the opaque material, JASPER, but which, like the chalcedonies, must be traced back to their origin in entirely common flint.

SECTION II

(Nos. 31 to 40: with supplementary series C. 1 to C. 15.) Illustrating the relations of QUARTZ *and* CHALCEDONY *to native metals and common metallic ores.*

31. Before, however, going on to study the jaspers, although it is out of scientific order, I think it will be well, and certainly it will be refreshing, to observe the conditions under which pure quartz is associated with *native* metals; of which it must be observed in the outset, that the native metals are scarcely ever—I do not myself recollect a single instance of finding them—associated with pure and clear *crystals*, either of quartz, or any other mineral. Both quartz and calcite, when they contain native gold, silver, or copper, are normally opaque, and only semi-crystalline. This generalization is bold, and must, of course, be taken only as the expression of my personal experience, not of a scientific fact; but it may be certainly stated positively, that in the countries hitherto known as mining districts, this law of the opacity of gold-vein-stone holds, with rare exception. (See one in No. 34, and another in 35.) This specimen, No. 31, is entirely characteristic of the kind of quartz in which gold is usually found, whether in America or Australia (see the long note on No. 40), and it is also entirely characteristic of the mode in which the greatest *quantity* of gold occurs in those countries; dispersed in plates of varying thickness, but still generally describable as laminæ, through what are apparently fissures in the imperfectly crystalline and feebly translucent quartz. This is a very rich and pretty specimen from Nova

Scotia, of which, as of every other specimen of the kind I ever saw, I can tell you,—absolutely nothing, more than that the gold is pure, and the quartz very nearly so; and that I have not the slightest idea how the gold got into it, or where it was before it got into it, unless perhaps in a place which I do not care to mention!

You will observe that I italicize above the word "native," meaning "pure" metals. Both quartz and calcite crystallize superbly in association with the *ores* of metals, especially with the oxides and sulphurets of copper and iron, and the oxide of titanium: and sometimes even native copper will form beautifully in the interior of calcite crystals. On the other hand, tellurium in any state is peculiarly destructive of crystalline power in its matrix. (See additional observations at pp. 498, 499, below.)

32. Native gold, in the same kind of quartz, but itself more massive, and partially crystalline. I believe this Australian; but it represents the kind of gold, all over the world, which forms the nuggets in alluvial deposit. That is to say, the flawed and more or less brittle quartz gets knocked away by attrition, while the gold, in grains and masses of various sizes, falls to the bottom of the stream. This specimen has itself been a little rolled and battered on the outside, but shows the fibrous crystalline character of the gold, beautifully, in its cavities. It was cut when I bought it, and in the section obtained shows the exact lines of contact between the gold and quartz, in the interior of the stone, where the metal is compact; while it effervesces in the cavities into moss.

33. Slice of quartz-rock from New Zealand, with native gold dispersed through the body of it, in extremely fine particles. This form is peculiar to New Zealand, and as I never use a microscope, I cannot say how far the gold is crystalline. I have added, therefore, the extremely rich smaller piece, for experimenting upon at leisure, by cutting slices off the edges of it, as thin as it is possible now to cut quartz,—and there is scarcely any limit to the fineness obtained by Mr. Sorby[1] and other microscopic observers. A great deal might, however, be found out by the schoolboys of St. David's, very interesting to older people as well as themselves. In the large specimen, No. 33, the gold is in somewhat finer grains and fibres, but my impression is that in both it is in a state very *closely corresponding to that of the chlorite in moss agate.*

34. Native gold from Transylvania, in a vein of rather finely crystallized quartz, traversing a grey rock, which is, I believe, called " Psammite." This form of gold is entirely peculiar to Transylvania, a country (you will know where it is, but I don't) which, it is well to observe in passing, has always taken the greatest pains to crystallize its minerals prettily and delicately, so that whenever you

---

[1] [See above, p. 207 n.]

see an extremely charming and peculiar specimen of anything, you may almost guess it Transylvanian, or if it be quartz only, to be of Schemnitz in Hungary, which, I believe, is somewhere thereabouts. This Transylvanian gold is beautiful in itself, and exceptional, as compared with American or Australian specimens, in being set on definitely crystalline, though minute, quartz.

35. I am very sorry to part with this specimen, of which I have no duplicate, and shall not easily find the like; but it is so instructive and beautiful, that I can't keep it any more idle in my cabinet. It is perfectly crystalline arborescent native gold, associated with perfectly formed, though minute crystals of carbonate of lime, and if it is not Transylvanian, I can only say I am very glad there is any other country that can work its gold in that way, but at least it is not American, nor Australian.

36. Crystalline gold in a cavity in quartz, more or less stained with oxide of iron. I do not know the locality of this specimen; it might be Australian, South American, or perhaps African; the form both of gold and quartz being the general one common to all; but it is comparatively rare to find the gold thus crystallized in the cavities only, and not dispersed through the body of the stone. Compact oxide of iron is also mixed among the stained quartz, the surface of which is very curiously crystallized round the cave, though not quite in Transylvanian style; and one could fancy rather a pretty fairy tale told about the cave.

37. Part of the wall of a large vein of quartz, I should guess from California, traversed by another small vein of whiter quartz, and having its porous cavities feebly crystalline, and in some places curiously cellular; filled with dispersed moss of finely crystalline gold, every particle of which becomes interesting under a powerful lens. Entirely wonderful and beautiful, though showing the always rude and feeble crystallization of gold-bearing quartz, everywhere except in Transylvania. The wall of solid quartz will be seen to be partially divided by a flaw passing into a fissure, which, with others parallel to it, seems to indicate the further rending asunder of the original vein as the mountain mass consolidated itself. This specimen is unique in my experience, both in beauty and phenomena, illustrating the occurrence of gold in quartz.

38. Detached examples of crystalline gold, chiefly, I believe, African, broken away from the quartz by the miners, but each in itself interesting.

39. Crystalline gold, I have no doubt Transylvanian, showing in one piece all the principal forms which the metal takes. It belongs to what mineralogists call the cubic system, and in any orthodox mineralogy, various pictures will be found of cubes, octahedrons,

dodecahedrons, and other charmingly symmetrical figures, as representative of those in which it is scientifically required to crystallize. But it never *does* crystallize in any one of them! I have one small octahedron (kept in a separate bottle), which I don't believe is genuine, and I had once two cubes, which certainly weren't. These are all I have been able to find in forty years' experience. The real crystals of gold, of which you may see plenty on one side of this specimen, are entirely indescribable by any human language, and I leave you to make what you can of them. It is impossible to see sharper or better examples, for gold has a curious way of never crystallizing brightly or neatly, but always as if it had been a little beaten about. The ground of the solid crystals, beautifully seen on the opposite side, is a flattish plate, in most places apparently made up of triangular small ones which project in little triangular pyramids. In other places, the plate will be seen to be woven out of fine fibres, and here and there apparently to be spun into wires. The laminated form of plate composed of triangles is the most frequent condition; the fibrous and arborescent lace, next commonest; massive crystals as fine as those in this example are extremely rare, and I think couldn't be done anywhere but in Transylvania, to this bewildering extent. The pale colour of the specimen is owing to a mixture of five per cent. of silver, which, however, as silver also belongs to the cubic system, rather helps than hinders the gold in its crystallization.

40. Massive gold, more or less fibrous, with interspersed quartz, but in several parts quite solid gold; I believe Californian, and representative of what is meant by a nugget, when it has not been much rolled. Even the richest nuggets are seldom wholly free from quartz. There is nothing particularly characteristic about this one, but it is nice to feel the weight of it in the hand, and the gold is extremely pure.

   As representative of native gold in commerce, the pinch of gold dust in the glass vial is enough to show what people carry about their waists and sell their lives for in various countries: but this little sprinkling is, I believe, Scotch, and the like of it obtainable only in very small quantities. Seen with a lens, it will be found composed of more or less flattened fragments, which are obtained by washing the sand of streams,—an extremely wicked waste of time. I do not give this gold-dust a separate number as it should be kept with No. 40, to show how gold is commonly found.

   We will not at present go on to silver, because I have allowed the gold to come into this place in the catalogue, chiefly for the illustration of quartz; and native silver differs from native gold in this particular (one of many), that, as a rule, silver likes limestone, and won't crystallize in quartz; while, as a rule, gold loves quartz, and hates limestone. The single specimen of the above ten, in which it is associated with calcite, was also unique in my collection. Looking back over the other nine specimens, they will all be found to exhibit the same characters of quartz, more or less compact and opaque, irregular in fracture, not like glass, nor yet like common stone: like, in fact, nothing but itself; one can only say of the fracture that it is "quartzose."

This kind of quartz forms an immense mass of the rocks of the world, but *in no place of all the world* does it form *globular concretions* like those of chalcedony. The two minerals, though the same in material, are everywhere and always distinct, though you will find it stated in mineralogical books (I quote Miller as most authoritative), that "chalcedony appears to be an intimate mechanical mixture of crystalline and amorphous quartz, botryoidal, reniform, stalactitic. It is called carnelian when of a red, yellow, or brown colour; plasma when dark-green; chrysoprase when of an apple-green colour, produced by the admixture of one per cent. of oxide of nickel."[1] This is all that science can make of the matter, but common observation will show you these further interesting particulars, that though chalcedony may be produced *with* either flint or quartz, it is a substance totally distinct from both, and belongs generally to a later period of the world's history, that of our own chalk, or of still more recent volcanic mountains. Further observe, *that no particle of gold, or of any other native metal,* has yet, so far as I know, been found in *chalcedony,* but the oxides of iron and manganese, and the sulphuret of iron, are frequently and beautifully associated with it. It will therefore be perhaps best to show, in connection with the quartz containing gold, the finest states of chalcedony containing iron, before we go on to the jaspers. I have arranged therefore, by themselves, a series of twelve examples of cut chalcedony, chosen of the finest kinds, and placed so as to form a pretty ornamental cross. The stones in this group containing brown and tree-like formations, are usually called Mocha stones,[2] being first got where we get our coffee; and I believe these Arabian ones are still the finest kinds found. Those with apparently green seaweed are called moss agates; but I believe there is not a fibre of moss in them, but only the green mineral called chlorite, of which we shall afterwards see tangible examples. The delicate brown fibres, I can state positively, have nothing whatever to do with vegetables. They may be imitated, sometimes very prettily, by dropping any watery dark colour into pasty white pigment, and they are certainly produced in these stones by partially crystalline ramifications through them, while they were still in a gelatinous state, of the oxide of iron or manganese. The substance of the stone, throughout, is pure chalcedony: the two little pieces cut into the shape of hearts are also finely agatescent, and show small white globular concretions on one of the stones, which we shall see better examples of afterwards.

It may be as well to number this group of stones C. 1 to 12, C. standing for Cross.

[1] [*Elementary Introduction to Mineralogy*, by the late William Phillips, new edition by Brooke and Miller, 1852, p. 250.]
[2] [On Mocha stones and "moss agates," see above, p. 377.]

C. 1. Mocha stone of the finest possible structure, arborescence radiating from points; and in clusters of two kinds, black and brown.

C. 2. The same, but with the black branches better defined at one part of the stone.

C. 3. Moss agate, fine, but of the usual structure.

C. 4. Moss agate, of less usual, and finer, delineation. The chlorite half washed away into it like weed.

C. 5 and 7. White agates, before described.[1]

C. 6. A little Mocha stone, merely put for a centre.

C. 8. A moss—or it might better be called a weed—agate of highest possible quality.

C. 9. Moss agate, nearly as fine, but richer and darker.

C. 10. An extremely unusual form, the chlorite seeming much mixed with iron.

C. 11. Mocha stone of the very highest quality, but less valuable to the public and the jeweller, because it can't be seen without a lens. If it hadn't been one of the cross stones, I wouldn't have given it away.

C. 12. Mocha stone, in which arborescence itself is of common character, but of great mineralogical interest, because formed in one of the zones of a well-banded agate, and bent under the point of the white zone at the top to right and left, like the branches of trees in a high wind.

With these twelve mossy chalcedonies, we had better number at once, as C. 13 to 15, three illustrative stones, which will make the system of arborization entirely intelligible.

C. 13. Fine chalcedonic agate, with arborescences and spots, of oxide of iron, running between its zones, the spots being entirely independent of the arborescence. An extremely rare example (for the sake of getting which, and one or two others, I bought a whole collection). I should like the reference to it in my old collection, 1292, preserved in this catalogue.

C. 14. White chalcedony, obscurely agatescent, formed on what I believe to be oxide of manganese with arborescences, which, in that case, would be of manganese, not of iron, and producing themselves, not between its bands, but on the exterior surface. A rare condition.

---

[1] [See above, p. 492 (No. 16).]

C. 15. The best piece I ever saw of the shaly limestone of the hills near the Bagni di Lucca, with arborescences of oxide of iron on both sides, originally, of course, formed in the fissures of the rock.

These fifteen examples being first carefully examined, we shall now be able more or less to understand the similar structures which take place in the finest jasper.

SECTION III

Nos. 41 to 60 : illustrating the nature and forms of JASPER.

JASPER is an opaque and compact stone, in substance extremely common, but which owes its occasional preciousness to its colour, its easy susceptibility of fine polish, and the mossy or flame-like delineations which sometimes traverse its mass. The scientific people only tell one that "it is a variously-coloured mixture of quartz with alumina, lime, carbon, oxides of iron, manganese,"[1] etc.; which somewhat vague statement I will venture partly to contradict, and partly to narrow. Jasper is no more a mixture of quartz with anything else, than it is of carbon with anything else. *Quartz* is properly the *crystalline* state of siliceous earth; chalcedony, the *semi-crystalline;* jasper, the compact *non-crystalline,* more or less mixed with either clay or lime. The mineralogists, in their high-mightiness, don't condescend to say with which;[2] but this, at least, you may depend upon, that jasper *never crystallizes into any definite form whatever,* and is only found in extremely rare cases in globular concretions, like chalcedony. The coarse forms of it, of a dull yellow, and only taking an imperfect polish, are extremely common. The finer qualities are either of a more glowing yellow, or brick-red, passing in the finest stones into scarlet, and even crimson. These colours are, I believe, without exception, given it by the red and yellow oxides of iron; but it is well worth notice, that the red oxide, although in a separate state, often pronounced enough in colour to have given occasion to its usual scientific name, Hæmatite (blood-stone), *never by any chance reaches its purest hues until it is mixed with jasper or agate.* I begin with an example which, though it carries us back to our old friends the flints, yet shows the colouring element of jaspers in perfection.

41. Portion of an imperfectly brecciated flint, coloured by red oxide of iron. I bought the whole piece out of a heap of flints, in a dealer's back-shop, catching sight of its red edge,—for sixpence; cut it, and

---

[1] [*Elementary Introduction to Mineralogy,* by the late William Phillips, new edition by Brooke and Miller, 1852, p. 251.]

[2] [For "... clay or lime. The mineralogists ...; but this, at least . . ," ed. 1 reads:—

"... clay or lime, and if the mineralogists were worth as much of either of them as would bury or burn them, they would have told us long ago with which. But this, at least . . ."]

have kept the biggest piece, which I would not part with for five pounds. (I have now, however, given it to the British Museum, No. 28 of the illustrative Siliceous series.[1]) It is the finest example of pure red jasperine colour I have ever seen in silica. The white portions might be called white jasper, if they were a little less brittle, and the grey parts might be called grey chalcedony, if they were a little more clear. But on the whole, it is better to call it a high-caste flint, traversed by fibres of red jasper.

42. Blue agate, surrounded by a coat of scarlet jasper. It will be seen
A.   that the globular concretions of the agate are independent of this
and   exterior coat; while yet the jasper is itself partly agatescent,
B.   and both are bent about by a brown concretion at the edge of the stone. The specimen is full of all kinds of puzzling interest.

43. An extremely beautiful fortification agate, formed of dark chalce-
A.   dony, with zones of white and red jasper. In all such cases *the*
and   *chalcedony is the active and formative element, and the jasper submissive*
B.   *to it.*

44. Spherical agate in zones of chalcedony, with fillings of quartz, the chalcedony traversed by mossy flakes of fine crimson jasper. On the fracture the stone looks almost like a compact jasper. Extremely fine.

45. Three portions of a deeply interesting agate, formed as a ball in
A.,   the hollow of a volcanic rock, and traversed, from its centre to
B.,   the rock, by opposite tubular veins. The quartz crystals, in which
and   the exterior band terminates, are coated with orange jasper.
C.   Their own sparkling cleavages seem like a diamond sand on the smaller surface of the middle slice. The whole thing is a trifid lump of wonder, and I wouldn't give it away, if it didn't always split my head into three slices to look at it. It is only half of the original stone after all, but I never had the other half, and found I had quite enough of it when I cut this into three.

46. Acicular agate passing into jasper. What I mean by acicular agate, you will see by looking at it with a lens, but I don't in the least know how it came to be like that. I was always afraid of breaking the specimen, but think that a thin slice might be taken off the flat surface, which though broken, would be very marvellous under the microscope.

47. Blue and purple agate, partly acicular, partly zoned, partly anyhow, with a jasperine coat; but these examples are all hitherto given for types of colour, and for illustration of the connection of jasper

[1] [See above, p. 402.]

with agate, rather than for specimens of jasper itself. We shall now pass on to examine its more compact forms.

48. Violet agate, enclosing bands of jasper, *with earthy walls in the centre*
A.   *of them.*
and   It is generally to be observed, that all the films, walls, and
B.   rods in the middle of bands or stalactites of agate, are of some substance less pure than the agate, which seems to throw everything that it doesn't like, and has got mixed up with, into the smallest compass that it can either of rod or film; and then to form itself round, or on the surface of, these rods or films. It will thus throw into its centre, and cover with its own pure substance, either the green earth called "greeny"! ("chlorite") or black oxide of iron, or sulphate of iron, or even the rough substance of the surrounding rock; but the *surface* of chalcedony is almost always pure, and the sprinkling of dust upon it, which takes place in one of the specimens in the first group of this collection, No. 24, is a phenomenon of extremely rare occurrence.

Now, herein is a specific, and I really think I have experience enough to justify me in saying, exceptionless, distinction between chalcedony and quartz in the process of their formation. *I have never in my life seen a crystal of quartz with a central rod, either of earth or iron.* They may be mixed with, or suspended in its substance in variously confused positions, but they are never thrown systematically to the centre, and, in very greatly the plurality of cases, they are consistently and steadily thrown to the *surface*, or to planes immediately within the surface. All the oxides of iron are either placed in spots upon the superficial bands of the crystal, or radiate from them towards the interior. Amianthus often roots itself on the interior surface, and shoots towards the inside; and mixed impurities encrust, or disturb the surfaces, leaving the centre of the crystal as far as possible pure. It is impossible to have a more positive proof of the distinct nature of the two minerals.

The particular specimen we are examining throws the rejected earth into curved lines, more or less resembling broken shells. I believe I was the first mineralogist who described this species of agate,[1] and though I have never been careful about precedence, intend in future quietly to state the things that I have myself discovered, leaving future investigators to point out, with any satisfaction it may give them, that they had been discovered before me, if they can.

There is one point of curious interest connected with this form of siliceous concretion,—phenomena exactly resembling it occur in agates which *have* really been formed on broken shells; and it is matter of extreme difficulty to determine, in many instances, whether the structure was not first organic. It has only been the attentive study of brecciated jasper which enabled me to establish the distinction. See the next following specimens.

---

[1] [In the papers in the *Geological Magazine*: see above, pp. 68, 69.]

49. Brecciated jasper, in variously interrupted bands apparently floating in grey chalcedony. An example of extreme interest connected with the shell-like jaspers on the one side, and with brecciated agates on the other.

50. Egyptian jasper, properly moss-jasper, casually stained red or yellow, by iron, yet with this distinction, that the red portions are finer in grain, and less ridiculous in wriggle, than the rest; while the yellow bits form a kind of soup of frantic vermicelli, unaccountable for on any principles of motion or construction hitherto exhibited on the face of the earth. I don't think the agatescent vegetarians will venture to claim any part of this stone as really botanic; but the name "moss-jasper" is the most conveniently descriptive that can be given. I am bound, however, to place beside it a specimen—

51. Submitted for examination to the greatest authority on muscose matter, Mr. Bowerbank;[1] who sent me word back that it was a charming specimen; and that the embedded forms were all real mosses. I am myself, nevertheless, still under the somewhat contrary impression, that Mr. Bowerbank's mosses must be all real minerals! At all events, the specimen is a superb, though a small one, and two or three film-slices might be cut for the microscope from its roughest end, without the least injury; and the question finally settled at St. David's.

52. Yellow and red jasper, thrown into bands by the force of the containing agate, which, however, is so far bothered by the stolidity of the jasper that it can't form its own orbs and angles properly; but wriggles and loops itself about, partly in the style of the jasperine vermicelli.

Small; but extremely valuable and interesting.

53. Perfect example of what, for distinction's sake, I have called conchoidal jaspers,[2] exhibiting all the previously described phenomena in perfection, yet with one added circumstance, which must be carefully noted. In all ordinary cases of the formation of rods and bands with earthy centres, the successive deposits or crystallizations on each side of the central film are alike: but in certain cases, not otherwise distinct from these, the deposit on one side of the central film is different in number, or in depth, of bands, from that on the other. This little agate, though not so grotesque as many others, is really more curiously unreasonable, in the apparently motiveless effort to make one side of its white promontories always deeper in bands than the other; though the red encircling jasper is practically of the same thickness throughout,

[1] [See above, p. 208 n.]
[2] [See above, p. 445 (A. 71).]

except where it clogs up in the bays. The pear-shaped or kite-shaped section of this agate is also extremely unusual, and, as far as my knowledge goes, unaccountable, all the more that the irregular white promontories would seem the result of forces which must also have produced an irregular surface.

54. Red jasper, in rough grey quartz, with inlets of chalcedony—two of them in perfectly straight lines, with central films, and very definitely unequal in the thickness of external deposit, while the material of the jasper seems to flow in between them, like a tide filling a harbour through a flood-gate. All these appearances of flux, as well as of fracture, in agates, are, I believe, pure and wilful imposture on the part of those unprincipled stones. But as I have never seen one of them making itself, I withhold at present any final expression of opinion.

55. Red jasper in bands, some straight, some bent, and some apparently not merely broken, but chopped up; the whole mixed with sandy quartz and gritty earth, in a manner, I believe, which no one could explain but Lord Dundreary.[1]

56. Red and yellow jasper variously coiled and squeezed, with veins of imperfect quartz,—a rare kind, which the traveller who gave it me told me he had ridden three days in a savage country to get. As I don't care about savage countries, nor whether a jasper, which one must ride three days to get a bit of (and can't make anything of when one has got it), is to be found in New Guinea or Old Guinea (this is, I believe, from New), I have no reluctance in parting with this specimen: but I believe it could not be easily matched.

57. We return to the regions of propriety and common sense, in a slice of pebble of the ordinary kind, known as Egyptian jasper. These pebbles are, I believe, found loose in the sand, and I do not find the mineralogists give us any account of where they come from; perhaps, however, the first question ought to be, where the *sand* they are found in comes from, since we get too easily into the habit of thinking that two-thirds of Arabia and Africa were originally manufactured of sand. The pebble itself is very characteristic in its substance,—jasper properly so called, without any admixture of chalcedony. Though so thin, it is perfectly opaque, and though apparently containing a considerable proportion of clay or lime, takes a quite lustrous polish. Its bands appear to be the result of the same process which takes place in banded flints.

---

[1] [The chief character in Tom Taylor's play, *Our American Cousin* (1858). The part was created by Sothern, whose drawling delivery of phrases such as "Don't you see? that's the idea," and, "That's a thing no fellow can understand," is well remembered by playgoers at the Haymarket Theatre.]

58. Our three last examples of the jasperine group sum its peculiarities, and exhibit them in the finest materials. This specimen, jasperine agate, with central quartz, is beautifully parallel and fine in the agatescent bands, which, please notice in passing, are also remarkable for the bastion-like points of their angles, to which the term "fortification," as descriptive of agate,[1] is properly limited. These angles are, in this example, only the re-entering ones of large arches, but we shall see them in others, formed by straight crystalline planes. The flammeate, jasperine, red and grey stains,—the latter especially, where they cross the white band, on the broadest polished side,—are of precision and beauty certainly in their kind, not to be surpassed, and in my own collection unrivalled.

59. Slice of a jasperine agate of the finest quality, showing agatescent formation, inside as well as outside its quartz. The external bands, variously arbitrary, and unmanaged, or unmanageable, but chiefly notable, because, on the interior of the narrow and consistent band of red jasper, which falls away at one end like the loose thread of a skein, there will be found a series of little circular domes, built into the quartz; white on the surfaces, and cut through into pretty violet sections, which perfectly illustrate the white spots in the two small agates cut into heart-shape, Nos. 5 and 7, of the "Cross" series, p. 500.

60. Pure scarlet jasper, with lateral yellow zones, entirely representative of the stone in its finest condition, and ordinary flammeate structure. Its scarlet colour under these conditions has through many ages maintained its position as the real representative of that colour among gems, connected always in the minds of the ancients with the more translucent types of the same colour in chalcedony, which we now call carnelian, but they, "Sardius"; whence in the Apocalypse, with the intention of describing the most beautiful colour of the immortal body, "He that sat was to look upon like a jasper and a sardine stone."[2]

SECTION IV

Nos. 61 to 80: illustrating the nature and forms of QUARTZ

61. Quartz vein in Coniston Grit. We now enter upon the study of the entirely separate and most perfect state of silica, which has permanently retained the German miner's name of "quartz." As the

---

[1] [See above, p. 438 (A. 40).]
[2] [Revelation iv. 3: see above, p. 184.]

first essential component part of granite and gneiss, it forms at least a third of the mass of the non-volcanic mountain chains of the world, and the threads or veins of it passing through softer rocks, are the repositories, as we have seen, of gold; and in non-auriferous districts of the greater number of the most interesting metallic minerals. It is to be noticed in passing, that although veins of quartz occur in granite itself, they are usually non-metallic; and that the rich mining districts are all in comparatively softer, and in most cases apparently sedimentary, rock.

The condition of quartz in an ordinary vein is mostly that with which we have been made familiar in studying gold; that is to say, solid, and opaque, with occasional cavities towards the centre of the vein, in which it takes more or less crystalline form. This piece of the rock on which the house of Brantwood is built, is a perfectly characteristic example of this general structure. The quartz is secreted out of a grey sandstone passing into slate, and is itself somewhat coarse and impure, so that in its cavities the forms of the crystals are still rude, but extremely illustrative of the first development of crystalline force.

62. Part of a vein of quartz, a little finer than the last example, or at least with more energetic crystalline power. The reduction of the six-sided prisms into gradually tapering pyramids is a somewhat unusual circumstance, usually regarded by mineralogists as an accident, but in my own catalogues I always class these tapering quartzes as a peculiar variety. Locality unknown, and of no consequence.

63. Fine white quartz, partly forming a vein, partly disseminated through a black rock, I believe an indurated slate,—the quartz forming fine sugary crystals at its surface, and set with small crystals of sulphuret of iron, and calcite.[1] One of these calcite crystals (all of which are excellent examples of the most characteristic form of calcite, an hexagonal prism terminated by trigonal pyramids, which are set the opposite way at each end), is farther interesting from having got hold of a crystal of iron, and swallowed it up, all to itself; while another at the edge of the specimen seems to have taken three or four, like pills.

64. Detached wall of solid quartz, passing into small sugary crystals on one side, and well-formed larger crystals on the other: sprinkled on the larger side with pearl spar (carbonate of combined lime and iron), which itself is sprinkled or peppered with sulphuret of iron, and tiny calcite confused among the quartz, a single cube of pretty green fluor spar setting off the whole very daintily.

[1] [Ed. 1 reads, ". . . of iron; and carbonate of lime, which, in future, I shall call by the shorter name of 'calcite.' One of these . . ."]

65. Fine compact quartz, developing into crystalline quartz, which is coated with minor crystalline quartz, which is coated again with sugary crystalline quartz, which is finally dropped over by chalcedonic crystalline quartz.

Entirely dainty, puzzling, and charming. Cornish,—and it may be noted in passing, that as Transylvania does the neatest things in gold, Cornwall does the neatest things in chalcedony.

66. Well-developed, full-sized crystals of quartz, coated with sugary quartz, which seems at first to have rather bothered them: together with impertinent sulphuret of iron, probably the cause of general discomfiture and distraction in the early life of the crystals. The natural bridge, or viaduct, from the smaller to the larger crystal, is extremely curious; but on the whole I am rather glad to get rid of the specimen, which always bothers me too much to find out what has been the matter with it. The full-sized crystals are extremely good specimens of quartz form, bringing all their sides neatly to the point, which they are very apt to fail of doing, and to look like an ill-cut pencil.

67. Fine quartz, which has meant to crystallize with great precision, but has been nearly tormented out of its life by some nasty earthy sulphuret of iron. The forms of the crystals at last achieved are of extreme interest, just because they are only half built. This form of quartz showing external layers is, however, peculiar to some localities. I think the best *un*fretted ones are found at Schemnitz, in Hungary, before mentioned.[1]

68. Compact quartz, vexed quite out of its moral character by sulphuret of iron, with, I think, a little zinc (the black glittering part at the bottom), and nasty carbonate of iron all over. This specimen is described in the *Ethics of the Dust*,[2] as illustrative of the temper of minerals that don't get on together. We will now take up a series of examples showing how quartz gets on with minerals whose company it likes.

69. Pure white agatescent quartz, terminated in crystalline surfaces, sustaining beautiful cubes of golden green fluor spar, which the quartz sugars all over to finish; paying partly the same attention to a large and well-behaved cube of galena (sulphuret of lead), the tiny fragment of sulphuret of iron not having been able to cause any dissension between these friends. The quartzose base of this grand specimen is of extreme beauty and interest, and the whole fine in its kind.

70. Part of a quartz vein, with a bit of the surrounding slate: the central cavity, lined by fine crystals, sprinkled with black oxide of

---

[1] [See above, p. 497 (No. 34).]
[2] [§ 67 (Vol. XVIII. p. 283).]

iron. These two minerals—quartz and the oxide of iron—are great friends, and there is no end to the pretty things they will do with and for each other. In this instance, the quartz crystals take the oxide just within their outer surface, and then crystallize over it, as if to take care of it; the iron being always allowed to take its own proper flaky crystalline forms with perfect precision. A piece of sulphuret of iron, apparently impressed by the good example, behaves itself very properly in the corner.

71. Quartz with Tourmaline. I observed, some time since, that quartz never allowed other minerals to get into the centre of it as a nucleus for its crystals, and, as a rule, kept all extraneous minerals, even those it was very fond of, either at its surface, or just under the surface, as in the last specimen: but with its closest personal friends it relaxes so far as to allow them to crystallize all through it,[1] and do whatever they like, and take whatever room they want, going on with its own crystals meanwhile, in perfect tranquillity, while the friends amuse or accommodate themselves by shooting through it in all directions. This piece is a fragment of a large crystal, traversed, I believe, by black Tourmaline. We won't stop just now to ask what Tourmaline is, especially as I am not sure that this *is* Tourmaline.

Polished on one side, fractured on the other, and on the longest edge; but extremely pretty. I cannot make out the granular black nests on the rough edge.

72. Slice of a quartz crystal, portions of the sides of which are seen at the edges, traversed by sheaves, of, I believe, actynolite (another friend admitted to much familiarity), too complex to be describable.

73. Group of three quartz-crystals, one nearly perfect, traversed by straight silky amianthus, which runs straight through the crystals without in the least disturbing them, and seems very nearly coming out on the other side.

74. Part of a quartz crystal traversed by white amianthus, which begins all of a sudden, before one knows that it has got in, and takes a delicate little curve at its own fancy, wholly careless about the intentions of the crystal. Polished one side, fractured on two, the rest crystalline.

---

[1] [Ed. 1 reads :—

". . . crystallize all through it, though, in most cases, they may still have to shoot from the surface inwards. Often, however, it seems to let favourite friends do whatever they like, going on with its own crystals meanwhile, exactly as if they did not exist, though they may amuse themselves by shooting through it in all directions. This piece . . . that this *is* Tourmaline. being only concerned with the fact that it is one of the intimate friends of quartz, and although crowded mostly at the surface, is allowed to shoot as far as it has a fancy to do into the interior. Polished . . ."]

75. Section of a quartz crystal containing white amianthus, which can be seen only in one light. Hold the piece with its rough side to the window, the figures on the ticket upright, and look with a magnifying glass through the polished plane. The threads of amianthus will then be seen on the internal surface, though only one or two are visible on looking through from the outside. Extremely mysterious, and looking like a piece of intended jugglery.

76. Quartz in perfect crystallization, with a friend whom it respects, but nevertheless keeps outside,—topaz. This mineral is found continually embedded in quartz, at the surface, in this manner, *but I have never seen a single instance of its getting inside it!*

77. Pure quartz crystal, with cavities, left, I believe, by rutile (oxide of titanium), but I have not the least notion how that mineral either got in or out; only quartz is always ready to let it do whichever it likes.

78. Crystal of the purest quartz, a little interrupted by calcareous earth, and showing structural flaws, therefore, of extreme interest and beauty. There is no end to the things that may be found out, and to the other things past finding out, gradually traceable in it.

79. Perfect quartz crystal, showing its mode of growth, by the accident of a pause when it had got half way, during which the surface of the then existent crystal was covered with mossy chlorite; all the planes in this specimen are genuine, none polished, and the example is extremely rare and good.

It is most singular that no mineralogist of any country on earth has ever brought up a school of miners, to take care of a good crystal when they had got it! This example has originally been as perfect as anything could be, and has been only spoiled, as single crystals always are, by the miner's throwing it into his bag with other stones, and banging them about at his leisure.

80. Perfect quartz crystal of Dauphiné, as good as well can be, and in the inside of it, fit for spectacles or telescopes or what not;—of course a little spoiled at the point in the manner above described, but I have not above four bits in my whole collection that are better, and can't spare any of them.

SECTION V

Nos. 81 *to* 100: *illustrating the states of siliceous minerals used in jeweller's work, and their grandest forms of combination*

81. White quartz coated with amethystine quartz. This example begins a short series of twelve specimens, representing the conditions of silica, which permit its employment in jeweller's or engraver's

work. The principal of these, both for its frequency of occurrence and ancient classic reputation, is the amethyst; it is also extremely beautiful, and stands alone as a purple gem; for although between the sapphire and the ruby there are gradations of violet colour of extreme loveliness, these stones of intermediate tint are never dark; you may find a dark red ruby, or a dark blue sapphire, but never a dark purple intermediate stone. The amethyst, on the contrary, is characteristically dark purple, and when coloured uniformly, of extreme beauty. In material it is only quartz, coloured, by—the mineralogists don't exactly know what, but it is so far specifically different from common quartz, that when the two, which very frequently happens, are found together, the white quartz is always inside, and the amethyst outside. I have seen exceptional cases, but they are very rare. In this example the purple colour only develops itself at the extremities of crystals: it does so, however, whether they are directed up or down!

82. Pale amethyst, on slaty sandstone, very pretty, and geologically interesting. I believe also rare: it is at all events the only example I have seen of this mode of occurrence.

83. Detached crystal of dark amethyst, very fine, though the amethystine layer is not above a quarter of an inch deep, over brown quartz. Had it been all amethyst, no jeweller would ever have let it come into a mineralogical collection.

84. The next seven specimens are examples of the rarest and most beautiful form of silica, the opal,—a stone, however, belonging more to modern romance than to classic history. The most precious forms of it are found somewhere in Hungary, I suppose under the influence of our old acquaintance, Transylvania. I can't spell the name of the place where they are found, and nobody could pronounce it if I did, and it doesn't matter. The other varieties of opal, some only discovered recently, are Brazilian and Australian. This No. 84 is rather a specimen of the Hungarian rock than of the opal, which, however, is seen beginning to develop itself in colour, out of the pale bluish white substance which forms the vein. All of it properly called opal, and known by its peculiar gelatinous fracture; the colour comes where its structure is perfectly developed, being dependent on structure only.

85. Hungarian precious opal, of finer quality, in a thicker vein. I broke away two bits of it on speculation, thinking to come at better colour, but I never have any luck in speculation, and pasted the bits in again. If they come loose they may be reattached better than I have done it, this specimen being one of considerable interest.

86. Brazilian opal in a narrow vein, which shows straight longitudinal banding. It is one of the principal distinctions between agate and opal, that when the latter is banded, the bands are always straight, never chalcedonic, nor is it ever thrown into chalcedonic globules. This quite simple, though singular distinction, was, so far as I know, first observed and stated by myself.

87. Brazilian opal, of the kind called hydrophane, which absorbs water on being dipped into it, and only then shows its perfect colour ; it can't be too often dipped into water, but neither it nor any kind of precious opal should be exposed to strong heat of fire or sunshine.

88. Precious opal, Hungarian, the best kind, *diffused through the stone*, not forming veins in it. The opals used in jewellery are pieces cut, as large as possible, out of this kind of rock and afterwards rounded and polished. *I never heard of a rolled opal being found in any stream, or a round pebble of it found in any rock.*

89. Australian opal, in veins, but very lovely. It is more prismatic than the Hungarian, and would form lovely gems, but that its colours are usually confined to narrow bands. It also shows well by candle light, by which most Hungarian opals are spoiled.

90. Australian opal, diffused through coarse brown jasper, and itself extremely beautiful and full of colour in the blue tones. It is more rarely found with lights of vivid green, but seldom with red, which is the colour most prized by jewellers. This quality of blue opal, however, is gradually establishing itself in commercial esteem, having been, I believe, only known since the year 1878.

91 are two examples of a stone difficult to place. It belongs properly
and   to the jaspers, but while the red jaspers are of universal occur-
92   rence, this green variety is found only in India, and is highly esteemed in commerce, under the name of "blood-stone," very truly descriptive of its character, showing, in fine specimens, spots of deep crimson on a green ground. It is much used for seals, vases, and other such ornamental work, forming a beautiful contrast in its deep green with chalcedony, gold, or any of the brighter gems. Exquisite fifteenth-century vases of it may be seen in the Louvre.

93. Purple Cornish chalcedony, of the finest kind. This, with the remaining seven large specimens, which complete the collection to the number of a hundred, are examples of the larger and grander forms assumed by siliceous minerals, and all exhibit structural phenomena too complex for description. No. 97 is remarkable for the extreme delicacy of its agate banding, enclosed within a double rank of

quartz crystals. No. 98 is an extremely interesting example of earthy stalactites in massive chalcedony; and No. 100, as complete and striking a piece of agate concretion as could be found in any national collection.

To this series of agates are added a few flint-fossils of which I really know nothing worth telling, except that the two in chalk, marked F. 12 and F. 13, were greatly esteemed by my old collector;—that the dainty little scallop Stylites (not its specific name,—but expressing its curious abode on the top of a column), F. 14, has been a great pet of my own, together with F. 15, which is so delicate as to look more like a fracture than a shell; —and that the pretty little hinge of the Terebratula is very charming in its transmutation to a loop of chalcedony. Don't mistake for fossils the two remnants on the other side of the brown card on which my old friend and Knight of the Hammer, Mr. Simon,[1] had pasted it for its glorification. I used also to value the little sponges, F. 4, 5 and 6; but the big one, F. 1, reminds me disagreeably of cold mornings. F. 3 is in three pieces, which fit into each other like weights, and contain about three thousand nondescript beasts, a nebula of siliceous vermin. The beast in the form of a figure of 8, No. 9, is, I believe, some acquaintance of the beast composed of a pentagon with two legs, No. 10; and the broken flat thing, No. 19, is a piece of somebody's shell, who was called an Inoceramus. I cut the little Echinus, F. 20, through the middle to see into his mind, but made nothing of it, and only lost the other half of him. Finally, the shattered conical one 22, and cast of a spine, 33, will join on properly to the finer Echini which the St. David's Museum already possesses.

JOHN RUSKIN.

BRANTWOOD, 27th February, 1883.

---

[1] [No doubt Mr. (afterwards Sir) John Simon, F.R.S., with whom Ruskin was at one time much in Switzerland: see Vol. VII. p. xlvi., and Vol. XVII. pp. xliii. n., 450.]

CATALOGUE OF MINERALS PRESENTED
TO THE CONISTON INSTITUTE (1884)

[*Bibliographical Note.*—This Catalogue, written by Ruskin, is reprinted from the following publication :—

    Catalogue | of the | Ruskin Museum, | Coniston Institute. | Price Three-pence.

The minerals are arranged in Cases XII. and XIII. (upper shelves), and the descriptions of them occupy pp. 31–35 of the Catalogue.]

CATALOGUE OF MINERALS PRESENTED TO THE CONISTON INSTITUTE (1884)

1. Carbonate of Lime from the Jura, nearest the Alps, showing fossils and cracks resulting from desiccation.
2. Carbonate of Lime with Iron Pyrites crystallized on Quartz.
3. Carbonate of Lime with Sulphides of Lead and Iron on Quartz.
4. Carbonate of Lime in large and small Crystals with sulphide of lead.
5. Carbonate of Lime in hexagonal plates.
6. Carbonate of Lime in small acicular and large twinned crystals, on Galena.
7. Carbonate of Lime in modified plates.
8. Carbonate of Lime in acicular and radiating columnar crystals.
9. Carbonate of Lime in acicular and radiating columnar crystals.
10. Carbonate of Lime in acicular and radiating columnar crystals.
11. Fluor spar in cubes beautifully modified on the angles.
12. Fluor spar in beautifully modified cubes.
13. Fluor spar, angle fractured, planes beautifully composed of low pyramids with inlaid Quartz.
14. Fluor spar on Quartz, two cubes curiously combined.
15. Blue Fluor spar on Quartz, truncate on the angles.
16. Blue Fluor spar on Quartz, cubes purple in the interior.
17. Fluor spar and Iron Pyrites on Quartz, cubes purple in the interior and beautifully modified on the angles.
18. Fluor spar, octahedric, transmuting itself into Chert.
19. Fluor spar much modified, its substance entirely replaced by Chert.
20. Large Agate formed in Basaltic rock, and containing crystals of Carbonate of Lime.
21. Agate showing involutions, natural colour.
22. Agate showing involutions and tubular centres, natural colour.
23. Agate showing large involutions, coloured.
24. Agate, coloured.
25. Amethyst, with coating of chalcedony on one side.
26. Chalcedony, becoming agatescent.
27. Chalcedony assuming Agate form.
28. Agate in its most complex manner of concretion.
29. Smoky Quartz (Switzerland).
30. Smoky Quartz, distorted Crystals, very fine.
31. Felspar, irregular white crystals.
32. Felspar, coated with specular Iron and Mica.
33. Felspar.
34. Felspar showing laminar structure.
35. Green Felspar, Amazon Stone.
36. Oxide of Iron, feathery crystals.

37. Ore of Antimony, scopiform crystals.
38. Ore of Antimony, acicular and fibro-laminar crystals.
39. Galena and prettily crystallized Carbonate of Lime.
40. Quartz, of a singular waxy lustre in long hexagonal prisms, coated with Iron Pyrites.
41. Galena on Chalcedonic Quartz.
42. Mica, showing one side of large crystal.
43. Mica and Felspar in beautiful tabular crystals.
44. Mica, radiating crystals attached to Rutile.
45. Asbestos attached to rock.
46. Amianthus intimately traversing Quartz, very beautiful.
47. Tourmaline with Quartz and Mica.
48. Schorlaceous rock.
49. Garnets (Cumberland).
50. Garnets, dodecahedral crystals (Airolo, St. Gothard).
51. Gneiss with Quartz and Topaz.
52. Actinolite (St. Gothard).
53. Green Mica, hexagonal tables.
54. Kyanite, called by Saussure "Sappare."
55. Quartz rock with Gold (White Hills, Bendigo. 1856).
56. Quartz rock from between the cheeks of a fissure in Kaolin rock, showing incipient and complete crystallization.
57. Thompsonite, Lime-Soda Zeolite (Old Kilpatrick, Scotland).
58. Prehnite (Old Kilpatrick, Scotland).
59. Natrolite (Aussig, Bohemia).
60. Natrolite (Aussig, Bohemia).
61. Albite and Natrolite (Aussig, Bohemia).
62. Albite, Soda-Felspar (Mahren).
63. Amazon-stone (Colorado).
64. Felspar (Baveno, Lago Maggiore).
65. Albite (Aussig, Bohemia).
66. Idocrase (Banat, Hungary).
67. Garnet (Hull, Canada).
68. Section of Garnet rock (Labrador).
69. Quartz upon Pearlspar Dolomite (Schemnitz, Hungary).
70. Quartz crystal exhibiting conchoidal fracture, with curved and cross striæ.
71. Quartz crystal containing specular Iron and minute white specular crystals; erosion where apparently Iron oxide has been deposited.
72. Quartz crystal with curious composite terminal plane and indentation showing internal structure.
73. Quartz containing red Hæmatite and specular iron; triangular facets on pyramidal planes.
74. Quartz crystal with curious thickening of one half of the pyramid (Tavetsch-thal).
75. Quartz crystal with enclosures and one large extra plane (Herkimer Co., New York).
76. Quartz crystal with one prism deflected.
77. Capped Quartz (Beeralston).
78. Doubly terminated Quartz crystals with Goethite (Bristol, England).

79. Milky Quartz containing a Zeolite.
80. Babel Quartz.
81. Lithomarge.
82. Flint coating Chalcedonic nucleus (Chalk).
83. Five Flints on Ventriculites (Chalk).
84. Flint containing crystallized Quartz (Chalk).
85. Siliceous cast of Ventriculite (Chalk, England).
86. Siliceous cast of Ventriculite (Chalk, England).
87. Choanites, polished cavity left by Ventriculite (Chalk, S. England).
88. Breccia of Quartz; Felsite cemented by Calcite.
89. Wollastonite, Table-spar (Diana Co., New York).
90. Silicate of Zinc (Altenburg, Belgium).
91. Hornblende (Bohemia).
92. Diaspore.
93. Tridymite (Drachenfels, Rhine).
94. Manganophyllite (Sweden).
95. Ludlamite (Cornwall).
96. Igloite (Styria).
97. Lettsomite (Var, France).
98. Noumeite (Noumea, New Caledonia).
99. Augite (Sussex Co., New York).
100. Talc (Tyrol).
101. Talc (Tyrol).
102. Tourmaline, Schorl, in Mica (Zieserthal).
103. Tourmaline in Mica schist (Norway).
104. Tourmaline (Czozlowa, Banat).
105. Sphene, Silico-titanite of Lime (Ala, Piedmont).
106. Sphene on Adularia Felspar (St. Gothard).
107. Sphene (Renfrew, Canada).
108. Chabasite, hydrated Silicate of Ammonia and Lime (Bohemia).
109. Cobalt Bloom, hydrated Arseniate of Cobalt (Saxony).
110. Tantalite (Chili).
111. Linarite, Cupreous Anglesite (Redgill, Cumberland).
112. Sulphate of Barytes (Cumberland).
113. Sulphate of Barytes, Heavy-spar (Cleator Moor, Cumberland).
114. Sulphate of Barytes, Cawk (Derbyshire).
115. Selenite, crystalline Gypsum (Switzerland).
116. Bromlite, Alstonite (Alston, Cumberland).
117. Calcite, Calc-spar (Cumberland).
118. Malachite, green Carbonate of Copper (Burra-Burra, Australia).
119. Carbonate of Iron (Cornwall).
120. Antimony (Felsöbanya, Hungary).
121. Plumose Antimony (Malazcka, Hungary).
122. Antimonite (Hungary).
123. Galena with Blende, Lead Sulphite and Zinc Sulphide (Alston, Cumberland).
124. Wood Opal (Tasmania).

Also sixty-five specimens from Mr. Ruskin's collection to illustrate his papers in the *Geological Magazine* on Banded and Brecciated Concretions (given by W. G. C.).

CATALOGUE OF MINERALS SHOWN
AT EDINBURGH

[*Bibliographical Note.*—The complete MS. of the sixth of the Catalogues of Minerals enumerated by Ruskin (above, p. 387) has not been found by the editors, nor was the collection of specimens, which it described, ultimately placed at Edinburgh, as he had proposed. The following Catalogue is put together from three sources :—

(1) A pamphlet of ten octavo pages (stitched, without wrappers). There is no title-page or drop-title; the headlines are "Catalogue" throughout. The general appearance is similar to that of the Kirkcudbright and Reigate Catalogues. The only copy which the editors have seen is in the possession of Mr. W. G. Collingwood. This Catalogue comprises Nos. 1-60. On p. 523, line 6, "Nectanebus" is here a correction for "Nectabenes."

(2) A sheet of MS. at Brantwood (Nos. 61-70). And

(3) Descriptions in the Edinburgh lecture (Nos. 71-80).]

A CATALOGUE OF MINERALS SHOWN
AT EDINBURGH

GROUP I.—1-20

Introductory Examples, ascending from COMMON FLINT *to* PERFECT
ENDOGEN [1] AGATE

1. Amber chalcedony with flint.

2. Quartzite chalcedony with cycloidal superstructure.

3. Cycloidal chalcedony in films on flint.[2]

4. Flint with chalcedonic fissure.

5. Nodule of red flint-chalcedony.[3]

6. Chalcedony in stalactitic coats, varying in their lines of current.[3]

7. Finest state of pure chalcedony, in irregularly combined rods and films.

8. Chalcedony on hæmatite.[3]

9. Spheroidal hæmatite in quartz.[4]

10. Grey spheroidal agate, partly stalactitic.[5]

11. Perfectly formed agatescent stalactite, of white chalcedony, with external quartz.

12. Agatescent stalactites of pure chalcedony, with green, more or less tubular, centres.

13. Filiform chalcedony with green tubular centres, its threads collecting gradually into a solid mass.

[1] [For the terms "endogen" and "exogen" applied to agates, see "Distinctions of Form in Silica," § 15, p. 378.]
[2] [For this specimen, see *ibid.*, § 18, p. 379.]
[3] [See *ibid.*, § 19, p. 380.]
[4] [See *ibid.*, § 20, p. 380.]
[5] [See *ibid.*, § 21, p. 381.]

14. Filiform chalcedony, with green centres, not tubular.

15. Pisolitic heliotrope (showing the spherical structure at the edge).

16. Pisolitic heliotrope in parallel bands.   Look at it by transmitted light.

17. Tubular agate, the common form, but a singularly fine example.

18. Pure spheroidal agate, developing itself round an earthy stalactitic centre, with external quartz (magnificent).

19. Half of a perfect geode of spheroidal agate, with external quartz.

20. Portion of a nodule of irregularly muscose and spheroidal agate, with external quartz.
    Note especially in this example, on its polished side, the portion of the external salmon-coloured band of the agate, which has been separated from the rest and carried up into the body of the crystallizing quartz.

GROUP II.—21–30

Exogen Agate

21. Geode of pure chalcedony, changing internally into crystalline quartz, and then filled with white chalcedony in bands, modified by spheroidal action, and leaving finally three small cavities between spheroidal zones, of which the parallelism is rigidly exact.   The small black spot formed by a local stain in the external bands is abnormal, and of extreme interest.

22. White exogen agate (partly lake agate), for comparison with 21, the beds absolutely unaffected by spheroidal action!   Superb.

23 Two pieces of a nodule of exogen agate tending to arrange itself in
and   crystalline planes towards the centre; the pure chalcedonic bands
24.   are intercalated with one coloured brown by iron, and singularly interrupted in its course.

25. Exogen agate of the finest quality, and most delicate colours, violet, green-grey, and white; its beds parallel round the central quartz, but suddenly contracting at the extremity, where they have the deceptive aspect of being bound by a ligature into the form of agate which I have hitherto been in the habit of calling "folded,"[1] but shall be most happy to adopt any other term better descriptive of the structure, when once that structure is understood.

[1] [See "Banded and Brecciated Concretions," § 26, pp. 64-65.]

26. Part of a nodule of folded agate of the rarest form, showing on its smaller polished side a cavity separated from the interior quartz by a narrow group of grey beds which, till then, concurrent in curvature with the external white ones, at this point suddenly quit them, and form a level plane of lake agate for the roof, or floor, of the cavity.

27 Two parts of a perfectly formed almondine nodule of exquisitely pure
and folded agate, composed of variously-transparent and translucent beds
28. of chalcedony, allowing the structure of folded agate to be studied in its simplicity.

29. Half of a large almondine nodule of common agate, with local interferences of the folded structure.

30. Folded agate of duplicate structure at two different periods (?), compare No. 45, filling the hollows of a geode of tubular chalcedony. Superb.

GROUP III.—31–40

LAKE AGATE

31. Accurately level lake agate deducing itself in a geode from surrounding beds of spheroidal chalcedony.

32. Accurately level lake agate, formed by grey chalcedony on the surface of an interferent level bed of dark brown chalcedony, and studded on its surface with spheroidal chalcedonic dew. Part of a geode in volcanic rock.[1]

33. Partially developed and imperfectly levelled lake agate between two masses of quartz.

34. White lake agate, with green surface films, the whole imperfectly levelled, and traversed by more or less vertical columns of irregularly outlined green chalcedony.

35. Perfectly levelled lake-chalcedony, with transverse green tubular stalactites.

36–38. Three unsurpassable examples of almond agate, showing the transition from the lacustrine structure to the ordinary "fortification."[2]

39 Two slices of an almond agate in which the obscurely banded semi-
and crystalline centre changes gradually into the fine encompassing bands.
40. See especially darker side of thick slice.

---

[1] [For a reference to this specimen, see "Distinctions of Form in Silica," § 22, p. 382.]

[2] [For this term, see A. 40 (above, p. 438).]

GROUP IV.—41–50

Segregate Agate

The ten following examples are chosen to illustrate the formation of agatescent rocks which have been carelessly confused with conglomerates, but are no more conglomerates than common volcanic rocks containing agate pebbles. The most beautiful existing example of them, so far as I know, is also one of the noblest monuments of Egyptian art, the sarcophagus of Nectanebus in the British Museum.[1] The examples given in the group here selected are, I believe, with two exceptions, English.

41. A common banded flint, showing, however, the secretion of the bands in curious variety of breadth, a group of ten narrow ones being suddenly developed in the space of a quarter of an inch, in the middle of a band of broad ones.

42 and 43. Sliced flint-pebble showing the secretion of its external coat, with unconformable bands in the interior.

44. Fragment of weathered surface of flint, bringing out in relief the harder portions of an interior banded concretion, which might be easily mistaken at first sight for a fossil.

45. Duplicate (one within the other, compare above, No. 30), almondine concretions, finely banded, in hornstone; incipiently brecciate in the centres, and in the brown concretions of the gangue.

46. Amygdaloidal concretion of black and grey chalcedony, out of Jura limestone. (Mont Salève, Savoy.[2])

47. Amygdaloidal concretion, banded, of red and yellow jasper, with rectilinear transverse bands and secondary concretions of white chalcedony. (Bought at Geneva, locality uncertain.)

48. Siliceous, partly banded, out of quartzite, becoming jasperine by irregular staining. A piece of the common rock of Sidmouth, Devon.

49. Fully developed grey "pudding-stone" (so called), probably of Hertfordshire.

50. Common Hertfordshire pudding-stone; a fragment showing in two places the intrusion of the quartzite gangue into the substance of the jasperine concretions. See the account of this beautiful siliceous formation, given in the postscript to the explanatory lecture.[3]

---

[1] [For other references to this sarcophagus, see *Ethics of the Dust*, § 101 (Vol. XVIII. p. 332), and *Fors Clavigera*, Letter 64, § 10.]

[2] [For this specimen, see "Distinctions of Form in Silica," § 22, pp. 381–382.]

[3] [See above, p. 389.]

GROUP V.—51–60

INLAID AGATES

A series illustrating the forms of agate usually called brecciate; but to which, for reasons stated in my Catalogue of British Museum Select Silicas, I give the safer name "inlaid."[1]

51. Slice of a pebble from the beach at Southwold, Suffolk,* so far as my experience reaches, unique.

52–55. Four crucial examples of inlaid structure.

56. A perfect example of the group of agates which I have called conchoidal,[2] and respecting the origin of which I am still uncertain.

57. A piece of the well-known Kunersdorf agates, in which the apparently disrupted structure is made most beautiful by the extreme fineness of its jasper secretions and amethystine gangue.

58–60. These three last examples are the pride of Brantwood, in their illustration of the finest possible states of jasper under chalcedonic action. All the three are—again speaking within the limits of my experience—matchless; but the last, the small oval, will best reward attention, in its determinate and lovely exhibition of the laws of agatescent structure. Under the lens, on the convex side, the flamboyant branchings are seen in pure section with their infinitely subtle bands, shooting from spherical nuclei:—on the flat side, they are seen in oblique section, swept into curves like the descending ridges of an Alp. No. 59, remarkable, first, for the band of iron oxide between the external quartz and belt of jasper, and secondly, for the orbicular-radiate structure of that jasper itself, broken up by the circumfluent chalcedony, which was described with care in my first papers on agates in the *Geological Magazine*.[3]

GROUP VI.—61–70

DENDRITIC AGATES

The general term Dendritic may rightly include both Mocha stones and moss agates, but see the distinctions between them defined at page 9 of the Explanatory Essay.[4]

* Found by the Rev. E. O. Morgan.

---

[1] [See above, p. 407 *n*.]
[2] [On this term, see above, p. 445.]
[3] [See above, p. 46; and compare p. 439.]
[4] [See above, p. 377.]

61. Moss agate, in straight chloritic layers of pale green with interfused chalcedony. Extremely rare.

62. Moss agate, in curved and broken chloritic layers with interfused chalcedony, and throwing out lateral muscose growths. Extremely rare.

63. Moss agate, common dark green fibrous, in clear white chalcedony. Very beautiful.

64. Moss agate, common dark green fibrous, in white chalcedony, clouded in the spaces between the fibres.

65. Moss agate, in dark green, pale green, and brown fibres of exquisite delicacy, with interclouded chalcedony becoming agatescent; a slab 6 inches by 15. The companion piece is in the St. George's Museum, Sheffield.[1]

66. Moss agate, in green, yellow, and red fibres, with interclouded chalcedony, becoming agatescent, the whole in state transitional to that of jasper. Superb.

67. Moss jasper, being a state of moss agate in which the fibres are set so close as to render the stone practically opaque. Purple, yellow, and blue, of extreme beauty.

68. Mocha stone, the common kind, but very finely developed.

69. Mocha stone, the arborescences formed at right angles to the zones of its chalcedony. Rare.

70. Mocha stone, in radiating spherical clusters. Unique in my experience

———

71–77. *Examples of* OPAL *to illustrate the Modes of Siliceous Solution and Segregation* [see above, pp. 382–383].

71, 72. Australian. Nodular and hollow concretion.

73. Normal state of Australian opal.

74. Arrangement in straight zones transverse to the vein.

[1] [Apparently A. 52; p. 444.]

75. Stellar crystallization.

76. Like a lake agate.

77. Hydrophane.

78–80. *Geological Specimens* [see above, p. 383].

78. Undulated jasper.

79. Hornstone.

80. Gneiss.

NOTES ON MINOR COLLECTIONS

[In addition to the six important collections mentioned by Ruskin above (p. 387), he presented smaller collections to schools, colleges, and individuals—to Miss Bell's school at Winnington, for instance, to Harrow School, and to Balliol College, Oxford (see the Introduction, p. xlviii.). For the collections just mentioned, he wrote no catalogues. The notes which follow here refer to collections (1) at the White-lands Training College, Chelsea; and (2) at the High School for Girls, Cork.

The first set of Notes are here reprinted from p. 21 of an octavo pamphlet (pp. 24) entitled :—

"The Ruskin Cabinet. Whitelands College. 1883."

For the Bibliographical Note on the Cork Collection, see below, p. 530.]

(1.) NOTES ON A COLLECTION OF TEN FINE SPECIMENS OF AGATE, ETC., GIVEN TO THE COLLEGE BY PROFESSOR RUSKIN

1. Cubic colourless Fluor, coated with opaque carbonate of lime, and formed on a base of curved plane quartz.

The whole on base of green fluor, with crystals of galena.

(Note the rounding off like the division of the margin of a leaf—instead of straight superimposed layers, the common form.)

2. Curved plane Quartz (? I am not sure if each small plane is curved or only the general form arrived at), formed in a nodule of agate and amethyst quartz.

3. Right plane quartz—aggregate in crusts (unusual).

4. Common Quartz in a close crust, with block Tourmaline and apatite (the hexagonal white thing)—phosphate of lime.

5. Common Globular agate, throwing itself into zones with Quartz outside (dyed brown artificially and spoiled in colour, but the lines of it better seen).

6. Common globular agate (variety I have called folded) with quartz inside.

7. Common agate—neither globular nor folded, but even, all round, and only following irregularities of matrix.

8. Purple agate with several interferent concretions. Very pretty.

9. Common Grey Chalcedony.

10. Purple Chalcedony on Quartz. (Cornwall and Cornwall only in this form.)

[At a later date (1885) Ruskin presented some further stones to Whitelands College, and wrote the following MS. notes explanatory of them. The numbers are either those which the examples bore in some arrangement of his own collection, or those in the dealer's list.]

"PIECES TO STAY AT WHITELANDS"

322. Mocha Stone and Quartz. It is very rare to get a bit of merlin-stone (the milky layer with brown moss-like traceries on it) in its matrix. This white kind seems to be generally thus in layers, forming *straight* veins in quartz on the basis of less regular crystallization. There is a piece of this same substance in the middle of the quartz.

262. Quartz with Ruhle (locality, Lukmanier). Extremely pretty.

325. Quartz with viscous fluid in triangular cavity (locality, Schemnitz). Very rare. I suppose the viscosity is proved by the slow motion of bubble. *I* can see no evidence of triangular cavity.

326. Quartz with small bubble of viscous fluid (Schemnitz). Here I can't make out the fluid at all, but the quartz itself is interesting in mark of division of summit.

332. Smoky Quartz, growth partially arrested. Curious, and not usually to be had—the like of it—for eighteen pence.

335. Quartz with gold. Still less the like of this for 2s. 6d. It is an extremely good and picturesque group of arborescently crystallized gold.

324. Quartz with extra planes (locality, Maderanerthal). *Warped* quartz, beautifully simple and complete in type. The larger one, exquisitely clear in substance, which I present to the College out of my own collection, will show by what infinitude of masonry such a curved crystal must be constructed.

325. Quartz and Calcite (locality, Alston).

328. Quartz (bipyramidal and Calcite), made of quartz aggregation. Notable.

329. Stalactitic Quartz (Bombay). And yet more so in this—a small central rod of agate.

336. Quartz, doubly terminated (locality, Cleator Moor, Cumberland).

A rare form, quartz being almost always  thus, not 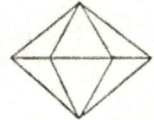 ;

but the little spangles of specular iron on the surface, not named by Mr. Butler,[1] are also interesting.

331. Fibrous Quartz (locality, Parma). Rare. Ask Mr. Butler, one day, what the green stuff is. The fibres are, I believe, amianthoidal.

[1] [Mr. Francis Butler, mineralogist, of the Brompton Road: see a letter of January 18, 1885, to the Rev. J. P. Faunthorpe, printed in a later volume of this edition.]

(2) MINERALS GIVEN TO THE CORK HIGH SCHOOL FOR GIRLS[1]

1. The best beginning of Mocha stone I ever saw, but they may, perhaps, be common at this locality. You can find out at leisure.

2. Agate, interrupted by quartz veins, which I described at greater length somewhere.[2] It cannot be too carefully looked at with pocket lens, and may some day be a classical stone.

3. Jasper with green coating. I believe Scottish, of quite infinite interest, and infinitely multiplied into infinite interest. By the time the youngest pupil in the school is ninety, she may know something about it.

4. Banded agate and jasper. Scotch; beat it in Ireland if you can.

5. Jasper passing into lake agate, an orbicular agate. Scotch also; but perhaps you may beat it at the Giant's Causeway.

6. Undulating jasper. I never thought to part with it, but it will be better at Cork.

7. Common black-banded flint. A rolled pebble.

8. Uncommon-banded flint, price 1s. 6d.; but I don't think you'll get the like of it for 2s. 6d.

9. Globular mica, the American fashion; but it will never make such good mountain as the old-fashioned mica.

10. Straight amianthus in quartz. Pretty, but the value of the specimen is in the three unpolished plains, with endlessly complex and with extremely minute cavities looking like spots.

11. Five stories of fairy amethyst mountain. Extremely rare and beautiful.

12. The last specimen I have of Sidmouth rock chert, becoming jasper by infusion of colour, red and yellow oxides of iron. Everywhere a beautiful enigma.

---

[1] [This list was copied into the *Pall Mall Gazette* of September 7, 1889, from an Irish newspaper. The *Pall Mall* had previously published a general account of Ruskin's various gifts to the School (July 8, 1889). Portions of the list appeared also in an article, by Miss Harriett A. Martin, entitled "Mr. Ruskin and the High School for Girls, Cork," in the *Ruskin Reading Guild Journal*, September 1889, No. 9, vol. i. p. 288. The list was next printed in *Igdrasil*, June 1890, vol. i. pp. 218–219, and was thence reprinted in the privately-issued *Ruskiniana*, vol. i., 1890, pp. 36–37. For a further account, see above, p. lx. In No. 5, "orbicular" is here a correction for "articular," and in No. 9, "globular" for "globulæ."]

[2] [Possibly on p. 76 above.]

VII

THE GRAMMAR OF SILICA

[This piece, referred to by Ruskin in one of the Catalogues already given (see above, p. 422), has not hitherto been published. It is here given from a printed proof found among Ruskin's papers. In § 30, line 1, "general" in the proof is here corrected to "regular."]

THE GRAMMAR OF SILICA

I. FLINT

1. Flint is an impure, but quite definite and distinct condition of silica, collected or secreted out of chalk rocks or the earthy beds connected with them; sometimes in large flat beds or masses, but for the most part in irregular knots and lumps of which the secretion seems in many instances to have been originally provoked by some organic substance.

2. Flint is characteristically black, pale, dull yellow, or grey; the paler kinds opaque, the dark feebly translucent, and these are to be considered the true substance of flint (the paler varieties containing some admixture of clay).

3. The material of flint is so delicate in grain or texture that it can take casts of the most delicate organic structures, but the modes of its secretion and deposition are hitherto undetermined.

4. Its substance is entirely singular among minerals in being so intensely *tough* in its coherence, though brittle, that it strikes fire easily against steel, and when violently broken it shows no flaws in its mass, and the fractures have curved and zoned surfaces, which from their resemblance to shells are called "conchoidal." Beautiful examples of entirely conic fracture of this kind will be found in the British Museum. I. F. 9 at Sheffield is entirely illustrative.[1]

5. The broken *surface* of perfect flint is feebly lustrous, passing into entirely dull smoothness on one side, and into a moderately bright polish on the other; but it is never perfectly dim, as jasper is, nor perfectly lustrous, as glass is. For its capacity of artificial polish, see British Museum 1, and Sheffield I. F. 12.[2]

6. The forms of knot which it characteristically assumes in its chalk matrix are extremely irregular and grotesque, but it shows a distinct tendency occasionally to throw itself into more or less spherical masses, usually hollow, but sometimes separating into inner and outer portions. The less regular masses are often hollow in the centre, probably, in many instances, having been formed round some organic substance which has disappeared; but really fine and representative flint is solid throughout, purest and hardest towards the centre, while the external coating is white, or porous, and granular, the flint seeming at its exterior to be connected with the chalk by a pulverulent state of its own surface. Sometimes, however, it is

[1] [See above, p. 422.]
[2] [See above, pp. 399, 423.]

quite well defined, and has then a most singular way of lapping and slopping itself round the substances it encloses, like dropped plaster. See Sheffield, I. F. 9.[1]

7. Many of the opaque and crumbling external conditions of agates seem owing to decomposition, but the actual inner substance of fine flint, such as is used for a building stone in Kentish churches, presents no appearance of undergoing change from exposure to the air and weather.

8. The hollows in flint are often lined with small crystals of quartz, but rarely, if ever, *filled* by quartz.   On the other hand, they are often thickly lined, and sometimes entirely filled, by chalcedony, but the method of the formation or introduction of this chalcedony is no more ascertained than that of the secretion of flint itself.

9. Some kinds of grey flint are banded, often with complexity and precision, somewhat in the manner of agates, but not in compliance with the outline of the external surface, which, for the most part, the bands of agate, if they do not follow, at least acknowledge as influencing their course.   But flint bandings are *characteristically transverse* to the surface, or occupy irregular spaces in the mass of the stone ; rudely concentric zones occur constantly towards the outer surfaces (see the Hertford pudding-stone, British Museum [2]), but these zones are never defined with agatescent precision, while the interior bands often approximate to, and sometimes reach it.   The translucent black varieties are, however, seldom, if ever, banded in this manner ; and perhaps we may consider the substance capable of banding as transitional between flint and jasper.

In some localities flints are coloured more or less vividly red, or touched with stains of red.   These in like manner may be regarded as transitional between flint and carnelian.

10. Chert and hornstone are siliceous concretions formed in rocks lower than the chalk, of importance only in relation to the histories of those rocks : they are not represented in the British Museum illustrative series, as they take no place in the transitions from flint to chalcedony.

11. The *colour* of perfect flint is the dark grey of this specimen, passing into jet black on one side, and dull white on the other.

The external coat of common flint is white, more or less thick.   I have chosen F. 9 for the extreme simplicity of its substance, and thinness of its coat.

The habit of fine flint to lap and flow (apparently), like dropped plaster, about the bodies it encloses is perfectly seen in this example.

II. JASPER

12. Jasper is an opaque, but extremely fine and homogeneous state of siliceous substance (slightly mixed with alumina), hard enough to take perfect polish, or give edge to the finest engraving, and characteristically of a vivid red, yellow, or green colour.

---

[1] [See above, p. 422.]
[2] [Nos. 20 and 21 : see above, p. 401.]

13. The white varieties seem connected with states of decomposing or ill-consolidated chalcedony, the yellow and red ones appear to be refined and compacted by the addition of the iron ochres which stain them. The manner and extent of this staining is extremely arbitrary, being sometimes in spots, sometimes in flamy undulations (peculiar to this mineral), sometimes in patches which colour, in common, portions of the stone of quite different substance; as, for instance, at once the paste and the enclosed pebbles of an apparent (or, possibly, true) conglomerate. Minute spheres of intensely scarlet, or crimson, floating in a paste of pure chalcedony, form many of the most beautiful hues and picturesque gradations of fine agates, but perhaps the globules in these stones are merely hæmatite in a peculiar state produced by suspension in gelatinous silica.

14. Jasper is either compact, flammeate (writhed in flamy curves), pisolitic, or mossy. It is never reniform[1] like chalcedony, some rare states of transition between jasper and sard presenting only languid and imperfect resemblances to chalcedonic form. Its fracture is smooth, but dull, and not conchoidal.

15. The green varieties called heliotrope are often minutely pisolitic, and divided by veins filled with chalcedony or minutely crystalline quartz.

16. The reticulated and weed-like fibres or films of green substance which traverse the clear chalcedony of some agates, and the yellow or red fibres resembling organic structures in *moss* agates, properly so called, have not yet been sufficiently examined.

III. CHALCEDONY

17. Chalcedony is a form of silica which appears to have been consolidated under various conditions from gelatinous states, or liquid solutions of the mineral. It is known and seen to be deposited by warm springs in Iceland and in North America, and is continually found in stalactitic groups which appear to have been deposited by falling water under the ordinary action of gravity. But all formations of the mineral whatsoever present phenomena different from those presented by ordinarily accumulative deposits, and partly at least dependent upon, and expressive of, the power of the mineral, if it has time enough allowed it to crystallize in reniform masses like those of wavellite and malachite. The radiant fibres of which these reniform masses are composed are indeed too delicate in chalcedony to be visible (even under the highest powers of the microscope), but there cannot be any doubt respecting the principle of crystallization, for the following perfectly simple reason, that while any ordinary accumulative deposit chokes up its primary hollows and irregularities of surface as thicker beds are superimposed, chalcedony *only increases as it thickens the precision and regularity of its spherical crystallization*. The cases in which a definitely fluent action, covering the interior substance over which the chalcedony has been formed, with a coating of irregular thickness like that formed by the guttering of a candle on the candlestick, are extremely

---

[1] [But see Ruskin's note on p. 48, above.]

rare. I have presented one such specimen to Sheffield,[1] and have two or three in my own collection; but the British Museum, in its entire series of silica, does not possess a single characteristic example. It would be extremely desirable for this reason to obtain for the National collection some large masses of the chalcedonic crust of the Geysers.

18. The ordinary forms of chalcedony, then, are

(A.) Parallel beds of beautifully equal thickness throughout, either level, or following the contours of the irregular surfaces on which they are laid (in which cases be it remembered that the equal thickness of a single bed of the substance over an irregular surface at once *does away with all possibility of supposing it a merely accumulative deposit*). And

(B.) Reniform or stalactitic processes, beautifully composed of aggregate spherical surfaces, and in the plurality of instances constructing themselves, like other crystals, without any distinct acknowledgment of the law of gravity. But as these deposits have necessarily been formed under many totally different conditions of chemical solution, rate and manner of supply, and time allowed for congelation, it is scarcely possible to find one example of chalcedony, in all particulars like another, and the problems presented by the varying phenomena of the mineral are virtually inexhaustible.

19. But whatever the form or manner of the deposit, the substance of the mineral singularly remains the same. There are no fine-grained or coarse-grained chalcedonies, as there are fine-grained and coarse-grained marbles; there are no compact-silky and fibrous chalcedonies, as there are compact and fibrous hæmatites and malachites. Mixtures of calcareous or aluminous substance may occasionally cause greater opacity of the chalcedonic substance, or interfere with the precision of its structure, but in all moderately good specimens, whatever their external structure, the material is the same; a translucent (*never* transparent), homogeneous, somewhat gelatinous-looking substance, if anything a little finer in grain than flint, so that it will engrave better, and less brittle than flint; its fracture duller, and not conchoidal, but resembling that of a limestone rock reduced in scale; its reniform surface *never* brightly lustrous, rarely lustrous at all, in ordinary cases delicately smooth without polish, in fine examples bloomed almost like the skin of a plum; its colour ordinarily warm grey, passing into yellow and various tones of pale red, more rarely cool grey, grey blue, and even full blue, the surfaces sometimes beautifully purple and blue, but it has never been found of amethyst purple in the interior. The varieties of deeper and purer red are always *warm* red, and commonly known as sard, or carnelian; but it is almost impossible now to distinguish the true natural stones of this colour from the artificially dyed ones which are among the worst disgraces of lapidary commerce.

20. In its less developed forms, chalcedony is constantly associated with common flint, and some of its aspects are altogether limited to its associations with that mineral, which for the most part forbid the idea of fluent deposit, and appear to result rather from slow processes of secretion and purification. Bedded chalcedony is the principal constituent of agates, and therefore peculiarly characteristic of the secretions of the minerals in volcanic rocks. The beautiful and perfect developments of its substance in

---

[1] [C. 1 in the Catalogue: see above, p. 424.]

flammeate masses, which I will ask leave to call Flamboyant Chalcedony, are, I believe, found only in cavities of metamorphic rocks; those of Trevascus in Cornwall are hitherto unrivalled. Peculiar guttate forms are associated with comparatively recent lavas in Auvergne and India; and incrustations, passing into semi-opal, are formed, as before noticed, by the hot springs of Iceland.

21. In no case has the mineral been found as a constituent part of any massive rock, nor has it ever been found containing, or associated with, any native metal, though its stalactites have very frequently centres of oxide or sulphuret of iron, and beautifully dendritic traceries are often constructed in it by the oxide of manganese. It never contains titanium, tourmaline, amianthus, nor any of the other minerals which are so frequently crystallized with quartz; nor is any filiform condition even of the oxides of iron possible in it; nor does it ever receive the flammeate stains of hæmatite which are characteristic of jasper, the colouring matter of hæmatite contained in it being exclusively in spots or globules.

22. On the other hand, it is continually associated with chlorite, which produces in it the structures valued by lapidaries, under the name of "moss" agates, of which scarcely anything is yet accurately known.

23. In the illustrative series at the British Museum the states of chalcedony associated with flint are shown in the flint compartment; then follow those of chalcedony proper; and then its associations with jasper and quartz, under the title of "agate."

The volcanic chalcedonies of Auvergne and India are not represented in this series, as being of local and accidental character, the illustrative series being intended to represent only general or constant forms.

IV. OPAL

24. This wonderful mineral is a hard siliceous jelly, usually filling all the pores and veins of the rocks in which it is found, but in some cases lambent on their surfaces, as if thinly poured over them. It is never stalactitic, never reniform, and except rudely under abnormal conditions, never concentrically zoned; but when it fills veins it is often straight zoned with clearness and decision in a direction transverse to the vein.

25. Its fracture is finely conchoidal, often striated with extreme delicacy, when the striæ are not concentric, and therefore do not produce the resemblance to casts of shells which originally suggested the term conchoidal.

26. Its lustre is vitreous, becoming sometimes slightly dull or gelatinous; its proper colour warm grey, passing into amber and brownish red (fire opal); and even the palest varieties always more or less yellow or amber coloured by transmitted light. Usually translucent only, but often approaching much nearer than chalcedony to transparency.

27. In some localities it is found traversed by a multitude of minute structural fissures, which enable it to separate the light it reflects into all colours of the prism, in a purity such as can only elsewhere be seen in clouds, or in the plumes of birds, and a few butterflies. The prismatic

colours of mother-of-pearl and shell nacre in general partially resemble those of opal, but are neither so intense nor so pure; the ordinary colours of the prism seen in the flaws of rock-crystal are on an entirely different scheme of division, much more vulgar than that of opal; and all metallic iridescences are impure in comparison. But the differences between pure and impure, between vulgar and graceful colour, have never been yet scientifically defined; and the definition cannot be arrived at without more knowledge of the relations of colour to sound than we yet possess. The varieties of opal which possess this iridescent power have always been highly valued in jewellery; the most beautiful stones cut out of the pores of a decomposing porphyry in Hungary are in the ground of a more or less clear milky white, traversed by scintillations of red, purple, blue, and green. But they never reflect yellow rays, except in combination with the red and blue, and the entire freedom of the purples and violets from any stain of yellow is one of the great sources of beauty in the stone.

28. Opals have been lately discovered in Australia in the pores of a highly ferruginous brown jasper, which with less varied scintillations possess deeper constant hues of blue and purple than had ever before been seen in the mineral.

V. HYALITE

29. Hyalite, properly so called, is an entirely transparent siliceous glass, which seems to have been dropped upon or poured over certain volcanic rocks, generally lavas, and is on the surface left in forms of concretion peculiar to itself, imperfectly represented by my woodcut, Fig .[1]

30. It is never regular, and never reniform in the accurate meaning of the word; but it is associated with conditions of silica which seem to have been deposited nearly in a similar manner, and which are, some of them, concentric in external concretion, and others pass gradually into stellar quartz.

31. The most interesting of these connected minerals are the guttate chalcedonies on the lava of Auvergne, feebly translucent or opaque, and resembling the droppings of a wax candle, associated with bitumen, and often passing into beautiful stars of imperfectly formed quartz; and the more or less flatly circular tablets of chalcedony sugared over with minute quartz and passing into stellar groups of well-developed and pure quartz crystals, for which I can at present give no narrower locality than "India." Lately, geodes of nearly similar chalcedony have been found in Uruguay, containing water, and brightly crystalline in the interior of their cavities, while on the external surface they present concentric groups of concretions more nearly approaching the forms of ordinary chalcedony than Hyalite; but, again, peculiar to themselves.

32. All these anomalous forms of chalcedony, I class under the general head of Hyalite, for the reasons stated at length in the eleventh chapter of the first volume of *Deucalion*.[2]

---

[1] [Left blank in the proof; perhaps Ruskin intended to insert Fig. 8 on p. 48, above.]

[2] [See above, p. 237.]

VI. QUARTZ

33. When pure silica is hardened in a compact mass it is white, opaque, feebly translucent on the edge, irregular and dim in fracture, and destitute of any apparent tendency to crystallize in one part more than in another. But if cavities are formed or left in this compact mass, its substance may produce, or seem to produce, well-formed crystals, usually vertical to the surfaces, or such crystals may occupy, or occur in, cavities in other rocks. Grey or dark varieties of compact quartz occur, more translucent than the white.

34. Absolutely compact pure quartz is, I believe, seldom found but in veins, rarely constituting large masses of mountain. The bulk of quartz substance is more or less granular, passing into sandy, variously associated with mica, and felspar which always subject and modify the quartz by their own crystallization, so that fine crystals of mica and micaceous minerals and of felspar and felspathic minerals continually occur imbedded in quartzose gangue; but never, so far as I know, fine crystals of quartz imbedded in micaceous or felspathic gangue. In order to crystallize, quartz requires a cavity; but the question of modes in which such cavities are primarily produced are hitherto unknown, or at least undetermined, as indeed also the entire production and construction of mineral veins.

35. From these general conditions of quartzose material it follows that rock-crystals are, as a rule, erect; that is to say, more or less vertical to the surface on which they are formed, and terminated finely only at their summits. The primary laws of their crystallization are stated in the appendix,[1] and descriptions of their actual forms, in special cases, given at length in my catalogues above referred to.

36. I believe that compact masses of pure rock-crystal, either filling cavities in rocks or where they show surfaces, exhibiting a sort of conchoidal crystallization (compare the apparent fractures of imperfect diamonds), are found in Madagascar, India, and Brazil; but I have never found any account of their occurrence in works on mineralogy. The general statement by Miller, p. 250,[2] that quartz is found "in attached and imbedded crystals, globular, reniform, and stalactitic, fibrous, compact," includes under the name of quartz all the varieties of jasper, chalcedony, and flint. The term "globular" can only refer to the structure of agate, and to the stellar associations of imperfect crystals; no such thing was ever seen as a globular surface of pure quartz, or a reniform one, nor is quartz ever fibrous with the visible fineness of malachite or hæmatite. In becoming reniform the radiating structure at once becomes invisibly subtle, all power of crystallizing in the form proper to the mineral is lost, and the substance is no more to be called quartz but chalcedony.

37. The varieties of colour attributed to quartz by Miller[3] are also inaccurately general, and belong to different substances. His sentence is, "colourless, white, violet-blue, rose-red, clove-brown, apple-green." By

---

[1] [Not written.]

[2] [*Elementary Introduction to Mineralogy, by the late William Phillips*, new edition by Brooke and Miller, 1852.]

[3] [*Ibid.*, p. 249.]

colourless, he only means transparent white, and by white, opaque white, and this is the proper aspect of quartz, as it is of ice. The other colours are all dyes, properly to be described as purple, rose-pink, amber-yellow, and clove or smoke-brown, and very feeble pale green. I do not know why called apple-green more than pear-green or bottle-green. It is scarcely ever deep enough to be called anything but greenish white, and that only in rock-crystal. Massive quartz is never green at all. Quartz is never blue at all, nor even violet-blue; all amethysts are of warm purple, passing rarely into violet, and never into blue. The pink varieties never reach a full red, and are properly to be described as rose-quartz. The finer states of yellow quartz, cairngorm, are pure full yellow, and very beautiful; the brown varieties pass from extremely pale smoke colour into the darkest browns, sometimes becoming opaque, nearly into black. Immense masses of granitic or gneissitic mountains are formed by a dull grey quartz, feebly translucent, grouped with felspar and other minerals, but it is remarkable that this grey quartz is rarely associated with metals.

38. As a rule, native gold is only found in white and extremely opaque quartz. I have never seen an instance of the occurrence of gold either with amethyst, rose-quartz, grey translucent quartz, or pure rock-crystal. Titanium associates itself most frankly with rock-crystal, and prefers it pure, but enters it only under the form of Rutile—never of Brookite, or of Anatase, which are both always superficial. Oxide of iron may be found associated with it in every condition except, so far as I am aware, that of rose quartz.

39. At Isella, in the ravine of Gondo, on the pass of the Simplon, British travellers have been, time out of mind, imposed upon by the sale of green bottle-glass as rock-crystal; but the glass thus sold has been cooled slowly, I suppose, at the bottom of disused furnaces, and it is peculiar, and extremely interesting, in containing white stellar and reniform crystals of, I suppose, true silicate of potash * or lead and potash; but I can find no account of the real crystalline form of this mineral. In one of my own specimens it is simply radiant globular; in another it presents the—to me elsewhere unknown—form of *stalked* crystals, forming a star like this [figure not given]; the actual form is of no consequence, but it would be extremely desirable if among all our glass furnaces we could produce a few museum specimens which should show the student, indisputably, the nature of crystallization from the hot fluent state of a mineral † as opposed to that from aqueous or other solution.

40. It is not, I believe, thought disputable by the mineralogists of

---

* "Globules of Devitrified Glass," Dana, p. 417. Why couldn't he have said "unglassed glass," and shown the real extent of his information?[1] In Miller's curious index I find "potasse nitratée" and "potasse solfatée," and "Glaskopf"; but not potassium, potash, potash-mica, nor glass. Glasskopf appears to be pure hydrate of red oxide of iron.

† "After a thunderstorm there was found in a meadow between Mannheim and Heidelberg a glassy mass, consisting of silicate of potash, upon the spot where a hayrick had stood that was struck by lightning" (Bischof, *Chemical Geology*, chap. xxvi.).

---

[1] [Compare the remark on "reptation," above, p. 317 *n.*]

to-day that granite is a cooled rock, but the manner of crystallization of quartz in it seems to be still at issue. It is not the business of this grammar to chronicle debates or signalize mistakes, but the general reader may be interested in hearing that the production of a crystal of quartz an inch in diameter from aqueous solution would, according to Bischof, require about a century, and the production of large crystals in lodes of granite an interval far greater than the age of the earth according to the Mosaic history (Bischof, chap. xxv.).

41. Questions especially to be submitted to mineralogists respecting the colours of siliceous mineral :—

1. Why pure crystalline quartz is never rose-colour blue nor green, except by mechanical admixture of chlorite or other green minerals, but only topaz colour, chocolate colour, or amethystine, while compact quartz is never amethystine, and rarely chocolate colour, but only rose colour, white, or grey.

2. Why jasper is characteristically red or deep green.

3. Why chalcedony is characteristically bluish grey, passing into blue and purple, but never into rose colour, and what green mineral it is which forms the greater part of the fibrous structures in moss agates.

4. The reason of the difference in *selection* of hue between the colours of opal and fissured quartz.[1]

---

[1] [Here the proof breaks off, Ruskin adding in MS., "End at present."]

APPENDIX

LETTERS, ADDRESSES, AND NOTES

I

NOTICE RESPECTING SOME ARTIFICIAL SECTIONS ILLUSTRATING THE GEOLOGY OF CHAMOUNI. BY JOHN RUSKIN, ESQ. COMMUNICATED IN A LETTER TO PROFESSOR FORBES[1]

(From the Proceedings of the Royal Society of Edinburgh)

[1858]

In the *Proceedings of the Royal Society,* vol. iii., p. 348, an account has been given by Professor Forbes[2] of the discussions which had then taken place as to the geological constitution of the chain of Mont Blanc, and as to the reality of the alleged superposition of the primary rock (gneiss) to the secondary (limestone), near Chamouni, and at Courmayeur.

In order to clear up any remaining doubt, Mr. Ruskin caused sections to be made, laying bare the junction at several points of the Valley of Chamouni.[3] The results, which are perfectly accordant with the conclusions of the above-cited paper, have been kindly communicated by Mr. Ruskin to Professor Forbes, and are described and sketched by him in the following note. The order of the sections is from the head of the Valley of Chamouni towards its lower or south-western extremity.

---

[1] [*Bibliographical Note.*—This "Notice" was read at the meeting held on February 15, 1858, and appeared in the *Proceedings of the Royal Society of Edinburgh,* 1857–1858, vol. iv., No. 48, pp. 82–84. It was published while Ruskin was abroad, and he says to his father in a letter from Bellinzona (June 18, 1858): "The Geological paper is very nicely done; the woodcuts carefully copied from my sketches."

It was reprinted from the *Proceedings* on a fly-sheet, octavo, pp. 2. The title given above occupies the upper portion of p. 1, the words "(From the Proceedings of the Royal Society of Edinburgh)" following. There is no imprint, and there are no headlines, p. 2 being numbered centrally.]

[2] [Forbes's paper was entitled "On the Geological Relations of the Secondary and Primary Rocks of the Chain of Mont Blanc," by James D. Forbes, D.C.L. An abstract of it was given in the *Proceedings of the Royal Society of Edinburgh,* 1856, and the paper was printed in full in the *Edinburgh New Philosophical Journal,* New Series, vol. iii. pp. 189–203.]

[3] [Ruskin's excavations and observations were made during his stay at Chamouni in 1856; details of them fill several pages of one of his diaries. See the Introduction, above, p. xxvi.]

Specimens of the more important rocks have been placed in the Museum of the Royal Society :—

"1. At Crozzet de Lavanchi, on road to Argentière, under the Aiguille de Bochard.

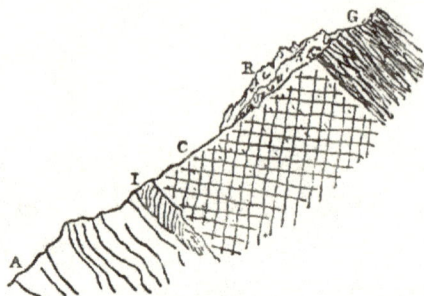

"A. Black calcareous rocks of the Buet, with belemnites, a good deal contorted (the same rock as at Côte des Pigets).

"I. Imperfect cargneule (porous limestone), about 2 feet thick.

"C. Common cargneule, used for limeworks, etc. (about 50 feet thick at the utmost).

"R. Débris concealing junction with gneiss.

"G. Gneiss laid bare, striking N. 50 E., and dipping 36° S.E., an unusually small angle, quite accidental and local, the average dip south being much steeper.

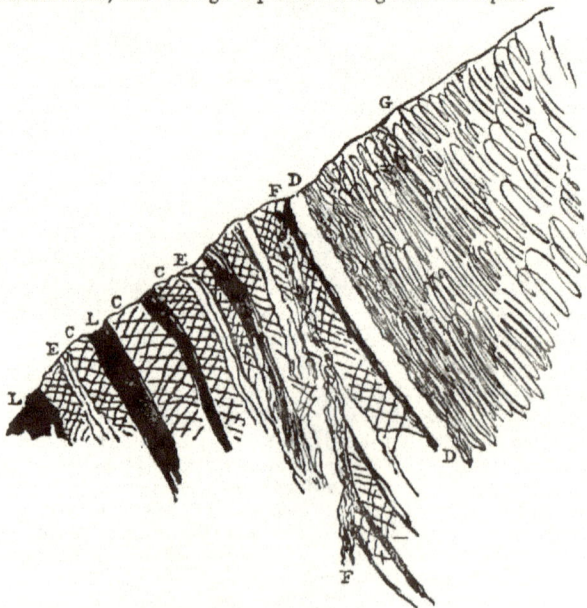

"L. Brown limestone, a form of the cargneule.

"C. Cargneule, generally inclosing fragments of the browner limestone, and with bands of greasy green earth, E, E, in the middle of its beds.

"F. Fault filled with fragments of clay and cargneule.[1]

"D. Decomposing white gneiss.

"G. Hard grey gneiss of Montanvert.

[1] [For this term see *Modern Painters*, vol. iv. (Vol. VI. p. 256 *n.*), where similar observations of the dip of the limestone under the gneiss are noted.]

"2. On the road to Chapeau, the same succession of beds takes place, the dip being greater (about 50°); the Buet limestones lower down dipping still more (about 65°). I say 'about,' not as guessing the angle, but giving the average of many accurate measurements.

"3. Junction opposite Prieuré of Chamouni, at my excavation.

"4. At Les Ouches, in the ravine under the Aiguille du Goûter.

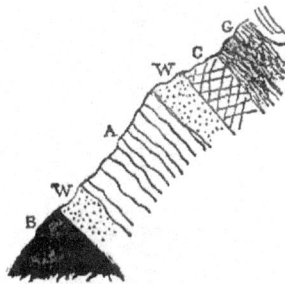

"B. Black slates of the Buet.
"W. Pure white fine-grained gypsum.
"A. Buet limestone (A of first section).
"W. Gypsum.
"C. Cargneule (C of first section).
"G. Gneiss."

LETTERS ON THE CONFORMATION OF THE ALPS[1]

(1864)

1

THE CONFORMATION OF THE ALPS[2]

DENMARK HILL, 10th *November*, 1864.

1. MY attention has but now been directed to the letters in your October numbers on the subject of the forms of the Alps.[3] I have, perhaps, some claim to be heard on this question, having spent, out of a somewhat busy life, eleven summers and two winters (the winter work being especially useful, owing to the definition of inaccessible ledges of strata by new-fallen snow) in researches among the Alps,[4] directed solely to the questions of their external form and its mechanical causes; while I left to other geologists the more disputable and difficult problems of relative ages of beds.

I say "more disputable," because, however complex the phases of mechanical action, its general nature admits, among the Alps, of no question. The forms of the Alps are quite *visibly* owing to the action (how gradual or prolonged cannot yet be determined) of elevatory, contractile, and expansive forces, followed by that of currents of water at various temperatures, and of prolonged disintegration — ice having had

---

[1] [The sections in this series of letters are here, for convenience of reference, numbered consecutively throughout.]

[2] [This letter first appeared in the *Reader*, November 12, 1864, and was reprinted in *Arrows of the Chace*, 1880, vol. i. pp. 255–258.]

[3] [The *Reader* of October 15 contained an article "On the Conformation of the Alps," to which in the following issue of the journal (October 22) Sir Roderick Murchison replied in a letter dated "Torquay, 16th October," and entitled "On the Excavation of Lake Basins in Solid Rocks by Glaciers," the possibility of which he altogether denied.]

[4] [On this passage, see the Introduction, above, p. xix. and *n*.]

small share in modifying even the higher ridges, and none in causing or forming the valleys.

2. The reason of the extreme difficulty in tracing the combination of these several operative causes in any given instance, is that the effective and destructive drainage by no means follows the leading fissures, but tells fearfully on the softer rocks, sweeping away inconceivable volumes of these, while fissures or faults in the harder rocks of quite primal structural importance may be little deepened or widened, often even unindicated, by subsequent aqueous action. I have, however, described at some length the commonest structural and sculptural phenomena in the fourth volume of *Modern Painters*,[1] and I gave a general sketch of the subject last year in my lecture[2] at the Royal Institution (fully reported in the *Journal de Genève* of 2nd September, 1863), but I have not yet thrown together the mass of material in my possession, because our leading chemists are only now on the point of obtaining some data for the analysis of the most important of all forces—that of the consolidation and crystallization of the metamorphic rocks, causing them to alter their bulk and exercise irresistible and irregular pressures on neighbouring or incumbent beds.

3. But, even on existing data, the idea of the excavation of valleys by ice has become one of quite ludicrous untenableness. At this moment, the principal glacier in Chamouni pours itself down a slope of twenty degrees or more over a rock two thousand feet in vertical height; and just at the bottom of this ice-cataract, where a water-cataract of equal power would have excavated an almost fathomless pool, the ice simply accumulates a heap of stones, on the top of which it rests.

The lakes of any hill country lie in what are the isolated lowest (as its summits are the isolated highest) portions of its broken surface,[3] and ice no more engraves the one than it builds the other. But how these hollows were indeed first dug, we know as yet no more than how the Atlantic was dug; and the hasty expression by geologists of their fancies in such matters cannot be too much deprecated, because it deprives their science of the respect really due to it in the minds of a large portion of the public, who know, and *can* know, nothing of its established principles, while they can easily detect its speculative vanity. There is plenty of work for us all to do, without losing time in speculation; and when we have got good sections across the entire chain of the Alps, at intervals of twenty miles apart, from Nice to Innsbruck, and exhaustive maps and sections of the lake-basins of Lucerne, Annecy, Como, and Garda, we shall have won the leisure, and may assume the right, to try our wits on the formative question.

J. RUSKIN.[4]

---

[1] [Compare above, pp. 98, 288.]

[2] ["On the Forms of the Stratified Alps of Savoy"—delivered on June 5, 1863: see above, pp. 3–17.]

[3] [Compare *Deucalion*, i. ch. i. § 11 (above, p. 106).

[4] [In reply to this letter, the *Reader* of November 19, 1864, published one from a Scottish correspondent, signed "Tain Caimbeul," the writer of which declared that, whilst he looked on Ruskin "as a thoroughly reliable guide in all that relates to the external aspects of the Alps," he could not "accept his leadership in questions of political economy or the mechanics of glacier motion." Ruskin thereupon wrote the letter which follows.]

2

CONCERNING GLACIERS[1]

DENMARK HILL, *November* 21.

4. I am obliged to your Scottish correspondent for the courtesy with which he expresses himself towards me; and, as his letter refers to several points still (to my no little surprise) in dispute among geologists, you will perhaps allow me to occupy, in reply, somewhat more of your valuable space than I had intended to ask for.

I say "to my no little surprise," because the great principles of glacial action have been so clearly stated by their discoverer, Forbes, and its minor phenomena (though in an envious temper, which, by its bitterness, as a pillar of salt, has become the sorrowful monument of the discovery it denies[2]) so carefully described by Agassiz, that I never thought there would be occasion for much talk on the subject henceforward. As much as seems now necessary to be said I will say as briefly as I can.

5. What a river carries fast at the bottom of it, a glacier carries slowly at the top of it. This is the main distinction between their agencies. A piece of rock which, falling into a strong torrent, would be perhaps swept down half a mile in twenty minutes, delivering blows on the rocks at the bottom audible like distant heavy cannon,* and at last dashed into fragments, which in a little while will be rounded pebbles (having done enough damage to everything it has touched in its course)—this same rock, I say, falling on a glacier, lies on the top of it, and is thereon carried down, if at fullest speed, at the rate of three yards in a week, doing, usually, damage to nothing at all. That is the primal difference between the work of water and ice; these further differences, however, follow from this first one.

6. Though a glacier never rolls its moraine into pebbles, as a torrent does its shingle, it torments and teases the said moraine very sufficiently, and without intermission. It is always moving it on, and melting from under it, and one stone is always toppling, or tilting, or sliding over another, and one company of stones crashing over another, with staggering shift of heap behind. Now, leaving out of all account the pulverulent effect of original precipitation to glacier level from two or three thousand feet above, let the reader imagine a mass of sharp granite road-metal and paving-stones, mixed up with boulders of any size he can think of, and with wreck of softer rocks (micaceous schists in quantities, usually), the whole, say, half a quarter of a mile wide, and of variable thickness, from mere skin-deep mock-moraine on mounds of unsuspected ice—treacherous,

---

* Even in lower Apennine, "Dat sonitum saxis, et torto vertice torrens" [*Æneid*, vii. 567].

---

[1] [This letter appeared in the *Reader*, November 26, 1864, and was reprinted in *Arrows of the Chace*, 1880, vol. i. pp. 259–267.]

[2] [See below, pp. 559 *seq.*]

shadow-begotten—to a railroad embankment, *passenger*-embankment, one eternal collapse of unconditional ruin, rotten to its heart with frost and thaw (in regions on the edge of each), and withering sun and waste of oozing ice; fancy all this heaved and shovelled, slowly, by a gang of a thousand Irish labourers, twenty miles downhill. You will conjecture there may be some dust developed on the way?—some at the hill bottom? Yet thus you will have but a dim idea of the daily and final results of the movements of glacier moraines,—beautiful result in granite and slate dust, delivered by the torrent at last in banks of black and white slime, recovering itself, far away, into fruitful fields, and level floor for human life.

7. Now all this is utterly independent of any action whatsoever by the ice on its sustaining rocks. It *has* an action on these indeed; but of this limited nature as compared with that of water. A stone at the bottom of a stream, or deep-sea current, necessarily and always presses on the bottom with the weight of the column of water above it—plus the excess of its own weight above that of a bulk of water equal to its own; but a stone under a glacier may be hitched or suspended in the ice itself for long spaces, not touching bottom at all. When dropped at last, the weight of ice may not come upon it for years, for that weight is only carried on certain spaces of the rock bed; and in those very spaces the utmost a stone can do is to press on the bottom with the force necessary to drive the given stone into ice of a given density (usually porous); and, with this maximum pressure, to move at the maximum rate of about a third of an inch in a quarter of an hour! Try to saw a piece of marble through (with edge of iron, not of soppy ice, for saw, and with sharp flint sand for felspar slime), and move your saw at the rate of an inch in three-quarters of an hour, and see what lively and progressive work you will make of it!

8. I say "a piece of marble"; but your permanent glacier-bottom is rarely so soft—for a glacier, though it acts slowly by friction, can act vigorously by dead-weight on a soft rock, and (with fall previously provided for it) can clear masses of that out of the way, to some purpose. There is a notable instance of this in the rock of which your correspondent speaks, under the Glacier des Bois. His idea, that the glacier is deep above and thins out below, is a curious instance of the misconception of glacier nature, from which all that Forbes has done cannot yet quite clear the public mind, nor even the geological mind. A glacier never, in a large sense, thins out at all as it expires. It flows level everywhere for its own part, and never slopes but down a slope, as a rapid in water. Pour out a pot of the thickest old white candied, but still fluent, honey you can buy, over a heap of stones, arranged as you like, to imitate rocks.[1] Whatever the honey does on a small scale, the glacier does on a large; and you may thus study the glacial phenomena of current—though, of course, not those of structure or fissure—at your ease. But note this specially:—when the honey is at last at rest, in whatever form it has taken, you will see it terminates in tongues with low rounded edges. The possible height of these edges, in any fluid, varies as its viscosity; it is some quarter of an inch or so in water on dry ground; the most fluent ice will stand at about a hundred feet. Next, from this outer edge of the stagnant honey,

---

[1] [Ruskin reverted to this experiment in *Deucalion*: see above, p. 162.]

delicately skim or thin off a little at the top, and see what it will do. It will not stand in an inclined plane, but fill itself up again to a level from behind. Glacier ice does exactly the same thing; and this filling in from behind is done so subtly and delicately, that, every winter, the whole glacier surface rises to replace the summer's waste, not with progressive wave, as "twice a day the Severn fills";[1] but with silent, level insurrection, as of ocean-tide, the grey sea-crystal passes by. And all the structural phenomena of the ice are modified by this mysterious action.

9. Your correspondent is also not aware that the Glacier des Bois gives a very practical and outspoken proof of its shallowness opposite the Montanvert. Very often its torrent, under wilful touch of Lucina-sceptre,[2] leaps to the light at the top of the rocks instead of their base. That fiery Arveron, sometimes, hearing, from reconnoitring streamlets, of a nearer way down to the valley than the rounded ice-curve under the Chapeau, fairly takes bit in teeth, and flings itself out over the brow of the rocks, and down a ravine in them, in the wildest cataract of white thunder-clouds (endless in thunder, and with quiet fragments of rainbow for lightning) that I have ever blinded myself in the skirts of.[3]

These bare rocks, over which the main river sometimes falls (and outlying streamlets always) are of firm-grained, massively rounded gneiss. Above them, I have no doubt, once extended the upper covering of fibrous and amianthoidal schist, which forms the greater part of the south-eastern flank of the Valley of Chamouni. The schistose gneiss is continuous in direction of bed, with the harder gneiss below. But the outer portion is soft, the inner hard, and more granitic. This outer portion the descending glaciers have always stripped right off down to the hard gneiss below, and in places, as immediately above the Montanvert (and elsewhere at the brows of the valley), the beds of schistose gneiss are crushed and bent outwards in a mass (I believe) by the weight of the old glacier, for some fifty feet within their surface. This looks like work; and work of this sort, when it had to be done, the glaciers were well up to, bearing down such soft masses as a strong man bends a poplar sapling; but by steady push far more than by friction. You may bend or break your sapling with bare hands, but try to rub its bark off with your bare hands!

10. When once the ice, *with strength always dependent on pre-existent precipice*, has cleared such obstacles out of its way, and made its bed to its liking, there is an end to its manifest and effectively sculptural power. I do not believe the Glacier des Bois has done more against some of the granite surfaces beneath it, for these four thousand years, than the drifts of desert sand have done on Sinai. Be that as it may, its power of excavation on a level is proved, as I showed in my last letter, to be zero. Your correspondent thinks the glacier power vanishes towards the extremity;

---

[1] ["There twice a day the Severn fills;
    The salt sea-water passes by,
    And hushes half the babbling Wye,
    And makes a silence in the hills."
                                —*In Memoriam*, xix.]

[2] [Lucina, the goddess of childbirth ("she that brings to the light"): see Horace, *Carmen Seculare*, 15.]

[3] [See the descriptions given in Vol. VI. pp. xxii.–xxiii., and above, p. xxiii.]

but as long as the ice exists, it has the same progressive energy, and, indeed, sometimes, with the quite terminal nose of it, will plough a piece of ground scientifically enough ; but it never digs a hole : the stream always comes from under it full speed downhill. Now, whatever the dimensions of a glacier, if it dug a big hole, like the Lake of Geneva, when it was big, it would dig a little hole when it was little—(not that this is *always* safe logic, for a little stone will dig in a glacier, and a large one build ; but it is safe within general limits)—which it never does, nor can, but subsides gladly into any hole prepared for it in a quite placid manner, for all its fierce looks.

11. I find it difficult to stop, for your correspondent, little as he thinks it, has put me on my own ground. I was *forced* to write upon Art by an accident (the public abuse of Turner) when I was two-and-twenty ;[1] but I had written a *Mineralogical Dictionary* as far as C, and invented a shorthand symbolism for crystalline forms, before I was fourteen :[2] and have been at stony work ever since, as I could find time, silently, not caring to speak much till the chemists had given me more help. For, indeed, I strive, as far as may be, not to speak of anything till I know it ; and in that matter of Political Economy also (though forced in like manner to write of that by unendurable circumfluent fallacy), I know my ground ; and if your present correspondent, or any other, will meet me fairly, I will give them uttermost satisfaction upon any point they doubt. There is free challenge : and in the knight of Snowdoun's vows (looking first carefully to see that the rock be not a glacier boulder),

> "This rock shall fly
> From its firm base, as soon as I."[3]

J. RUSKIN.[4]

3

ENGLISH VERSUS ALPINE GEOLOGY [5]

DENMARK HILL, 29th Nov.

12. I scarcely know what reply to make, or whether it is necessary to reply at all, to the letter of Mr. Jukes in your last number. There is no antagonism between his views and mine, though he seems heartily to

---

[1] [See on this subject Vol. I. p. xxxiii., and Vol. III. p. xviii.]
[2] [See above, p. 97.]
[3] [*Lady of the Lake*, canto v. stanza 10.]
[4] [Following this letter in the same number of the *Reader* was one from the well-known geologist, Mr. Joseph Beete Jukes, F.R.S. (1811–1869), who, writing from "Selly Oak, Birmingham, Nov. 22," was described as the "originator of the discussion." He therefore was no doubt the author of the article in the *Reader* alluded to above (p. 548 *n.*). Ruskin thereupon wrote the further letter which follows.]
[5] [This letter appeared in the *Reader* on December 3, 1864, and was reprinted in *Arrows of the Chace*, 1880, vol. i. pp. 268–273.]

desire that there should be, and with no conceivable motive but to obtain some appearance of it suppresses the latter half of the sentence he quotes from my letter.[1] It is true that he writes in willing ignorance of the Alps, and I in unwilling ignorance of the Wicklow hills; but the only consequent discrepancy of thought or of impression between us is, that Mr. Jukes, examining (by his own account) very old hills, which have been all but washed away to nothing, naturally, and rightly, attributes their present form, or want of form, to their prolonged ablutions, while I, examining new and lofty hills, of which, though much has been carried away, much is still left, as naturally and rightly ascribe a great part of their aspect to the modes of their elevation. The Alp-bred geologist has, however, this advantage, that (especially if he happen at spare times to have been interested in manual arts) he can hardly overlook the effects of denudation on a mountain-chain which sustains Venice on the delta of one of its torrents, and Antwerp on that of another; but the English geologist, however practised in the detection and measurement of faults filled in by cubes of fluor, may be pardoned for dimly appreciating the structure of a district in which a people strong enough to lay the foundation of the liberties of Europe in a single battle,[2] was educated in a fissure of the Lower Chalk.

13. I think, however, that, if Mr. Jukes can succeed in allaying his feverish thirst for battle, he will wish to withdraw the fourth paragraph of his letter,[3] and, as a general formula, even the scheme which it introduces. That scheme, sufficiently accurate as an expression of one cycle of geological action, contains little more than was known to all leading geologists five-and-twenty years ago, when I was working hard under Dr. Buckland at Oxford;[4] and it is so curiously unworthy of the present state of geological science, that I believe its author, in his calmer moments, will not wish to attach his name to an attempt at generalization at once so narrow, and so audacious. My experience of mountain-form is probably as much more extended than his, as my disposition to generalize respecting it is less;[5] and, although indeed the apparent limitation of the statement which he half quotes (probably owing to his general love of denudation)

---

[1] [The following is the sentence from Mr. Jukes' letter alluded to: "Therefore when Mr. Ruskin says that 'the forms of the Alps are quite visibly owing to the action of elevatory, contractile, and expansive forces,' I would entreat him to listen to those who have had their vision corrected by the laborious use of chain and theodolite and protractor for many toilsome years over similar forms." For the passage curtailed by Mr. Jukes, see § 1 (above, p. 548).]

[2] [Either the battle of Morgarten (1315) or that of Sempach (1386): for references to them, see Vol. V. p. 415, and Vol. XVIII. p. 538.]

[3] [To the effect that "the form of the ground is the result wholly of denudation." For the "scheme," consisting of ten articles, see note 5, below.]

[4] [For Dr. William Buckland, see the Introduction, above, p. xx.]

[5] [This and the following sentences allude to parts of the above-mentioned scheme. "The whole question," wrote Mr. Jukes, "depends on the relative dates of production of the lithological composition, the petrological structure, and the form of the surface." The scheme then attempts to sketch the "order of the processes which formed these three things," in ten articles, of which the following are specially referred to by Ruskin. "1. The formation of a great series of

from my last letter, to the chain of the Alps, was intended only to attach
to the words "quite visibly," yet, had I myself expanded that statement,
I should not have assumed the existence of a sea, to relieve me from the
difficulty of accounting for the existence of a lake;—I should not have
assumed that all mountain-formations of investiture were marine; nor
claimed the possession of a great series of stratified rocks without inquiring
where they were to come from. I should not have thought "even more
than one" an adequate expression for the possible number of elevations
and depressions which may have taken place since the beginning of time
on the mountain-chains of the world; nor thought myself capable of compressing into Ten Articles, or even into Thirty-nine, my conceptions of the
working of the Power which led forth the little hills like lambs,[1] while it
rent or established the foundations of the earth; and set their birth-seal
on the forehead of each in the infinitudes of aspect and of function which
range between the violet-dyed banks of Thames and Seine, and the vexed
Fury-Tower of Cotopaxi.

14. Not but that large generalizations are, indeed, possible with respect
to the diluvial phenomena, among which my antagonist has pursued his—
(scarcely amphibious?)—investigations. The effects of denudation and deposition are unvarying everywhere, and have been watched with terror
and gratitude in all ages. In physical mythology they gave tusk to the
Graiæ, claw to the Gorgons, bull's frontlet to the floods of Aufidus and Po.[2]
They gave weapons to the wars of Titans against Gods, and lifeless seed
of life into the hand of Deucalion.[3] Herodotus "rightly spelled"[4] of
them, where the lotus rose from the dust of Nile and leaned upon its
dew;[5] Plato rightly dreamed of them in his great vision of the disrobing
of the Acropolis to its naked marble;[6] the keen eye of Horace, half poet's,
half farmer's (albeit unaided by theodolite), recognized them alike where
the risen brooks of Vallombrosa, amidst the mountain-clamours, tossed their

---

stratified rocks on the bed of a sea. . . . 3. The possible intrusion of great masses
of granitic rock" in more or less fluent state; and 6, 7, 8, 9, which dealt with
alternate elevation and depression, of which there might be "even more than one
repetition."]

[1] [Psalms cxiv. 4.]

[2] [On the "physical mythology" of the Graiæ and the Gorgons (the soft rain-
clouds and the storm-clouds), see *Modern Painters*, vol. v. (Vol. VII. pp. 182, 183),
where Ruskin refers to the "beak or tooth" given to the Graiæ as illustrating
"the way they tear down the earth from the hillsides"; and to the "brazen hands"
of the Gorgons, in a like connexion. For the "bull's frontlet" of the Aufidus
("tauriformis Aufidus," Horace, *Odes*, iv. 14, 25) and for the Eridanus (Po), see
Ruskin's letters on inundations, Vol. XVII. p. 547.]

[3] [For "diluvial phenomena" in connexion with the wars of Titans against
Gods, see Hesiod's *Theogony*, 674 *seq.*; and for the legend of Deucalion, see the
Introduction, above, p. xlvi.]

[4] [See Milton : *Il Penseroso*, 170.]

[5] [Herodotus, ii. 92 (a chapter describing the habits of those who live in the
fen-land of Egypt).]

[6] [*Critias*, 112 : "The Acropolis was not as now. For the fact is that a single
night of excessive rain washed away the earth and laid bare the rock; at the same
time there were earthquakes, and then occurred the third extraordinary inundation
which immediately preceded the great destruction of Deucalion."]

champed shingle to the Etrurian sea,[1] and in the uncoveted wealth of the pastures,

> "Quæ Liris quietâ
> *Mordet* aquâ, taciturnus amnis."[2]

But the inner structure of the mountain chains is as varied as their substance; and to this day, in some of its mightier developments, so little understood, that my Neptunian[3] opponent himself, in his address delivered at Cambridge in 1862, speaks of an arrangement of strata which it is difficult to traverse ten miles of Alpine limestone without finding an example of, as beyond the limits of theoretical imagination.[4]

15. I feel tempted to say more; but I have at present little time even for useful, and none for wanton, controversy. Whatever information Mr. Jukes can afford me on these subjects (and I do not doubt he can afford me much), I am ready to receive, not only without need of his entreaty, but with sincere thanks. If he likes to try his powers of sight, "as corrected by the laborious use of the protractor," against mine, I will in humility abide the issue. But at present the question before the house is, as I understand it, simply whether glaciers excavate lake-basins or not. That, in spite of measurement and survey, here or elsewhere, seems to remain a question. May we answer the first, if answerable? That determined, I think I might furnish some other grounds of debate in this notable cause of Peebles against Plainstanes,[5] provided that Mr. Jukes will not in future think his seniority gives him the right to answer me with disparagement instead of instruction, and will bear with the English "student's" weakness, which induces me, usually, to wish rather to begin by shooting my elephant than end by describing it out of my moral consciousness.[6]

J. RUSKIN.

---

[1] [*Odes*, iii. 29, 33 *seq.* :—

> "Cetera fluminis
> Ritu feruntus, nunc medio alveo
> Cum pace delatentis Etruscum
> In mare, nunc lapides adesos
> Stirpesque raptas et pecus et domus
> Volventis una non sine montium
> Clamore vicinæque silvæ
> Cum fera diluvies quietos
> Irritat amnes."]

[2] [*Odes*, i. 31, 7. For another reference to the passage, see Vol. XVII. p. 494.]
[3] [For this term, see above, p. 117 *n.*]
[4] [The address was delivered by Mr. Jukes as President of the Geological Section of the British Association for the Advancement of Science, which met in 1862 at Cambridge. See the *Report of the Association*, vol. xxxii. p. 54.]
[5] [See *Redgauntlet*, Letter 13, and chaps. i. and ii. of the Narrative.]
[6] [Mr. Jukes' letter had concluded by recommending English geologists to pursue their studies at home, on the ground that "a student, commencing to learn comparative anatomy, does not think it necessary to go to Africa and kill an elephant." In the following number of the *Reader* (Dec. 10) Mr. Jukes wrote, in answer to the present letter, that he had not intended to imply any hostility towards Ruskin, with whose next letter the discussion ended.]

4

CONCERNING HYDROSTATICS [1]

NORWICH, 5th December.

16. Your pages are not, I presume, intended for the dissemination of the elements of physical science. Your correspondent "M. A. C." has a good wit, and, by purchasing any common treatise on the barometer, may discover the propriety of exercising it on subjects with which he is acquainted.[2] "G. M." deserves more attention, the confusion in his mind between increase of pressure and increase of density being a very common one. It may be enough to note for him, and for those of your readers whom his letter may have embarrassed, that in any incompressible liquid a body of greater specific gravity than the liquid will sink to any depth, because the column which it forms, together with the vertical column of the liquid above it, always exceeds in total weight the column formed by the equal bulk of the liquid at its side, and the vertical column of liquid above that. Deep-sea soundings would be otherwise impossible. "G. M." may find the explanation of the other phenomena to which he alludes in any elementary work on hydrostatics, and will discover on a little reflection that the statement in my last letter [3] is simply true. Expanded, it is merely that, when we throw a stone into water, we substitute pressure of stone-surface for pressure of water-surface throughout the area of horizontal contact of the stone with the ground, and add the excess of the stone's weight over that of an equal bulk of water.

17. It is, however, very difficult for me to understand how any person so totally ignorant of every circumstance of glacial locality and action, as "G. M." shows himself to be in the paragraph beginning "It is very evident," could have had the courage to write a syllable on the subject. I will waste no time in reply, but will only assure him (with reference to his assertion that I "get rid of the rocks," etc.), that I never desire to get rid of anything but error, and that I should be the last person to

---

[1] [This letter appeared in the *Reader* of December 10, 1864, and was reprinted in *Arrows of the Chace*, 1880, vol. i. pp. 274–276.]

[2] ["M. A. C." wrote "Concerning Stones," and dealt—or attempted to deal—with "atmospheric pressure" in addition to the pressure of water alluded to in Ruskin's letter of November 26. The letter signed "G. M." was entitled "Mr. Ruskin on Glaciers": see next note. Both letters appeared in the *Reader* of December 3, 1864.]

[3] [Not in the "last letter," but in the last but one: see above, p. 551, "A stone at the bottom of a stream," etc. The parts of "G. M.'s" letter specially alluded to by Ruskin are as follows: "It is very evident that the nearer the source of the glacier, the steeper will be the angle at which it advances from above, and the greater its power of excavation. . . . Mr. Ruskin gets rid of the rocks and *débris* on the under side of the glacier by supposing that they are pressed beyond the range of action in the solid body of the ice; but there must be a limit to this, however soft the matrix."]

desire to get rid of the glacial agency by friction, as I was, I believe, the first to reduce to a diagram the probable stages of its operation on the bases of the higher Alpine aiguilles.[1]

Permit me to add, in conclusion, that in future I can take no notice of any letters to which the writers do not think fit to attach their names. There can be no need of initials in scientific discussion, except to shield incompetence or license discourtesy.

J. RUSKIN.

[1] [See *Modern Painters*, pt. v. ch. xiii. ("Of the Sculpture of Mountains"); in this edition, Vol. VI. p. 197.]

III

JAMES DAVID FORBES[1]

(1874)

THE incidental passage in *Fors*, hastily written, on a contemptible issue,[2] does not in the least indicate my sense of the real position of James Forbes among the men of his day. I have asked his son's permission to add a few words expressive of my deeper feelings.

For indeed it seems to me that all these questions as to priority of ideas or observations are beneath debate among noble persons. What a man like Forbes first noticed, or demonstrated, is of no real moment to his memory. What he was, and how he taught, is of consummate moment. The actuality of his personal power, the sincerity and wisdom of his constant teaching, need no applause from the love they justly gained, and can sustain no diminution from hostility; for their proper honour is in their usefulness. To a man of no essential power, the accident of a discovery is apotheosis; to *him*, the former knowledge of all the sages of earth is as though it were not; he calls the ants of his own generation round him, to observe how he flourishes in his tiny forceps the grain of sand he has imposed upon Pelion. But from all such vindication of the claims of Forbes to mere discovery, I, his friend, would, for my own part, proudly abstain. I do not in the slightest degree care whether he was the first to see this, or the first to say that, or how many common persons had seen or said as much before. What I rejoice in knowing of him is that he had clear eyes and open heart for all things and deeds appertaining to his life; that whatever he discerned, was discerned impartially; what he said, was said securely; and that in all functions of

---

[1] [*Bibliographical Note.*—This piece appeared at pp. 205–207 of *Theory of the Glaciers of Savoy,* by M. le Chanoine Rendu, *translated by Alfred Wills, Q.C., to which are added the Original Memoir, and Supplementary Articles by P. G. Tait, M.A., and John Ruskin, D.C.L., Professor of Fine Arts in the University of Oxford.* Edited, with Introductory Remarks, by George Forbes, B.A. : London, Macmillan and Co., 1874. The origin of the book was Tyndall's suggestion (in his *Forms of Water*, 1872) that James Forbes, in his theory of glacier motion, had been anticipated by Bishop Rendu. Ruskin had noticed Tyndall's insinuation in *Fors Clavigera*, Letter 34 (October 1873), and the passage from that letter was reprinted in the book by James Forbes's son, Professor George Forbes (pp. 199–205), the supplementary passage (here given) following the extract. The passage was reprinted in *Arrows of the Chace*, 1880, vol. i. pp. 277–282, under the heading "James David Forbes : His Real Greatness." On the whole subject, see the Introduction, above, pp. xxxiii. *seq.*]

[2] [*i.e.*, the personal question of the credit due to Agassiz, Rendu, and *Forbes* respectively.]

thought, experiment, or communication, he was sure to be eventually right, and serviceable to mankind, whether out of the treasury of eternal knowledge he brought forth things new or old.[1]

This is the essential difference between the work of men of true genius and the agitation of temporary and popular power. The first root of their usefulness is in subjection of their vanity to their purpose. It is not in calibre or range of intellect that men vitally differ; every phase of mental character has honourable office; but the vital difference between the strong and the weak—or let me say rather, between the availing and valueless intelligence—is in the relation of the love of self to the love of the subject or occupation. Many an Alpine traveller, many a busy man of science, volubly represent to us their pleasure in the Alps; but I scarcely recognize one who would not willingly see them all ground down into gravel, on condition of his being the first to exhibit a pebble of it at the Royal Institution. Whereas it may be felt in any single page of Forbes's writing, or De Saussure's, that they love crag and glacier for their own sake's sake; that they question their secrets in reverent and solemn thirst: not at all that they may communicate them at breakfast to the readers of the *Daily News*,—and that, although there were no news, no institutions, no leading articles, no medals, no money, and no mob, in the world, these men would still labour, and be glad, though all their knowledge was to rest with them at last in the silence of the snows, or only to be taught to peasant children sitting in the shade of pines.

And whatever Forbes did or spoke during his noble life was in this manner patiently and permanently true. The passage of his lectures in which he shows the folly of Macaulay's assertion that "The giants of one generation are the pigmies of the next,"[2] beautiful in itself, is more interesting yet in the indication it gives of the general grasp and melodious

---

[1] [Matthew xiii. 52.]

[2] [For the reference to this saying, see Vol. XVI. pp. 154–155 *n.* Forbes's criticism of it and of the whole address may be found in a lecture introductory to a course on Natural Philosophy, delivered before the University of Edinburgh (Nov. 1 and 2, 1848), and entitled *The Danger of Superficial Knowledge;* under which title it was afterwards printed, together with a newspaper report of Macaulay's address (London and Edinburgh, 1849). The following are parts of the passage (extending over some pages) in Forbes's lecture alluded to by Ruskin: "How false, then, as well as arrogant, is the self-gratulation of those, who, forgetful of the struggles and painful efforts by which knowledge is increased, would place themselves, by virtue of their borrowed acquirements, in the same elevated position with their great teachers—nay, who, perceiving the dimness of light and feebleness of grasp, with which, often at first, great truths have been perceived and held, find food for pride in the superior clearness of their vision and tenacity of their apprehension!" Then, after quoting some words from Whewell's *Philosophy of the Inductive Sciences*, vol. ii. p. 525, and after some further remarks, the lecturer thus continued: "The activity of mind, the earnestness, the struggle after truth, the hopeless perplexity breaking up gradually into the fulness of perfect apprehension, —the dread of error, the victory over the imagination in discarding hypotheses, the sense of weakness and humility arising from repeated disappointments, the yearnings after a fuller revelation, and the sure conviction which attends the final advent of knowledge sought amidst difficulties and disappointments,—these are the lessons and the rewards of the discoverers who first put truth within our reach, but of which we who receive it at second-hand can form but a faint and lifeless conception." (See pp. 39–41 of *The Danger of Superficial Knowledge.*)]

tone of Forbes's *reverent* intellect, as opposed to the discordant insolence of modernism. His mind grew and took colour like an Alpine flower, rooted on rock, and perennial in flower; while Macaulay's swelled like a puff-ball in an unwholesome pasture, and projected itself far round in deleterious dust.

I had intended saying a few words more touching the difference in temper, and probity of heart, between Forbes and Agassiz, as manifested in the documents now[1] laid before the public. And as far as my own feelings are concerned, the death of Agassiz[2] would not have caused my withholding a word. For in all utterance of blame or praise, I have striven always to be kind to the living—just to the dead.[3] But in deference to the wish of the son of Forbes, I keep silence: I willingly leave sentence to be pronounced by time, above their two graves.

JOHN RUSKIN.

The following letters,[4] one from Forbes to myself, written ten years ago, and the other from one of his pupils, received by me a few weeks since, must, however, take their due place among the other evidence on which such judgment is to be given.

J. R.

"ST. ANDREWS, 2nd Dec., 1864.

"MY DEAR SIR,—Having almost retired from the din of controversy, it is a pleasurable surprise to me to find now and then some old friend who has not forgotten me, and who has still spirit and fidelity enough to break a lance in my favour.

"For about three years I have felt disheartened and hopeless as to stemming the popular tide of prepossessions against my 'Theory of Glaciers,' or, I should rather say, against me personally as its author. The article in the *Quarterly*[5] seemed to me to complete the triumph of Tyndall within the limits of his influence. Its dulness and stupid reiteration of exploded statements seemed to me to give the best promise of its speedy oblivion. Advancing years and a permanently depressed state of health have taken the edge off the bitterness which the injustice I have experienced caused me during many years. But, as I said, the old fire revives within me when I see any one willing and courageous, like you, to remember an old friend, and to show that you do so.

"I have just read ———'s paper in the *Phil. Mag.* for October,[6] which is better written than he usually does, and he makes some good points. But what amuses me is the *necessity* to which glacialists are driven (whatever opinions they may profess) to treat glaciers as plastic, in the most uncompromising sense of the term.

---

[1] [In the edition of Rendu's *Glaciers of Savoy* already alluded to.]
[2] [Forbes died December 31, 1868; Agassiz in 1873. Saussure's death had taken place in 1845.]
[3] [With this passage, compare Vol. XVI. p. 32 n.]
[4] [The letter from Forbes to Ruskin (dated December 2, 1864) was presumably elicited by the allusions to Forbes in Ruskin's letter to the *Reader* of November 26, 1864 (see above, p. 550; and compare the Introduction, p. xxxviii.).]
[5] [July 1863: see again, above, p. xxxviii.]
[6] ["On the Conformation of the Alps," by John Tyndall, F.R.S., in the *London, Edinburgh, and Dublin Philosophical Magazine*, 4th series, No. 189, October 1864.]

"I hope your health is good and that you still enjoy frequent trips to Switzerland. I am so bad a traveller now, that even the railway journey is quite a barrier, especially were my family to accompany me. I hear much of the increased obstacles to large parties moving about since I was last abroad, and the utterly vulgarized state to which Chamouni is reduced.

"Ever yours sincerely and obliged,
(Signed) "James D. Forbes."

The following are portions of a letter from Mr. Robert B. Watson, dated :—

"Madeira, *Dec.* 13, 1873.

"The Nat. Phil. Class was an essential preparation for the study of Divinity in the Free Church of Scotland, of which I am now a minister. Thus I attended Forbes. That was in the session '44–45. Glaciers, of course, came prominently into the course, but of Agassiz we learned only as a great and honoured name. In January 1846 (I think on the 2nd of the month), going from Geneva to Berne in bitter cold and deep snow, alone in the diligence, there entered to me at Neuchâtel, as the early dark settled down upon us, a cheery fellow-traveller, who drew out the last moments in eager talk with a parting friend—one with a certain stamp of greatness on him—and their talk was all of glaciers. On that subject—I kindled ; and when at length Desor, my new fellow-traveller, told me that the Agassiz of whom I spoke was none other than he who had stood beside me at the coach door, *all my boyish enthusiasm was stirred,*[*] and I told all I had learned of him, and from whom. Then it was Desor told me he and others of Agassiz's friends—their leader himself, about to start for America, being unable to come—were going to the Aar Glacier, to measure for the first time the actual movement of the ice. Very soon it was arranged that I should go too. Next day we were joined by M. Dolfuss-Ausset (Papa Dolfuss) from Mulhouse. The following day we reached Meiringen, and the day after we walked to the Grimsel. There we spent several days in bright fierce cold, making various theodolite measurements at many points, digging a deep hole into the glacier beyond the influence of superficial cold there, by coloured water to test the percolation susceptibility of the ice. Desor communicated the results to a Berne paper ; probably they have been published elsewhere, but they amounted to little more than was known before, viz. that the ice throughout, even in winter, was drenched with water, and that the movement was as regular from day to day, though a little slower than in summer.

"Forgive these personal details—at least I have not thrust them on you—but they serve to illustrate these two points—1st, that Forbes taught us to reverence Agassiz ; and 2nd, that Forbes's theory of viscosity, being the final explanation of glacier motion, forced itself on me, in spite of much concurring to the contrary, and specially of my being taught Agassiz's views, *in situ*, and by his own friends. Neither then or since could I make out what Agassiz's antagonistic view was, and time has only matured the conviction that when Forbes uttered the word 'viscosity,' he had said the very last that could be said on the subject ; and that Thomson's experiments on melting under pressure, and regelation, explain the manner of the viscosity—the manner, not the fact. With Rendu it was a guess among guesses too—with others it was a poetical metaphor, that a glacier flowed like a river. Forbes proved that it was a fact, and Thomson confirmed his proof by showing whence viscosity comes."

---

* *The italics are mine.* I think this incidental and naïve proof of the way in which Forbes had spoken of Agassiz to his class, of the greatest value and beautiful interest.— J. R.

IV

A LECTURE ON STONES[1]

(1876)

On commencing, Mr. Ruskin at once put himself on good terms with his audience by a few friendly words, and then introduced his subject by calling attention to a sovereign and a pebble.[2] Of the first, he remarked that he knew, to some extent, what a pound would do, and what it was good for, but there was much to be learnt about the sovereign which very few people were acquainted with. For instance, it had stamped on it a man on a flying horse, stabbing at an impossible dragon in a manner in which it would be impossible to kill him.[3] They called the man St. George, but people did not know much about St. George, or why he should be on the sovereign. Then as to the pebble, any apothecary could tell them of what it was made, and, for that matter, could tell, too, what the lecturer and his hearers were made of; but in doing so, he would not tell them what a boy or man was, and still less what such an inexplicable thing as a girl or a woman was. To know what a thing was, they must know what it could do and what it was good for.

They knew that a piece of flint, when struck by steel, would produce fire, but there was more to be learned about it than that. In some places it fortified the cliffs of England, and on the very brow of that cliff the flints were taken up in the softest form by the grasses that grew there. This had been examined by one of his pupils—a student at Oxford, who had taken the stalk of a particular kind of grass, and had, by subjecting it to the influence of an acid, obtained pure flint—the fact being, that there was not a bit of grass grown that did not make its skeleton of flint, just as human beings' bones were built up of lime.

To understand the stones, it was necessary to consider not only what they could do, but what they could suffer, and whether or not they were

---

[1] [*Bibliographical Note.*—This report of an address to the boys of Christ's Hospital appeared in the *City Press* of April 15, 1876, and was reprinted in the privately-issued *Ruskiniana*, 1891, Part ii., pp. 225–226. On p. 564, in the fifth line from the bottom, "dear" is here substituted for "dead."]

[2] [It would thus appear that this lecture was in part the same as the one delivered at the London Institution on February 17 and March 28, 1876, and incorporated in *Deucalion:* see above, i. ch. vii. §§ 4–9, pp. 167–168.]

[3] [Compare the criticism of Pistrucci's design in *Fors Clavigera*, Letter 26; and for the story of St. George, see the same letter.]

amiable things—whether they had ever been loved.[1] Now, it was easy to understand the love which men had for sovereigns, and their love for the bright and beautiful amethyst, and opal, and onyx; but was that little black pebble cared for? And yet there was beauty in it, and it was the right understanding of this beauty that brought us into somewhat of sympathy with the power that made us, and preserved us, and gave us the dust of the earth even for our delight. The Bible mentioned specially amongst the minerals, gold, crystal (or bdellium), and onyx; and these three might be regarded as types—the gold standing for all metals, the crystal for all clear, brilliant stones, and the onyx for all stones in which the colours were arranged in veins or bands. Gold suggested thought of the powers exercised by the metals, and what had been done by them for and against man; and the crystal reminded them of the influence of jewels over the mind, and of the ill that had been wrought in connection with them by man in error, and by woman in vanity; and the question arose whether these stones were blessings or not? They were spoken of as blessings throughout the Scriptures, but it was doubtful if we had not again and again made them curses, and sacrificed them to devils rather than to God.

Then there came the inquiry, How these stones were made for us? We were told, "There is a vein for the silver and a place for the gold, where they find it"; but no geologist could say how that gold got there —they could only tell that there it was. Another mystery was that of crystallization, and another the growth of a pebble. On every hand, the student was reduced to a continual asking of questions, and the wiser he was the more he asked. The man who knew but little was satisfied that he was very wise; but as he progressed, he found himself floating on the top of a sea of knowledge, and was delighted to find how deep the waters were.

Alluding to the museum which is in course of formation at Christ's Hospital, Mr. Ruskin observed that it was important, for the purposes of study, that it should not contain too much.[2] With a small amount to look at, they would be able to give it better attention. There might be a thousand beautiful specimens of crystallization, but they would not be able to grasp them all at once, and it was better to study three or four than to attempt to make the acquaintance of so many; they could no more look at twenty things worth looking at in an hour, than they could read twenty good books in a day. It might be asked, "What were precious stones made for?" They could only have a moral value, for one could not eat diamonds; and though one might drink pearls,[3] the effect would only be to make the vinegar dear. A pig would not care to have his food served in a silver trough, but noblemen and gentlemen liked silver troughs; they preferred their food off plate, thereby showing their superiority over the pig, as seeing something better than the pig does. The love of the beautiful was associated with the use of colours; and, in connection with

---

[1] [Here, again, compare *Deucalion*, i. ch. vii. pp. 168, 169.]

[2] [Compare *Deucalion*, i. viii. § 11, p. 203.]

[3] [The reference is of course to the tale of Cleopatra dissolving her pearl ear-drop in vinegar and drinking to Antony.]

this point, the lecturer explained, somewhat in detail, the symbolical effect of colour in heraldry, from which some interesting lessons were drawn.

Mr. Ruskin concluded by reminding his hearers that central truths were always beautiful and lovely, though half-way truths might be exceedingly ugly.[1] The symbolical aspect of stones suggested important reflections; as, for instance, when men were found worshipping a black stone because it fell from heaven,[2] they were doing a thing which was not wise, but it was half-way to wisdom. The lecture throughout was copiously illustrated with diagrams, and a collection of stones, to which frequent reference was made by the speaker.[3]

[1] [Compare *Deucalion*, p. 187.]

[2] [*Ibid.*, p. 193.]

[3] [The report thus concludes : "At the close, a vote of thanks was proposed by Mr. J. D. Allcroft (the Treasurer), and seconded by the Rev. G. C. Bell, the boys endorsing it with three ringing cheers, which, however, hearty as they were, could hardly have more fully indicated their satisfaction and enjoyment than did the close attention with which the lecturer's remarks were followed. Mr. Ruskin, in responding, expressed his pleasure at being present, and promised to repeat his visit as opportunity might offer."]

V

THE ALPINE CLUB AND THE GLACIERS[1]

(1878)

BRANTWOOD, CONISTON, LANCASHIRE.
August 27, 1878.

MY DEAR SIR,—I have only this morning taken up the August number of the *Alpine*, and should have before thanked you for the candid and exhaustive history of the Buet, and its just notices of dear old Saussure and Bourrit.

No less for the courteous paper on Alpine art, the most sensible I have ever seen.

I should like to send you a few words on the matter for your October number, if September 12 will be in time for it, of which the gist will be an affirmative answer to your question whether I have ceased to hope for Alpine art, and a courteous reproach to the writer of the article for supposing snow paintable.

I have told my publisher to send you the back numbers of *Deucalion*, and to continue it for the Club library. I hope that some day the members of the Club may desire to gather together their knowledge of glaciers, and make a wholesome end of all glacier theories, by due acknowledgment of James Forbes's conclusive ascertainment of glacier facts.[2]

They owe this duty to science, and should, it seems to me, take honourable pride in fulfilling it. Always believe me, my dear Sir,

Your faithful servant,
J. RUSKIN.

---

[1] [The following letters are reprinted from the *Alpine Journal*, No. 148, May 1900, vol. xx. p. 129. They were included in an "In Memoriam" notice of Ruskin, and were addressed to Mr. Douglas Freshfield, who was at the time editor of the *Journal*. The references in Ruskin's first letter are to two papers by Mr. Freshfield, in the number for August 1878 (vol. ix. pp. 6–31, 37–45). In the article on "Alpine Art in the Exhibitions" Mr. Freshfield referred to Ruskin's statement in the "Turner Notes" of 1878 (see Vol. XIII. p. 509) that the upper snows are unpaintable, and contrasted it with a passage in the first volume of *Modern Painters* (see Vol. III. p. 449), where Ruskin bids the painters "refresh us a little among the snow." Mr. Freshfield combated the idea that the snows are unpaintable, and argued that Turner did not paint the high Alps, only because he did not climb them.]

[2] [In a discussion at the Royal Geographical Society, upon the question "Do Glaciers Excavate?" on March 27, 1893, Mr. Freshfield cited this letter, and said,

BRANTWOOD, CONISTON, LANCASHIRE,
September 1, 1878.

"MY DEAR SIR,—I am greatly delighted with your letter, and most glad of all the suggestions in it, especially of that about the spirit of climbing and travelling in Switzerland. It will be very refreshing to me to think over all these once so much loved subjects, and I hope to be able to send you an interesting, though a very quiet, paper, for I can't let myself get excited in writing since my illness without too much fatigue.

Ever most truly yours,
J. RUSKIN.

[Undated, but written about the end of the year 1878.]

"DEAR MR. FRESHFIELD,—I have at least ten times set myself to do that paper for you, and ten times have been unable to write a word for sorrow, as I thought of the wasted pride and energy of our youth and the total destruction of the beauty of Switzerland. . . . I find myself still unable to write, and cannot venture to think on a subject to me so appalling. But as to the possibility of Art for alps I may say, merely for your own guidance in what you write in future, that, if an artist could paint an icicle or an opal, he might in time paint an alp. But if he will first try a branch in hoar-frost, and succeed—I shall like to see it! Calame[1] and that man—I forget his name—are merely vulgar and stupid panorama painters. The real old Burford's work[2] was worth a million of them. In true sorrow for my failure,

Faithfully yours,
J. RUSKIN.

"I do not think the Alpine Club can be accused of having failed in doing its part in the work Mr. Ruskin proposed for it; in bringing, that is to say, recent geological theories into close contact with geographical facts. Mr. Whymper, as we all know, in his book on the Alps, entered largely into the question of glacier action; and since that time four Presidents of the Club—our late respected and beloved Fellow and Councillor, Mr. John Ball, Mr. William Mathews, Mr. Bonney (whom you have heard to-night), and last and least, myself—have done our best to show that the geological theory of glacial excavation is inconsistent with the topographical facts as we and others have seen them, and that it is supported mainly by appearances which I may fairly call superficial" (*Geographical Journal*, June 1893, vol. i. p. 500). It may be added that another President of the Alpine Club, Mr. T. W. Hinchcliff, was one of the earliest of those who defended Forbes's theory against the criticisms of Tyndall: see the appendix to his *Summer Months among the Alps* (1857).]

[1] [For this painter, see Vol. III. p. 449, Vol. VI. p. 101, and Vol. XV. p. 112.]

[2] [Robert Burford (1791–1861) exhibited panoramas in Leicester Square. Ruskin refers to them with commendation in *Præterita*, i. § 136.]

VI

"THE LIMESTONE ALPS OF SAVOY"

(1884)

INTRODUCTION [1]

1. The following book is the fulfilment, by one of the best and dearest of those Oxford pupils to whom I have referred in the close of my lectures given in Oxford this year,[2] of a task which I set myself many and many a year ago, and had been obliged, by the infirmities of age, with deep regret to abandon. The regret is ended now, for the work is here done in a completeness which, among my mixed objects of study, it could never have received at my hands.

2. The subject of the sculpture of mountains into the forms of perpetual beauty which they miraculously receive from God, was first taken up by me in the fourth volume of *Modern Painters*;[3] and the elementary principles of it, there stated, form the most valuable and least faultful part of the book. They had never been before expressed, or even thought of, for the simple reason that no professed geologists could draw a mountain,[4] nor therefore *see* the essential points of its form. So that at this very time being, the large model of the Valley of Chamouni exhibited in the library of the British Museum,[5] is a disgrace not only to the Museum first,

---

[1] [*Bibliographical Note.*—This Introduction occupies pp. ix.–xxiii. of a work with the following title-page :—

> Deucalion.—First Supplement. | The | Limestone Alps of Savoy ; | A Study in Physical Geology. | by | W. Gershom Collingwood, | M.A., Oxon. | with an Introduction by | John Ruskin, D.C.L., LL.D., | Honorary Student of Christ Church, Honorary Fellow of Corpus Christi | College, and Slade Professor of Fine Art, Oxford. | George Allen, | Sunnyside, Orpington, Kent. | 1884.

The paragraphs are here numbered for convenience of reference. In § 6, line 4, "1380" is here a correction for "1881."]

[2] [This Introduction, though dated at the end January 1884, must have been begun in 1883, as the reference here is to the last but one of the lectures on the *Art of England* (§ 154), delivered in November 1883.]

[3] [See above, p. 549.]

[4] [Compare *Deucalion*, p. 161.]

[5] [This model has now for some years been removed from the King's Library, and consigned to obscurity in a far corner of the Map Department. The printed key to the model is entitled, *Tableau geographico-topographique pour servir d'explication au Relief du Mont Blanc et des contrées voisines construit par Auguste Köhler, d'après les données du Dr. W. Pitschner* (Berlin, 1862). The Geological Society was not concerned in its production or exhibition. Ruskin's remark is explained by a passage of the same date in *Fors Clavigera* (Letter 95, § 16): "the Geological

and the Geological Society next, but actually it is a libel on the ordinary intelligence of human nature. For if people resolutely refuse to look at things in the right way, the law of their nature is, they come to look at them exactly in the wrong.

The only member of the Geological Society, since its energies were diverted to palæontology, who could draw a mountain in outline, was James Forbes, and even *he* could not draw in light and shade ; but his outlines were precise and lovely. And it was the accuracy of observation directed by this practice that enabled him to recognize the lines of flux in glaciers, which no previous (nor subsequent, for that matter) geologist had so much as a glimpse of ;—to this day most of them remaining as incapable of tracing the linear indices of motion of glacier waves as the puces-de-glace that live in them.

3. After comparing notes with James Forbes at the village inn of the Simplon (see *Deucalion*, chap. x.[1]), in 1849,[2] I went up to the Bell Alp, then totally unknown, and drew the panorama of the Alps, from the Fletschhorn to the Matterhorn, which is now preserved in the Sheffield Museum.[3] Then, going up to Zermatt I took the first photograph * of the Matterhorn (and, I believe, the first photograph of any Alp whatever) that had then been made.[4] On the work done in Zermatt at that time, the mountain section of *Modern Painters* was principally based ; but in 1861 I went into Savoy, and spent two winters on the south slope of the Mont Salève, in order to study the secondary ranges of the Alps, and their relation to the Jura.[5] I quickly saw that the elements of the question were all gathered in the formation of the mountains round the Lake of Annecy : and, at Talloires, in the spring of 1862, made a series of studies of them, which only showed me how much more study I wanted.[6]

Being called to England, I left the light blue lake with resolution of swift return ; and the time of Troy-siege passed by, before I stood again upon its brink among the vineyards.[7]

4. In the meantime I had been able to do some collateral work that was of use. The autumn and half winter, till Christmas, of 1862, were spent at Lucerne and Altorf, in examining the relations of the limestone of Uri with the Northern Nagelfluh and Molasse. The summer of 1866, though principally given to *Proserpina*, yet allowed me time, at Brientz and Inter-lachen, to trace the lines of Studer's sections across the great lake-furrow of central Switzerland. I learned enough geological German to translate for myself the parts of his volumes which relate to the Northern Alps, and wrote them out carefully, with brilliantly illuminated enlargements of

* Properly daguerreotype—photography then being unknown.

Society should, for pure shame, neither write nor speak another word, till it has produced effectively true models to scale of the known countries of the world."]

[1] [Above, p. 220.]
[2] [Here Ruskin confuses the dates. It was in 1844 that he met Forbes (see the Introduction, p. xxi.); his work at Zermatt was in 1849.]
[3] [See above, p. 222.]
[4] [Compare above, p. 97.]
[5] [Compare above, p. 548.]
[6] [For these visits, see the Introduction, above, pp. xxvii., xxviii.]
[7] [In April 1872 Ruskin again visited Annecy : see Vol. XXII. p. xxvi. *n.*]

his tiny woodcuts, proposing the immediate presentation of the otherwise somewhat dull book to the British public in this decorated form. A letter, bringing me bad news, interrupted me one bleak wintry day; and the since untouched manuscript, with its last drawing only half coloured, remains on the library-shelf behind me, like an inoffensive ghost.[1]

5. It chanced, or rather mischanced, also, that having written *Unto this Last*, in the Valley of Chamouni in 1860, and *Munera Pulveris* at Mornex in 1861,[2] I was eager at Lake Lucerne in 1862 to translate into pithy English the two first books of Livy,—a design which broke down like that for Mr. Studer, after I had lost a whole lovely day of clear frost at Altorf in cataloguing the forces of the preposition *ob*, as a prefix to verbs.[3] In the course of these desultory efforts, however, I ascertained that the essential facts of Alpine construction remained to be detailed: that Studer had given only superficial examination to a far too widely extended surface; and while he had spent years of unremitting labour in partially determining the conditions of form in the Jura, the Apennines, and the entire length of the Alpine chain between Savoy and the Tyrol, had never given the exhaustive attention to any single valley, which was needed to ascertain the scope and results of the metamorphic, as distinguished from mechanical, changes of feature in its secondary strata.

6. I took up this subject with renewed eagerness in the first leisure given by retirement from my Oxford Professorship in 1879, and received from time to time the kindest assistance and coadjutorship from Mr. Clifton Ward,[4] whose lamented death in 1880 deprived modern science of one of her most patient, powerful, and candid observers; and left me again discouraged, and at pause, in the presence of questions which had become by his help more definite, but in that very distinctness, less assailable.

Feeling also that my strength would no more permit me the climbing of Swiss hills, I resigned hope of doing more among the precipices of the Buet or the Jungfrau; and began, as better suited to my years, the unadventurous rambles by the streams of Yewdale, whose first results were given in my Kendal lecture (*Deucalion*, chap. xii.[5]).

7. But here again I soon was in need of help. Though still able easily enough to get to the top of Wetherlam or Silver How, on occasion, I had no time for such survey of the country in all the lights of evening and morning as I felt to be necessary for the understanding of its essential forms: and I entreated Mr. Collingwood, who had before been at work for me on the bed of the retreating Glacier des Bossons, to come to my assistance at Coniston, and make me a perfect model of the mountain group which was within the day's walk of Brantwood.[6]

He gave a summer to this task; and completed his model to a scale of six inches to the mile,—the best, I am bold to say, yet made of any

---

[1] [The book, a quarto account-book bound in vellum, is at Brantwood. It is labelled by Ruskin "Rock Book," and dated, not 1866, as here stated, but "1861-2."]

[2] [See Vol. XVII. pp. xxi. *seq.*]

[3] [In one of Ruskin's note-books there are many pages of such grammatical notes. The diary shows that his studies in *ob* were made on November 26, 1861. For his interest in Livy at the time referred to, see Vol. XVII. p. xlvi.]

[4] [For his contributions to *Deucalion*, see above, pp. 267 *seq.*]

[5] [Above, pp. 243 *seq.*]

[6] [This model is now in the Coniston Museum.]

part of the Lake District. But, before we had settled the colouring of it, the usual malignity of my fairy godmother (or gnome godfather) interfered; and Mr. Collingwood had to leave his whole summer's work—like my former ones, *re infecta;* and to come with me to Italy, more in the capacity of physician than geologist.[1] His watchful care of me had such good results that before recrossing the Alps, I had formed the hope of returning to my duties in Oxford; and in a newly active frame of mind, asked my friend, while yet the snows were high, to review with me some of the old problems in the much loved recesses of the Dorons and Tournette.

To my (somewhat unreasonable) surprise I found his instinct for the lines expressing the action of the beds far more detective than my own; and felicitous beyond my hopes, in that he was fettered by no scientific theory, and saw the most wonderful group of mountains in Europe with entire freshness of mind and eye.

8. But he had another advantage over me, in his glance over strata, which I was not prepared for, and which not a little provoked, while it mightily assisted me. All through France and Italy, where we had been drawing Gothic sculpture, Mr. Collingwood, trained in recent science of anatomical draughtsmanship, had been putting me continually in a passion by looking for insertions of this and the other tendon and gut, instead of the general effect of his figure; but when we got to the hills, I saw that this habit of looking for the insertion of tendon and gut was of extreme value in its way, and often enabled him to see the real direction of original movement in the mountain mass, where *I* saw only the effects of time and weather on the superinduced cleavages, often opposed altogether to, and always entirely independent of, the lines indicating primary motion. On the other hand, as I read over the sheets now ready for the press, I find he has not at all dwelt on one of the questions respecting this motion itself, which I thought to have indicated to him as one of the most needful subjects of inquiry, namely, the relation to it of joints, as distinguished from cleavage.

9. True cleavage never pays the smallest attention to the fluctuation, involution, or any other caprice of the several strata; but assuredly in any substance not fluid, nor elastic, nor capable of easy molecular adjustment, fluctuation cannot take place without fissures; and I greatly marvel to see my enthusiastic friend shaking his mountains up and down as a terrier shakes a rat, or a rug,* without ever telling us in what state of cohesion

---

* "Or switching them about like a whiplash into loops and curls," p. 63.[2]

---

[1] [In 1882. Mr. Collingwood has given an account of the tour in Chapter IV. ("Ruskin's Old Road") of his *Ruskin Relics* (1903).]

[2] [The full passage in *The Limestone Alps* is as follows: "In these figures [between pp. 64, 65] the upper broad layer, divided by a fine line, crossed by vertical fissures, and lettered U, is the Urgonian; playing everywhere a very prominent part; not only piled in sturdy precipices, but switched about like a whiplash into loops and curls. At first, indeed, it seems almost too hard to believe that so solid and rigid a stuff should lend itself to such flowing and fantastic lines; and quite too hard to imagine what process could have produced them from original flat layers. But there are places where the twist is indubitable to the naked eye in the naked rock; notably in the Jurassic beds at the Nant d'Arpenaz."]

their substance must have been at the time of the operation, or seeming to remember that though one can wave a flag, or wreathe hot iron, one can't wrinkle a deal board, or pucker a *baked* pie-crust. It did not, of course, enter into the design of this volume to touch on the structural phenomena of metamorphism or any others connected with the baking of the earth's crust;* and it is wise in the author, on the whole, to have restricted himself absolutely to the description of existing forms, and the abolition of recklessly adopted explanations of them by glacial or pluvial agency. But I regret that in his enumeration of component rocks, he should have taken notice only of the changes in one of them, the Lias, where it approaches the Central Alps; and tell us nothing about the differences of aspect or structure traceable in others of the formations under the same condition. In reviewing my own experience in this matter, it becomes more and more wonderful to me how little the rocks seem to modify each other at their actual junctions. Gneiss runs into Protogine at Chamouni in tongues and veins,[1] without the slightest loss of its own character or pardonablest proclivities to the Protoginesque; and oolite lies flat upon granite at Avallon, with no apparent discomfort or objection, and without allowing the slightest change in its own shaly and crumbly substance, till within a few feet of the actual junction; while the metamorphism which in other localities affects these, or even more recent formations, appears, as for instance in the crystalline marbles of Tuscany, the result of the equable diffusion of heat and distribution of pressure for myriads of years through the entire mass of the substance under modification.

10. I might have easily prevented the appearance of neglecting these and some other connected difficulties, had I thought it right to interfere in any way with the natural impulse of the author's thoughts. But I was, on the contrary, so anxious not to disturb—and above all not to check— the direct energy which was doing such good work in its chosen field, that I not only refrained, in looking over the manuscript, from making any suggestions, except in matters of mere arrangement; but took great pains when we were at Geneva to prevent Mr. Collingwood from getting hold of Professor Favre's elaborate analysis of the same district.[2] This I did for two reasons; the first, that I greatly feared Mr. Collingwood might give up the whole design, if he saw to what precision and extent

---

* Geologists seem satisfied, nowadays, that the whole globe is a sort of flying haggis, or lava pudding, out of which, I see by Mr. Ball's lecture on the Corridors of Time, the Moon got pinched at the baker's.[3]

---

[1] [On this subject, see *Modern Painters*, vol. iv. ch. xv. § 15, and Fig. 55 (Vol. IV. p. 253). Ruskin stayed for some days at Avallon during the tour of 1882; his Preface to the 1882 edition of *Sesame and Lilies* is dated thence (see Vol. XVIII. p. 52).]

[2] [Professor Favre's analysis of the district is contained in two works—(1) *Considérations géologiques sur le mont Salève et sur les terrains des environs de Genève*, by Alphonse Favre, Geneva, 1843; (2) *Recherches Geologiques*, etc., 1867 (as noted above on p. xxviii.).]

[3] [*A Glimpse through the Corridors of Time: Lecture delivered at the Midland Institute, Birmingham, October 24, 1881*, by Professor R. S. Ball, 1882. See pp. 17 *seq.* for the "moon as a fragment torn off."]

Professor Favre's study of Savoy had been carried out; in the second place, I was extremely desirous to see how far the conclusions of Professor Favre would be confirmed by an independent observer.

Accordingly, I assured my friend, when I had got him with his full sketch-book into a quiet corner of the Hotel des Bergues, that if only he would go on preparing his drawings for the wood blocks, I would myself ransack the libraries of Geneva for whatever geological works could be of the smallest assistance to him. And so I did: but I only gave him those whose assistance to him *was* "of the smallest!"—and locked Professor Favre carefully up in my own portmanteau. The result was absolutely satisfactory, and the corrections of his own views in points of detail which Mr. Collingwood afterwards found necessary on comparing Professor Favre's sections with his own, were easily made and collected in the postscript to the third chapter.[1]

11. The drawings which I was so eager to see in progress quite deserved my solicitude, being indeed much better than any by which the volume is now illustrated. Made on the spot, or from immediate memory, they were vivid and expressive in the extreme; but, of course, in many points inaccurate or incomplete. The correction and finishing, with continual hesitation as to what could or could not be expressed in wood-engraving, has taken half the life out of the first drawings; and I shall take good care in any future geological expeditions with the author, to lock up his own drawings in my portmanteau, as well as other people's, and not let him meddle with them afterwards, till I can get them engraved. On the other hand, the extreme fidelity and skill with which Mr. W. Hooper has facsimiled the final states of the sketches, deserve the author's best thanks, and the public's also; for truly, whatever their shortcomings, the figures in this book are quite the most illustrative of mountain form in its wide symmetries that have yet been contributed to the syntax of constructive geology.

12. I have but a word more—partly in modification,—partly in support, —of the author's remarks on the influence of Rudisten-kalk in the production of hermits.[2] In modification, that the Eremitic character sometimes takes

---

[1] [See pp. 101–105 of *The Limestone Alps*, where Mr. Collingwood says: "It was only after completion of the foregoing remarks that Professor Alphonse Favre's works were put into my hands. The object of his *Recherches Géologiques* has been exhaustively to survey and describe in detail those appearances of nature which I have sought to grasp as a whole, and to give in rough generalization. They embrace, with much country beyond my range, the northern part of my district, which I name after the old duchy of Genevois. In much eagerness, and more fear, I compared his masterpiece with my school sketch from the same model; and discovering that in general I have not misrepresented the facts which he has so ably and so minutely investigated, I leave my own work as it was; choosing rather to point out my mistakes than to tacitly correct them."]

[2] [The first portion of the first chapter of Mr. Collingwood's book is entitled "The Homes of the Hermits," the author remarking (pp. 7, 8): "It is interesting, though perhaps impertinent, to remark how intimately limestone mountains are connected with the typical development of Christian asceticism. To the limestone caves and coves of Yorkshire we owe the English hermits, of whom Richard Rolle of Hampole was the chief, and whose influence on our own religious history is more real than recognized. And the Savoy limestone mountains were the home

the less recognized form of Rousseau's retirement on St. Peter's Island,[1] or Byron's by the Bosquet de Julie,[2] while St. Bernard of the crusades, himself the product of a French monticule instead of a Swiss mountain, "discoursing of the lake—asked where it was."[3] Something also in this kind may be insinuated for the marshes of Croyland and Citeaux;[4] nevertheless I am thankful to be able to associate with this pretty opening chapter of my friend's book, my reminiscence of a real live hermit, whom I found in a cave two thousand feet or so above the valley of the Rhone, —alas! now fifty years ago;—and the expression of my reverent sense of the wisdom of his mossy and cressy retirement (his little garden had flowers in it also), as compared with the tormented existence of the modern travelling Eremite, in caves which he has paid millions of money to dig, that he may not see the Alps when he gets to them.

To such better sympathy as may yet be found of Benedictine, Carthusian, or Augustine gentleness, in the hearts of pilgrim folk, I commend this book, for their mountain guide among some of the fairest scenes that ever were formed by earth, or blessed by Heaven.

JOHN RUSKIN.

BRANTWOOD, 10th January, 1884.

---

of St. Germain of Talloires, St. Bernard of Menthon, and St. Francis of Sales; not to mention the monks of the Grande Chartreuse, and a host of lesser saints whose memory still survives, and whose names remain; as St. Jeoire or Jorioz, St. Ruph or Rodolph, the brother of Germain,—with many who in their age exerted a power of light and leading which the superficial reader of history attributes to kings and warriors. But the kings and warriors were made or unmade by the monks. And the monks were made by the mountains. And the mountains? It is our business to see how these were made. For we do geology a wrong if we think its tale ends where history begins. Wars and rumours of war, revolutions and reformations, they do not make history; the people and their feelings, aspirations, passions, culture, they make history; of which, battles and sieges, councils and codes, are only as the barometer is to the weather,—merely the indications. And, once again, the people are swayed by the thinkers, and the thinkers by nature —little though they know it; and specially by the sublime in nature; and most vigorously of all by these limestone mountains."]

  [1] [See Vol. XVIII. p. xxxviii.]
  [2] [At Clarens. See Byron's *Letters and Journals*, 1899, vol. iii. p. 352.]
  [3] [See Vol. V. p. 363.]
  [4] [See, in a later volume of this edition, the lecture entitled "Mending the Sieve."]

VII

THE GARNET[1]

(1885)

1. In the last chapter,[2] I distinguished granulated rocks from composite rocks, of which the most important to the general reader are of course those in which one of the component substances may be sometimes separable in the form of a gem. Of these, one of the most interesting masses in the world is the mountain with the pretty name, Adula, which rises in the midst of the St. Gothard Pass, above the plain, I believe, on which the old Hospice still stands. In the substance of that single mountain are found, in confused crystals, some fifteen or twenty (I will count presently) different minerals, all of them interesting, and *five* precious; namely, first, the one which takes its name from the mountain, Adularia—in the finest conditions of it, used by jewellers under the name of moonstone; secondly, the red garnet, which is the subject of our immediate inquiry; thirdly, the most beautiful rock-crystal that can be found in the world; fourthly, the jewel described by Saussure under the name of Sappare,[3] as blue as a pale sapphire, and much brighter than any sapphire, if left in its natural crystal; and, lastly, the mineral called, I know not why, but very prettily, Tourmaline,[4] sounding as if it were the Tower of Mechlin, and indeed forming towers, when perfectly crystallized, which uninformed fairies might take for the Tower of Giotto built of ruby.

2. In this chapter I describe only the occurrence of the best known gem—the garnet, which is divided by jewellers into Oriental and common garnet, not on account of any real difference in the mineral, but because, like all other jewels, the Eastern examples are the best crystallized, while in Europe a perfectly clear crystal is so seldom found that it cannot become an article of commerce.

3. In the first place, then, understand of garnets this much, that they are always roughly round things embedded in the rock in which they occur, never forming themselves on the outside of it. Note this very

---

[1] [Printed from a MS. (in Mrs. Severn's hand) headed "In Montibus Sanctis. Part III."]

[2] [That is, Chapter III., with a postscript as printed in *In Montibus Sanctis*: see now Vol. VI. pp. 144, 145.]

[3] [*Voyages dans les Alpes*, § 1901, vol. iv. p. 83 (1796 edition).]

[4] [Compare *Ethics of the Dust*, § 96 (Vol. XVIII. p. 325).]

particularly, because a great many other jewels are found in cavities of the rock, and sometimes in large caverns, so that, in a rich mine, you might suddenly break your way into a cave glittering all over roof and floor with topazes, beryls, sapphires, amethysts, or cairngorms.   But you never could have a roof of garnets, diamonds, or emeralds—those stones invariably forming themselves in the mass of the rock, gravel, or clay in which they are found.

4. Farther, garnets never form themselves in groups; they are the sulkiest of precious stones; and, however crowded in the rock where they form, never unite their crystals, and therefore never shoot into rods.   Nearly all long crystals are produced by the building together of smaller crystals, and so also the beautiful star-like groups of chalcedony, quartz, or zeolite; but a garnet insists upon keeping its symmetrical and stupid crystal of untrue sides, and, roughly and practically speaking, is never long, never flat, never irregularly grouped, and never star-like.   It has no more idea of arranging itself decoratively than the plums have in a pudding, and cannot even become massive, without losing the beauty of its colour; but, under these limitations, it adds greatly to the beauty, weight, and perhaps strength, of the grandest rocks in the Alps, and fine specimens of it are among the most interesting that present themselves to the mineralogist. Although I have never neglected an opportunity of securing them,[1] there is only one in my collection of more than three thousand crystalline minerals which can be considered as perfect.

5. I believe we may most simply describe the varieties of the garnet by naming them from their localities—as their different aspects belong much more distinctly to places than to rocks.   For the general student they may be summed under four heads—(1) the Scawfell garnet, occurring in Scawfell greenstone; (2) the St. Gothard garnet, occurring in St. Gothard mica slate; (3) the Monte Rosa garnet, occurring in Monte Rosa protogine; and (4) the Bohemian garnet, the only one fit for jewellery, occurring I know not yet in what crest of the Bohemian Mountains.

6. First, then, the Scawfell garnet is interesting in the strict limitation of its size, never exceeding that of small bird shot; and in the perfection of its form, being always neatly crystallized all round, but never attaining purity of colour.   It is of a dark purplish red, not transparent, but contrasting agreeably with the olive-green of its matrix.   I shall in future keep the name of greenstone for this rock, wherever it occurs, having a close-grained Basalt or rather Basaltic paste—so far as I know, never forming columns, and confined by the Scotch with the greyer rocks, known generally as whinstone.   I suppose the colour of the Scawfell formation to be owing to dispersed chlorite; at all events, it is definite enough to justify the name I reserve for it.[2]

---

[1] [Ruskin refers to some of his specimens in *Ethics of the Dust*, as quoted above.]

[2] [The MS. adds:—

   "The annexed illustration will enable the reader to reflect for himself upon the strangeness of the crystalline action, which withdraws the material of the garnet from the mass, and imbeds it, in perfect form, without the slightest cavity or flaw in the substance of the rock."

The "annexed illustration" does not, however, appear on the MS., and the fragment here breaks off.]

VIII

A GEOLOGICAL RAMBLE IN
SWITZERLAND[1]

ONLY half of the town of Lucerne remains, the rest having been destroyed, and a mass of hotels built on its ruins. The huge barracks beside the Reuss stand on the site of one of its most picturesque groups of gate and wall. But by a strong effort of fancy the traveller may still imagine the old town running round the extremity of the lake, with narrow quays before its old-fashioned houses, connected, by the long covered bridge which yet remains, with the large suburb on the other side of the Reuss, and by another, now destroyed, with the eastern shore near the cathedral; the whole group defended from an attack on the land side by the chain of walls and towers, of which a portion yet remains on the grassy hill behind the town (the walks about this and the immediately succeeding hills between it and the Rothsee used to be among the most exquisite in Switzerland). The eminence which carries the towers is caused by a sharp elevation of the beds of the molasse sandstone, as may be clearly seen in the quarry on the other side of the Reuss, on the old Berne road. That road, about a mile and a half from Lucerne, crosses a picturesque dingle in this same sandstone, by which the torrent from the northernmost spurs of Pilate joins the Emme. There is a road through the ravine, and the opening view of Mount Pilate from the top of the hill is of singular loveliness. Then the torrent cuts itself a narrow cleft in the sandstone, and falls into it in a pretty cascade; and a walk up through this dingle, then up to the little chapel of Herr Gotts Wald, and back by Kriens, will give some idea of the character of Swiss landscape on the outer edge of the Alps, and of the nature of this molasse formation, quite worth the day's pause.

In the ravines of the rifle-ground behind Lausanne, by the river-side beyond the bosquet of Vevay, or in any of the valleys near Fribourg, the same formation may be seen exposed in large masses; but on the whole the most interesting fragment of it that I know is just behind the town of Bonneville, on the road to Samoens. There is a small quarry to the right

---

[1] [Printed from one of Ruskin's note-books (see the Introduction, pp. xviii., xxviii.). The passage is headed "1st Route. From Lucerne over St. Gothard to Locarno," but it only takes us a short way.]

of it, in which, unless by this time the rocks are blasted away and the site overgrown with vegetation, the molasse may be seen in sheets inclined at a steep angle, forming a dark grey or brown building stone, on the surface of which the ripple-marks of the waves which deposited it are as fresh in form as on the day when the last of those waves retired from them.

Now of this sandstone these following points are to be noted.

First. It occupies the whole space of the great lowland valley of Switzerland between Alps and Jura, lying partly under superficial gravels, partly showing itself in rounded hill masses, or in cliffs at the edges of streams. And it is the cause that throughout all this great Swiss valley there are no very beautiful pieces of broken crag scenes, nor any white cliff scenery like that of our Derbyshire and Yorkshire hills. Wherever the rock rises, it is grey or brown; there is none even of the fine red colour of the sandstone of Warwickshire or South Scotland.

Also, there being a great deal of mud mixed in the mass of this sandstone, there are no pure lowland streams of any importance, for every shower fouls them, and all the beautiful glacier streams are spoiled as soon almost as they leave the lakes. The Aar is fouled in the 17 miles between Thun and Berne; the Reuss used, indeed, in old times to be nearly pure as far as Bremgarten, and the Limmat as far as Baden, but the Sarine and other such minor streams were always more or less opaque.

A great many moral phenomena in the character of the Swiss are necessarily connected with the monotony and comparative uniformity of this great sandstone formation, as opposed to the fantastic though diminutive variety of the rocks and brooks on the outskirts of our own mountains. Where, however, the molasse is thrown into bold undulations, well cultivated, and near enough to the higher range to be joined with them in general landscape effect, there are some scenes in lowland Switzerland to which the pensiveness of the grey rock colour, and the softness of the slopes into which it falls, only add to the subdued charm.

Next, for the geology of the matter. Observe you have here a vast valley covered with a mass of grey sand, of which the average depth is at least five hundred feet. What is this sand, and where did it come from?

It is a fresh-water deposit. It is one of the first which was formed after the Alps rose out of the sea; and it shows that for some time after this elevation the valley of Switzerland and many neighbouring districts formed a vast fresh-water lake (perhaps connected here and there with the sea?) into which this mass of sand settled, more or less quietly. Now, where did this fresh water flow from, bringing this peculiar sand with it? What were the circumstances of the Alps themselves after their elevation, which permitted a basin so near them to be filled only with grey sand? One would have supposed that immediately after such a convulsion the waters flowing from or off the newly-created mountains would have been surcharged with every kind of rock ruin—we shall see afterwards what evidence there is making such a conclusion all but inevitable; but here is the fact—there is no variety of ruin at all, but an even deposit of grey sand containing a few fresh-water shells and fern leaves.

And what is the origin of this peculiar deposit itself? When the waters brought no other disintegrated rocks, where did they get this mass of micaceous sand and clay? The sea sands are the remains of the endless grinding of quartz to flint pebbles by the shore waves; but what rock was ground or dissolved into this dark, peculiar lake sand, and where was the grinding done?

Lying level, mostly in the centre of the great Swiss valley, the molasse beds are sharply elevated and distorted near the Alps. And this distortion which threw the molasse beds upright under Lucerne towers was the last but one on a great scale which the Alps sustained. And it gave rise to another formation altogether, above this fresh-water one, for which we must keep a separate chapter.

INDEX

INDEX I

OF SUBJECTS, MAINLY IN "DEUCALION"[1]

Æras of mountain formation, in sum, three, 117, 118

Agate, 177, 178. See Chalcedony; also, if possible, the papers on this subject in the "Geological Magazine," vol. iv., Nos. 8 and 11; v., Nos. 1, 4, 5; vi., No. 12; and vii., No. 1 [in this volume pp. 37–84]; and Pebbles

Ages of rocks, not to be defined in the catalogue of a practical Museum, 203

Alabaster, sacred uses of, 172

Alabastron, the Greek vase so called, 172, 183

Alpine Club and Glaciers, 566

Alps, general structure of, 102, 275; are not best seen from their highest points, 103; general section of, 105; violence of former energies in sculpture of, 112; sum of traceable former history of, 112; Bernese chain of, seen from the Simplon, 225; sections of, given by Studer, examined, 279

Amethyst, 186; and see Hyacinth

Anatomy, study of, hurtful to the finest art-perceptions, 102; of minerals, distinct from their history, 241

Angelo, Monte St., near Naples, 122

Angels, and fiends, contention of, for souls of children, 263

Anger, and vanity, depressing influence of, on vital energies, 95

Argent, the Heraldic metal, meaning of, 186

Arrangement, permanence of, how necessary in Museums, 204

Artist, distinction between, and man of science, 116; general description of an artist, 116; how to make one, 173

Athena, her eyes of the colour of sunset sky, 185

Author, the, gives account of his rest in the Valley of Cluse, 152; of his studies on the Simplon, 219–225; holds himself a brother of the third order of St. Francis, 225; his dispositions not saintly, 236; his character, practical, 166; *not a philosopher, 333; his natural theology, 334; summary of his geological pursuits, 569, 570*

---

[1] [This is for the most part a reprint (with altered references to pages) of the Index compiled and printed by Ruskin at the end of *Deucalion*, vol. i. A reader who should compare it with the original edition of that volume would find several additions. Many of these are printed from Ruskin's list of *addenda* in his own copy. Other additions are distinguished by being printed in *italics*. These supply references to the second volume of *Deucalion* and to other papers in the present volume. For notice of other alterations now made in Ruskin's Index, see Bibliographical Note (above, p. 92). As explained in the Introduction, the Index is purposely kept very short (see above, pp. xlvii.–xlviii.).]

[1] [In the Magazine for January 1870.]

Rocks, wet and dry formation of, 207
Rood, Professor, Author receives assistance from, 164 *n*.
Rosa, Monte, the chain of Alps to the north of it, 222
Rose, the origin of the Persian word for red, 183
Rossberg, fall of, how illustrating its form, 107

SABLE, the Heraldic colour, meaning of, 186
Salève, façade of, 6
Scarlet, the Heraldic colour, meaning of, 184
Science, modern, duties of, 117, 244; modern vileness and falseness of, 263; true, how beginning and ending, 266; scientific mind cannot design, 174
Scientific persons, how different from artists, 116
Sealskins, use of, in the Jewish Tabernacle, 189
Selfishness, the Author's, 236 and *n*.
Sense, in morals, evil of substituting analysis for, 115
Senses, the meaning of being in or out of them, 115
Sensibility, few persons have any worth appealing to, 102
Sentis, Hoche, of Appenzell, structure, of, 104, 109
Sheffield Museum, 166, 234
Silica in lavas, 233 *seq.*; varieties of, defined, 235; *distinction of form in, 373. See also Index II.*
Siliceous minerals, arrangement of, 200
Simplon, village of, 219; Hospice of, 227
Sinai, desert of, coldness of occasional climate in, 170 and *n*.
Slate, cleavage of, generally discussed, 279 *seq.* Compare "Modern Painters," Part V., chapters viii.–x. [Vol. VI. pp. 128 *seq.*]
Sloth (the nocturnal animal), misery of, 264
Snakes, index to the contents of lecture on, 301. See also 342–343
Snow, Alpine, structure of, 128, 131–133
Sorby, Mr., value of his work, 207. *See also 354*
Sovereign (the coin), imagery on, 168
Squirrel, beauty of, and relation to man, 264
Stalagmite, incrustation of, 205–206
Standing of aiguilles, method of, to be learned, 113
Stockhorn, of Thun, structure of, 104
Stones, loose in the Park, one made use of, 167; precious, their real meaning, 193
Streams, action of, 28, 249. See Channels; and compare "Modern Painters," vol. iv. ch. vii. [Vol. VI. pp. 121 *seq.*]
Studer, Professor, references to his work on the Alps, 25, 109, 278
Sun, Heraldic type of Justice, 182, 183

TABERNACLE, the Jewish fur-coverings of, 189; the spiritual, of God, in man, 195
"Téméraire," the fighting, at Trafalgar, 182
Tenny, the Heraldic colour, meaning of, 184
Theory, mischief of, in scientific study, 99, 205; the work of "Deucalion" exclusive of it, 112
Thinking, not to be trusted, when *seeing* is possible, 130

Thoughts, worth having, come to *us;* we cannot come at *them,* 150

Thun, lake and vale of, 106; passage of the lake by modern tourists, 110; old-fashioned manners of its navigation, 111

Time, respect due to, in forming collections of objects for study, 203–204

Topaz, Heraldic meaning of, 182

Torrents, action of, in forming their beds, debated, 120

Town life, misery of, 265

Travelling (in Switzerland), old ways of, 104, 111

Truth, ultimate and mediate, differing character of, 187

Turner, J. M. W., Alpine drawings by, 103

Tylor, Mr. Alfred, exhaustive analysis of hill curves by, 290. *See also 365*

Tyndall, Professor, experiments by, 130; various reference to his works, 139, 144, 161, 227, 280, 285

Tyrwhitt, the Rev. St. John, sketches in Arabia by, 170

VALLEYS, lateral and transverse, of Alps, 105; names descriptive of, in England how various, 244; *not excavated by ice, 549*

Valtelline, relation of, to Alps, 106

Vanity of prematurely systematic science, 197

Vert, the Heraldic colour, meaning of, 184–185

Via Mala, defile of, 105, 112

Viollet-le-Duc, unwary geology by, 223; real grasp and faculty of, 223

Viscosity, definition of, 141, 157; first experiments on viscous motion of viscous fluids by Professor Forbes, 139

Volcanoes, our personal interest in the phenomena of, in this world, 262

WAVES of glacier ice, contours of, in melting, 231; *of mountain form,* 6, 7, *29*

Weathering of Coniston slate, 255 *n.,* 256

Willett, Mr. Henry, investigations of flint undertaken by, 206; proceeded with, 212

Woman, supremely inexplicable, 167

Wood, the Rev. Mr., method of his teaching, 264 *n.;* and compare "Fors Clavigera," Letter 51, § 22[1]

Woods, free growth of, in Savoy, 153

Woodward, Mr. Henry, experiment by, on contorted strata, 109

YELLOW, how represented in Heraldry, 182

Yewdale, near Coniston, scenery of, 247, 252, 254

Yewdale Crag, structure of, 254; a better subject of study than crags in the moon, 262

[1] [Where the reference is to popular works on Natural History by the Rev. W. Houghton.]

INDEX II[1]

LIST OF MINERALS MENTIONED IN THIS VOLUME, AND
OF *TECHNICAL* TERMS USED BY THE AUTHOR IN
THEIR DESCRIPTION

References to Catalogues (p. 387 and *n.*): *A.* = *British Museum.* *C.* = *Coniston.*
D. = *St. David.* *E.* = *Edinburgh.* *K.* = *Kirkcudbright.* *S.* = *Sheffield.*

*The numbers of specimens in these different catalogues are enclosed in brackets,
the other numbers being those of the pages in this volume.*

---

[1] [This Index is compiled on the lines indicated in notes among Ruskin's MSS.
(see the Introduction, p. lx.). In these notes he had begun to index the several
Catalogues of Minerals; the Index is here extended, so far as *mineralogical* matter
is concerned, to all the contents of the present volume. General topics mentioned
in the volume are included in the General Index to this edition.]

END OF VOLUME XXVI

Also from Benediction Books ...
Wandering Between Two Worlds: Essays on Faith and Art
Anita Mathias
Benediction Books, 2007
152 pages
ISBN: 0955373700

Available from www.amazon.com, www.amazon.co.uk

In these wide-ranging lyrical essays, Anita Mathias writes, in lush, lovely prose, of her naughty Catholic childhood in Jamshedpur, India; her large, eccentric family in Mangalore, a sea-coast town converted by the Portuguese in the sixteenth century; her rebellion and atheism as a teenager in her Himalayan boarding school, run by German missionary nuns, St. Mary's Convent, Nainital; and her abrupt religious conversion after which she entered Mother Teresa's convent in Calcutta as a novice. Later rich, elegant essays explore the dualities of her life as a writer, mother, and Christian in the United States--Domesticity and Art, Writing and Prayer, and the experience of being "an alien and stranger" as an immigrant in America, sensing the need for roots.

About the Author

Anita Mathias was born in India, has a B.A. and M.A. in English from Somerville College, Oxford University and an M.A. in Creative Writing from the Ohio State University. Her essays have been published in The Washington Post, The London Magazine, The Virginia Quarterly Review, Commonweal, Notre Dame Magazine, America, The Christian Century, Religion Online, The Southwest Review, Contemporary Literary Criticism, New Letters, The Journal, and two of HarperSanFrancisco's The Best Spiritual Writing anthologies. Her non-fiction has won fellowships from The National Endowment for the Arts; The Minnesota State Arts Board; The Jerome Foundation, The Vermont Studio Center; The Virginia Centre for the Creative Arts, and the First Prize for the Best General Interest Article from the Catholic Press Association of the United States and Canada. Anita has taught Creative Writing at the College of William and Mary, and now lives and writes in Oxford, England.

www.anitamathias.com,
christiancogitations.blogspot.com
wanderingbetweentwoworlds.blogspot.com

www.ingramcontent.com/pod-product-compliance
Lightning Source LLC
Chambersburg PA
CBHW020751300326
41914CB00050B/63